THE SONGS OF THE
GRASSHOPPERS AND CRICKETS
OF WESTERN EUROPE

THE
SONGS OF THE GRASSHOPPERS
AND
CRICKETS
OF
WESTERN EUROPE

D. R. RAGGE & W. J. REYNOLDS

The Natural History Museum, London

in association with The Natural History Museum

Harley Books (B. H. & A. Harley Ltd)
Martins, Great Horkesley,
Colchester, Essex CO6 4AH, England

Text set in Photina by
Ann Buchan (Typesetters), Shepperton, Middlesex
Text printed by St Edmundsbury Press Ltd,
Bury St Edmunds, Suffolk
Colour reproduced and printed by
Hilo Colour Printers Ltd, Colchester, Essex
Bound by Woolnough Bookbinders Ltd, Irthlingborough, Northants

Designed by Geoff Green

The Songs of the Grasshoppers and Crickets of Western Europe
published by Harley Books in association with
The Natural History Museum, London

British Library Catalogue-in-Publication Data applied for

ISBN 0 946589 49 6

Contents

Preface

by the Senior Author

This book was probably conceived during two summers in the late forties, when I was camping in the New Forest in southern England. These summer camps formed part of my university course and I had decided to make a special study of grasshoppers and crickets, which are more richly represented in this region than in most other parts of the British Isles. I soon found that, although some of the British grasshoppers are quite difficult to identify from their appearance, their conspicuous and contrasting songs made field identification extremely easy. By 1952, when I began work on these insects at the Natural History Museum in London, I had become convinced that these songs should be exploited in taxonomic research on the group. Unfortunately, because of the lack of suitable facilities in the Museum, the pursuit of such a project had to remain dormant for some twenty years.

The serious application of these songs to taxonomic research requires two essential facilities: the means of making high quality recordings of the sounds and the means of analysing their rhythmic patterns. In the early fifties such facilities were beginning to be used by ethologists and physiologists, but their studies were usually in universities and few of them had any interest in taxonomy. In museums, where most taxonomic research is based, it has been traditional to work on dead specimens, and facilities for studying insect behaviour are generally lacking. It was not until the early seventies, when there was a lucky coincidence of a suitable site and available funds, that a properly equipped acoustic laboratory could be built at the Natural History Museum. Dormancy thus gave way to gestation, and serious progress began on the field and laboratory work that made this book possible.

The book has been written during the five years following my retirement from official work in 1990, when the necessary time became available. Although I have written and illustrated it myself, I have thought it entirely appropriate to include Jim Reynolds, my research assistant throughout the project, as a co-author. He made about half the song recordings and was largely responsible for compiling the lists of references to past work on the songs that are given for each species. His continual availability as a consultant in dealing with the numerous problems that have arisen has been of immense value to me. My retirement sadly brought to an end any official research on Orthoptera in the Natural History Museum, and Jim Reynolds now works on butterflies. I am most grateful to Dr R. P. Lane, Keeper of Entomology at the Museum, for allowing Jim to help me, after my retirement, in the preparation of the master tapes for the companion CDs; after these were completed the Acoustic Laboratory effectively ceased to exist, being converted for quite different use.

In order to maximize the geographical coverage, Jim and I have carried out our field-work independently of each other. On my own field trips I have always been accompanied by my wife and, on the first three in the seventies, my two young daughters. While it has always been a great pleasure to me to have my family with me on these excursions, I have come to realize that it was not always the unalloyed pleasure for them that I imagined – or at least hoped – it was at the time. There were occasions, I now appreciate, when they found my single-minded enthusiasm rather hard to take. Nevertheless, their reservations were seldom expressed and I am immensely grateful to them for their companionship, support and real help during these travels, and greatly appreciate the sacrifices they made from time to time in the interests of entomological success. To my wife in particular, who has accompanied me during some 70 000 km covered by car on these trips, I owe an enormous debt of gratitude. In order to reach localities for species of particular interest, we have had to drive up mountains along precipitous and rocky tracks that sometimes seemed never-ending. My wife, no doubt silently praying with eyes tightly closed, seldom complained during such ordeals, though sometimes looking rather shaken when we finally reached the right spot.

The book is based mainly on the 1200-odd song recordings made by Jim Reynolds or myself, either during field excursions or in the Natural History Museum Acoustic Laboratory. All these recordings were analysed oscillographically and used for the song descriptions; about half of them were used for the oscillograms published in this book or for the song excerpts on the compact discs. We have also been able to use 51 recordings made by the following recordists, who very kindly made them available to us: H. C. Bennet-Clark (*Gryllotalpa vineae*), M. F. Claridge (*Lyristes plebejus*), M. Duijm (*Barbitistes obtusus, Polysarcus scutatus, Pholidoptera littoralis, Callicrania selligera*), K.-H. Garberding (*Acheta domesticus*), J. A. Grant (*Cicadetta montana*), J. C. Hartley (*Polysarcus scutatus*), K.-G. Heller (*Isophya kraussii*), P. A. D. Hollom (*Locustella luscinioides*), M. Iannantuoni (*Caprimulgus europaeus*), S. Ingrisch (*Isophya pyrenaea, I. kraussii, Barbitistes serricauda, B. fischeri, B. obtusus, Yersinella raymondii, Y. beybienkoi*), N. D. Jago (*Omocestus viridulus*), R. Margoschis (*Locustella naevia*), D. J. Robinson (*Acrometopa servillea*), I. Robinson (*Meconema thalassinum*), J.-C. Roché (*Bufo viridis*), P. Rudkin (*Acheta domesticus*), M. J. Samways (*Decticus verrucivorus* f. *monspeliensis, Platycleis falx*), R. Savage (*Conocephalus dorsalis*), L. Svensson (*Locustella lanceolata*), P. Szöke (*Locustella fluviatilis*), F. Willemse (*Acrometopa servillea, Poecilimon ornatus, Tettigonia caudata, Decticus verrucivorus assiduus, Platycleis nigrosignata, Pholidoptera fallax, P. femorata, Rhacocleis germanica, Antaxius spinibrachius, Ephippigerida areolaria, Uromenus elegans, U. andalusius, Baetica ustulata, Platystolus martinezii, Pycnogaster inermis, Gryllus campestris, Euchorthippus pulvinatus pulvinatus*). I am particularly indebted to my friend and colleague Fer Willemse who, in addition to making available the recordings mentioned above, has generously provided copies of many of his other song recordings for the Natural History Museum collection.

I should also like to thank the following, who have kindly provided livestock (either from laboratory cultures or, more often, collected by them personally) from which we have made song recordings: D. W. Baldock, B. H. Betts, P. Bragg, M. Bugren, J. P. Carvalho, R. Cumming, M. Chinery, R. W. Crosskey, C. F. Dewhurst, R. Ehrmann, D. Elias, K. Guichard, E. C. M. Haes, J. C. Hartley, D. H. Harvey, K. Harz, T. M. Howard, N. D. Jago, W. Latimer, W. B. Lee, J. A. Marshall, G. Maybury, K. Nicolle, J. Paul, L. M. Pitkin, R. Ranft, S. Robilliard, M. C. D. Speight, P. Thorens, B. C. Townsend and W. G. Tremewan. Of these, two deserve a special mention:

Chris Haes, whose many personally collected specimens have over the years much enriched the collection of the Natural History Museum in London, and Colin Hartley, who has generously provided many live Tettigoniidae, some collected personally and some from the cultures that he so successfully established at Nottingham University. Among the others, Theresa Howard, Linda Pitkin, Bruce Townsend and Gerry Tremewan were members of my staff at one time or another during the period of the project, and Judith Marshall for the whole of it; I should like to thank all of them, and especially Judith, for their invaluable support over the years. I am also most grateful to John Huxley, another member of my staff in the early days of the project, for his help in designing the data sheet that accompanies all our song recordings.

For many years it has been a pleasure to collaborate with the National Sound Archive (NSA), a subdivision of the British Library. As soon as our collection of song recordings reached a significant size, Jim Reynolds made copies of them for what was then the British Library of Wildlife Sound (BLOWS), now simply the Wildlife Section of the NSA. This was arranged through the good offices of Ron Kettle, then curator of BLOWS, who provided the necessary tape and has been a helpful consultant from time to time throughout the project. Ron's successor, Richard Ranft, has been equally helpful, and we are especially grateful to him for providing copies of song recordings, held by the NSA, of most of the non-Orthoptera that form the subject of Chapter 10 and whose calls are included on the second companion CD; the recordists, who kindly agreed to this use of their recordings, have already been acknowledged above. Richard also very kindly arranged for the NSA to provide, as a safety precaution, both analogue (tape) and digital (CD) copies of the master tapes for both companion CDs; we are most grateful to the NSA for making their facilities available for this purpose and to Nigel Bewley, who made the copies. We are also grateful to the NSA for permission to use the recording of the song of *Caprimulgus europaeus* from their cassette *The sounds of Britain's endangered wildlife*.

The earliest song recordings used for this book were made in 1962 during a memorable excursion, on behalf of the British Broadcasting Corporation, to the Isle of Purbeck and New Forest in southern England. The participants in this enterprise were my long-standing friend John Burton, who organized the trip, the late Bob Wade and myself. John was at that time Librarian of the BBC Natural History Unit, and Bob a BBC Mobile Engineer with long experience of recording animal sounds. The results of our efforts were issued as a long-playing disc (Ragge *et al.*, 1965) and, much later and augmented by other recordings, a cassette tape (Burton & Ragge, 1987). Although none of these recordings have been used for the CDs issued with this book, I have, with the kind permission of the BBC, used oscillograms taken from two of them; these are shown in Figs 171 (*Conocephalus dorsalis*) and 348 (*Metrioptera brachyptera*).

Most of the photographs of live Orthoptera used for the three colour plates were very kindly provided by the following photographers: A. Beaumont (*Pholidoptera griseoaptera*), P. A. Bowman (*Meconema thalassinum, Metrioptera brachyptera*), C. G. Butler (*Leptophyes punctatissima, Gryllus campestris, Nemobius sylvestris*), P. T. Chadd (*Conocephalus dorsalis, Decticus albifrons, Platycleis sepium, Anonconotus alpinus, Arcyptera fusca, Euthystira brachyptera, Omocestus viridulus, Euchorthippus declivus*), M. Chinery (*Tylopsis lilifolia, Eupholidoptera chabrieri, Antaxius pedestris, Ephippiger ephippiger*), R. & C. Foord (*Tettigonia viridissima, Decticus verrucivorus, Metrioptera roeselii*), E. C. M. Haes (*Gomphocerus sibiricus*), the late D. E. Kimmins (*Chorthippus parallelus*), B. C. Pickard (*Gryllotalpa gryllotalpa*), K. G. & R. A. Preston-Mafham (*Phaneroptera falcata, Gomphocerippus rufus*), M. J. Skelton (*Polysarcus denticauda, Stethophyma grossum*) and P. H. Ward (*Chorthippus biguttulus*). The remaining photographs were taken by the following members of the Natural History Museum Photographic Unit: P. Crabbe (*Poecilimon ornatus*), F. Greenaway (*Stenobothrus lineatus*), E. Millar (*Stauroderus scalaris*) and H. Taylor (*Ruspolia nitidula, Platycleis tessellata, Uromenus stalii, Oecanthus pellucens*).

I am much indebted to the Natural History Museum, London, and in particular to Dr L. A. Mound and his successor Dr R. P. Lane, Keepers of Entomology during the five years following my retirement, for providing me with working space during this period and the facilities of the Acoustic Laboratory for as long as they were needed. For the first two of these years I was awarded an Emeritus Fellowship by the Leverhulme Trust, to whom I am most grateful for funding two further field excursions and the cost of expendable materials used in the preparation of the book.

My publisher, Basil Harley, has been unfailingly helpful during the preparation of this book. Like all his other products, it has benefited greatly from his careful thought and high standards, and I am most grateful to him and his wife Annette for their friendly collaboration throughout its production.

Finally I should like to express, on a more personal note, my warm appreciation of the help of my co-worker Jim Reynolds. Always conscientious, patient and hard-working, never complaining, he has made a major and wholly indispensable contribution to this book.

David R. Ragge
Purley, Surrey, 1997

The companion compact discs

High quality recordings of the songs of all 170 species of Orthoptera included in this book are available in a companion boxed set of two compact discs entitled *A sound guide to the grasshoppers and crickets of western Europe*. Also included on the discs are the songs (described in Chapter 10) of four cicadas, two toads and five birds, all of which could be confused with the songs of Orthoptera. The discs include a total of 352 song excerpts, with a playing time of 120 minutes.

Authors' note

We have found the terms 'syllable' and 'echeme' to be indispensable in the song descriptions and elsewhere in this book. Readers unfamiliar with the use of these terms in describing grasshopper and cricket songs are recommended to read the section on song terminology (p. 23) before consulting the song descriptions or attempting to use the identification key (p. 81).

Chapter 1

Introduction

General

No summer in the European countryside is complete without the chorus of grasshopper and cricket songs that fills the air on sunny days and fine nights. These sounds are usually taken for granted; like bird song they are the expected accompaniment to any open-air country occasion on a fine summer's day. Yet these sounds provide a wealth of information to the trained ear: in particular, most of the 170 species included in this book can be identified instantly from their songs – and often more reliably than from their appearance. Indeed, there are several groups in which identifying the species by morphological characters alone is almost impossible, although they can be recognized at once by their songs.

European grasshoppers and crickets are used widely in ecological, cytogenetic and bioacoustic research, as well as, more recently, in evolutionary biology. They are being increasingly used in the conservation field as indicators of undisturbed habitats, and include about 30 agricultural pests, of which about a third are regarded as major ones. Accurate identification is vital in all these fields and there has been a growing need for a comprehensive account of the songs that would provide the means of reliably identifying even the more difficult species. The main aim of this book is to provide such an account for western Europe, with the emphasis on the use of these songs in taxonomy and identification. Most of the previously published studies on the songs of European Orthoptera have been in German, and we feel that an account in English will have the added advantage of making information on these songs much more accessible to English-speaking readers. We also hope that the compact discs of recorded songs will be useful not only to serious students of insect sounds but also to anyone who happens to notice one of these songs, perhaps while on holiday, and is curious to know which insect is producing it.

The insects that form the subject of this book are the sound-producing members of the Orthoptera s. str., i.e. the Tettigoniidae (bush-crickets), Gryllidae (true crickets), Gryllotalpidae (mole-crickets) and Acrididae (grasshoppers) (we have found it convenient to use a very conservative classification, discussed briefly on p. 9). The only other European insects that make comparable sounds are cicadas (Hemiptera: Cicadidae); these belong to a totally unrelated group and we have not described their songs in the main text of the book, but we have given in Chapter 10 oscillograms and brief descriptions of the songs of four species, *Cicada orni*, *Lyristes plebejus*, *Tettigetta argentata* and *Cicadetta montana*, and have included recordings of their songs on the second compact disc. We have given similar treatment to seven vertebrate animals whose songs could easily be confused with those of Orthoptera: *Bufo viridis* (Green Toad), *Alytes obstetricans* (Midwife Toad), *Locustella naevia* (Grasshopper Warbler), *L. luscinioides* (Savi's Warbler), *L. fluviatilis* (River Warbler), *L. lanceolata* (Lanceolated Warbler) and *Caprimulgus europaeus* (Nightjar).

The 170 species of Orthoptera whose songs are described in this book include all the common ones and most of the rarer ones occurring in western Europe. The singing species not included are mostly local forms, often of doubtful status, whose songs are unknown to us. The songs we have described are primarily the calling songs of isolated males, but we have also included the special courtship song produced by the males of some species when next to a female and, where it is diagnostically useful, the rivalry song produced when males interact with one another. We have not included the rather nondescript stridulatory sounds produced by many locustine grasshoppers, since these are of little or no diagnostic value.

The geographical area we have included in our study is all of Europe west of, and including, Scandinavia, Germany, Austria and Italy (Fig. 1). Also included are the Mediterranean Islands on which we have been able to carry out field-work: the Balearic Islands (Majorca, Ibiza, Formentera), Corsica, Sardinia and Sicily. North Africa is in general excluded but, because of its special interest, we have included the endemic North African grasshopper *Chorthippus marocanus*.

It should be emphasized that our account is based entirely on sounds audible to man, i.e. below about 20 kHz in frequency. The songs of gomphocerine grasshoppers and, especially, tettigoniid bush-crickets, contain much sound that is too high-pitched to be audible to even the keenest of human ears. Although such ultrasonic frequencies often have a communicative value to the insects producing them, they have little effect on the rhythmic pattern of the song (and hence on the oscillograms portraying it), and it is this pattern, easily appreciated from the audible part of the song, that is most useful in the taxonomy and identification of these insects. Some of the smaller European Tettigoniidae have songs that are almost entirely ultrasonic; however, by using an ultrasound detector or even good audio equipment without such an aid, it is nevertheless possible to make tape recordings of them that can be used to produce good oscillograms. The songs of the true crickets (Gryllidae) and mole-crickets (Gryllotalpidae) are usually much lower-pitched than those of gomphocerines and tettigoniids and are more easily audible to man. In these groups the dominant frequencies of the calling songs are often low enough to form a musical note, the pitch of which can itself be useful in identification.

Although clear descriptions of the songs, illustrated by oscillograms, can give quite an accurate impression of their component sounds, and especially their rhythmic patterns, such accounts inevitably fall far short of good quality recordings of the songs themselves. The compact discs issued in conjunction with this book are intended to provide such good quality recordings and thus to enable the user to make direct comparisons between them and a song that needs to be identified. These days, when highly portable CD players are widely available, a recorded song can easily be played back in the field, thus providing a particularly effective means of identifying a singing insect.

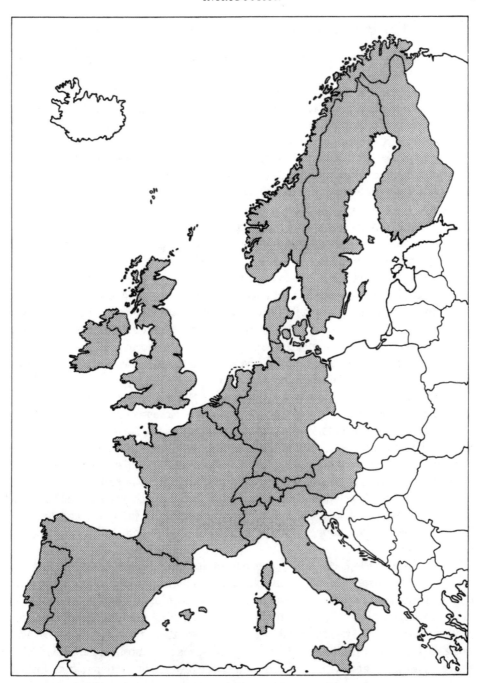

Figure 1 Map of western Europe, showing the main area of study (shaded).

An introduction to the European Orthoptera

The Orthoptera, in the sense used here, include all the saltatorial orthopteroid insects, i.e. those typically with the hind legs enlarged for jumping and hence sometimes known as the Saltatoria. These insects are found throughout the warmer parts of the world: they are richly represented in the tropics, rather less so in temperate regions, and few penetrate the Arctic or Antarctic Circles. The total number of species known is at present about 20 000, of which some 800 occur in Europe.

The Orthoptera are relatively primitive insects. They have biting mouthparts, and the legs, wings and antennae are usually of simple structure. The hind wings, when present, have a large, foldable fan; when flexed they are at least partly protected by the usually tougher fore wings. Apart from this fan, the venation of the fore and hind wings is fairly similar and rather primitive. Many species of Orthoptera are brachypterous, with both pairs of wings too small for flight; some normally brachypterous species occasionally occur in a fully winged (macropterous) form, usually capable of flight. The pronotum is much enlarged in Orthoptera, extending backwards to cover the bases of the fore wings and downwards on each side of the thorax in the form of lateral lobes. Female bush-crickets have a conspicuous ovipositor, which is laterally flattened, often upcurved, sometimes straight and occasionally slightly downcurved; it is a unitary structure, although composed of six closely integrated valves. In the true crickets the ovipositor is usually needle-like, with a slight thickening at the tip; it is four-valved but the valves are again closely integrated to form a unitary structure. The mole-crickets have no ovipositor. Female grasshoppers have an ovipositor composed of four quite separate valves; it is usually rather inconspicuous (sometimes only the tips of the valves can be seen), but in *Euthystira brachyptera* the valves are long and always completely visible. The Orthoptera are divided into two clearly distinct groups that are not closely related: the Ensifera (bush-crickets, true crickets, mole-crickets, etc.) and Caelifera (grasshoppers, ground-hoppers, etc.). The sound-producing and hearing organs are quite different in these two groups and are described in Chapter 3. It should also be noted that, although well-developed calling songs are produced by almost all bush-crickets, true crickets and mole-crickets, many grasshoppers lack a stridulatory apparatus and in the European grasshoppers calling songs are largely confined to the subfamily Gomphocerinae.

These insects develop gradually from egg to adult, without a sudden metamorphosis or pupal stage; because of this the young stages are usually known as nymphs rather than larvae. The nymphs resemble miniature adults, but lack wings. In the fully winged species the wings develop gradually, in the form of wing-pads, during successive moults, becoming obvious during the last two nymphal stages; at this time they are directed upwards to meet along the mid-line of the body, with the hind wing-pads outermost. At the final moult the emerging wings expand to their full size and are then reversed in position so that what were the dorsal margins are directed downwards and the fore wings cover the hind wings. Sexual maturity is usually reached a few days after the final moult, and it is at this time that the males begin to sing.

As explained in Chapter 4, the main function of the songs is to bring the two sexes together. When this has been achieved mating usually follows, sometimes preceded by a special courtship song. The usual mating position in the Ensifera, at least initially, is with the female above the male, whereas in the grasshoppers the male mounts the female. In both groups the position may change during mating so that the male and female face in different directions, remaining in contact only at the abdominal tips. In the bush-crickets the male usually grips the tip of the female abdomen with his cerci, a sharp point on each cercus often engaging in a special socket on the female. Male crickets sometimes secrete an attractant that encourages the female to adopt, and remain in, the correct mating position. In all these groups the sperm is transferred to the female in a spermatophore. This can be easily seen in the Ensifera, as the spermatophore emerges externally from the male abdomen, either before or at the time of its attachment to the female. In the bush-crickets the spermatophore is usually accompanied by a conspicuously large, gelatinous mass (the *spermatophylax*) on which the female feeds after mating. In the grasshoppers the spermatophore is not produced until mating has begun and is thus not visible externally; part of it remains in the male when the sexes separate. After mating, females of all these groups are unresponsive to the male calling song until they have laid at least some eggs.

Egg-laying usually begins within a day or two of mating and often continues, at intervals, through most of the remaining life of the female. Bush-crickets and true crickets lay their eggs, usually in small batches, in a wide variety of sites. Many species lay them in the ground, but many others lay them on or in various forms of vegetation; in *Phaneroptera* they are inserted along the edges of leaves, between the upper and lower surfaces. *Gryllotalpa gryllotalpa* lays its eggs in an underground nest-chamber, where they are tended by the mother until they hatch; this is probably also true of the other European mole-crickets. Almost all grasshoppers lay their eggs in the ground or in the bases of grass-tufts. They are laid in a group and then a frothy liquid is poured over them; this rapidly solidifies to form a spongy protective case called an egg-pod. The egg-pods of *Euthystira brachyptera* are laid on or between the leaves of grasses or other plants and are sometimes quite openly visible. *Chrysochraon dispar* is exceptional in laying its eggs in plant stems, often in the pithy interior exposed at the end of a broken stem.

In the European bush-crickets and grasshoppers it is usual for the eggs to undergo an obligatory dormant phase (*diapause*) so that they cannot hatch at an unfavourable time of the year. This phase is brought to an end by a cold period lasting for at least a few weeks (normally provided by the winter), after which development continues as soon as the temperature is high enough to permit it and hatching takes place in the spring. Some bush-crickets have more than one diapause during egg development and in many species the eggs do not hatch until at least the second spring after they were laid. A few European bush-crickets (e.g. *Phaneroptera nana*) have no egg diapause and the eggs seem to be prevented from hatching in the late summer only by an insufficiently high temperature; as far as is known at present, no European Tettigoniidae have more than one generation per year. It seems likely that the eggs of most European grasshoppers hatch in the first spring after they were laid, as is true of all the British species. In southern Europe some species have two generations per year (e.g. *Chorthippus jacobsi* and *Omocestus raymondi* in Spain) and, conversely, we suspect that the eggs of some European grasshoppers, like those of so many bush-crickets, have slow developments extending through more than one winter. In most European bush-crickets and grasshoppers with one generation per year, maturity is reached by July and the adults die off when the cold autumn weather arrives. However, some non-gomphocerine grasshoppers overwinter as nymphs or adults in southern Europe and this may be true of some gomphocerines.

The European true crickets and mole-crickets show a greater variety of life-cycles. Some overwinter as eggs, like most bush-crickets and grasshoppers, but many overwinter as nymphs or adults, so that singing and breeding take place earlier in the year, from May onwards in the north and April onwards further south; in some of these species (e.g. *Gryllus campestris*, *Gryllotalpa gryllotalpa*) the eggs have no diapause and hatch within a few weeks of being laid. A few have two-year life-cycles (e.g. *Nemobius sylvestris*, *Gryllotalpa gryllotalpa*), at least in the more northerly part of their range.

The European Orthoptera are most abundant in open habitats and seldom occur in dense forest. The clearing of the natural forest by man during the past 2000 years has thus made the central and northern parts of Europe much more favourable to Orthoptera, in spite of the agricultural use of many areas of cleared land. These insects may be found in a wide range of open habitats, from wet bogs to the driest steppe and semi-desert. Species that typically occur on shrubs and trees, or in the leaf-litter beneath them, are not normally found in unbroken forest, but rather in open woodland, by forest clearings or borders, or in hedgerows. Some crickets (e.g. *Gryllus campestris*) spend much of their lives in or near burrows, and the mole-crickets are mainly subterranean insects, seldom being seen above ground. The gomphocerine grasshoppers are typically grassland insects, though some are associated more with low shrubs and a few occur on taller ones.

The feeding habits of Orthoptera show a similar wide variety. Phaneropterine bush-crickets are almost exclusively phytophagous, *Meconema* is almost entirely carnivorous, and most tettigoniids feed on a mixture of plant and animal matter. Crickets are generally opportunistic feeders, eating whatever suitable plant or animal foods are readily available. The gomphocerine grasshoppers are exclusively phytophagous, usually feeding entirely on grasses or sedges.

The Orthoptera include the locusts, which rank among the world's most destructive pests. These insects are simply grasshoppers that periodically form migrating swarms – the *gregarious* phase – and can cause immense damage to crops; at other times – when in the *solitarious* phase – they behave like other grasshoppers and present no major problems. Most locust damage occurs in tropical countries and locusts are seldom a problem in western Europe, where the last major outbreak was of the Migratory Locust (*Locusta migratoria*) in France during the years 1946–48. The Italian Locust (*Calliptamus italicus*) and Moroccan Locust (*Dociostaurus maroccanus*) are both common in southern Europe but again seldom cause serious damage to crops in the area of our study. Several species of ordinary grasshopper are quite serious pests in eastern Europe and temperate Asia, where they often occur in high population densities, but this seldom happens farther west.

The European Ensifera include a number of pests, mostly of quite minor importance. Several bush-crickets (e.g. *Barbitistes fischeri* and several species of *Ephippiger*) cause damage to vines and fruit-trees when they are exceptionally abundant, and others (e.g. *Decticus albifrons*, *Platycleis albopunctata*, *P. intermedia*) sometimes damage cereal crops. The House-cricket (*Acheta domesticus*) is occasionally a nuisance in heated buildings or on rubbish tips, and the Mole-cricket (*Gryllotalpa gryllotalpa*) sometimes causes damage to root-crops.

The Ensifera show a strong tendency to be nocturnal: even the species that are active during the day are also active during at least part of the night. Some species (e.g. *Meconema thalassinum*) are strictly nocturnal, remaining completely dormant during the daytime. The grasshoppers, by contrast, are typically diurnal, although a few species remain active for a time after dark if the temperature is sufficiently high. Even during their more active periods many Orthoptera are rather static insects, tending not to move far unless disturbed. Some of the fully winged species, although well equipped for flight, seldom seem to fly spontaneously, and even the species that fly more readily often limit themselves to flights of only a few metres. Spontaneous flight becomes much more likely when population density is high, and sufficiently dense populations sometimes disperse themselves by long-distance flight. The most striking example of this is provided by the locusts, but many of the larger fully winged species are capable of long flights when a population explosion occurs.

The European Orthoptera have many natural enemies. They are attractive items of food to a number of vertebrate predators, including small mammals, birds, reptiles and amphibians, and also fall victim to spiders, beetles and even predatory bush-crickets. They are parasitized by various flies and wasps, some of which specialize in using Orthoptera as their hosts. Some of these predators and parasitoids are known to be attracted by the calling songs of their orthopteran prey (see p. 43). Orthoptera are also attacked by nematode worms and various fungal, bacterial and viral diseases. Mites, especially the blood-sucking larvae of the Trombidiidae, are often found as external parasites on grasshoppers, but they do not appear to do their hosts any serious harm.

Presentation

The main purpose of this book is to provide a comprehensive account of the songs of the western European Orthoptera, illustrated by oscillograms (see p. 12) and accompanied by excerpts from recorded songs on the companion compact discs (see p. 13). The descriptions of songs are given, species by species, in Chapters 7–9. For each species the information is given in a standard sequence, headed by the currently valid scientific name (see p. 9) followed by a reference to the original description of the species. The English vernacular name is then given, in bold type, followed by vernacular names in the principal western European languages in which such names exist (see p. 12). There is then a list of references to any significant past descriptive accounts of the song, classified according to whether they include oscillograms (including sound-level tracings), diagrams (hand-drawn or symbolic representations) of the song, sonagrams (audiospectrograms), illustrations of stridulatory wing- or leg-movements, frequency information, musical notation, or verbal description without any of these additions. Any commercially issued disc or cassette recordings of the song are also listed; long-playing discs are indicated by '(LP)' after the reference, compact discs by '(CD)'. These references are not intended to be exhaustive – there are many brief statements about the songs, especially in the earlier literature, that do not warrant inclusion; our aim has been to list all sources (including some unpublished theses) that the reader might find useful to refer to for additional or confirmatory information on the songs. Sources in which the specific name used is not the currently valid one are annotated by quoting the specific name used; if such a name is repeated under other references, it is abbreviated, usually to the initial letter. No annotation is given for the use of other *generic* names. We have not usually distinguished in these references (or any other references in the text) between different authors with the same surname; in almost every case the year of publication is sufficient to enable the reader to find the full bibliographical information, including the authors' distinguishing initials, in the list of references beginning on p. 543. The authors are arranged alphabetically under each subhead.

Under the heading *Recognition* brief notes are given on recognizing the species by morphological characters, and there is often also a brief reference to the diagnostic value of the song. The full song description is given under the heading *Song*, with references to the oscillograms and relevant compact disc. The description is primarily of the calling song of an isolated male, but in appropriate cases we have also described the special courtship song produced when the male is next to a female and, where it is diagnostically useful, the rivalry song produced when two or more males are together. We have not, however, separately described the songs produced by female gomphocerine grasshoppers when in a sexually receptive state (see p. 41); these songs are seldom heard in the field and are in any case broadly similar in pattern to the calling songs of conspecific males. The rhythmic pattern of the songs can be seen at a glance from the oscillograms, and so the descriptions are concise and the emphasis is on the measurements of various song characters and the variation that can be assessed from studying all the available recorded songs rather just than those chosen for the oscillograms and compact discs. Durations are usually expressed in seconds (s) or milliseconds (ms), i.e. thousandths of a second. In resonant songs the frequency of the carrier wave (see p. 28) is expressed in kilohertz (kHz), i.e. thousands of cycles per second. Repetition rates, on the other hand, are simply expressed as so many per second (/s). For the species *Psophus stridulus* and *Bryodema tuberculata*, in which the only significant sound production is during flight, the heading *Crepitation* is used instead of *Song*. A summary of the distribution of each species is given under the heading *Distribution*, including a brief mention of the distribution in the British Isles for each British species.

In Chapter 2 we describe our methods of recording and analysing the songs, as well as storing, documenting and copying the recordings. We have presented this information in such a way that it can, we hope, be used as a guide by anyone wishing to record the songs of Orthoptera. We also discuss the terminology that has been applied to the songs and define the terms that we have decided to use ourselves. Chapter 3 describes the methods of sound production used by the European Orthoptera and includes a brief account of the hearing organs. In Chapter 4 we give a general account of the songs, their functions and the extent to which they vary, both individually and geographically. We attempt to classify the songs and speculate on their evolution.

In Chapter 5 we turn to the use of the songs in taxonomy, for which we coin the new term *phonotaxonomy*. After a fairly comprehensive review of past work we apply the songs to a number of taxonomic problems among the western European Orthoptera. The chapter concludes with a brief introduction to the use of the songs in identification, leading, in Chapter 6, to an identification key, based primarily on the songs, to all the species included in this book. The key has to depend on verbal description and some of the song distinctions used are inevitably difficult to describe, but we hope the key will be of some help to readers trying to identify an orthopteran singer.

After describing the song of each species of Orthoptera in Chapters 7–9, we give in Chapter 10 a brief account of the songs of four cicadas, two toads and five birds, all of which could quite easily be confused with the songs of ensiferan Orthoptera. The book concludes with a check-list of all the species included, a list of new changes to scientific names (largely resulting from applying the songs to taxonomy), a glossary of technical terms and a list of all the literature, discs and cassettes referred to in the text.

Classification

We have found it convenient to use a simple, conservative classification of Orthoptera in this book, as set out in the check-list on p. 485. Higher ranks are often given to many of the group-names listed, but we have considered it inappropriate to do this here. We have retained the subfamily Decticinae purely for convenience, while recognizing that there is no clear distinction between this group and the Tettigoniinae when the world fauna is taken into account (Rentz, 1985); in Europe, where the two groups are fairly distinct, it is still customary to give them separate status, and we have found it helpful to be able to use the adjective 'decticine'. The bush-crickets we have included in the genus *Platycleis* are sometimes treated as belonging to a number of distinct genera, following Zeuner (1941), but we prefer to treat these genera as subgenera of *Platycleis* and have not considered it even worth mentioning them here (see Ragge, 1990); the same applies to the genus *Metrioptera*. We have indeed used no subgeneric names in this book, considering such names to be of doubtful value in the classification of the European Orthoptera. We have similarly ignored most of the numerous subspecific names that are currently in use for European Orthoptera; many of these subspecies are of doubtful validity and would in our opinion be better treated informally as, at best, local forms. The only subspecies mentioned in the check-list are those with distinctive songs. The sequence in which we have arranged the species is largely that used by Harz (1969, 1975b), with a few minor modifications.

Unlike some authors we have preferred to treat the two rather similar grasshoppers *Chrysochraon dispar* and *Euthystira brachyptera* as generically distinct. The striking difference in the ovipositor (associated with quite different egg-laying habits) is much greater than, for example, the ovipositor difference used to separate *Omocestus* and *Stenobothrus*.

The grasshopper *Stauroderus scalaris* is now included by some authors in the genus *Chorthippus*, but we prefer to keep it generically distinct. This view is lent some support by the work of Corey (1933, 1937) on the chromosomes.

Pfau (1996) has recently included the bush-cricket *Callicrania selligera* (together with the other species of *Callicrania*) in the genus *Platystolus*, but this change was published too late for us to be able to implement it in this book.

A few taxonomic problems at specific and subspecific level are discussed on pp. 69–79.

Scientific names

The Latin or 'scientific' names used for animals and their groups are governed by the *International Code of Zoological Nomenclature*, currently in its third edition (1985). The scientific names we use in this book conform with the requirements of this *Code* (but see *Conocephalus discolor* below) and occasionally differ from those used by some workers in continental Europe. The difference is often merely in spelling, but we sometimes use a different generic or specific name from those customarily used in continental publications. We have therefore thought it desirable, in all such cases of divergence, to give in this section brief explanations for the spellings and names we have adopted.

For one subfamily of grasshoppers the alternative names Locustinae and Oedipodinae have both been in recent use, and at the time of writing the International Commission on Zoological Nomenclature is considering which of these names should be given precedence. We have used the simpler name Locustinae pending the Commission's ruling.

Tylopsis lilifolia (Fabricius)

The specific name is often spelt *liliifolia*, but the original spelling used by Fabricius (1793) was *lilifolia* and this must stand under Article 32 (b) of the *Code*.

Isophya pyrenaea (Serville)

Serville (1838) spelt the specific name *pyrenea* in the heading to the original description (p. 481) and *pyrenæa* in the Index (p. 766); both these spellings have been widely used in the subsequent literature. Heller (1988: 187) discusses this problem and concludes that, since Serville used the more correctly latinized spelling in the Index, the spelling *pyrenea* was an inadvertent error that must be corrected under Article 32 (c, d) of the *Code*. We agree with this view.

Platycleis veyseli Koçak

This species was known until recently as *P. vittata* (Charpentier). However, it was originally described as *Locusta vittata*, and this name is a junior homonym of *Locusta vittata* Thunberg and thus invalid under Article 52 of the *Code* (see Ragge, 1990: 6, 28).

Metrioptera roeselii (Hagenbach), *Yersinella raymondii* (Yersin), *Uromenus stalii* (Bolívar), *U. perezii* (Bolívar), *U. martorellii* (Bolívar), *Platystolus martinezii* (Bolívar), *Omocestus uhagonii* (Bolívar), *Stenobothrus bolivarii* (Brunner)

All these specific names are often spelt with a single terminal '-*i*', but the original spellings were with the ending '-*ii*' in every case and must stand under Article 32 (b) of the *Code*.

Conocephalus discolor (Thunberg)

Until Roberts (1941) suggested otherwise, this species was referred to as *C. fuscus* (Fabricius), and this name is still occasionally used in the European literature. Fabricius (1793: 43) originally described the species under the name *Locusta fusca*, and Roberts argued that this was 'preoccupied by *Gryllus Locusta fuscus* Pallas, 1773' [now *Arcyptera fusca*] and that it should therefore be replaced by its first available synonym, *Conocephalus discolor* (Thunberg).

At the time of Roberts' work the only edition of the *Code* available was the original one, published in 1905. Article 35 of that edition, which deals with the question of specific homonymy, makes no mention of subgenera and so does not make it clear whether these two specific names should be treated as homonyms. However, following discussions at the 1948 Paris meeting of the International Commission on Zoological Nomenclature, it was decided to insert a provision into the *Code* stipulating that subgeneric names were to be disregarded in determining specific homonymy. This was done when the second edition of the *Code* was published in 1961 (Article 57 (a)), and it then became clear that *Locusta fusca* Fabricius and *Gryllus Locusta fuscus* Pallas were not homonyms.

In the meantime many authors had followed Roberts in adopting *C. discolor* as the valid name for the species, as has Harz (1969) in the definitive work on the European Orthoptera, and this name is currently in almost universal use. In view of this we are reluctant to revert to the name *C. fuscus*, even though there is no doubt as to its validity under the current *Code*. It may now be in the better interests of nomenclatural stability to ask the Commission to set aside this earlier name.

Modicogryllus bordigalensis (Latreille)

This species, with its specific name spelt thus, was described from the vicinity of Bordeaux, known to the Romans as Burdigala. To be etymologically correct the name should have been spelt *burdigalensis*, and this spelling is now almost universally used. However, Article 32 of the *Code* makes it absolutely clear that the original spelling must be retained, and so we have used it in the present work.

Pteronemobius heydenii (Fischer)

This species often appears in the European literature under the name *P. concolor* (Walker). The two names are generally considered to refer to the same species, but *P. heydenii* is the earlier name and so must be used as the valid name under Article 23 of the *Code*. The spelling *heydeni* is often used, but the original spelling was *heydenii* and must stand under Article 32 (b) of the *Code*.

Stethophyma grossum (Linnaeus)

In continental Europe this species has usually been referred to as *Mecostethus grossus*. However, as Roberts (1941) pointed out, Kirby (1910) designated *Gryllus grossus* as the type species of *Stethophyma* and *Gryllus alliaceus* Germar as the type species of *Mecostethus*. The effect of this is to make *Stethophyma grossum* and *Mecostethus alliaceus* the currently valid names for these two species.

Omocestus rufipes (Zetterstedt)

When Zetterstedt (1821: 89, 90) first described this species he named the female *Gryllus ventralis* and the male, on the following page, *Gryllus rufipes*. It was soon realized that both names referred to the same species and at first *rufipes* was given priority, probably because it was based on the male sex. Then, during the present century, *ventralis* began to be preferred on the grounds that it had 'page priority' in the original publication. This view was endorsed by the 1948 Paris meeting of the International Commission on Zoological Nomenclature, at which it was decided that all cases of this kind should be settled on the basis of page priority. However, this ruling was reversed at the 1953 Copenhagen meeting of the Commission, when it was decided that priority in such cases should be given to the name favoured by the 'first reviser', in this case Borck (1848: 124), who had chosen *rufipes*. The first reviser principle, which formed part of the original 1905 edition of the *Code* and has been retained in the second (1961) and third (1985) editions, seems likely to prevail and establishes beyond doubt the validity of the name *rufipes* for this species. The name *ventralis* is nevertheless still often used in the European literature.

Gomphocerus sibiricus (Linnaeus)

This species often appears in the European literature under the name *Aeropus sibiricus*. However, Roberts (1941) pointed out that Brullé (in Audouin & Brullé, 1835) had designated *Gryllus sibiricus* as the type species of *Gomphocerus*, thus making *Gomphocerus* a senior synonym of *Aeropus*, which also has *sibiricus* as its type species.

Gomphocerippus rufus (Linnaeus)

The effect of Brullé's designation of *sibiricus* as the type species of *Gomphocerus* (see above) was to leave *rufus* without a valid generic name. Roberts (1941) therefore proposed the name *Gomphocerippus* with *rufus* as its type species. This has therefore been the correct generic assignment for this species since 1941, although the combination *Gomphocerus rufus* still often appears in the European literature.

Vernacular names

The English names used for the broad groups of singing Orthoptera have now been used fairly consistently in England for over half a century: bush-cricket (Tettigoniidae), cricket (Gryllidae), mole-cricket (Gryllotalpidae) and grasshopper (Acrididae). The name 'bush-cricket' was invented by Burr (1936) in a successful attempt to supplant the older name 'long-horned grasshopper', on the grounds that the Tettigoniidae are much more closely related to the Gryllidae than to the Acrididae. 'Bush-cricket' has recently been given formal recognition by *The New Shorter Oxford English Dictionary* (Brown, 1993) and is now used widely in continental Europe when texts are published in English. The only significant departure from these names in the English-speaking world is the use by Americans of 'katydid' instead of 'bush-cricket', but this American term has so far made little headway in Europe.

In the main descriptive part of this book (Chapters 7–9) we have listed, under the Latin-name heading of each species, vernacular names in the principal western European languages in which such names exist. Where one species has more than one vernacular name in the same language, we have chosen the one that seems to be definitive or most appropriate. The English names for the species occurring in the British Isles were first firmly established by Kevan (1952) and have been consistently used since then in all significant works on the British Orthoptera. For the non-British northern European species we have adopted the English names used in Bellmann (1988); these were specially coined by us for that work, with the present book in mind. This left some 90 southern European species treated in this book for which no English names existed; for these we have coined new ones.

The French names are taken from Luquet (1993) and the German ones from Bellmann (1993a). The Dutch names are from Beukeboom (1986), with one addition (zuidelijke veldkrekel) from Duijm & Kruseman (1983) and another two (bruin schavertje and gaspeldoornsprinkhaan) from Kleukers *et al.* (1997). For the Scandinavian names (including Danish and Finnish) we have generally followed Holst (1986), but have occasionally given preference to Wallin (1979) for Swedish names.

In Spain, Portugal and Italy there are very few vernacular names that can be confidently applied to particular species of Orthoptera. Following advice kindly given by our colleagues Eugenio Morales Agacino, José Quartau and Baccio Baccetti, we have cited vernacular names in Spanish, Portuguese and Italian for only two species, *Gryllus campestris* and *Acheta domesticus*. There are vernacular names for *Gryllotalpa* in all three languages but, because there are several morphologically similar species of this genus in these southern peninsulas, we cannot apply such names to particular species.

The English vernacular names are given first in bold type, followed by those of the other languages (when represented) in the following sequence: French (F), German (D), Dutch (NL), Danish (DK), Swedish (S), Norwegian (N), Finnish (SF), Spanish (E), Portuguese (P) and Italian (I).

The oscillograms

The songs of each species of Orthoptera are illustrated by oscillograms at two, three or sometimes four different speeds, produced by the method described on p. 23. The speeds chosen (as indicated by the scale lines) are standard for each major group, but differ from group to group in order to show most clearly the characteristic features of the songs in each group. The slowest speed gives the best impression of the rhythmic pattern of the song as heard by the human ear; the faster speeds reveal features of the song that cannot be detected by the unaided ear. For the Tettigoniidae the speeds used are 20 mm/s, 160 mm/s, usually 640 mm/s, and occasionally a fourth speed of 2560 mm/s, 6400 mm/s or 12 800 mm/s in order to show the tooth-impacts in a non-resonant song or the carrier wave in a resonant song. For the Gryllidae the slowest speed is also 20 mm/s, but the higher speeds are 320 mm/s and 6400 mm/s (showing the carrier wave of the calling songs), with the fourth speed of 25 600 mm/s in three species to show the carrier wave of the high-frequency components of their courtship songs. For the Gryllotalpidae the two slower speeds are 20 mm/s and 320 mm/s, as for the Gryllidae, but the highest speed of 3200 mm/s was found to be better for showing the carrier wave of the rather lower pitched calling songs. The rhythmic pattern of the songs of the Acrididae were found to be better shown by slower oscillograms: 10 mm/s, 80 mm/s, usually 320 mm/s and occasionally 1280 mm/s; for the unusually long songs of *Chorthippus cazurroi* and *C. mollis* we have included an extra slow speed of 5 mm/s so that the whole song could be shown in one oscillogram. Unless otherwise stated the oscillograms of gomphocerine grasshoppers are taken from 'two-legged' songs, but for a number of species we have included oscillograms taken from the songs of males with only one hind leg; such oscillograms often show the syllables more clearly (e.g. Figs 1125, 1265, 1352, 1504) and are also useful in showing the individual tooth-impacts from a single stridulatory file (e.g. Figs 871, 960, 1031, 1241).

The oscillograms of cicada songs are at the single speed of 10 mm/s; higher speeds were found to show little of interest in comparing these songs with those of Orthoptera. The oscillograms of toad calls are at 20 mm/s, 320 mm/s and 1600 mm/s, and those of bird songs are at 20 mm/s, 160 mm/s, 640 mm/s and 3200 mm/s, with the extra speed of 320 mm/s for the Nightjar (*Caprimulgus europaeus*) to facilitate comparison with the songs of mole-crickets.

For each species of Orthoptera the number of calling songs illustrated by oscillograms depends mainly on how variable the song is, both individually and geographically. We have, for example, included 41 oscillograms taken from 17 songs of the interesting grasshopper *Chorthippus yersini*, which shows much individual variation as well as having a number of local 'song-dialects'. At the other extreme we have sometimes given oscillograms from only one song when the species concerned shows little variation or is so rare that we have recorded songs from only one male. For some widespread species we have felt it desirable to give oscillograms of songs from widely scattered localities in order to demonstrate that there is *no* detectable geographical variation.

The compact discs

The two companion compact discs include excerpts from the songs of all 170 species of Orthoptera whose songs are described in this book, as well as four cicadas, two toads and five birds whose songs could be confused with those of Orthoptera. The excerpts from the songs of Orthoptera have been selected as being typical of each species and, in some cases, showing how the songs vary geographically.

Each track on the compact discs begins with an announcement of at least the Latin and

English names of the species (except for the cicadas, which have no generally agreed English names). We have given much thought to the pronunciation used for the Latin names. In the English-speaking world there are two main systems for pronouncing Latin: the traditional pronunciation used by gardeners and many biologists, and the 'reformed' academic pronunciation used by classical scholars. The traditional pronunciation simply treats Latin as if it were English, and thus uses the long vowel sounds (as in *mate, mete, mite, mote, mute*) so characteristic of spoken English since the Great Vowel Shift of the 15th and 16th centuries; speakers of other European languages, in which these vowels are usually pronounced quite differently, often find the traditional English pronunciation of Latin unintelligible. The academic pronunciation of Latin is nearer to the pronunciation used by Continental Europeans, but differs in making *ae* sound like a long English *i* (as in *mite*) and *v* like an English *w*; in addition, *c* and *g* are always hard (as in *cap, gap*), not being softened before *e, i* and *y* as in the Romance languages and, at least in words of Romance origin, in English.

Our compact discs are intended for use throughout Europe and so we have adopted a pronunciation for scientific names that should be easily intelligible to all Continental Europeans. The vowels are pronounced as in *far, met, fit, on* and *pull*, becoming more or less extended according to whether they are long or short (the long *i* is as in *machine*, the long *u* as in *brute*); *y*, when used as a vowel, is pronounced in the same way as *i*. The intrusive *y* sound that the English frequently use when pronouncing a long *u*, as in *mute* ('myoot'), *tune* ('tyoon'), is eliminated completely. The combinations *ae* and *oe* are pronounced as 'ee', *au* as 'ow' (in *cow*) and *ei* as in *rein*. The consonants *c* and *g* are pronounced hard before *a, o* and *u* (as in *cap, gut*), soft before *e, i* and *y* (as in *cent, gin*); *ch* is pronounced as 'k'; *j* is pronounced as a consonantal 'y'; *s* is always voiceless, as in *sit*; and *v* is given its usual English pronunciation (as in *van*).

Scientific names based on personal and modern geographical names are treated as exceptions: the name of the person or place is pronounced as it is in its country of origin, with the addition of the Latin ending and with any modification that such an addition requires. Hence, the *ch* in *chopardi* (named after the Frenchman Lucien Chopard) is pronounced 'sh', as in French, and the *ci* in *cialancensis* (named after Cialancia in the Piedmont Alps) is pronounced like an English 'ch', as in Italian, but the *in* in *yersini* (named after the French Swiss Alexandre Yersin) loses its nasal French pronunciation (as it would do in French) because it is no longer the terminal syllable.

In names with several syllables there is also the question of which syllable to stress. We have followed the usual rules for the accentuation of Latin words; these are outlined by Stearn (1992), who also offers useful general guidance on the pronunciation of Latin.

We are confident that the pronunciations we have adopted will be easily comprehensible to Continental Europeans. Some of them will sound strange to English ears, but the English vernacular names, always given in addition, together with the accompanying written list of species, should soon dispel any doubt.

Chapter 2

Acoustic methods

In the course of assembling the data for this book we have made over 1200 tape recordings of the songs of western European Orthoptera, about half of them in the field and most of the remainder in the Acoustic Laboratory of the Natural History Museum in London. For some common and widespread species we have made song recordings from over 40 males from many different localities in western Europe and, at the other extreme, we have sometimes recorded the song of only one male belonging to a rare and very local species. All these recordings have been analysed oscillographically and used in the preparation of the song descriptions; about half of them have been used for the oscillograms published in the book; and about a quarter for the song excerpts on the compact discs.

As emphasized in the Introduction this book is concerned solely with *audible* sounds and this chapter deals only with our methods of recording and analysing sounds of frequencies below about 20 kHz. It is quite possible to record ultrasound, using more specialized equipment, as is discussed, for example, by Pye (1992), but such techniques are seldom needed in using orthopteran sounds for identification and taxonomic research.

For such purposes it is also unnecessary to make stereophonic recordings, and the techniques described here are purely for recording monophonically with a single microphone. Information on the stereophonic recording of wildlife sounds is provided by Tombs (1974, 1980), Margoschis (1977) and Simms (1979).

Finding the insects

Most singing grasshoppers and crickets can be heard by human ears from a distance of several metres. Given suitable weather at an appropriate place and time, finding males is simply a matter of listening for the song and then tracking it down. Trying to find them in wet weather is a waste of time, and even in dry weather the day-singing species seldom sing when it is cold and dull. When the insects are abundant they are usually easy to find, but in sparser populations, particularly of bush-crickets, it may be necessary to track down a single singing male. The key to doing this successfully is to move slowly and cautiously, and this often requires a lot of patience. Most of these insects will stop singing if they are disturbed, and some are much more easily disturbed than others. As soon as the singing stops one must remain completely motionless until it begins again. Bush-crickets are generally more difficult to track down than true crickets and grasshoppers, partly because their songs have remarkably ventriloquial properties and partly because they often sing from less accessible places. Moreover, if they are sufficiently disturbed they may move deeper into the undergrowth and it may be some time before they begin to sing again. Ideally, one should approach such insects so stealthily that they never stop singing until one can actually see them. A bush-cricket song can be more accurately pin-pointed if two people approach it from different directions.

Some of these insects (including most of the grasshoppers) sing only during the day, many bush-crickets and true crickets sing both during the day and at night, and some are strictly nocturnal singers. Night-singers are often easier to track down (with the help of a torch or head-mounted light) than day-singers; they seem less easily disturbed and often continue to sing when illuminated.

Some quiet singers, especially many of the Phaneropterinae and some of the smaller Decticinae and Conocephalinae, have songs that are so faint or high-pitched that they cannot be easily tracked down even by the best of unaided human ears; they are indeed completely inaudible to some people. In such cases an ultrasound detector (bat detector), which converts the frequencies that are normally too high to hear into easily audible sounds, can be very useful, both by day and at night.

The basic data for the recordings used for the oscillograms reproduced in this book and for the excerpts included on the companion compact discs are given in Appendix 3 (pp. 491–537). Further information on our own recordings is given in the following pages.

Recording the songs in the field

We have heard it said, even by bioacousticians, that it is impossible to make good recordings of orthopteran songs in the field. With this we totally disagree. There are some quiet singers for which a quiet indoor environment is necessary, and it is of course essential to use a laboratory for much experimental work on the songs, but when we embarked on this project in the seventies we soon learnt that it was quite possible to obtain high quality field recordings of the songs of most of the European Orthoptera. Indeed, in some ways such recordings are preferable in that one can feel more confident that the insect is producing a typical song, uninfluenced by the artificial conditions of a cage and studio. We have also found that some species are reluctant to sing in the unusually quiet environment of a sound-proof room. About a third of the recordings used for the companion CDs were made in the field.

For field use recording equipment has to be portable and battery-operated. Although the miniaturization of such equipment usually results in some loss in recording quality, the best pocket-sized cassette recorders will now produce better recordings than the heavy mains-operated open-reel recorders of 30 years ago, and recordings of excellent quality can be obtained with equipment that is not much larger than this. The choice of a tape recorder for field use has to be a compromise between weight and quality, quite apart from considerations of cost and the idiosyncracies of personal preference. We have used the Uher Report series of open-reel recorders (4000 L, 4200, 4200 IC or 4200 Monitor) for all our field-work. Although portable cassette recorders are now able to approach these machines in quality, the limitations of narrower tape and slower tape speed prevent them from quite equalling the quality of open-reel recorders. The facility with which the tape speed can be reduced on the Uher recorders has also lent itself admirably to our method of analysis (see below). The original recordings were, however, all made at a tape speed of 19 cm/s in order to achieve the best possible quality. The tape used was BASF LG35, LP35, LP35LH or LPR35LH. Although most of our recording equipment is stereo, almost all our recordings are mono; there is little point in using stereo when recording the song of a single male in a quiet environment.

The famed Kudelski Nagra III and IV open-reel recorders are of significantly higher quality than the Uher Report series, but are about twice as heavy and we found them to be too cumbersome for convenient use in the field. The Nagra SN, on the other hand, is much lighter than the Uher Report, but we considered its drawbacks, including the use of non-standard tape, to outweigh this advantage.

These high quality analogue recorders have recently been rivalled by portable digital recorders using rotary recording heads and digital audio tape (DAT) in very compact cassettes. Such machines are as light as a portable analogue cassette recorder of professional standard and can produce recordings of significantly higher quality. They are much more expensive at present, but are likely to become more widely used in the recording of wildlife sounds.

The microphone we have used for all our field-work is the AKG D202. This well-known dynamic microphone is unusual in having separate low and high frequency capsules, with a change-over frequency of about 800 Hz. We have had our microphones modified so that the low frequency capsule can be switched off. Most orthopteran songs are above 800 Hz in frequency, and so by recording with the low frequency capsule switched off one can exclude much unwanted background noise (which is mostly below 800 Hz), while leaving the song unaffected. In general, dynamic microphones, with a cardioid response (most sensitive in the direction in which they are pointed), are best for field recording. Capacitor microphones tend to be less robust and are sometimes affected by changes in humidity.

Whether headphones are used to monitor field recordings is a matter of personal preference. They are particularly helpful to beginners in wildlife recording as they quickly reveal a number of causes of poor quality recordings, e.g. over-modulation (see below), handling noises and other background sounds that one might not have otherwise detected. Headphones are of course especially useful when off-tape monitoring is possible. They should be well padded and of the closed back kind to exclude directly heard sound. (The loudspeaker on the recorder cannot be used for monitoring in the field because the sound from it is likely to be picked up by the microphone, resulting in a feedback 'howl'.)

Given the appropriate equipment, all one needs to make a field recording of a diurnal singer is a calm, sunny day in a fairly quiet place. Such conditions are common enough in continental Europe, but much less so in Britain, where it is often windier and usually subject to more noise from aircraft and road traffic. The effect of light winds on the microphone can be largely eliminated by fitting it with a windshield, but when the wind is strong one should seek a sheltered spot, perhaps in the lee of a hill, or resort to recording indoors from a captive insect. If there is no sunshine the insects will probably not sing at all and even if they do the song is likely to be much slower than usual and not really typical of the species.

Recording from nocturnal singers is usually easier: it is usually quieter and calmer at night and the insects are less easily disturbed. It is helpful to use a head-mounted light or to have a second person to shine a torch on the insect while one positions the microphone.

It is generally best to have the microphone close to the singer, so that the song will be

Figure 2 The senior author making a field recording.

recorded at a much higher level than any other sound the microphone may pick up. For a quiet song a microphone distance of 10 cm or even less may be desirable, and even for a loud song the microphone should be no more than a metre from the insect. Holding the microphone in one's hand should be avoided whenever possible, as there is a high risk of handling noises being recorded. If the insect is sitting on or near the ground, the microphone can usually be rested on the ground near to it. If the vegetation makes this impossible the microphone can sometimes be suspended over the insect by its lead. If the microphone has to be hand-held, great care should be taken to keep the hand and fingers still and also to arrange the microphone lead so that it is not brushing against anything. Whenever possible, the recorder should be on the opposite side of the microphone from the singing insect, so that any noise it makes while running is not recorded (Fig. 2). Similarly, it is best to position the microphone so that any obtrusive background sound (e.g. running water from a stream or a distant agricultural tractor) is behind it, in the direction of its lowest sensitivity.

Parabolic or conical reflectors have been used with some success in recording insect sounds, but we have not felt the need for such aids and have not used them for our recordings.

Manual control of recording level ('gain') is almost essential when recording orthopteran songs, and the level used for each recording is crucial. The aim should be to achieve as high a signal/noise ratio as possible without using such a high level as to cause over-modulation distortion. For recording many sounds, such as speech and music, the level meter on the recorder is a reliable guide in setting the level control. However, these meters are rather insensitive to the high frequencies of grasshopper and bush-cricket songs and are consequently of limited use in determining the best level setting for recording them. If the recorder allows off-tape monitoring, the level can be adjusted while recording so that it is as high as possible without any audible distortion. If off-tape monitoring is not possible, the level should be set so that a grasshopper or bush-cricket song gives no more than a small deflection on the level meter; playing back trial recordings will then allow further adjustments to the level if necessary. The songs of the true crickets and mole-crickets contain much less high frequency sound and so a level setting giving a higher reading on the meter can be used.

For the present study our aim has usually been to make recordings of the songs of single males, ideally with no other nearby Orthoptera either singing themselves or influencing the subject male in any way. Recordings made in such conditions are ideal for analysis and are likely to be of typical calling songs. When two or more males are together they often influence one another and none of them can be relied on to produce a typical calling song. In the grasshoppers and crickets the presence of a female will often result in a male producing a special courtship song, which, although of interest in its own right, may be quite different from the calling song. If a population is so dense that an isolated male is difficult to find, it may be necessary to capture a male and release it in a more suitable place for a song recording.

A frequent problem in field recording, especially with grasshoppers, is that males are disinclined to sing from one spot for any length of time. Just as one has all the equipment set up for a perfect recording the male will move to another spot several metres away! It is in fact normal behaviour for the males of some species (e.g. *Omocestus viridulus* and *Stauroderus scalaris*) to produce only one calling song each time they settle in one place and then to fly off to another spot before singing again. In dealing with such species in the field patience and perseverence are essential. Fortunately, *O. viridulus* sometimes sings more than once from one spot. *S. scalaris* is less likely to do this, but we have found one technique that is usually successful. One follows a male making its post-song flight and disturbs it slightly when it lands to prevent it producing a calling song immediately; the equipment is then positioned appropriately and set to record, after which there is a good chance that the male will produce a typical calling song before moving again.

A quite different problem is presented by a burrowing insect that sings from the mouth of its burrow, such as *Gryllus campestris*. This insect retreats into its burrow at the slightest disturbance and it can be some time before it re-emerges and begins to sing again. The best technique to use in this case is to mount the microphone in a suitable stand (or simply rest it on the ground) so that it is pointing at the mouth of the burrow, and then to use a long microphone lead so that one can sit with the tape recorder well away from the burrow. Then, when the insect resumes singing, it is particularly easy to make a good recording without disturbing it any further.

An essential part of our field technique for the present study has been the capture of the singing insect after a satisfactory recording of its song has been made. Each of our recordings is thus normally associated with a specimen in the reference collection of the Natural History Museum in London.

Recording the songs of captive insects

Most of our recordings of captive insects were made in the purpose-built Acoustic Laboratory of the Natural History Museum, London; this no longer exists, having been reconstructed for completely different purposes since the completion of work on this book. It contained an inner studio (with thermostatically controlled heating) which could be acoustically isolated from an outer laboratory. The insect to be recorded was placed in a cage in the studio with the microphone, mounted on a suitable stand, pointing at it; the lead from the microphone passed through a dividing wall into the outer laboratory, where the tape recorder and all other acoustic equipment was housed. A double-glazed observation window enabled the singing insect in the studio to be seen from the outer laboratory. Both the studio and the outer laboratory had carpeted floors, acoustic tiles on the ceiling, and walls lined with 25 mm thick padding under 25 per cent perforated hardboard; there were no external windows.

We used simple cages of various sizes, with the walls made of acoustically transparent netting. A bench lamp was used to provide diurnal singers with light and radiant heat. The air temperature for a diurnal singer was usually about 25°C, but a lower temperature was used for most nocturnal singers.

Almost all our studio recordings were made with a Nagra IV tape recorder and Sennheiser MKH405 condenser microphone. The Nagra recorder was modified from its standard specification in two ways: the slowest standard tape speed of 9·5 cm/s was replaced by an extra high speed of 76 cm/s, and the recording amplifier circuit was fitted with a switchable low-pass heterodyne filter with a corner frequency of 20 kHz and slope of 24 dB/octave. The extra high tape speed has the potential to improve recordings of high frequency sounds, and we found it helpful for a few difficult subjects. The filter was particularly useful in recording bush-cricket songs and enabled undistorted recordings to be made at a higher level than is possible with the unmodified circuit.

Most of our studio recordings were made at a tape speed of 38 cm/s, some at 19 cm/s and a few at 76 cm/s. The tape used was BASF SP52, SPR50LH or 911.

As with field recordings the microphone should be close to the singing insect and preferably pointing straight at it. The recording level setting is again important and should be just below the point at which distortion begins. The level meter of the Nagra, like that of other tape recorders, is of little value in determining the best setting, and the level should be set, whenever possible, by monitoring off-tape with a good quality amplifier and speaker. If the monitoring speaker is acoustically isolated from the microphone, headphones are not needed.

When indoor recordings of captive insects were necessary during our field excursions, we made them in a suitable quiet room and used the same recording equipment as for field recordings. In a small room with reflective walls and ceiling the microphone is liable to pick up the mechanical sound made by the tape recorder, but this can be minimized by using a long lead so that the tape recorder can be as far from the microphone as the room will allow. As in studio recording, the cage used always had walls made of acoustically transparent netting. An electric light bulb, for example in a table lamp, was used to provide diurnal singers with light and radiant heat. Nocturnal singers were usually recorded in total darkness, often in the early hours of the morning.

Storage and documentation of tape recordings

Almost all our recordings have been made on 13 cm spools of tape, mainly because this is the largest size that can be used on the Uher 4000 series of recorders. We have found it convenient to store these spools in the plastic boxes made for them by BASF or, since BASF stopped making these boxes a few years ago, the very similar boxes made by the French firm

Posso. Our tapes are all kept in their original form and are simply numbered serially as they are added to the collection. The tapes of field recordings normally include a number of different species; we have not reassembled such recordings into 'species tapes' since for our purposes there would be little to gain from the extra labour of doing so.

The tapes have been stored at a fairly even temperature, averaging about 20°C, and we have not detected any deterioration in well modulated recordings made up to thirty years ago. However, we have one or two early recordings of quiet tettigoniid songs in which the signal/noise ratio, poor when they were originally made, is now noticeably worse. Only two or three tapes out of the 800-odd we have used have shown any signs of being physically defective: these were shedding an excessive amount of their oxide coating and the recordings on them, which seemed to be unaffected, were copied as a safety precaution.

All our recordings were made for scientific research and so needed to be fully documented. The information was entered on a standard data sheet, of which a typical example is shown in Fig. 3. This sheet, 24 × 20 cm in size, was designed so that it could be folded to fit comfortably inside the plastic boxes in which our tapes are kept. The details were typed manually on to the sheet, using a small-sized type (16 characters/inch) and taking a carbon copy on an identical sheet. The top copy went (unfolded) into our file catalogue and the carbon copy was put with the tape. We used our oscillograph (see below) to make a routine analysis, at several different paper speeds, of a typical part of each recording, and the resulting oscillograms were filed in transparent folders with the top copy data sheets. These sheets are filed systematically by major group and then alphabetically by genus and species, so that they provide both a catalogue and an index to the collection. By virtue of the full data and analyses that these files contain, they provide a most useful data bank in themselves, often enabling us to obtain all the information we needed about the songs of particular species without needing to take a tape off the shelf.

Analysing the songs

In order to make use of song differences in taxonomic research it is particularly important to be able to analyse the sounds so that objective comparisons may be made. By far the most taxonomically useful characteristic of orthopteran songs is their rhythmic pattern, and so what is needed is an objective representation of this pattern against a time scale. During the period of our study the two main methods of producing such a graphic representation have been by the use of (1) an oscilloscope or oscillograph, producing an oscillogram, i.e. a plot of the waveform (and hence amplitude) of the sound on a time scale, and (2) an audiospectrograph (or sonagraph), producing an audiospectrogram (or sonagram), i.e. a plot of the frequency spectrum of the sound against a time scale. Where there is a clear dominant frequency in the song an oscillogram also enables that frequency to be accurately measured. Conversely, a sonagram gives a rough idea of the amplitude of the sound by the darkness of the trace.

Among the Orthoptera, frequency seems to be of relatively little taxonomic importance in the Acrididae and Tettigoniidae, becoming more significant in the Gryllidae and Gryllotalpidae, which produce more musical sounds of well defined pitch. Sonagrams are thus quite useful in the analysis of cricket songs, but are much less appropriate for portraying the songs of grasshoppers and bush-crickets. Oscillograms give an excellent picture of

Figure 3 A typical completed data sheet for a tape recording in the Natural History Museum, London.

the rhythmic pattern of the sound in all these groups and, at an appropriate speed, will also show quite clearly any dominant frequency in the song. In order to have a uniform method of graphic representation for all these groups, we therefore decided to use oscillograms.

In the early seventies when our studies began, choosing the best method of producing oscillograms was not easy. We decided against the use of an oscilloscope camera or ultra-violet oscillograph because of the difficulties of using light-sensitive film or paper and the problems of reproducing the results adequately in a publishable form. We eventually decided in favour of the Siemens Mingograf 34T ink-jet recorder because of its extreme ease of use, low running-cost and ability to produce instant oscillograms (on either grid or blank paper) for permanent record or publication without any further treatment. After allowing for tape speed reduction to bring the full audio frequency range within the Mingograf's capabilities, the wide choice of paper speeds (ranging from 2·5 mm/s to 500 mm/s) proved to be precisely what we needed. Unfortunately the Mingograf 34T is no longer made and the nearest equivalents produced by Siemens at the time of writing are much less suitable for analysing insect songs.

The latest method of producing graphic analyses of orthopteran and other sounds is by using computer-generated plots from digital or digitized recordings. Although at present this method has certain limitations in resolution and in the duration of song that can be continuously analysed, its ease of use and versatility are likely to make it a widely used technique for many years to come.

Copying recordings

We have deposited copies of our recordings in the Wildlife Section of the National Sound Archive (NSA), a subdivision of the British Library. The high frequencies occurring in grass-hopper and, especially, bush-cricket songs cause problems when copying analogue recordings, just as they do when making recordings direct from the singing insect (see above, p. 18). The level meter of the copying tape deck is again a poor guide to the recording level at which the copy should be made, and off-tape monitoring (or, if that is not available, trial and error) should be used to achieve the best possible signal/noise ratio without distortion. When making copies of our recordings for the NSA we found that a significantly better signal/noise ratio could be achieved without distortion by filtering off frequencies above 15 kHz. When these recordings are used in taxonomy and identification, or simply listened to for interest, these high frequencies seldom serve any useful purpose.

Song terminology

It is hardly an exaggeration to say that there are almost as many systems of orthopteran song terminology as there have been bioacousticians working on these insects. Several attempts have been made to establish a consistent set of terms for general use by orthopterists (Chavasse et al., 1955; Broughton, 1963a; Alexander, 1967a); they did not agree with one another and each failed to gain general acceptance. Broughton's attempt, based on a draft circulated for comment to 36 workers in the field, perhaps deserved more attention than it got; following the German use of 'Silbe', he proposed the English term 'syllable' for the sound produced by one to-and-fro movement of the sound-producing apparatus, but he failed to provide a satisfactory term for a group of syllables. He remedied this in 1976, introducing the new term 'echeme' (pronounced eck´eem) for a first-order assemblage of syllables, and thus provided the basis for the terminology we have adopted here.

In the English-speaking world, North American workers on orthopteran songs have adopted a terminology of their own, with little regard for European practice. The term 'phonatome', used by the French for a simple wave-train (equivalent to the English 'pulse') has been adapted by some Americans or Canadians working on tettigoniid songs (e.g. Walker & Dew, 1972; Morris & Walker, 1976) to mean the same as the syllable of British bioacousticians; this agrees with the French use of phonotome for the songs of Gryllidae and Gryllotalpidae, but not for those of Tettigoniidae and Acrididae. The term 'echeme' has apparently not yet crossed the Atlantic.

In deciding on the best song terminology to use in this book our main choice was between terms based on the movements of the sound-producing apparatus and terms based on the nature of the sounds themselves without regard to how they are produced. The second alternative needs a term for a sound that seems unitary to the human ear, such as 'chirp' in the sense of Broughton (1963a and 1976, but not 1952b). Such a term presents two difficulties. First, whether a sound seems unitary is clearly a subjective matter likely to give rise to disagreement. Second, a sound that seems unitary when the insect is singing in warm conditions may well become split into several clearly distinct sounds when the conditions are cooler and the sound-producing apparatus is moving more slowly; for example, the diurnal song of a bush-cricket may have to be described in quite different terms from the song of the same insect singing on a cool night.

Adopting a terminology based on the leg- or wing-movements of the singer overcomes these difficulties and is completely objective. It does, however, require some knowledge of how the sounds are associated with these movements. In many cases this is quite obvious, but when the movements are very rapid it may be difficult to determine, even by close observation, exactly how they correspond with the elements of the sound produced. Fortunately this problem has been solved by the use of a Hall-generator (Elsner, 1970, 1974a) or opto-electronic camera (Helversen & Elsner, 1977; Elsner & Popov, 1978; Heller, 1988) to record these movements in many European Acrididae and almost all the European Tettigoniidae. The calling songs of the European Gryllidae and Gryllotalpidae are easily correlated with the wing-movements producing them, and the more complex wing-movements used in the courtship songs of Gryllus campestris and G. bimaculatus have been studied in detail by the use of miniature angle detectors (Koch, 1980; Koch et al., 1988). In those Gomphocerinae whose stridulatory leg-movements have not yet been studied we have seldom had any difficulty in determining how they are related to the songs, either by direct observation of singing males or by oscillographic analysis.

The term 'syllable' lends itself well to such derivatives as diplosyllable, hemisyllable, microsyllable, monosyllabic, disyllabic, etc., and even for this reason alone we would prefer it to such alternatives as 'phonotome' in the sense of Walker & Dew (1972) and Matthews & Matthews (1978). We have found Broughton's term 'echeme' for a group of syllables to be indispensable in describing orthopteran songs, and we prefer it to the term 'mode' coined at about the same time by Morris & Walker (1976) and defined slightly differently. These terms do not of course preclude the use of simple descriptive terms such as 'buzz', 'tick', etc., where these are helpful in conveying an impression of an insect song.

The main terms we have used in describing songs are defined below, beginning with a brief discussion of the use of the term 'song' itself. Note that the French and German equivalents given have not always been used in the senses defined here; for example, Dagmar & Otto von Helversen, in their excellent studies on the song of *Chorthippus biguttulus*, have used the terms 'Vers', 'Silbe' and 'Puls' for the song-elements we refer to as echeme-sequence, echeme and hemisyllable, respectively.

Song. In bioacoustics this term is used in two main senses: in the broadest sense it is applied to the deliberate acoustic output of animals (or a group of animals) in general, and in a more restricted sense it is applied to the acoustic output of a particular species or individual. In this book the term is almost always used in the latter, more restricted sense. For species producing relatively short bursts of sound, usually lasting less than 30 s, separated from other similar bursts of sound by intervals of more than about 10 s, we refer to such bursts as songs (as well as, more specifically, echemes or echeme-sequences,

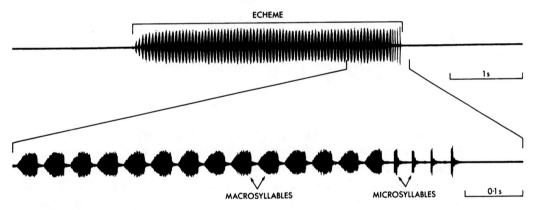

Platycleis falx

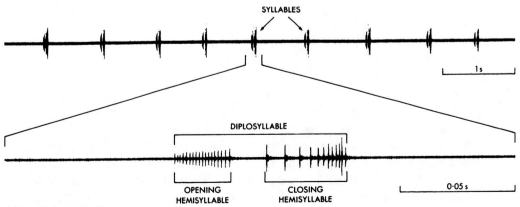

Platycleis veyseli

Figure 4 Oscillograms of the calling songs of two species of bush-cricket, showing the terminology used in this book. (From Ragge, 1990)

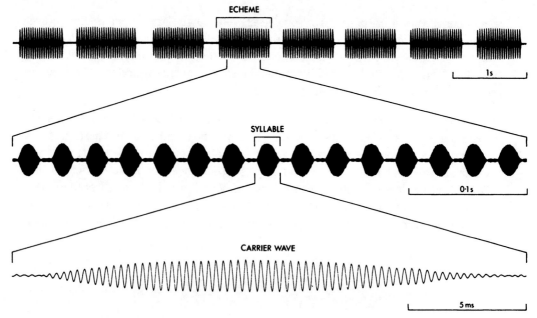

Oecanthus pellucens

Figure 5 Oscillograms of the calling song of a cricket, showing the terminology used in this book.

as appropriate). For species producing continuous sound, or regularly and rapidly repeated short bursts of sound, in each case for long periods of indefinite duration, we
refer to the whole of such an emission as a song. Thus we refer to the song of *Chorthippus
vagans* as a single, relatively long echeme, but the song of *C. dorsatus* as a long series of
short echemes. The distinction is one of degree only, and when the intervals between
repeated units of sound are of the order of 10 s one can apply the term either way. (French:
chant; German: *Gesang*.)

Calling song. The song produced by an isolated male. (French: *chant ordinaire, chant d'appel*;
German: *gewöhnlicher Gesang, Lockgesang, Rufgesang*.)

Courtship song. The special song produced by a male when close to a female. (French: *chant
de cour*; German: *Werbegesang*.)

Rivalry song. The special song produced by two or more males reacting to one another.
(French: *chant de rivalité*; German: *Rivalengesang*.)

Syllable (Figs 4–7). The sound produced by one to-and-fro movement of the stridulatory
apparatus. In the Ensifera we have always considered a syllable to begin with the opening stroke of the fore wing (except when this stroke is silent) and to end with the closing
stroke. In the gomphocerine grasshoppers we have usually considered a syllable to begin
with the upstroke of the hind femur and to end with the downstroke (assuming that
both strokes produce sound); in some species, however, we have found it more convenient to consider a downstroke followed by an upstroke to constitute a syllable. (French:
syllabe, accent; German: *Silbe*.) (See below.)

   The songs of a number of European bush-crickets (especially Decticinae) include two
contrasting kinds of syllable, differing markedly in duration. In such songs we have termed
the longer, more normal ones *macrosyllables* and the shorter ones, usually lasting less

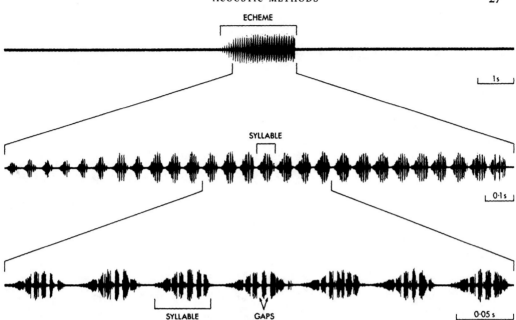

Stenobothrus stigmaticus

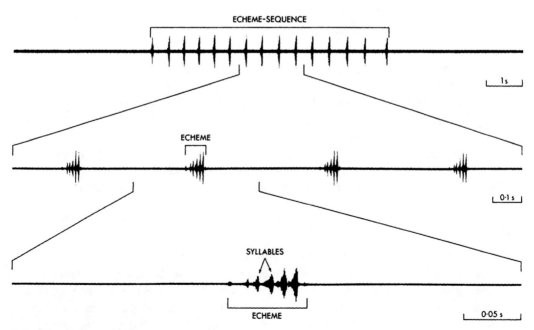

Stenobothrus grammicus

Figure 6 Oscillograms of the male calling songs of two species of grasshopper, showing the terminology used in this book. (From Ragge, 1987a)

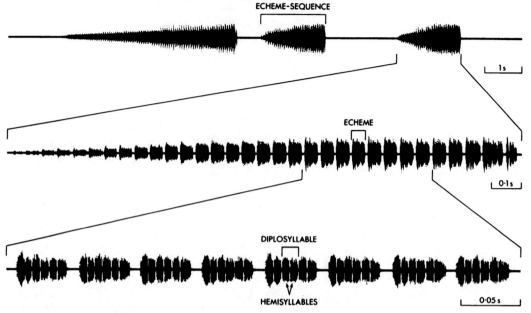

Chorthippus biguttulus

Figure 7 Oscillograms of the male calling song of a third species of grasshopper, showing the terminology used in this book. The oscillograms are taken from the song of a male with only one hind leg, so that the song-pattern can be seen more clearly.

than 10 ms, *microsyllables* (Fig. 4).

In gomphocerine grasshoppers there are often momentary breaks in the sound during the louder, downstroke part of the syllable, and occasionally during the upstroke. We have referred to such breaks, of at least 1·25 ms duration, as *gaps* (Fig. 6).

Diplosyllable (Figs 4, 7). A syllable in which sound is produced by both opening and closing wing-strokes in the Ensifera, and both upward and downward leg-strokes in gomphocerine grasshoppers.

Hemisyllable (Figs 4, 7). The sound produced by one unidirectional movement (opening, closing, upward or downward) of the fore wings or hind legs. (See below.)

Echeme (Figs 4–7). A first-order assemblage of syllables. (French: *phrase*; German: *Vers.*) (See below.)

Echeme-sequence (Figs 6, 7). A first-order assemblage of echemes. (French: *strophe*; German: *Strophe.*)

Carrier wave (Fig. 5). In the context of this study, the carrier wave is the fundamental wave of a resonant song, i.e. a song with an almost pure dominant frequency. Resonant songs are characteristic of the true crickets and mole-crickets, and it is the carrier wave that gives rise to the musical pitch of these songs. Some bush-cricket songs are also resonant, but the frequency of the carrier wave in these cases is usually too high to produce a musical pitch.

The stridulatory movements of the fore wings and hind legs are sometimes sufficiently complicated to cause difficulties in applying some of these terms. For example, in the course of one song the wing-strokes or leg-strokes may vary greatly in length and the wings or legs may be vibrated in higher or lower positions. In such cases we have used the term syllable for the sound produced by any significant to-and-fro movement, irrespective of whether the movement is long or short, or of the position in which the vibrating wings or legs are held. The wings or legs may pause, sometimes several times, in the course of a stroke, but we have always treated the sound produced by such a stroke as a single hemisyllable, irrespective of these interruptions in movement and sound. Sometimes there are *slight* (usually silent) reversals in the direction of movement during a largely unidirectional stroke of the wings or legs; these we have generally ignored in applying our terminology. In deciding when to treat such reversals as sufficiently significant to affect the terminology we have used common sense and analogy to the songs of related species.

In gomphocerine grasshoppers a further complication arises from the fact that there is a paired stridulatory organ: both hind legs are normally used to produce sound simultaneously. Even when the left and right legs perform the same movements they are often not quite in phase, and sometimes their movements are significantly different. Occasionally they vibrate in opposite phases, so that the upstroke of the left leg coincides with the downstroke of the right leg, and vice versa (e.g. the calling songs of *Myrmeleotettix maculatus* and *Chorthippus albomarginatus*). These complexities are of considerable ethological and physiological interest, but are of less importance taxonomically and it is usually quite easy to decide the best way to apply the terms we have adopted.

For convenience in describing the songs of Tettigoniidae the definition of the term hemisyllable given above is slightly looser than that given by Broughton (1963b: 853). Broughton took the view that this term should be used only in referring to each of the two components of a diplosyllable, i.e. when sound is produced by both the opening and closing movements of the fore wings; he suggested the term 'haplosyllable' for a syllable in which sound is produced by only one of the two movements of the fore wings, usually the closing stroke. In many Tettigoniidae the opening strokes of the fore wings sometimes produce sound but sometimes do not; using Broughton's definition the sounds produced by the closing strokes would sometimes be hemisyllables and sometimes haplosyllables, and description would become too cumbersome. The definition of hemisyllable given above removes this difficulty. In songs containing both micro- and macrosyllables we have further simplified the terminology by sometimes referring to 'closing macrosyllables' when 'closing macrohemisyllables' would be strictly more accurate.

There are occasional difficulties in applying the term echeme. For example, the song of *Platycleis stricta* includes a dense series of equally spaced syllables in which every third one is louder than the rest (Figs 344, 346). It could be argued that these units of one louder and two softer syllables (or even just the two softer ones) are first-order assemblages of syllables and that they should therefore be termed echemes, but we have found it less cumbersome to regard the complete series of syllables as a single echeme. In effect, this interpretation requires the presence of distinct pauses as a criterion for separating groups of syllables into echemes, rather than the mere repetition of a pattern based solely on changes in amplitude. Similarly, the syllable-sequences of some grasshoppers (e.g. *Chorthippus pullus*, *C. dorsatus*) are divided, without pause, into two distinct parts, in which the legs are moved in a different way and produce a different kind of sound; we have found it convenient to treat the whole of such a sequence as a single echeme rather than a combination of two.

Chapter 3

Sound production and reception in the European Orthoptera

The songs of Orthoptera have attracted the attention of many insect physiologists and there is now a vast literature on the physiology of sound production and reception in these insects. The subject has been reviewed many times and we shall not attempt here to give more than a brief account, particularly of the main methods of sound production in Orthoptera. Readers needing more detailed information are referred to Bailey (1990, 1991a), Ewing (1989), Greenfield (1997) and, for crickets, the relevant chapters of Huber *et al.* (1989) and the more recent studies of Pfau & Koch (1994) and Stephen & Hartley (1995). In the Orthoptera as a whole many different methods of sound production have been developed (see especially Kevan, 1955), but we shall describe here only those methods used by the western European species included in this book.

Stridulation

This is the term generally used for deliberate sound production by the friction of one structure against another. Some authors (e.g. Broughton, 1963a; Leston & Pringle, 1963; Sales & Pye, 1974) have used the etymologically more appropriate term 'strigilation' in this sense, but it has not been adopted generally.

The stridulatory organ of the Tettigoniidae is at the base of the fore wings (Fig. 8); in some brachypterous species the stridulatory organ is the only part of the fore wings that remains. In the male, a toughened edge (plectrum) at or near the hind margin of the right fore wing is scraped against a file (formed along vein Cu_2) on the underside of the left fore wing. The stridulatory part of the left fore wing overlaps that of the right one when the wings are fully closed. When the male sings the fore wings are raised slightly from their position of rest to bring the plectrum into contact with the file (Fig. 13); each syllable of sound is produced by the fore wings moving apart slightly (the opening stroke) and then towards each other (the closing stroke). Most of the audible sound is usually produced during the closing stroke, but in some species the two strokes produce sounds of comparable intensity and in others all the audible sound is produced by the opening stroke (see p. 110).

The two male fore wings are usually noticeably asymmetrical in the stridulatory region. The right fore wing often has a small area of transparent membrane, the mirror, supported by a vein running round its rim; there is also often a redundant, usually poorly developed file – the homologue of the functional file on the left fore wing. The left fore wing usually

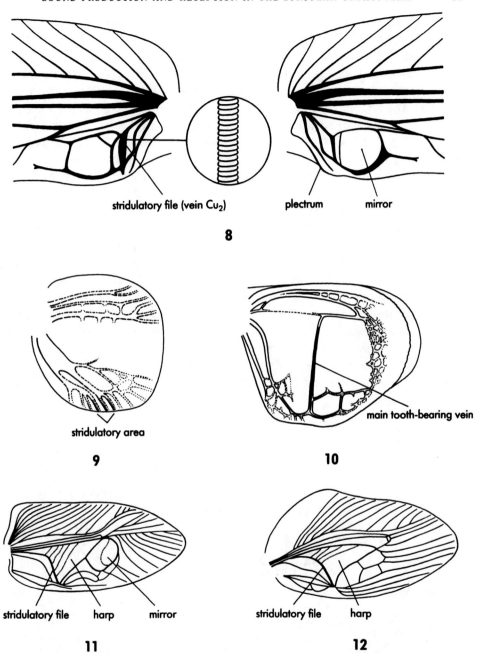

Figures 8–12 Stridulatory organs of bush-crickets, crickets and mole-crickets. **8.** The basal parts of the male fore wings of *Tettigonia viridissima*, with part of the underside of the stridulatory file enlarged to show the teeth (modified from Ragge, 1965). **9–12.** The right fore wings of (9) *Leptophyes punctatissima*, female; (10) *Ephippiger terrestris*, female; (11) *Acheta domesticus*, male; (12) *Gryllotalpa gryllotalpa*, male.

lacks a mirror and, because it overlaps the right one in this region, the mirror of the right fore wing is visible only when the fore wings are at least slightly raised or opened.

In most of the western European Tettigoniidae the females are silent. However, in the Phaneropterinae, Ephippigerinae and Pycnogastrinae the females can also stridulate, sometimes in response to the male calling song and sometimes when handled or threatened by a predator. The stridulatory organ is again at the base of the fore wings, but is quite different from that of the male and has clearly evolved independently. In the Phaneropterinae there are tooth-bearing veinlets along the more proximal part of the hind margin of the right fore wing (Fig. 9), which is again overlapped by the left one. The teeth are on the *upper* surface of the wing and are scraped by a plectrum on the left fore wing. The left fore wing lacks the stridulatory teeth and this asymmetry is occasionally enhanced by a difference in shape; when the females are very brachypterous (e.g. *Leptophyes*) the asymmetry is much less obvious. In the Ephippigerinae and Pycnogastrinae, in which the fore wings are always reduced to small flaps in both sexes, the female stridulatory apparatus is similar in principle (teeth on the upperside of veins on the right fore wing, plectrum on the left fore wing), but the position of the tooth-bearing veins is different (Fig. 10) and the apparatus probably evolved independently from that of female Phaneropterinae.

The male stridulatory organ of the Gryllidae and Gryllotalpidae is similar to that of the Tettigoniidae, except that the right fore wing normally overlaps the left one (Fig. 14). The file on the underside of the right fore wing is thus scraped by the plectrum of the left fore wing. Often all the audible sound is produced by the closing stroke, but in some species there may be quieter opening hemisyllables. During gryllid calling songs the fore wings are

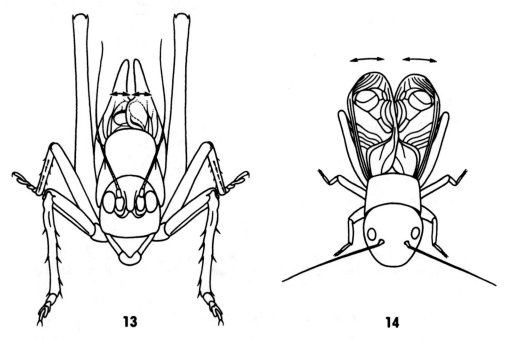

13   **14**

Figures 13, 14 Anterodorsal views of singing males of (13) *Tettigonia viridissima* and (14) *Gryllus campestris*, showing how the fore wings are moved during stridulation.

raised at an angle of about 30–50° to the abdomen in many species, and almost vertically in Oecanthinae; during the special courtship songs of *Gryllus* and *Acheta* the fore wings are held in a much lower position, closer to the position used by Tettigoniidae. On the rare occasions when male mole-crickets have been seen while producing a calling song, the fore wings have been held at a lower angle than in the true crickets.

The male fore wings of the Gryllidae and Gryllotalpidae are less asymmetrical than those of the Tettigoniidae. In many Gryllidae the left and right male fore wings are almost identical; they both have large areas of transparent membrane known as the harp and mirror (Fig. 11), and the files are equally developed in the two wings. In at least some of these species the males can still sing effectively if the fore wings are reversed in position, with the left one overlapping the right one, and indeed males with such wing-reversal occasionally occur in nature. However, the file of the left fore wing is reduced in many other species, and is sometimes completely lacking (Masaki *et al.*, 1987).

The male fore wings of Gryllotalpidae are less highly specialized for stridulation. They have stridulatory files of the gryllid type, but have a rather poorly developed harp and no mirror (Fig. 12); the left and right fore wings seem to be identical. The right fore wing usually overlaps the left one but they are often seen in the reversed position and some mole-crickets have been observed singing with the left fore wing overlapping the right one (Ulagaraj, 1976; Kavanagh & Young, 1989).

Female Gryllidae are normally silent, although Nickle & Carlysle (1975) have found structures on the dorsal surface of the female fore wings of *Gryllus rubens* that suggest a stridulatory function. Harz (1962: 66) observed a female of *Acheta domesticus* making a quiet sound by moving its fore wings in a similar manner to a singing male, but he suspected this insect was abnormal. Female Gryllotalpidae often have well-developed teeth on the underside of one or more veins in the cubito-anal region of the fore wings and can, in a number of species (including *Gryllotalpa gryllotalpa*), produce sounds by moving the fore wings in the same way as the males. These female sounds are usually short-lived and harsh, and seem to be produced when the female encounters another mole-cricket of either sex.

Some crickets make use of their immediate environment to improve the efficiency of their sound production. Walker (1969c) noticed that males of *Oecanthus jamaicensis* sometimes sang while holding the front part of their body in a small hole in a leaf, and males of several South African species of *Oecanthus* make holes in leaves that are specially shaped to fit the vibrating fore wings during singing (Prozesky-Schulze *et al.*, 1975). The American species *Anurogryllus muticus* sings from a self-made depression in the soil surface, and the common habit among ground-living crickets of singing at a burrow entrance may also improve the dissemination of the sound (Forrest, 1982). A number of species of mole-cricket, including *Gryllotalpa gryllotalpa* and *G. vineae*, construct carefully tuned singing burrows opening at the surface of the soil through horn-shaped entrances, and these have been shown to amplify the song (Bennet-Clark, 1970a, 1987; Nickerson *et al.*, 1979).

The rate at which the fore wings are vibrated in these groups varies enormously from one species to another. In the bush-crickets *Uromenus rugosicollis* and *U. elegans*, one to-and-fro movement of the reduced fore wings can take as long as 2 s on a cool night. At the other extreme, the syllable repetition rate of *Ruspolia nitidula* and *Metrioptera roeselii* can exceed 100/s in warm conditions, and in some tropical bush-crickets the rate is sometimes over 400/s (Heller, 1986). In crickets and some bush-crickets the song is produced by simple opening and closing movements of the fore wings, but studies with an opto-electronic camera have shown that the wing-movements of many bush-crickets are more complex (Heller, 1988, 1990). The amplitude of the movements often varies during the course of the song, and the closing strokes are sometimes interrupted by momentary pauses or even slight reversals of movement. The teeth on the tettigoniid stridulatory file are often shaped like a ratchet, so that the plectrum slips over them quite easily during the opening stroke but tends to be caught by them during the closing stroke. There is clearly the potential for sound to be generated in two ways: by the *plucking* action as the plectrum frees itself from a tooth, and by the *percussive* action as the plectrum hits the next tooth. Some of our high-speed oscillograms show double spikes, which could perhaps have been generated in this way (see, for example, Figs 530, 559; these are from field recordings and so echoes can be ruled out).

The songs of Gryllidae and Gryllotalpidae are resonant, i.e. they have an almost pure dominant frequency. This frequency is usually low enough to form a musical note and often seems to correspond with the tooth-impact rate of the stridulatory organ, as well as, at least in the few cases studied, with the natural frequency of vibration of the fore wing or part of it (usually the harp or, in the case of *Oecanthus*, the harp together with the greatly enlarged mirror). Stephen & Hartley (1995) have recently shown that in *Gryllus bimaculatus* the air space beneath the fore wings acts as a resonator, amplifying the sound produced and emphasizing its dominant frequency. Some Tettigoniidae have resonant songs (e.g. *Ruspolia nitidula*, *Decticus albifrons*), although the frequencies are too high to be musical and sometimes, instead of equalling the tooth-impact rate, seem to be a multiple of it.

Gomphocerine grasshoppers sing by rubbing their hind femora up and down against the flexed fore wings (Fig. 15). Each hind femur has a stridulatory file, consisting of a row of minute pegs, on its inner side (Fig. 16), and this is brought to bear against the radial vein of the fore wing. Sound is usually produced by both upstrokes and downstrokes, but the downstroke sound is almost always louder. Some of the membranous areas of the fore wings are often enlarged in the male, presumably to amplify the sound produced. The stridulatory pegs are usually much better developed in the male than in the female, but the females of

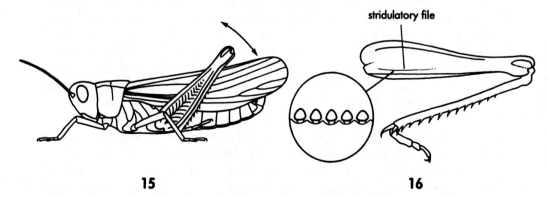

Figures 15, 16 Stridulatory organ of a grasshopper. **15.** Side view of a singing male grasshopper, showing how each hind leg is moved during stridulation. **16.** Inner view of a grasshopper hind leg, showing the position of the stridulatory file, with part of the file enlarged to show the pegs (modified from Ragge, 1965).

many species nevertheless sing when they are receptive to mating, either in response to the male calling song or sometimes spontaneously. The sound produced is usually similar in pattern to the male calling song, but generally quieter; it is seldom heard in nature as receptive females are usually mated very quickly, after which they cease to be receptive at least until they have laid an egg-pod.

An important aspect of gomphocerine stridulation is that the stridulatory apparatus is *double*: each of the two hind legs is engaged in independent sound production. In some species the movements of the two legs are synchronized and identical, but more often there are small but consistent differences between their movement patterns. A small time-lag between the movements of the left and right hind legs is very common; in one case, *Stenobothrus lineatus*, this can be easily observed because of the unusually slow leg-movements. In *Chorthippus mollis* the characteristic 'tick' that begins each echeme is produced by one hind leg only, although the two legs are moved in a very similar way during the rest of the song; this too can be seen by careful observation. Sometimes the two hind legs are completely out of phase, the left one moving upwards while the right one moves downwards (e.g. the calling songs of *Myrmeleotettix maculatus* and *Chorthippus albomarginatus*). The analyses of stridulatory leg-movements carried out by Elsner (1974a), using a Hall-generator, and Helversen & Elsner (1977), using an opto-electronic camera, have revealed further differences in the movements of the two legs of singing grasshoppers that would otherwise have been impossible to detect.

The rate at which the hind legs are moved up and down differs markedly from species to species. In *Stenobothrus lineatus* there is sometimes only one complete up-down movement per second, whereas a male of *S. nigromaculatus*, singing in hot sunshine, may reach a syllable repetition rate of 120/s. When the rate of leg-movement is within the range of about 5–20/s, it is common for the downstrokes to be interrupted by several momentary pauses of about 2–5 ms; the two legs usually pause synchronously, thus producing gaps in the downstroke hemisyllables that show up clearly in oscillograms (see, for example, Figs 927, 928). Occasionally such pauses occur in an upstroke, as in the rattling component of the calling song of *Stauroderus scalaris*.

The males of *Brachycrotaphus tryxalicerus*, a primarily African grasshopper that occurs very locally in southern Europe, have ridges on the inner side of the hind tibiae that are generally considered to be stridulatory in function. However, the usual gomphocerine stridulatory file is also present on the hind femora, and it has not yet been determined whether the tibial ridges play a significant part in producing the rather quiet, brief echemes of the song of this species.

The stridulatory files of all these groups vary greatly from one species to another in the number and density of the teeth or pegs; for the range of file structure among the European species see especially the illustrations given by Heller (1988) for Tettigoniidae and Jacobs (1953a) for Acrididae. There are often taxonomically useful differences in file structure in groups of species that are otherwise very similar morphologically. However, the rhythmic patterns of the songs are determined entirely by the stridulatory movements of the fore wings and hind legs, so that it is quite impossible to predict the characteristics of a song by examining the stridulatory file. Among the western European grasshoppers we know of only one species, *Stenobothrus lineatus*, that seems to show a clear correlation between between the structure of the stridulatory file and the way in which the stridulatory apparatus functions. In this species the stridulatory pegs are unusually dense and the file is exceptionally long (Jacobs, 1953a; Pitkin, 1976); during the calling song the hind legs are moved unusually slowly over an exceptionally wide angle. *Uromenus rugosicollis* is perhaps a similar case among the bush-crickets: its stridulatory file has an unusually large number of dense teeth (Heller, 1988) and the wing-movements are exceptionally slow. Heller (1988) has suggested that the bumps on the unusually complex stridulatory file of *Acrometopa servillea* play a part in producing the characteristic terminal click of its song.

In normally brachypterous bush-crickets and grasshoppers that sometimes occur in a fully winged form, the song seems to be affected remarkably little by the length of the fore wings. Oscillograms of the songs of such fully winged forms are given in this book for *Metrioptera roeselii* (Figs 405, 411, 417), *Euthystira brachyptera* (Figs 769, 772, 775) and *Chorthippus parallelus* (Figs 1518, 1525, 1532).

During the course of the adult lives of bush-crickets, crickets and grasshoppers, their stridulatory files undergo a considerable amount of wear. If one assumes that a male of, for example, *Metrioptera roeselii* sings with a syllable repetition rate of 100/s for a total of 6 hours per day for 46 days during its adult life, it would scrape its file about 100 million times! It is therefore not surprising that bush-cricket files often show signs of wear on the crests of the teeth. Hartley & Stephen (1989) have shown the effect of such wear on the quality of the song in *Poecilimon schmidti*, and Stiedl *et al.* (1991) have found that females of *Ephippiger ephippiger* can discriminate in their phonotactic response between calling songs made with unworn files and those made with worn ones. In male gomphocerine grasshoppers the stridulatory pegs are liable to be knocked out of their sockets by repeated impacts, and it is not unusual for the files of older males to lack several pegs.

Like the Gomphocerinae, locustine grasshoppers can stridulate by rubbing their hind femora against the fore wings, but the apparatus is reversed: there is an inner keel, without pegs, on each hind femur, and this is rubbed against a tooth-bearing additional vein, the medial intercalary, on the fore wing. The sounds produced are rather nondescript and of little or no diagnostic value; they are not therefore described in this book.

The grasshopper *Stethophyma grossum* produces its calling song by a method unique among the European Orthoptera: it flicks one of its hind tibiae backwards against the distal part of the flexed fore wings, making a loud 'ticking' sound. Tibial flicks of this kind also feature in the courtship songs of some grasshoppers (e.g. *Omocestus viridulus*).

Tremulation

While stridulating on trees and shrubs, bush-crickets cause the substrate to vibrate, and it has been shown in some species that this substrate-borne vibration plays a part in communication (see especially Keuper *et al.*, 1985). In some species of Ephippigerinae both sexes transmit substrate-borne signals by silent tremulation, during which the body is vigorously shaken. In at least one species, *Ephippiger perforatus*, tremulation is accompanied by a special quiet stridulation, quite different from the normal calling song.

Percussion

The bush-cricket *Meconema thalassinum* uses another method of transmitting substrate-borne signals: it drums one of its hind tarsi on the substrate. The loudness and quality of the airborne sound produced depends on the substrate; drumming on a twig is almost inaudible, but if a suitable leaf is used in a very quiet environment the drumming can sometimes be heard from a distance of a few metres.

Crepitation

A number of European grasshoppers crepitate, i.e. produce a rattling, whirring or buzzing sound in flight, or sometimes by vibrating the wings while sitting on the ground. A brief history of studies on grasshopper crepitation is given on p. 63; further useful observations are given by Otte (1970: 15). Exactly how the sound is produced is still unknown, but it seems to be generally agreed that only the hind wings are involved. According to Elsner (1974b) the gomphocerine grasshopper *Stenobothrus rubicundulus* produces its loud wing-buzz by beating the strongly sclerotized anterior margins of the two hind wings against each other. Locustine grasshoppers that crepitate in flight often have thickened veins in the anal fan of the hind wing, and Otte (1970) has suggested that the sound in these species may be produced by the more distal membrane between these thickened veins popping in and out. In many species both sexes can crepitate, although male crepitation is more frequent. Crepitation seems usually to be facultative, in that silent flight is also possible. In North America the males of many locustine grasshoppers have impressive display flights, involving loud crepitation, but in western Europe such behaviour is shown by only one species, *Bryodema tuberculata*.

Sound reception

All the Orthoptera included in this book hear by means of tympanal organs. There is a voluminous literature on the functional morphology of these organs and no more than a brief summary is given here; it should be noted that hearing has been studied in only a small number of species and the results of such studies do not necessarily apply to all the remaining ones.

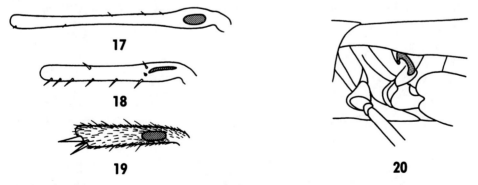

Figures 17–20 Tympana. **17–19**. Outer view of the left fore tibia of (17) *Leptophyes punctatissima*; (18) *Pholidoptera griseoaptera*; (19) *Acheta domesticus* (from Ragge, 1965). **20**. Side view of part of a grasshopper, showing the position of the tympanum. In each drawing the tympanum or tympanal aperture is shaded.

The Tettigoniidae, Gryllidae and Gryllotalpidae have tympanal organs at the proximal end of the fore tibiae. In the Tettigoniidae each fore tibia has a pair of tympana, which may be openly exposed or partially covered by flaps (Figs 17, 18). Closely applied to the inner side of each tympanum is a large trachea, which runs proximally down the leg, uniting as it does so with the trachea from the other tympanum; the resulting single trachea usually reaches the open air at a special acoustic spiracle, which can be very large, on the side of the prothorax. There are thus two sound inputs to each tympanum, one directly to its external surface and one through the acoustic spiracle and trachea to its internal surface. The relative importance of these two inputs to bush-cricket hearing has not yet been fully elucidated, but it is becoming clear that both play a significant part. Some bush-crickets are able to close the acoustic spiracle and, in at least one Australian species, the females do this when responding to the male calling song (Bailey, 1990); this suggests that the direct input to the tympana is particularly important during phonotaxis.

The tettigoniid tympanal organ is composed of three sensory parts (Fig. 21). The most proximal one is the subgenual organ, which responds primarily to vibrations transmitted through the substrate but also shows some response to airborne sound of relatively low frequency. Slightly distal to the subgenual organ is the intermediate organ, the function of which is at present unknown but is believed to be at least partly auditory. Continuing distally from the intermediate organ, and not clearly differentiated from it, is the main sensory organ for airborne sound, the crista acustica, which is composed of a long row of specialized

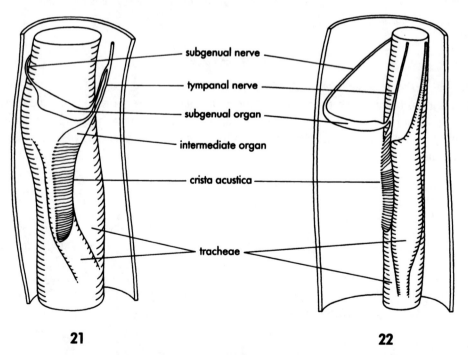

subgenual nerve

tympanal nerve

subgenual organ

intermediate organ

crista acustica

tracheae

21   **22**

Figures 21, 22 Part of the right fore tibia of (21) a bush-cricket and (22) a cricket, with the front cut away to show the tympanal organ. (Diagrammatic and much simplified from Lakes & Schikorski (1990) and Schwabe (1906), with Fig. 22, originally drawn from a left leg, reversed for easier comparison with Fig. 21)

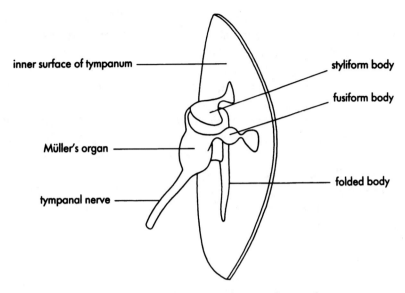

Figure 23 Inside view of the right tympanal organ of a grasshopper. (Diagrammatic and simplified from Gray, 1960)

sensilla (scolopidia) running along the dorsal wall of the anterior tibial trachea. The scolopidia become progressively smaller from the proximal to the distal end of the crista acustica, and Zhantiev & Korsunovskaya (1978) and Oldfield (1982) have shown that their peak sensitivities are to progressively higher frequencies. The tettigoniid tympanal organ can thus detect the pitch as well as the intensity of sounds.

The gryllid ear is basically similar to the tettigoniid one. There is usually a pair of tympana in each fore tibia, though the anterior one is smaller and often thicker; the tympana are usually openly exposed on the surface of the tibia (Fig. 19). The tracheal arrangement is similar, but the tracheae from the left and right fore legs become closely associated in the prothorax, so that acoustic input from the right trachea can reach the left tympanal organ and vice versa; this seems not to be true of most Tettigoniidae, but Zhantiev *et al.* (1995) have recently shown that there is acoustic contact (via tracheae) between the left and right tympanal organs of some Deracanthinae. Each trachea ends at a fairly normal spiracle, rather than a special acoustic one. There is a subgenual organ and crista acustica, each similar in function to that of the Tettigoniidae, but the intermediate organ seems to be lacking (Fig. 22). As in the Tettigoniidae, the crista acustica is composed of a row of scolopidia, and Zhantiev & Chukanov (1972) and Oldfield *et al.* (1986) have shown that the more distal the position of the scolopidium the higher the frequency of its peak sensitivity, so that the gryllid ear is also capable of pitch discrimination. In at least some species the ear is particularly sensitive to the carrier frequency of the calling song, and Nocke (1972) found a second peak of sensitivity in *Gryllus campestris* at the frequency of the 'ticks' in the courtship song. (For recent reviews see chapters 7 and 8 of Huber *et al.*, 1989.)

It should be noted that in both the Tettigoniidae and Gryllidae there are homologues of the tympanal organ, including the crista acustica, in the mid and hind tibiae. However, tympana are never present in the mid and hind legs, and there is no doubt that the fore tibial tympanal organ provides the main means for these insects to hear airborne sound.

The Acrididae have a tympanal organ on each side of the first abdominal segment, where it usually lies in a shallow depression and is often partly covered by the rim of the depression or an anteroventral flap (Fig. 20). On the inner side of the tympanum is a sensory complex known as Müller's organ, which is attached to the tympanum by three structures, the styliform, folded and fusiform bodies (Fig. 23). Just anterior to the tympanum is the first abdominal spiracle, which opens inwardly into an air cavity lying immediately under the tympanum. Müller's organ contains four groups of scolopidia, whose positions are detailed by Gray (1960), and Michelsen (1971) has shown that these groups differ in their frequency responses. Thus grasshoppers, like bush-crickets and true crickets, have the means to discrimate pitch, although using a very different kind of tympanal organ. In a recent study based on a sample of 20 European species, Meyer & Elsner (1996) have shown that the overall frequency response of the tympanal organ is generally well-matched to the frequency spectrum of the conspecific song.

Chapter 4

The nature and function of the songs

Types of song

The song most often heard in the field is the song produced spontaneously by an isolated male: the *calling* song (sometimes called the proclamation song). Given favourable weather, the right time of the year, and the right time of the day or night, mature males of most species need no further stimulus to produce their calling song. Indeed, they often sing at frequent intervals, or even continuously, for hours on end. Males of most species sing throughout their adult lives, but at least some species of *Gryllotalpa* seem to be exceptions. Males of *G. gryllotalpa*, although adult for the whole of the summer, seem to sing only from about mid-April to the end of June; the same is true of *G. vineae*, at least in France, though in this case singing usually seems to begin again in September.

The main function of the male calling song is to attract, or elicit a response from, conspecific females. In the Gryllidae and most Tettigoniidae the females are silent; when they are physiologically receptive they respond to a singing male by walking or flying towards him (*phonotaxis*, see the next section). In most Phaneropterinae and many Ephippigerinae the females produce a response song; ephippigerine females that respond in this way nevertheless still walk to the singing male, but in the Phaneropterinae it seems usually to be the male that walks part or all of the way to the responding female. In the Gryllotalpidae the females often have a disturbance call, but this plays no part in phonotaxis: receptive females are attracted to singing males. This also seems to be true of the North American 'true katydid', *Pterophylla camellifolia*, a pseudophylline bush-cricket (Caudell, 1906).

In gomphocerine grasshoppers the male calling song usually elicits a response song from a receptive conspecific female. The female song, usually similar in pattern to the male calling song, is seldom heard in the field because most females, having been recently mated, are unreceptive. However, females kept in captivity without access to males will not only sing in response to male calling songs but will also often sing quite spontaneously. Females that are too brachypterous to produce any audible sound (e.g. those of *Chorthippus parallelus*) nevertheless make stridulatory movements with their hind legs. In some species a female response results in the male walking to the female; in others both sexes walk towards each other, at least until visual contact is established. In brachypterous species with silent females, a responsive female has to approach the male until she is near enough to attract his attention visually.

The calling song of one male often acts as a stimulus to another nearby male (not necessarily conspecific) to sing. Sometimes singing males are attracted to one another, and when two conspecific male grasshoppers are sufficiently close to each other they often produce a modified song, known as the *rivalry* song. The rivalry songs of grasshoppers are often faster or abbreviated versions of the calling song, but sometimes they are rather different in character. Perhaps the most striking rivalry song among the European grasshoppers is that of *Chorthippus brunneus* (see p. 413). Some bush-crickets and true crickets also produce rivalry songs when two males encounter each other; these are often more vigorous or prolonged versions of the calling song. In some species loose groups of males sing in a chorus for short periods separated by intervals of silence (e.g. *Pholidoptera aptera*, p. 207). Interactions of other kinds (e.g. synchrony, alternation, spacing) also occur and have been recently reviewed by Ewing (1989) and Bailey (1991a).

In many crickets and gomphocerine grasshoppers the males produce a special song when next to a female, the *courtship* song. This may be a modified calling song, or a sequence that includes a calling song (modified or unmodified) as well as some new elements, or a song that is quite different from the calling song. In some gomphocerine grasshoppers courtship is quite complex; examples among the European species are *Myrmeleotettix maculatus* (p. 366), *Gomphocerippus rufus* (p. 374) and *Chorthippus albomarginatus* (p. 443). The function of such complex courtship behaviour is not clear, since mating sometimes occurs without it; for discussions of this problem see Otte (1972) and Bull (1979).

In addition to calling, rivalry and courtship songs (which are formally defined on p. 26), Orthoptera produce many other stridulatory sounds that are of little or no importance in taxonomy or identification. For example, ephippigerine bush-crickets of both sexes often make defensive disturbance sounds, as do some female mole-crickets. Male gomphocerine grasshoppers often make short-lived sounds when interacting with other individuals of either sex, and commonly make loud monosyllabic sounds immediately before an attempt to mate (the *Anspringlaut* of German workers). Such grasshopper sounds have been dealt with at length in the German literature (see especially Faber, 1953a; Jacobs, 1953a) and it would be inappropriate to describe them further here.

Female response and its specificity

Regen (1913), in his pioneering experiments on female phonotaxis in *Gryllus campestris*, was the first to show that females responded phonotactically to the male calling song, even when this was transmitted through a telephone so that the possibility of visual or olfactory attraction was eliminated. Similar responses were later demonstrated in *Ephippiger ephippiger* (Duijm & van Oyen, 1948; Busnel *et al.*, 1956) and *Oecanthus pellucens* (Busnel & Busnel, 1955). Jacobs (1944) first described the female songs of *Chorthippus brunneus* and *Gomphocerippus rufus*, produced in response to conspecific male calling songs, and later observed them in many gomphocerine grasshoppers (Jacobs, 1950a, 1950b, 1953a).

Although these studies demonstrated a female response to conspecific male calling song, none of them established how *specific* the response was, i.e. to what extent the female might respond to the calling songs of other species. Jacobs (1953a, 1953b) found that a receptive female of *Chorthippus montanus* responded much more strongly to the male calling song of its own species (64 per cent) than to the rather similar male calling song of *C. parallelus* (1 per cent). Perdeck (1957), in his classic study of the two species *Chorthippus brunneus* and *C. biguttulus*, was able to show not only that the female response (by singing) to the male calling song was highly specific in these two species, but also that the same applied to males responding to the spontaneous songs of females. At about the same time Walker (1957), in a study of nine species of oecanthine crickets, showed that, where two or more species were sympatric, the females were highly specific in their phonotactic responses to male calling song. These pioneering studies have been followed by a number of others, all confirming the specificity of the female response (e.g. Bailey & Robinson, 1971; Hill *et al.*, 1972; Zaretsky, 1972; Ulagaraj & Walker, 1973; Paul, 1976, 1977; Helversen & Helversen, 1981; Vedenina & Zhantiev, 1990; Doherty & Callos, 1991; Stumpner & Helversen, 1992, 1994). There is now ample evidence to support two conclusions, both likely to have general application to Orthoptera producing communicative sounds: (1) receptive females respond to conspecific male calling song, and (2) this response is highly specific when two or more species occur together in the same place at the same time of the year. It has also been established that in some phaneropterine bush-crickets the short *interval* between the male calling song (or a particular part of it) and the response song from a receptive female is highly specific; the male recognizes a conspecific female by the duration of this interval (often under 100 ms) and then finds her by phonotaxis (see especially Spooner, 1968; Heller & Helversen, 1986; Robinson, 1990). In such cases it is clearly unnecessary for the female to respond exclusively to the calling songs of conspecific males.

The attraction of predators and parasites

When using their calling songs to attract females, male Orthoptera, like other animals using such calls, run the risk of attracting the unwanted attention of predators and parasites. Many animals will eat Orthoptera when the opportunity arises, and at least some of these are known to be attracted acoustically to singing males. Walker (1964a) demonstrated experimentally that domestic cats (*Felis domestica*) can track down bush-crickets and true crickets by their calling songs, and Willey (1970) observed a dwarf weasel (*Mustela erminea muricus*) pursuing a crepitating grasshopper (*Arphia conspersa*) by sound. Bell (1979) showed that the heron *Florida coerulea* can locate crickets (*Anurogryllus celerinictus*) after dark by their calling songs, and an experimental study by Walker (1979) suggested that the spadefoot toad (*Scaphiopus holbrooki*) is attracted to singing crickets (*Anurogryllus arboreus*) during the early part of the night. Sakaluk & Belwood (1984) found that the insectivorous gecko *Hemidactylus tursicus* was attracted at night to the calling song of the cricket *Gryllodes supplicans*; in this case the male crickets were singing from burrows, where they were safe from attack, and the geckos were presumed to be feeding on conspecific female crickets that were also attracted by the calling males. Insectivorous bats often feed on Orthoptera and some have been shown to respond to singing males of various species of bush-cricket (Buchler & Childs, 1981; Tuttle *et al.*, 1985; Belwood & Morris, 1987; Belwood, 1990). Parasitoid flies belonging to the tachinid subfamily Ormiinae have been shown to locate their nocturnally active orthopteran hosts (bush-crickets, true crickets and mole-crickets) by homing in on the calling song (Cade, 1975; Mangold, 1978; Burk, 1982; Fowler & Garcia, 1987; Lakes-Harlan & Heller, 1992; Walker, 1993).

Individual and geographical variation

Although the songs of Orthoptera are highly stereotyped, there is naturally some variation from individual to individual of the same species. Such individual differences in song are usually quite small in comparison to the differences between the songs of different species, and so seldom detract from the value of the songs in taxonomy and identification. In the species of Tettigoniidae that can be either brachypterous or macropterous (e.g. *Metrioptera roeselii*) the degree of wing development seems to have little effect on the song, and the same is true of such species of Gomphocerinae (e.g. *Chorthippus parallelus*). Gomphocerine grass-hoppers singing with only one hind leg are as easily recognizable from their song as those singing with two; the pattern of the sound is not affected significantly although, not surpris-ingly, oscillograms of 'one-legged' songs usually show the elements of the song more clearly.

Geographical variation is more significant, although the evidence suggests that a long-lasting interruption in gene flow is necessary before a noticeable difference in song develops. From west to east across western Europe, north of the Pyrenees and Alps, we have been unable to detect any significant change in song. The songs of the British Orthoptera, which have been cut off from the Continental gene pools for some 8000 years, show no apparent differences from those of Continental populations of the same species. Recordings of the songs of the gomphocerine grasshoppers *Chorthippus biguttulus*, *C. brunneus* and *C. mollis* made in the vicinity of Kursk in Russia (kindly sent to us by Drs R.D. Zhantiev and V.Yu. Vedenina) show no apparent difference from the songs of these species in France.

It is in the southern peninsulas that the main examples of geographical variation in song are to be found. In the Iberian Peninsula song differences have developed in several species of grasshopper. The calling song of *Omocestus viridulus* tends to be a little shorter in Spain, where it occurs as the subspecies *O. v. kaestneri*. The echemes of the calling song of *Myrmeleotettix maculatus* tend to be longer and to be repeated more slowly in the Iberian Peninsula (and, to a lesser extent, in the Italian Peninsula). In the Sierra Nevada an 'aftersong' is usually added to the main echeme of the calling song of *C. vagans*. The Iberian form of *C. parallelus*, at present regarded as the subspecies *C. p. erythropus*, has a well-devel-oped courtship song that is completely lacking in typical *C. parallelus*.

Montane species, and montane populations of widespread species, are particularly likely to develop song differences. The Pyrenean species *Omocestus antigai*, for example, shows differences in courtship song in different, well-isolated populations. Although not varying much within the area of our study, the song of *Stenobothrus rubicundulus* is significantly different in Greece, where some of the downstroke hemisyllables are replaced by short bursts of wing-vibration (Elsner & Wasser, 1995a, 1995b, 1995c). *Chorthippus cialancensis*, known only from the Piedmont Alps, shows small but consistent differences in calling song be-tween the two localities where we have studied it. The calling song of the widespread species *C. biguttulus* tends to consist of a larger number of shorter echeme-sequences in parts of the southern Alps, and some other local variations in song have been noticed in this species (see p. 402). The Alpine form of *C. mollis*, *C. m. ignifer*, has a shorter calling song, with fewer echemes and a more abrupt ending. The calling song of *C. yersini* shows a number of local song differences in different upland areas of Spain, tending to lose its quiet 'aftersong' in southern Spain (and Sicily) and becoming more like the calling song of *C. biguttulus* in the north-eastern part of its range.

Rather surprisingly, we have not generally found significant differences in song in populations isolated on islands. An exception is the Sicilian form of *C. brunneus*, which has unusually long echemes in both its calling and rivalry songs. This seems also to be true of Finnish *C. brunneus* (Perdeck, 1957) and some Alpine populations of this species (Ingrisch, 1995).

Among the Tettigoniidae, *Ephippiger ephippiger* shows quite marked geographical variation in calling song in parts of southern France and Catalonia, where the normally monosyllabic song changes to polysyllabic echemes (see p. 71). However, the most striking case of apparent geographical variation in the song of a European bush-cricket is provided by *Decticus verrucivorus*. Although the song of this well-known species is very uniform in most of western Europe, a form occurs in the mountains of central and eastern Spain that has a markedly different song (see p. 158). This form shows no significant morphological differences from typical *D. verrucivorus* and is at present treated as the subspecies *D. v. assiduus*, but it would be most interesting to know whether the song difference provides any ethological barrier to interbreeding.

We have not so far noticed any significant geographical variation in the songs of the western European crickets.

The effect of temperature

Once an orthopteran is singing the main environmental factor that affects the song is temperature. Anyone studying these songs soon notices that, like most of the activities of a poikilothermic animal, they are speeded up by an increase in temperature. It is the body temperature that is important, and in sunny weather a diurnal singer is more likely to be influenced by solar radiation than by the temperature of the ambient air.

Although this effect is well known there have not been many carefully controlled studies. Early observations on *Oecanthus fultoni* (the Snowy Tree Cricket or 'thermometer cricket') are outlined on p. 62 and are reviewed in more detail by Frings & Frings (1962). Crickets lend themselves well to studies on the effect of temperature: they will usually sing without radiant heat, the songs are easily heard, tape recorded and analysed, and there are several easily measured song characters (e.g. syllable repetition rate, echeme repetition rate, carrier frequency). The more recent studies of Alexander (1956), Walker (1957, 1962a, 1962b, 1963, 1969a, 1969b, 1973), Sismondo (1979), Prestwich & Walker (1981) and Pires & Hoy (1992) on North American crickets have firmly established that the syllable repetition rate increases with ambient air temperature and that the relationship is largely linear. When the syllables are regularly grouped into echemes, Walker (1962a), Block (1966), Toms (1992) and Pires & Hoy (1992) have shown that the echeme repetition rate varies with ambient air temperature in the same way, although Block believed the relationship to be exponential rather then linear. The carrier frequency of the calling songs of these North American species often also increases with temperature, but not as much as the syllable and echeme repetition rates, and in *Gryllus rubens* and *G. firmus* only very slightly or not at all. Nielsen & Dreisig (1970) found that the echeme repetition rate in the North African cricket *Platygryllus brunneri* shows a linear relationship with ambient air temperature at night, but that the relationship is non-linear during the day. Kutsch (1969), Doherty & Huber (1983), Doherty (1985) and Koch *et al.* (1988) showed that the syllable and echeme repetition rates of the calling songs of the Old World species *Gryllus campestris* and *G. bimaculatus* also show a linear relationship with air temperature and that the carrier frequency, as in *G. rubens*, is only slightly affected. In these species of *Gryllus* the sound-producing closing stroke of the fore wings is hardly affected by changes in temperature, and hence the tooth impact rate, which seems to correspond with the carrier frequency, remains almost constant. The increase in syllable repetition rate that follows a rise in temperature is the result of a reduction in the interval between successive closing hemisyllables, and this is largely achieved by an increase in the speed of the opening stroke of the fore wings. Elliot & Koch (1985) have suggested that the relative constancy of the tooth impact rate and carrier frequency in *Gryllus* is achieved by a means analogous to the escape mechanism of a mechanical clock, the harp acting as a pendulum in limiting the tooth impact rate to its resonant frequency; Stephen & Hartley (1995), however, disagree with this analogy, believing that the carrier frequency in *Gryllus* is controlled by a system of auditory feedback. Sismondo (1993) has shown that the clock analogy does not in any case apply to *Oecanthus*, in which the carrier frequency rises steadily with temperature and the fore wings are able to resonate over a range of frequencies.

The earliest significant observation on the effect of temperature on a tettigoniid song seems to have been that of Hayward (1901), who counted the number of echemes per minute in the calling song of the North American 'true katydid', *Pterophylla camellifolia*, over a wide range of ambient air temperatures. No conclusions were drawn from these observations at the time but, as in crickets, they show a clear linear relationship when plotted on a graph. Allard (1929b) observed a similar effect in *Orchelimum agile*, and Hallenbeck (1949) claimed that the number of syllables per echeme increased with temperature in *P. camellifolia*. Alexander (1956) gave graphs demonstrating a linear relationship between syllable repetition rate and ambient air temperature in *Neoconocephalus ensiger*, *N. nebrascensis*, *Orchelimum vulgare* and *Atlanticus testaceus*. Frings & Frings (1957), however, considered this relationship to be exponential rather than linear in a fairly thorough study of *N. ensiger*. In 1962 these two authors published the results of another thorough study of the effect of air temperature on various aspects of the song of *O. vulgare*; they did not give a graph for syllable repetition rate (within the buzzing component of the song), but the figures they gave, when plotted, show a linear relationship agreeing well with Alexander's graph. In a comprehensive study of the acoustic behaviour of *Scudderia texensis*, Spooner (1964) demonstrated a clear linear relationship between syllable repetition rate and ambient air temperature for both the fast- and slow-pulsed types of song produced by this species. Walker *et al.* (1973) and Walker (1975b, 1975c) found the same linear relationship in more than 20 further species of North American Tettigoniidae.

There have been few comparable studies on European Tettigoniidae. Dumortier (1963c) gave a graph showing a decrease in the duration of a four-syllable echeme (i.e. an increase in the syllable repetition rate) as the ambient air temperature rises in the European bush-cricket *Ephippiger provincialis*, and Nielsen & Dreisig (1970) gave similar graphs for echeme repetition rate in *Tettigonia viridissima* and syllable repetition rate in *T. cantans*; in all three cases the relationship appears to be exponential rather than linear. We often mention the effect of temperature on bush-cricket songs in the descriptive part of this book, especially in species that sing both during the day and at night. In *Platycleis albopunctata*, for example, the echeme repetition rate is typically 2–4/s in warm daytime conditions, but is often below

1/s on cool nights. In *Tettigonia cantans* the nocturnal song not only has a much slower repetition rate, but continues for long periods instead of being broken up into the echemes typical of the daytime song.

It is worth noting that, even without solar radiation, the temperature of the body – or at least of the thorax – of a singing orthopteran is likely always to be higher than the ambient air temperature as a result of heat generated by the stridulatory muscles. This has been shown to be true of several Tettigoniidae: the North American species *Neoconocephalus robustus* (Heath & Josephson, 1970) and *N. triops* (Josephson, 1984), and the Oriental species *Euconocephalus nasutus* (Josephson, 1973) and *Hexacentrus unicolor* (Heller, 1986). In at least two of these species (*N. robustus* and *H. unicolor*) the males have a 'warm-up' period before singing, during which the thoracic temperature is raised by activity of the stridulatory muscles but no sound is produced.

The songs of gomphocerine grasshoppers are affected by temperature in a similar way to those of crickets and bush-crickets, but they do not lend themselves so well to carefully controlled studies. These insects are active by day and at normal air temperatures most species are reluctant to sing without direct solar radiation. In the laboratory they are similarly stimulated to sing by radiant heat from an electric lamp. In these circumstances the ambient air temperature cannot be used as an indication of the insect's body temperature, which is often mainly determined by radiant heat.

Helversen (1972) overcame this problem in a laboratory study by eliminating any source of radiant heat, and showed that the duration of the echemes and the intervals between them in the calling song of *Chorthippus biguttulus* were both reduced by increasing the ambient air temperature; she believed the relationship to be exponential rather than linear. Helversen (1979) and Helversen & Helversen (1981) demonstrated a similar relationship between syllable duration and temperature in the calling songs of both *C. parallelus* and *C. montanus*, and Skovmand & Pedersen (1983) showed the same effect in *Omocestus viridulus*. In further laboratory studies, Tschuch & Köhler (1990) showed that both syllable and echeme duration were similarly affected by temperature in *C. parallelus* and *C. montanus*, the echeme duration changing slightly more than the syllable duration. Defaut (1983, 1988a) attempted field studies on the effect of air temperature on song in *C. parallelus*, but found it was complicated by two factors: firstly, as noted by Yersin (1854b), the males produce mainly rivalry rather than calling songs in the early morning and, secondly, at a given air temperature (at least below 25°C) the syllable repetition rate of the calling song is higher in the first half of the day, before the temperature peak, than later in the day, when the temperature is falling. He nevertheless found the tempo of the calling song to be markedly affected by air temperature, the syllable and echeme repetition rates increasing and the duration of the echemes decreasing as the temperature rises.

Since the function of the male calling songs of Orthoptera as a mate recognition system is believed to be based mainly on such rhythmic characteristics as syllable and echeme repetition rates, and these rates are so markedly affected by temperature, how do the females manage to recognize and respond to conspecific songs over a wide range of temperatures? This question is particularly relevant when the calling songs of sympatric species differ mainly, if not solely, in their tempo. The answer, as far as can be judged from a very limited number of studies on this topic, seems to be that the female response system varies with temperature in a way that matches the changes in the male calling song. This was demonstrated in the two North American tree-crickets *Oecanthus quadripunctatus* and *O. nigricornis* by Walker (1957), in *Gryllus firmus*, also North American, by Pires & Hoy (1992) and in the Old World cricket *Gryllus bimaculatus* by Doherty & Huber (1983) and Doherty (1985). In the grasshopper *Chorthippus biguttulus* Helversen (1972) showed that the *optimum* response of the female changed with temperature in a way that matched the change in the male calling song, but that a female at 35°C would nevertheless respond to calling song patterns corresponding to a wide range of temperatures; she established that the most important characteristic of the song was the ratio between echeme duration and the interval between successive echemes, and that this ratio remained constant at different temperatures.

The two grasshoppers *C. parallelus* and *C. montanus* present a particularly interesting case. The main difference between their male calling songs is in tempo: at a given temperature the syllable repetition rate of *C. parallelus* is about double that of *C. montanus*. However, the calling song of *C. parallelus* at, say, 20°C could easily be confused with that of *C. montanus* at about 32°C. These two species quite often mingle in the same locality and so it is important that the females distinguish between the two calling songs. Helversen (1979), Helversen & Helversen (1981) and Bauer & Helversen (1987) have shown that, in both these species, the female response to the calling song at a particular temperature is quite closely tuned to the song tempo of a conspecific male at the same temperature, so that these species have an effective system for preventing cross-mating at any temperature.

The marked effect of temperature on the songs of Orthoptera can also cause problems when the song-patterns are being used for identification. In the descriptive part of this book we comment on the effect of temperature in many places, and especially for species that sing both during the day and at night. For the gomphocerine grasshoppers we always describe the song heard in full sunshine, when the temperature of the singing male is more dependent on solar radiation than on the ambient air temperature; most of these sun-loving insects in any case stop singing altogether in dull weather.

In concluding this section it is worth mentioning that Otte (1970) found no significant effect of ambient air temperature on the crepitation rates of several North American species of locustine grasshopper.

The songs of hybrids

Although seldom found in nature, hybrids between related species of Orthoptera have often been bred in the laboratory. The first investigation of the effect of hybridization on song was by Fulton (1933) on two North American crickets now called *Nemobius allardi* and *N. tinnulus*, whose songs differ mainly in syllable repetition rate. He found that the F_1 hybrid males had a syllable repetition rate intermediate between those of the parent species, and the same applied to most of the F_2 hybrids. Back-crosses with the parent species tended to produce syllable repetition rates intermediate between the hybrids and the parent species. Studies on the Old World species *Gryllus campestris* and *G. bimaculatus* by Hörmann-Heck (1957) are of less interest in this context as the acoustic behaviour of these two species is very similar. She did, however, find that the male hybrids were intermediate between the parent species in the sounds they made as a preliminary to the courtship song. Bigelow (1960) found the songs of male hybrids between the American species *G. rubens* and both *G. veletis* and *G. assimilis* to be intermediate in syllable repetition rate but not necessarily in their echeme pattern, which tended to follow the maternal species. The calling songs of hybrids between *Teleogryllus commodus* and *T. oceanicus*, and between *Gryllus argentinus* and *G. peruvianus*, were found by Leroy (1966) to be intermediate between those of the parent species in every respect. Bentley (1971), Bentley & Hoy (1972, 1974), Hoy & Paul (1973), Hoy (1974) and Hoy *et al.* (1977) made a particularly thorough study of the acoustic behaviour of hybrids between *T. commodus* and *T. oceanicus*, and found not only that the calling songs were intermediate in most respects between those of the parent species, but that the female hybrids responded more strongly to the calling songs of male hybrids than to those of either parent species. They also found that certain characters of the hybrid calling song were exceptional in tending to follow the maternal species rather than being intermediate (so that the songs of the *commodus* male × *oceanicus* female hybrids were significantly different from those of the reciprocal cross), and were even able to establish that the female hybrids were more attracted by the songs of their 'brothers' than by those of male hybrids from the reciprocal cross. Inagaki & Matsuura (1985a, 1985b) studied the songs of hybrids between *Teleogryllus siamensis* and *T. emma*, and between *T. siamensis* and *T. taiwanemma*; in both cases they found that the hybrid songs, although showing a general tendency to be intermediate, were variable, some being closer to one parent species and some, even from the same cross, closer to the other.

In a recent study on the bush-cricket *Ephippiger ephippiger*, Ritchie (1992a) obtained crosses between two populations with respectively monosyllabic and tetrasyllabic male calling songs. He found that the F_1 male calling songs were predominantly disyllabic and that the F_1 females preferred these songs to both the monosyllabic and tetrasyllabic calling songs of the parental males.

For the gomphocerine grasshoppers the first significant study of hybrid songs was by Perdeck (1957) on *Chorthippus biguttulus* and *C. brunneus*. He found that, whichever way the cross was made, the calling songs of male hybrids were intermediate between those of the parent species in both the duration and number of the constituent song-units (echeme-sequences in the case of *C. biguttulus*; echemes in *C. brunneus*). Receptive females of the parent species rarely responded to the male hybrid song, but receptive female hybrids tended to respond to the calling songs of one or both parent species as well as to those of male hybrids. Natural hybrids between these two species occasionally occur and can be easily recognized by their songs; oscillograms of the calling song of the only (putative) natural hybrid found by us is shown in Figs 24–32, in comparison with the calling songs of the parent species (see also Ragge, 1976). The calling song of this hybrid is included on Compact Disc 2 (track 53).

The acoustic behaviour of hybrids between *C. biguttulus* and *C. mollis* has been thoroughly studied by Helversen & Helversen (1975a, 1975b). They found that some aspects of the very variable hybrid calling songs were intermediate between those of the parent species, but others seemed to result from the superimposition of elements from both parents; there was a tendency for the song-pattern to be closer to the maternal than to the paternal species. Most of the female hybrids preferred the calling songs of one or both parent species to those of male hybrids. The male calling song of the Alpine form of *C. mollis* (*C. m. ignifer*) is remarkably similar to those of some of the male hybrids reared by the Helversens, particularly those from a *C. mollis* mother, and leads us to suspect that this form might have arisen by hybridization between *C. biguttulus* and *C. mollis* (Ragge, 1981, 1984).

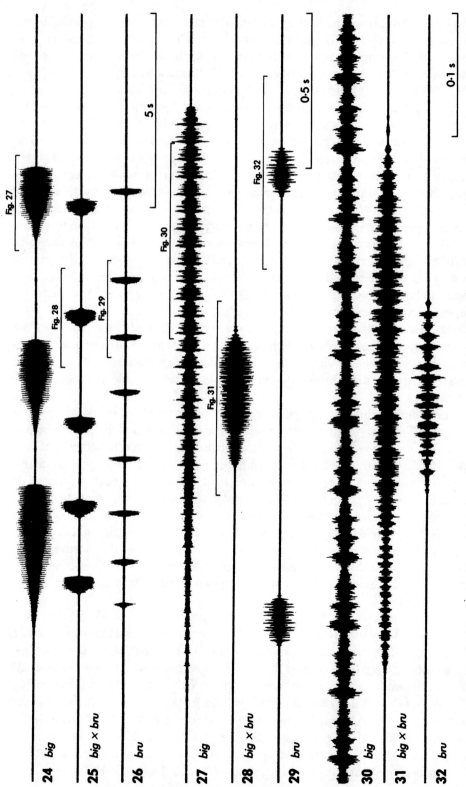

Figures 24–32 Oscillograms at three different speeds of the male calling song of a putative hybrid between *Chorthippus biguttulus* and *C. brunneus* (25, 28, 31) in comparison with the male calling songs of the parent species (24, 26, 27, 29, 30, 32).

Tschuch & Köhler (1990) found that two laboratory-reared male hybrids between *Chorthippus parallelus* and *C. montanus* had calling songs intermediate in syllable and echeme duration between those of the parent species.

Classifying the songs

The bewildering variety of song-patterns produced by the European Orthoptera comes near to defying classification. They range from the simple isolated syllables of such species as *Phaneroptera nana* to the complex echeme-sequences of *Conocephalus dorsalis* and such grasshoppers as *Chorthippus apricarius*, *C. cazurroi* and *C. mollis*. Between these extremes lie innumerable patterns of intermediate complexity. Among the many factors contributing to this variety are the duration and repetition rates of the syllables and echemes, the relative intensity of the component sounds and, in the resonant songs of the crickets, mole-crickets and some bush-crickets, the frequency of the carrier wave. In spite of this, most of the song-patterns fall into one of a small number of basic categories, and we feel it is helpful, if only as an aid to identification, to set these out here (see also Fig. 33). Some conclusions, and further remarks on the song-patterns, are given in the next section.

The broad categories that follow are concerned with the rhythmic patterns of the songs rather than the method of sound production, and apply only to the calling songs; some of the courtship songs of grasshoppers reach a degree of complexity that would put them in a category of their own. We have included crepitation in only two cases: the wing-buzz of *Stenobothrus rubicundulus*, which is associated with more normal femoro-alary stridulation (see category 10), and the display flight of *Bryodema tuberculata*, which falls into category 6 according to its rhythmic pattern (treating each burst of crepitation rather loosely as an echeme). The foot-drumming of *Meconema thalassinum* is also included in category 6, each burst of sound again being treated as an echeme. Some of the distinctions between the categories are of degree rather than kind, and in these cases we have had to draw the line at an arbitrary, though definable, point. Because of this, a number of songs fall into more than one category. Some songs are a mixture of different kinds of sound and we have gathered these together to form an extra, final category. The relevant species are listed, in systematic order, in each category.

1. *Continuous trains of rapidly repeated, ungrouped syllables, lasting indefinitely (often more than a minute)*

Whether a long series of syllables sounds like a continuous train depends on at least three factors: syllable repetition rate, syllable duration in relation to the intervals between successive syllables, and regularity of syllable repetition. If the syllable repetition rate in such a series is at least 5/s we always regard it as a continuous train. If this rate is below 5/s we regard it as a continuous train only if the intervals between the syllables are no longer than the syllables themselves; if this is not so we regard the syllables as isolated (see category 2, below). The syllable repetition rate need not be completely regular but the syllables must not fall into clearly recognizable groups (see category 3, below).

This category can be subdivided further, as follows:

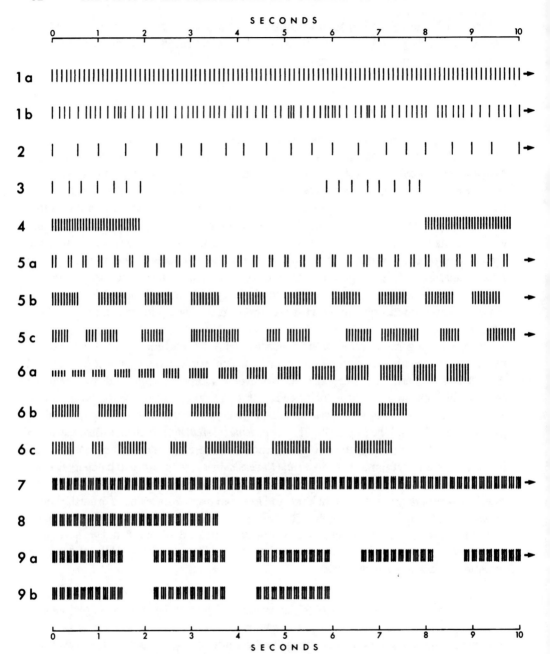

Figure 33 Diagrams illustrating the classification presented here of the song-patterns of European Orthoptera. Each vertical mark represents a syllable of sound. The arrows on the right indicate that the song lasts indefinitely (often more than a minute). Note that each category includes a range of time characteristics, so that the rhythmic pattern is more important than the time scale. See the text for details of category 10 and for precise definitions of all the categories.

(a) Trains with a regular syllable repetition rate

Isophya kraussii
Polysarcus denticauda
Ruspolia nitidula
Tettigonia cantans (night-time song)
Decticus verrucivorus assiduus
  (cool conditions)
Platycleis tessellata (sometimes)
Platycleis veyseli (sometimes)

Platycleis nigrosignata (sometimes)
Metrioptera roeselii
Gampsocleis glabra
Ctenodecticus siculus
Pycnogaster jugicola (rarely)
Uromenus rugosicollis (sometimes)
Gryllotalpa gryllotalpa
Gryllotalpa vineae

(b) Trains with an irregular syllable repetition rate

Barbitistes fischeri   *Decticus albifrons*

Note that, pending studies on the wing-movements, it is at present uncertain whether *Gampsocleis glabra* belongs to category 1 or category 7.

2. Isolated syllables

We regard syllables as being isolated if they are repeated less frequently than 5/s, the intervals between them are longer than the syllables themselves, and they are not formed into recognizable groups.

Phaneroptera falcata (sometimes)
Phaneroptera nana
Tylopsis lilifolia (night-time song)
Isophya pyrenaea
Leptophyes punctatissima
Leptophyes albovittata
Leptophyes boscii
Leptophyes laticauda
Poecilimon ornatus
Poecilimon jonicus
Conocephalus conocephalus (sometimes)
Platycleis tessellata (sometimes)
Platycleis veyseli (sometimes)

Platycleis nigrosignata (sometimes)
Pachytrachis striolatus (sometimes)
Eupholidoptera chabrieri
Yersinella raymondii
Yersinella beybienkoi
Antaxius spinibrachius
Thyreonotus bidens
Ephippiger ephippiger (sometimes)
Ephippiger terrestris
Ephippigerida taeniata
Uromenus andalusius
Eugryllodes pipiens (sometimes)

Note that in the song of *Conocephalus conocephalus* there is often a mixture of quieter and louder syllables.

3. Loose groups of syllables

A group of syllables is loose if the syllable repetition rate is less than 10/s and the intervals between the syllables are longer than the syllables themselves.

Tylopsis lilifolia (daytime song)
Platycleis tessellata (sometimes)
Platycleis veyseli (sometimes)
Platycleis nigrosignata (sometimes)
Pachytrachis striolatus (sometimes)

Stethophyma grossum
Ramburiella hispanica
Dociostaurus maroccanus
Dociostaurus jagoi

4. *Isolated dense echemes of limited duration (less than a minute)*

We regard echemes as isolated if they are separated by intervals of at least 5 s; however, in many species these intervals are very variable and it is an arbitrary matter whether the song is regarded as being in this category or the next one. We always regard an echeme as dense if the syllable repetition rate is at least 10/s. If this rate is below 10/s we regard the echeme as dense only if the intervals (if any) between the syllables are no longer than the syllables themselves.

Phaneroptera falcata (sometimes)
Acrometopa servillea
Tettigonia cantans
   (daytime song, sometimes)
Tettigonia caudata
Platycleis iberica (sometimes)
Platycleis falx (sometimes)
Platycleis montana (sometimes)
Pholidoptera griseoaptera (sometimes)
Pholidoptera fallax (sometimes)
Pholidoptera femorata (sometimes)
Pholidoptera littoralis
Anonconotus alpinus (sometimes)
Antaxius pedestris (sometimes)
Antaxius hispanicus (usually)
Ephippiger perforatus
Uromenus rugosicollis (sometimes)
Uromenus elegans
Uromenus brevicollis
Uromenus asturiensis
Uromenus stalii
Platystolus martinezii
Pycnogaster jugicola (sometimes)
Pycnogaster sanchezgomezi
Pycnogaster inermis
Eugryllodes pipiens (sometimes)
Brachycrotaphus tryxalicerus
Chrysochraon dispar (sometimes)

Euthystira brachyptera (sometimes)
Omocestus viridulus
Omocestus rufipes
Omocestus haemorrhoidalis
Omocestus raymondi (sometimes)
Omocestus panteli
Omocestus antigai (sometimes)
Omocestus bolivari (sometimes)
Omocestus uhagonii (sometimes)
Omocestus femoralis
Omocestus uvarovi
Stenobothrus lineatus
Stenobothrus fischeri
Stenobothrus festivus
Stenobothrus stigmaticus (usually)
Stenobothrus apenninus
Stenobothrus ursulae
Gomphocerus sibiricus
Chorthippus corsicus
Chorthippus nevadensis
Chorthippus alticola
Chorthippus vagans
Chorthippus reissingeri
Chorthippus binotatus
Chorthippus dichrous
Chorthippus parallelus (sometimes)
Chorthippus montanus (sometimes)

5. *Separately audible dense echemes in a series lasting indefinitely (often more than a minute)*

The echemes in such a series must be clearly audible as separate units and repeated less rapidly than 7/s.

This category can be subdivided further, as follows:

(a) Disyllabic echemes

Platycleis intermedia
Platycleis sepium (sometimes)

Ephippiger ephippiger (sometimes)
Acheta domesticus (sometimes)

(b) Polysyllabic echemes, of uniform duration

Polysarcus scutatus
Conocephalus discolor (sometimes)
Tettigonia cantans
    (daytime song, sometimes)
Platycleis albopunctata
Platycleis sabulosa
Platycleis iberica
Platycleis falx (sometimes)
Platycleis montana (sometimes)
Metrioptera brachyptera
Metrioptera saussuriana
Metrioptera buyssoni
Metrioptera caprai
Metrioptera abbreviata
Metrioptera burriana
Pholidoptera griseoaptera (sometimes)
Pholidoptera fallax (sometimes)
Pholidoptera femorata (sometimes)
Anonconotus alpinus (sometimes)
Rhacocleis germanica
Antaxius pedestris (sometimes)
Antaxius hispanicus (occasionally)
Ephippiger ephippiger (sometimes)

Ephippiger ruffoi
Ephippigerida areolaria
Uromenus catalaunicus
Uromenus perezii
Uromenus martorellii
Baetica ustulata
Gryllus campestris
Gryllus bimaculatus
Modicogryllus bordigalensis
Eugryllodes pipiens (sometimes)
Pteronemobius heydenii
Oecanthus pellucens
Chrysochraon dispar (sometimes)
Stenobothrus cotticus
Chorthippus apicalis
Chorthippus parallelus (sometimes)
Chorthippus montanus (sometimes)
Euchorthippus declivus
Euchorthippus pulvinatus
Euchorthippus chopardi
Euchorthippus albolineatus siculus
Euchorthippus sardous

(c) Polysyllabic echemes, of different duration

Decticus verrucivorus assiduus
    (warm conditions)
Platycleis affinis
Platycleis romana

Platycleis stricta
Acheta domesticus (sometimes)
Nemobius sylvestris

6. *Separately audible dense echemes in a series of limited duration (usually less than a minute)*
As in the last category, these echemes must be clearly audible as separate units and repeated less rapidly than 7/s.

This category can be subdivided further, as follows:

(a) Series with a crescendo, at least in the early part

Omocestus petraeus
Omocestus minutissimus
Myrmeleotettix maculatus
Chorthippus apricarius

Chorthippus cialancensis
Chorthippus cazurroi
    (often longer than a minute)
Chorthippus mollis

(b) Series without a crescendo (although there is often a crescendo within the constituent echemes); echemes of uniform duration

Pholidoptera aptera
Rhacocleis neglecta

Bryodema tuberculata
    (crepitation during display flight)

Euthystira brachyptera (sometimes)
Dociostaurus hispanicus
Omocestus raymondi (sometimes)
Omocestus antigai (sometimes)
Omocestus bolivari (sometimes)
Omocestus uhagonii (sometimes)
Stenobothrus nigromaculatus
Stenobothrus grammicus (sometimes
   with crescendo in early part)
Stenobothrus bolivarii (sometimes

   with crescendo in early part)
Stenobothrus stigmaticus (sometimes)
Stauroderus scalaris
Chorthippus brunneus
Chorthippus jacobsi
Chorthippus yersini (usually)
Chorthippus jucundus
Chorthippus albomarginatus
Euchorthippus angustulus

(c) Series without a crescendo; echemes of different durations

Barbitistes serricauda
   (echemes with few tick-like syllables)
Barbitistes obtusus
   (echemes with few tick-like syllables)

Meconema thalassinum (foot-drumming)
Conocephalus conocephalus (sometimes)
Chorthippus modestus

Note that in the song of *Conocephalus conocephalus* the later echemes in the series are often much louder than the earlier ones.

7. *Continuous dense echeme-sequences, lasting indefinitely (often more than a minute)*
A dense echeme-sequence is one in which the echemes are not clearly audible as separate units or are repeated at the rate of at least 7/s.

Conocephalus discolor (sometimes)
Tettigonia viridissima (occasionally)
Tettigonia hispanica (sometimes)

Decticus verrucivorus verrucivorus
Metrioptera bicolor (sometimes)

8. *Isolated dense echeme-sequences, of limited duration (less than a minute)*
We regard echeme-sequences as isolated if they are separated by intervals of at least 5 s. However, as in category 4 above, these intervals are so variable in some species that the song can be placed in both this category and the next one.

Gomphocerippus rufus
Chorthippus biguttulus (occasionally)
Chorthippus yersini (occasionally)

Chorthippus rubratibialis
   (with brief interruptions)
Chorthippus marocanus (occasionally)

9. *Series of dense echeme-sequences with intervals between them*
In such a series there are clear intervals between successive echeme-sequences, but not between successive echemes within each echeme-sequence.

(a) Series lasting indefinitely (often more than a minute)

Tettigonia viridissima (usually)
Tettigonia hispanica (sometimes)

Metrioptera bicolor (sometimes)

(b) Series of limited duration (less than a minute)

Chorthippus biguttulus (usually)
Chorthippus yersini (occasionally)

Chorthippus marocanus (usually)

Note that the song of *Chorthippus rubratibialis* comes into the above category if the characteristic interruptions are unusually long or if they are considered to divide the song into several echeme-sequences.

10. *Songs that are a combination of different kinds of sound and thus do not fall into any of the above categories*

(a) Alternations of echeme-sequences and simple trains of syllables
Conocephalus discolor (sometimes) Conocephalus dorsalis

(b) Echemes containing at least two different kinds of syllable
Callicrania selligera Chorthippus pullus
Platystolus faberi Chorthippus dorsatus
Uromenus asturiensis and sometimes U. stalii could also be included here, and so could the species of Platycleis and Metrioptera that have microsyllables (P. affinis, P. romana, P. iberica, P. falx, P. stricta, M. saussuriana, M. buyssoni, M. caprai, M. abbreviata and M. burriana); these species are listed in the appropriate categories above without taking into account any lack of uniformity in the syllables of their songs.

(c) Combinations of single syllables and dense echemes
Arcyptera fusca Arcyptera microptera
Arcyptera tornosi

(d) Combination of leg stridulation and wing-buzz
Stenobothrus rubicundulus

The two species of Polysarcus, listed in earlier categories according to the most long-lasting part of their songs, would also fall into category 10 if the full complexity of their songs were taken into account.

Several species (e.g. Uromenus asturiensis, Chorthippus yersini) could be included here on the basis of the occasional or frequent 'aftersongs' following (and often quieter than) the main component of their calling songs. The songs of such species have nevertheless been classified in the appropriate earlier categories without taking the aftersongs into account. The same applies to the song of Ephippiger perforatus, in which the main echemes of the calling song are often supplemented by bursts of tremulation during which much quieter sounds are produced.

The song-patterns and their evolution

A few general points emerge from the classification set out above. None of the European grasshoppers produce a train of rapidly repeated syllables lasting for an indefinite period, so that category 1 includes only bush-crickets and mole-crickets. Category 2, isolated syllables, includes only bush-crickets and one true cricket. A few grasshoppers produce regularly repeated echemes for indefinite periods of a minute or more (category 5), but this kind of song is again much commoner in bush-crickets and crickets; on the other hand, echemes in a series of *limited* duration (category 6) are much commoner in grasshoppers than in bush-crickets. Similarly, long-lasting dense echeme-sequences (category 7) are known only in bush-crickets, and short-lived dense echeme-sequences (category 8) are known only in grasshoppers. Short-lived isolated dense echemes (category 4) are common in the songs of both grasshoppers and bush-crickets.

A feature shown by the songs of many European Orthoptera is a crescendo during the course of an echeme. In a polysyllabic echeme this sometimes affects only the first two or three syllables, and suggests merely that it takes a few cycles of movement for the stridulatory organ to become fully functional. However, the crescendo quite often continues through a large proportion of a long echeme and in such cases is clearly a significant part of the song-pattern; examples include *Tettigonia cantans* (daytime song), *T. caudata*, *Pteronemobius heydenii*, *Omocestus viridulus* and *O. rufipes*. A crescendo during an echeme-sequence is found only in grasshopper songs (at least in western Europe); commonly heard examples include *Myrmeleotettix maculatus*, *Chorthippus apricarius*, *C. mollis* and, in dense echeme-sequences, *C. biguttulus*.

A common feature of grasshopper songs is the presence of momentary breaks in the sound during the course of a syllable. These are normally in the downstroke hemisyllable and are usually detectable only by oscillographic analysis (or at least by slowing down a recorded song), though they often impart a 'scratchy' quality to the sound. Such gaps are present in most grasshopper songs with syllables lasting more than about 30 ms, but particularly striking examples include *Omocestus uhagonii* (Figs 924–929) and *Stenobothrus apenninus* (Figs 1035–1040). In *Stauroderus scalaris* the gaps are just audible to the unaided ear: they are in the upstroke hemisyllables that form the rattling parts of the loud echeme-sequences produced by this grasshopper.

A number of bush-crickets belonging to the genera *Platycleis* and *Metrioptera* produce two kinds of syllable differing markedly in duration: macrosyllables, of normal duration, and very short-lived microsyllables, usually lasting less than 10 ms. The microsyllables are usually produced in a rapid sequence following a series of macrosyllables or, occasionally, a single macrosyllable (*Platycleis stricta*). *Callicrania selligera* and *Platystolus faberi* also produce echemes in which the syllables differ markedly in duration, but in the songs of these species the shorter syllables precede the longer ones and are usually longer than the microsyllables of *Platycleis* and *Metrioptera*. The echemes of *Uromenus asturiensis*, and sometimes *U. stalii*, also contain syllables of strongly differing durations, though showing less of a contrast than in the species mentioned above, and in the songs of many Orthoptera the syllables show some increase in duration during the course of an echeme. In the song of *Uromenus martorellii*, all the syllables are extremely brief, lasting less than 2 ms.

The songs of several European grasshoppers include syllables of markedly different kinds. In the genus *Arcyptera* dense echemes are accompanied by isolated syllables, and in *A. fusca* and *A. tornosi* the isolated syllables are of two different kinds, differing greatly in duration. The brief echemes of the song of *Stenobothrus bolivarii* consist of a longer syllable followed by a series of very short syllables, and are remarkably similar in pattern to the shorter of the two kinds of echemes in the song of *Platycleis stricta*. In the well-known song of *Chorthippus dorsatus* the first part of each echeme consists of syllables with gaps in the downstrokes and with the hind legs moved synchronously, and the second part is composed of syllables without such gaps and with the hind legs moving alternately.

In some grasshoppers the main echemes of the calling song are followed by a quieter and usually structurally different 'aftersong'. This occurs almost always in the song of *Gomphocerus sibiricus*, frequently in the song of *Chorthippus yersini* and occasionally in the songs of *Chorthippus mollis ignifer* and *C. rubratibialis*. In *C. vagans* it occurs as a local variant of the song in the Sierra Nevada in southern Spain. Quiet aftersongs are also a feature of the *courtship* songs of several species, including *C. biguttulus* and *C. mollis ignifer*.

Given the immense variety shown by these song-patterns, one cannot help wondering how they evolved – indeed, how sound production itself evolved in these insects. This is a topic that can never progress beyond speculation, but we think it appropriate to touch briefly on a few possibilites here.

Since the ensiferan Orthoptera (bush-crickets, true crickets and mole-crickets) stridulate with their fore wings, it is generally believed that sound production in these groups arose in association with flight, or at least with a 'warming-up' vibration of the fore wings preparatory to flight; Bailey (1991*a*), however, has suggested wing-vibration for dispersing a pheromone as an alternative precursor to bush-cricket stridulation. The sounds would at first have been produced accidentally, but would then have acquired selective value as an acoustic aid to pair-formation. These insects would of course have needed the ability to hear air-borne sound, and the communicative advantage of the sounds produced would have had to outweigh their potential for attracting predators and parasites. Gwynne (1995) has recently suggested that the tettigoniid stridulatory and hearing organs evolved quite separately from those of the Gryllidae and Gryllotalpidae, in spite of the close similarity in their structure.

In the gomphocerine grasshoppers, which use their hind legs as the active part of the stridulatory apparatus, it seems likely that sound production began in association with movements of the hind legs for locomotion, defence or visual signalling. However, Elsner (1968, 1983*b*) has suggested that, since the metathoracic muscles used for moving the legs are also used for moving the hind wings in flight, gomphocerine stridulation may have begun by the hind legs being vibrated at the frequency of wing-beat during flight – about 30–80/s, depending partly on the size of the grasshopper. He points out in support of this possibility that a number of European grasshoppers, when singing, vibrate their hind legs at frequencies of this order (Elsner & Popov, 1978). In species with a much slower syllable repetition rate, he suggests that the wing-beat frequency is reflected in the jerky movement that often occurs during the downstrokes (or, occasionally, the upstrokes) of the hind legs, resulting in brief gaps in the syllables. However, we have found the gap frequency to be generally higher than the frequency of wing-beat in the European Gomphocerinae: it is usually over 100/s in *Omocestus* and *Stenobothrus*, and reaches well over 200/s in some species of *Euchorthippus*. Elsner suggests that at least some of these discrepancies can be explained by the gap frequency being double (or occasionally half) the wing-beat frequency, but it would be difficult to account for all the gap frequencies of the European Gomphocerinae in this way. Helversen & Helversen (1994) have recently suggested that, since rapidly pulsed sound produces a stronger response in a grasshopper's auditory system than longer bursts of continuous sound, gapped syllables may have evolved simply as a more effective way of stimulating a female response. It is perhaps worth noting that the production of gapped downstroke hemisyllables is facultative in at least some species; for example, *Omocestus antigai* produces both gapped and continuous downstroke hemisyllables in its courtship song, and the echemes of the calling song of *Chorthippus dorsatus* begin with gapped downstroke hemisyllables and end with continuous ones.

Once the means of acoustic signalling had evolved it is not surprising that it should have become an important aid to pair-formation. It is also easy to understand that, following (or during) allopatric speciation, species isolated from one another geographically should develop differences in song, just as they often diverge morphologically. In this way a variety of different songs could soon develop. In the Ensifera it seems likely that the earliest songs were long, simple trains of syllables, showing a close analogy to flight and needing similar neural control. The breaking up of such trains of syllables into echemes (or isolated syllables) would have conserved energy and allowed the singing male to hear predators, other singing males and, in some cases, acoustically responding females. In the gomphocerine grasshoppers both sexes sing, producing similar song-patterns with the same apparatus, and we think it likely that singing evolved in the two sexes simultaneously. The primitive song-pattern was probably a simple echeme of limited duration, as in the songs of many present-day European grasshoppers; the females normally sing in response to the male calling song, and the need for the males to hear this response would have made continuous singing disadvantageous.

The development of the more complex rhythmic patterns, such as the dense echeme-sequences of such species as *Conocephalus discolor* and *Chorthippus biguttulus*, is less easy to explain. One possibility, currently in favour, is that females have sometimes preferred more elaborate songs and have thus effected, by sexual selection, a trend towards greater complexity. Another, more controversial, possibility is that selective pressure has favoured more distinctive songs in cases where different species with similar songs have been wasting resources by intercopulating, either fruitlessly or producing hybrids of reduced fertility.

Sporadic interbreeding between different species that have contrasting songs, but that are nevertheless completely interfertile, could provide a further method of song evolution, since the hybrids would be likely to have songs differing from those of both parent species. Because of the general effectiveness of songs as a mate recognition system, this is not likely to have happened often, but one possible case is provided by *Chorthippus mollis ignifer* (see p. 76).

Readers needing fuller discussions of the evolution of orthopteran song are referred to those of Heller (1990), Bailey (1991a, 1991b), Otte (1992) and Helversen & Helversen (1994). The evolution of the extraordinarily complex patterns of the courtship songs of some grasshoppers is discussed by Otte (1972), Bull (1979) and Helversen (1986).

Chapter 5

The value of the songs in taxonomy and identification

Historical review

It is a humbling thought that forerunners of our present-day crickets were singing to one another over 200 million years before there were any human ears to hear them. There are clear signs of a stridulatory organ on the male fore wings of fossil members of the ensiferan family Haglidae, which first appeared in the early Triassic Period, and Sharov (1968: 26–28) believed that even the more primitive fossil family Oedischiidae had an alary stridulatory organ during the preceding Permian Period. It was not until a mere 3 million or so years ago that early man appeared on the scene, thus providing the first opportunity for human interest in the sounds of these insects. Exactly when our forebears first noticed that particular sounds were characteristic of particular insects we shall never know, but early man would soon have discovered that certain large insects that were good to eat could be tracked down by their songs. We can be sure that human interest in insect sounds dates back at least to the earliest historical records, since there are a number of references to singing cicadas and crickets in literature from as long ago as the second millennium BC (Kevan, 1974). Indeed, according to Bodson (1976) the ancient Greeks had already discovered by 650 BC the location of the cicada's tymbal organ, a sound-producing apparatus much less obvious than those used by crickets and grasshoppers. Aristotle, the traditional founder of natural history, gave a brief account of insect sounds in about 330 BC, mentioning that the songs of grasshoppers are produced by rubbing with the hind legs (Thompson, 1910).

During the following two thousand years the subject hardly advanced. In the English literature Gilbert White's (1789) brief accounts of the songs of *Gryllus campestris* and *Gryllotalpa gryllotalpa* are perhaps worth a mention, but it was not until the nineteenth century that there were any significant developments. Goureau (1837*a*), in an essay on insect stridulation in general, described sound production in a number of Orthoptera, noting that in *Ephippiger* both sexes stridulated and observing, remarkably, that, although both males and females used their much reduced fore wings for this purpose, the stridulatory modifications of the female fore wings were quite different from those of the male. Shortly afterwards, Goureau (1837*b*) observed that males of *Leptophyes punctatissima* moved their fore wings as if stridulating but produced no sound that he could hear; he even speculated that they were nevertheless producing sound that could be detected by the appropriate equipment and thus came remarkably near to predicting the discovery, a century later, that these insects produce ultrasound. Siebold (1842) described the songs of several Orthoptera, including the characteristic hind tibial kick of *Stethophyma grossum*, and a little later (1844) gave a fuller account of sound-producing and, especially, hearing organs in Orthoptera. Soon afterwards, Fischer (1849, 1850, 1853) included a number of song descriptions in primarily faunal or systematic studies.

At about this time there was a major advance in the recognition of the diagnostic value of the songs of Orthoptera: Yersin (1852, 1853 and especially 1854*b*) produced a series of papers specifically on the songs of European Orthoptera and their value in identification. In 1852 he described clearly for the first time the striking differences between the songs of the closely similar species *Chorthippus brunneus*, *C. biguttulus* and *C. mollis*, and the paper of 1854, in which he described the songs of 38 species, included the first attempt to illustrate the songs graphically in the form of musical notation (Fig. 34). Yersin (1854*b*: 116–117) also described, probably for the first time, a nymph of *C. parallelus* making stridulatory movements with its hind legs typical of those made by adults, though producing no sound.

A few years later there was a strikingly parallel development in North America, where Scudder (1868*a*, 1868*b* and especially 1893) produced a series of papers on the songs of North American Orthoptera, in which he also used musical notation to illustrate their rhythmic patterns. In the 1893 paper he described the songs of some 50 species, drawing on the work of others (notably Riley, 1874; McNeill, 1891) as well as his own observations.

Both Yersin (1854*b*) and Scudder (1893) included descriptions of the flight sounds (crepitation) produced by some grasshoppers, though there is much greater emphasis on these sounds in Scudder's account simply because there are many more crepitating species in North America than in Europe. For similar reasons Yersin's account includes many gomphocerine grasshoppers and few crickets, whereas in Scudder's account the reverse is true. These two early observers were perhaps the first to recognize, in their respective hemispheres, the great value of the songs in the taxonomy and identification of Orthoptera. Scudder's work no doubt acted as an important spur to further studies in North America, but Yersin's work was largely ignored in Europe, perhaps because most of it was published in a local Swiss journal of limited circulation.

Among other significant nineteenth century studies on sound-production in Orthoptera, those of Graber and Landois perhaps deserve a special mention. In the works of both these authors the emphasis was more on the anatomy of the sound-producing organs than on the sounds themselves. Graber (1871, 1872*a*, 1872*b*) included accounts of the stridulatory files of female grasshoppers and of stridulatory structures on the female fore wings of bush-crickets belonging to the genera *Ephippiger* and *Odontura*. He also described the crepitation of *Bryodema tuberculata* during its impressive display flights (Graber, 1873). Landois' work was on insects in general (1867) or even animals in general (1874) and included few original observations on the songs of Orthoptera, but in the 1867 paper he made an attempt, using musical notation, to indicate the pitch of various insect sounds, including the 'Schnarrtöne' of grasshoppers and 'Schrilltöne' of *Acheta domesticus*.

Towards the end of the nineteenth century there was a flurry of interest in the effect of temperature on the song of the Snowy Tree Cricket or 'thermometer cricket' (*Oecanthus fultoni*) in the USA. This began with the observations of Brooks (1881), but it was only when Dolbear (1897) published his famous, but inadequate, formula for determining the night air temperature from the 'chirp rate' of this insect that widespread interest was aroused. A number of studies followed (Bessey & Bessey, 1898; Edes, 1899; Shull, 1907; Fulton, 1925; Allard, 1930), in which it was established that, although there was a clear relationship between chirp rate and temperature in this cricket (as is true of most singing Orthoptera), the effects of individual variation and other factors made it impossible to express the relationship accurately by a single formula.

The turn of the century was notable for the attempt by Kneissl (1900) to classify the songs of the Bavarian grasshoppers and crickets, giving descriptions of them in the form of an identification table, though not yet a fully developed key. He also gave a separate account of a number of courtship songs, including the highly developed ones of *Chorthippus albomarginatus* and *Gomphocerippus rufus*, and even noted the hind tibial kicks (of the kind used in the calling song of *Stethophyma grossum*) that form part of the courtship song of *Omocestus viridulus*.

In the following year Petrunkewitsch & Guaita (1901) published an account of the stridulatory structures of Orthoptera in which they described and illustrated such structures in the females of a number of species, including grasshoppers, bush-crickets and the European mole-cricket *Gryllotalpa gryllotalpa*. Regen (1902, 1903) developed this theme further and described (1903: 387), probably for the first time, the actual sounds produced by females of *G. gryllotalpa*. Shortly afterwards Baumgartner (1905 and, in more detail, 1911) published accounts of female stridulation in a North American mole-cricket currently (though rather doubtfully) identified as *Neocurtilla hexadactyla*.

The next significant development was the experimental discovery in Europe by Prochnow (1907: 159) and Karny (1908: 118–119), each unaware of the other's work, that the rattling sounds (crepitation) produced by flying grasshoppers were generated solely by the hind wings, with no involvement of the fore wings; Prochnow used *Psophus stridulus* for his experiments, and Karny used both this species and *Arcyptera fusca*. Stäger (1930) repeated these experiments, this time including *Stauroderus scalaris* in addition to the other two species, and came to the same conclusion; he even found that all three species could crepitate with only one hind wing, after the other hind wing, both fore wings and both hind legs had been removed! Fulton (1930: 615) concluded from experiments on some unspecified North American grasshoppers that loud crepitation was at least partly dependent on the fore wings, but Isely's (1936) experiments on seven named Texan grasshoppers seemed to confirm the earlier European studies; Isely quotes a Dr Gordon Alexander as having discovered that crushing the strongly developed veins in the anal fan of the hind wings prevented crepitation though flight was still possible.

In 1928 there was an important step forward in the use of the songs for identification when Faber produced his key to the German Orthoptera based entirely on their songs. This was a considerable advance on Kneissl's (1900) earlier attempt and was followed by two major studies on sound-production in gomphocerine grasshoppers (Faber, 1929b, 1932) and an equally comprehensive one on locustine grasshoppers (Faber, 1936). Another significant work produced at this time was Baier's (1930) study of stridulation and hearing in insects generally, notable not only for its scientific content but also for the fact that it was published in English, although in a German journal and based on work carried out at the University of Fribourg in Switzerland.

After World War II, Jacobs entered the field of insect acoustics, soon producing (1950b) a major study of grasshopper songs and associated behaviour, to be followed a few years later

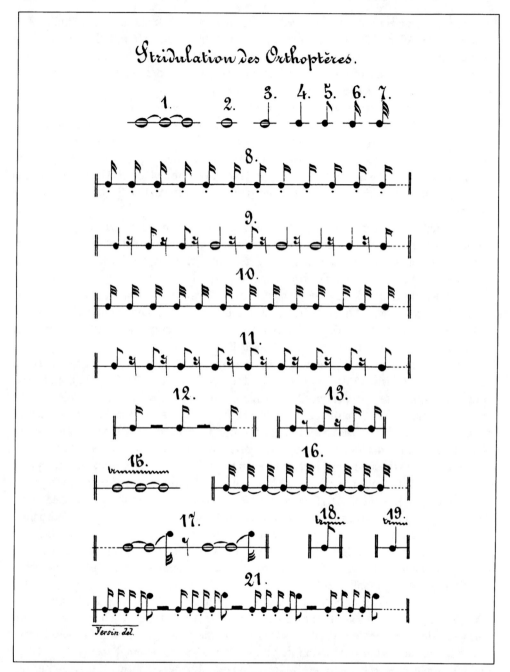

Figure 34 The remarkable attempt by Alexandre Yersin (1854*b*) to illustrate the songs of European Orthoptera by musical notation. 1. Three tied semibreves (whole notes), indicating indefinite duration. 2. Semibreve, indicating limited duration but at least 4–5 s. 3. Minim (half note), lasting 2–3 s. 4. Crotchet (quarter note), lasting about 1 s. 5. Quaver (eighth note), lasting no more than half a second. 6. Semiquaver (sixteenth note), 3–5/s. 7. Demisemiquaver (thirty-second note), very short sound, often repeated too rapidly to count. 8. *Gryllus campestris, Chorthippus vagans.* 9. *Nemobius sylvestris.* 10. Several species of Tettigoniidae, *Gomphocerus sibiricus.* 11. *Platycleis*

albopunctata. 12. *Platycleis sepium, Ephippiger terrestris, Pholidoptera griseoaptera*. 13. *Pholidoptera griseoaptera*. 14. *Metrioptera roeselii*. 15–17. *Polysarcus denticauda*. 18. *Euthystira brachyptera*. 19. *Euchorthippus declivus*. 20. *Chorthippus albomarginatus*. 21. *Chorthippus dorsatus*. 22. *Chorthippus parallelus*. 23. *Stenobothrus lineatus*. [24 is missing] 25. *Omocestus rufipes, Omocestus viridulus, Gomphocerippus rufus*. 26. *Chorthippus apricarius*. 27. *Stauroderus scalaris*. 28. *Chorthippus biguttulus*. 29. *Chorthippus brunneus*. 30. *Chorthippus mollis*. 31. *Myrmeleotettix maculatus*. 32, 33. *Arcyptera fusca*.

with an even more comprehensive work on the same subject (Jacobs, 1953a). Almost simultaneously, Faber (1953a) produced an equally significant book on the songs of both grasshoppers and phaneropterine bush-crickets. These works contained voluminous descriptions of the acoustic repertoire of the species included and together comprised by far the most comprehensive account of the songs of European grasshoppers to have appeared up to that time.

Jacobs (1950b, 1953a) used simple diagrams against a time scale to illustrate the rhythmic patterns of the songs. Ragge (1965) introduced a more sophisticated type of diagram, amounting to a simplified oscillogram, and diagrams of this kind have since been used by Holst (1970), Samways (1976e), Luquet (1978), Wallin (1979) and, in a slightly different form, by Duijm & Kruseman (1983) and Bellmann (1985a). Identification keys based on the songs were provided, for restricted parts of Europe, by Luquet (the Mont Ventoux area of the French Alps, Acrididae only), Wallin (Sweden), Duijm & Kruseman (Benelux), Weber (1984, Westphalia) and Bellmann (Germany). All these studies were concerned with regional faunas and contain no taxonomic research, although Luquet gives much emphasis to the importance of the songs in taxonomy.

Meanwhile parallel developments had been taking place in North America. From 1910 onwards Allard published many short papers giving descriptions of the songs of Orthoptera in the eastern USA, eventually producing a more comprehensive review, including cicadas (Allard, 1929a). His contemporary, Fulton, also described many songs during this period in papers that were primarily faunistic or taxonomic. In 1931 Fulton published a study of the North American species of the cricket genus *Nemobius* that was significant in two respects: it contained what was probably the first description of a new taxon (*tinnulus*) detected primarily because of its distinctive song, and it included the first identification key based entirely on songs to be produced in North America. In the following year Fulton (1932) published a key to the Tettigoniidae, Gryllidae and Gryllotalpidae of North Carolina, again based entirely on their songs; he was almost certainly quite unaware of the earlier keys of this kind produced by Kneissl (1900) and Faber (1928). In 1952 Fulton published an account of the field crickets of North Carolina, in which he showed that what had been regarded as a single species consisted of four different 'races' with different songs; he found experimentally that these forms would not interbreed but he stopped short of recognizing them formally as distinct species. Two further works from this era need to be mentioned: a useful review by Lutz (1924) of work on insect sounds in general, and Pierce's (1948) book on insect sounds (almost all of North American Orthoptera), in which he described the songs of some 40 species and illustrated them with tracings of level against time produced by a 'photographic recorder' of his own design.

Up to the nineteen-fifties work on the songs of Orthoptera – and other insects – had been severely hampered by the lack of a convenient means of making (a) recordings of the sounds being studied and (b) objective and detailed analyses of the sounds in the form of patterns against a time-scale. Insect sounds were probably first recorded on discs in the early part of the present century, through being accidentally included in recordings of bird song or other better-known wildlife sounds. The earliest deliberate recording of an insect sound known to us is one of *Gryllus campestris* made by Ludwig Koch in 1934 and kept in the BBC Sound Archives. This was followed shortly afterwards by a disc of American insect sounds prepared by Brand *et al.* (1937) for private circulation. In 1947 Ossiannilsson produced, again for private circulation, a disc of the sounds made by 14 Swedish species of Homoptera Auchenorrhyncha. Since World War II many commercial discs of insect sound recordings have been issued (see Kettle, 1984). The use of disc-recorded sounds in scientific studies has been very limited, but Broughton (1952b, 1954, 1955a) carried out some analysis from discs, both electronically and by direct photography of the disc grooves.

The advent in the fifties of the high quality portable tape recorder at last provided bioacousticians with a convenient means of 'storing' songs for repeated study and analysis. Audiospectrographic and oscillographic analyses became much easier, and from this time onwards bioacoustic research papers were almost always illustrated with audiospectrograms (frequency spectrum against time) or oscillograms (waveform, and hence amplitude, against time) (see Chapter 2). This technology was soon used in North America in the first serious application, by R. D. Alexander, of the songs of Orthoptera to their taxonomy, in a series of papers stemming from his doctoral thesis (Alexander, 1956). Perhaps the most significant of these was his account of the field crickets of the eastern USA (Alexander, 1957b), in which he recognized five species, including Fulton's (1952) four 'races'. In other papers he applied, sometimes jointly with E. S. Thomas, song characters to the taxonomy of North American Nemobius (Thomas & Alexander, 1957; Alexander, 1957a; Alexander & Thomas, 1959) and Orchelimum (Thomas & Alexander, 1962). He also published two general reviews of sound communication in Orthoptera and other insects (Alexander, 1960; 1967a), a useful discussion of the role of song studies in cricket taxonomy (Alexander, 1962), and a small book in which he discussed a number of the more interesting aspects of singing insects in North America (Alexander, 1967b). In 1972 Alexander et al. produced an account of the singing insects of Michigan in which they gave keys based mainly on the songs for the Tettigoniidae and Gryllidae.

Song characters were also applied to the taxonomy of American Orthoptera by T. J. Walker, beginning with his studies on Oecanthinae (Walker, 1962a, 1963). In 1964 he published a general discussion of 'cryptic' species among Gryllidae and Tettigoniidae, observing that about a quarter of the species of these groups in the eastern USA had been unrecognized or of doubtful status until the calling songs and life histories were studied (Walker, 1964b). He went on, sometimes with co-workers, to use the songs to elucidate the taxonomy of the American Orocharis (Walker, 1969a), Cyrtoxipha (Walker, 1969b), Orchelimum (Walker, 1971), Borinquenula (Walker & Gurney, 1972), Neoconocephalus (Walker et al., 1973; Walker & Greenfield, 1983), Anurogryllus (Walker, 1973), Hapithus (Walker, 1977) and Pictonemobius (Gross et al., 1990). Among Walker's many other papers on the bioacoustics of Orthoptera are a detailed study of the effect of temperature and other factors on cricket songs (Walker, 1962b), a discussion of character displacement in insect songs (Walker, 1974) and the first really accurate accounts (based on high-speed cinematography) of the stridulatory wing-movements of Gryllidae and Tettigoniidae in relation to the sounds produced (Walker et al., 1970; Walker & Dew, 1972; Walker, 1975a).

Among other significant North American studies of the songs of Orthoptera two are particularly noteworthy, although the sounds studied are not in either case applied to taxonomic problems: Spooner's (1968) account of the songs of phaneropterine bush-crickets and Otte's (1970) detailed study of communicative behaviour in gomphocerine and locustine grasshoppers. At a more popular level, Dethier (1992) has recently provided a nostalgic account of the time he spent in the late thirties working with G. W. Pierce, whose pioneering work on the songs of North American insects has already been mentioned (Pierce, 1948); Dethier includes identification keys, based on both morphology and song, to some of the more common Orthoptera of New England. Even more recently, Pelletier (1995) has produced a useful guide to the singing Orthoptera and cicadas of eastern North America, accompanied by a choice of cassette or CD of recorded songs.

In other parts of the world the serious taxonomic exploitation of the songs of Orthoptera has been slower to develop. Outside Europe taxonomic studies in which the songs have played a significant part include those of Nevo & Blondheim (1972) on Israeli mole-crickets, Bailey (1975, 1980) on African and Australian copiphorine bush-crickets, Pitkin (1977) on two sibling species of *Thyridorhoptrum* (Tettigoniidae) in Africa, Otte & Alexander (1983) on Australian crickets, Otte & Cade (1983*a*, 1983*b*, 1984*a*, 1984*b*, 1984*c*), Otte (1983, 1985, 1987) and Otte *et al.* (1988) on African crickets, Otte *et al.* (1987) on the crickets of New Caledonia, Rentz (1985, 1993) on Australian Tettigoniidae, Rentz (1988) on southern African Tettigoniidae, Ingrisch (1993) on the grasshoppers of Thailand and Otte (1995) on the crickets of Hawaii. Also worth mentioning is a guide to some of the singing insects of Western Australia, including an identification key based on song, produced by Gwynne *et al.* (1988).

The use of the songs in taxonomic research on European Orthoptera has begun only quite recently. This is largely because the taxonomic problems in need of bioacoustic solutions have been mainly in the southern peninsulas (Iberian, Italian and Balkan), where the songs have generally been ignored. In addition, the European cricket fauna is much poorer than that of North America and it seems unlikely that the discovery of cryptic species that has occurred on such a scale in the USA will be repeated in Europe. In the gomphocerine grasshoppers the reverse is true: the European fauna is much richer than the American one and the songs are already playing a vital role in solving taxonomic problems. Examples of such studies are those of Ragge & Reynolds (1984) on *Euchorthippus*, Helversen (1986) on the *Chorthippus albomarginatus* group, Ragge (1986) on *Omocestus*, Ragge (1987*a*) on *Stenobothrus*, and Ragge & Reynolds (1988), Helversen (1989) and Ragge *et al.* (1990) on the *Chorthippus biguttulus* group. Comparable studies on the Tettigoniidae include those of Heller (1984), Heller & Reinhold (1992) and Willemse & Heller (1992) on *Poecilimon*, Ragge (1990) on *Platycleis* and Pfau (1996) on *Platystolus*. Heller's (1988) comprehensive account of the bioacoustics of European Tettigoniidae should also be mentioned, although not primarily intended for taxonomic application (the two new subspecies described in this work are based on morphological rather than acoustic differences). Bennet-Clark (1970*b*) described a new species of mole-cricket from France partly on the basis of its distinctive song and there is undoubtedly scope for applying the songs to the taxonomy of the southern European mole-crickets in general; because of the difficulties in working with largely subterranean insects, we have not had the time to undertake such a study ourselves.

Most of the bioacoustic studies on the European Orthoptera during the past 40 years have been on the physiology of sound production and reception, or on the details of acoustic behaviour, and it would be inappropriate to review this work in any depth here. The many descriptions of the songs of particular species that have been published in the course of these studies are referred to where the songs of these species are described in this book. It suffices to mention here a few of these studies that have a more general relevance to our work. The special value of the songs of Orthoptera in taxonomy and identification stems from the fact that they form a mate recognition system and are thus likely to be a particularly reliable guide to the identity of the species producing them. We must therefore emphasize the importance of the much-quoted study of Perdeck (1957), in which he established experimentally that the songs of the well-known sibling pair of grasshoppers *Chorthippus brunneus* and *C. biguttulus* provided virtually the only barrier to interbreeding. Helversen & Helversen (1975a, 1975b, 1981) showed that this was also true of *C. biguttulus* and *C. mollis*, and outside Europe there have been comparable studies on several groups of crickets (e.g. Walker, 1957; Hill *et al.*, 1972; Zaretsky, 1972; Ulagaraj & Walker, 1973; Popov *et al.*, 1974; Paul, 1976, 1977).

The terminology we (and many others) use in describing the songs of Orthoptera depends on the way the stridulatory apparatus is moved in producing the songs. We have therefore found particularly valuable the development by Elsner (1970, using a Hall-generator) and Helversen & Elsner (1977, using an opto-electronic camera) of highly sophisticated methods of recording the leg-movements of singing grasshoppers; the opto-electronic camera has also been used by Heller (1984, 1988) for recording the wing-movements of singing bush-crickets, and miniature angle detectors have been developed by Koch and his colleagues for recording the wing-movements of true crickets during both calling and courtship songs (Koch, 1978, 1980; Koch *et al.*, 1988). These techniques have revealed that the two hind legs of singing grasshoppers are seldom exactly synchronized and often perform different patterns of movement (Elsner, 1974a).

Most of the other recent work on the songs of European Orthoptera is less relevant to their application in taxonomy. Access to much of this literature may be gained through a number of books and reviews, of which the most useful that have not already been mentioned are probably the following: Haskell (1961), Dumortier (1963b, 1963c, 1963d), Uvarov (1966: 176–195, 1977: 220–239), Sales & Pye (1974), Otte (1977), Elsner & Popov (1978), Zhantiev (1981), Elsner (1983a), Hutchings & Lewis (1983), Kalmring & Kühne (1983), Bennet-Clark (1984), Ewing (1984, 1989), Huber *et al.*(1989), Gribakin *et al.* (1990), Bailey & Rentz (1990), Bailey (1991a), Pye (1992), Greenfield (1997). For the earlier literature the excellent bibliography of Frings & Frings (1960) is frequently useful.

The phonotaxonomy of the western European Orthoptera

As indicated in the preceding pages, the songs of Orthoptera are now recognized as providing a vital aid in resolving taxonomic problems at the species level. We have thought it best not to encumber our descriptions of the songs with discussions of such taxonomic problems, of which there are many among the European Orthoptera, but rather to deal with them separately in this section. Our emphasis is naturally on the taxonomy indicated by the songs, for which we propose to coin the term *phonotaxonomy*.

As is only to be expected of a mate attraction and recognition system, the taxonomic usefulness of the songs is almost entirely at the species level. Species that appear to be closely related on all other criteria often have quite different songs. A marked contrast in song is a strong indication that species are *distinct*, but not necessarily that they are *unrelated*. The *Chorthippus biguttulus* group is a striking example of species that are closely similar in morphology and can interbreed, producing fertile hybrids, and yet have quite different songs. Indeed, when such interfertile species are sympatric it is clearly necessary for them to have some means of achieving reproductive isolation, and a contrast in calling song can provide such a means.

The songs are thus particularly useful in detecting and resolving complexes of sibling species, and in deciding on the status of allopatric populations showing small morphological differences. Their use in assessing relationships between species, and therefore as a tool in phylogenetics, is more limited.

We do not pretend that the songs are all-important in species-level taxonomy, or that they should always outweigh other criteria in making taxonomic judgements. We do, however, believe that the calling songs, in providing the basis for a mate recognition system, are a particularly reliable indication of species limits.

Particular species or groups that call for taxonomic discussion are dealt with below, following the same systematic sequence as in Chapters 7–9, and with cross-references to where the songs are described. The formal nomenclatural changes at the specific level that follow from our conclusions are listed on p. 490.

Decticus verrucivorus (Linnaeus) (p. 157)

As has been shown by Götz (1970) and Samways & Harz (1982), this species is very variable in southern Europe, where it occurs mainly in mountains. These southern populations often have unusually short wings and sometimes show other morphological differences; some of them are regarded as belonging to distinct subspecies. The calling song, however, is very uniform in most of western Europe, including the Pyrenees, Alps and Apennines, and lends strong support to the treatment of all these populations as a single species.

In the mountains of central and eastern Spain the opposite situation occurs: there is a form that is morphologically indistinguishable from typical *D. verrucivorus* (though having the rather short wings that are usual in montane populations), but in which the male calling song is strikingly different (cf. Figs 224, 229). Such a difference in song suggests that these Spanish populations belong to a distinct species, but in the apparent absence of any morphological support for this it would seem better to treat them as a form of *D. verrucivorus* until a thorough biological study can be made. Ingrisch *et al.* (1992) have treated them as the subspecies *D. v. assiduus*, and this seems a reasonable compromise for the time being. It would of course be most interesting to determine experimentally whether this form is interfertile with typical *D. verrucivorus* and whether the difference in calling song provides any kind of ethological barrier. If such a study were to establish that there is no barrier to interbreeding, this difference in calling song would be the most striking example of geographical variation in song known at present in the European Orthoptera.

Metrioptera saussuriana (Frey-Gessner), *M. buyssoni* (Saulcy) and *M. caprai* Baccetti (pp. 189, 191)

The members of this trio are so similar in morphology and song that we are not entirely confident that they are distinct species. *M. buyssoni* is so far known only from the départements of Haute-Garonne and Ariège in the French Pyrenees. The much more widespread species *M. saussuriana* also occurs in the eastern Pyrenees, including several localities in Ariège, though it has never been found at the same site as *M. buyssoni*. The morphological differences between these two species, purely in coloration and genitalia, are small but seem nevertheless to be clear and reliable. We have, however, been unable to find any difference in the calling song (see Figs 357–365). Dr B. Defaut has made a special study of these species and is convinced that they are distinct (Defaut, 1981, 1988b). We remain doubtful, but are content to preserve the status quo until stronger evidence either way is available.

M. caprai presents a rather different problem as it is completely allopatric with the other two species, occurring only in the Apennine range in Italy. The morphological differences are again slight, of a similar order to those between M. saussuriana and M. buyssoni. We have studied the song at the type locality, Terminillo in the Monti Reatini, and further south in the Gran Sasso d'Italia. At the type locality the calling song of the three males studied was consistently different from that of M. saussuriana, the number of syllables per echeme being higher (Fig. 369), but the calling song in the Gran Sasso (Fig. 368) was much closer to M. saussuriana. It seems likely that similar local variations in the song will be found in other parts of the Apennines, perhaps paralleling the morphological variants that have been formally recognized as subspecies. We are again doubtful about the status of M. caprai but, in the absence of any strong evidence to the contrary, prefer to leave it as a distinct species for the time being.

Metrioptera roeselii (Hagenbach) (p. 199)

In southern Europe this common species occurs in a number of local forms, most of which have been regarded at some time in the past as distinct species. Götz (1969) reduced the status of M. azami (Finot) and M. ambitiosa Uvarov to subspecies of M. fedtschenkoi (Saussure), and Canestrelli (1981) did the same with M. brunneri (Ramme). Finally, Heller (1988) reduced M. fedtschenkoi and M. bispina (Bolívar) to the status of subspecies of M. roeselii, thus treating this whole group of variants as a single species. In western Europe we have studied the song of M. r. fedtschenkoi in southern France, where it occurs very locally as the form azami, and have found it to be indistinguishable from the song of typical M. roeselii (see Figs 401–418). This form does, however, show clear morphological differences from typical M. roeselii in the shape of the tenth abdominal tergite and cerci of the male, and the subgenital plate of the female, and we are not entirely convinced that it is not a distinct species in spite of the apparent identity in song.

Ephippiger ephippiger (Fiebig) (p. 227)

This species is highly variable in size and coloration, and also to some extent in such morphological characters as pronotal texture and genitalia. Two variants occurring in particular areas have been regarded until recently as distinct species: E. cruciger (Fieber) in the Mediterranean region of France, and E. cunii Bolívar in the eastern Pyrenees. The song is predominantly monosyllabic in most of the range in western Europe, but changes to polysyllabic echemes in the French départements of Aude and Pyrénées-Orientales, and in Catalonia. The change is gradual and the number of syllables per echeme is variable throughout this region of southern France and north-eastern Spain; because of this we do not regard these song differences as indicating that more than one species is involved. Recent studies on interbreeding, enzyme analysis and song have resulted in a consensus that the variants occurring in these regions are best treated as no more than local forms of E. ephippiger, and we are in full agreement with this view (see especially Duijm, 1990; Hartley & Warne, 1984; Hartley & Bugren, 1986; Oudman et al., 1990; Ritchie, 1991, 1992a, 1992b).

Ephippiger terrestris (Yersin) (p. 231)

Like *E. ephippiger* this species is very variable, though much more restricted in distribution. Until recently the more easterly populations were regarded as a distinct species, *E. bormansi* (Brunner), but it is now generally agreed, on the basis of morphology, enzyme analysis and interbreeding experiments, that only one species is involved (see especially Duijm & Oudman, 1983; Landman *et al.*, 1989; Nadig, 1980; Oudman *et al.*, 1989). The song shows little variation throughout the range, being almost always monosyllabic, and thus supports this conclusion.

Ephippiger perforatus (Rossi) (p. 235)

Heller (1988) regards this species as a subspecies of *E. ephippiger*. However, there is one significant aspect of its acoustic behaviour that is not apparently shared with *E. ephippiger* and indeed seems to be unusual in the Ephippigerinae. When males of *E. perforatus* tremulate they produce a quiet but clearly audible stridulation (see p. 235 and Figs 539–546). We have not observed this 'tremulation-sound' in *E. ephippiger*, which also frequently tremulates, and consider this difference in behaviour to be sufficient reason for retaining *E. perforatus* as a distinct species.

Arcyptera microptera (Fischer de Waldheim) (p. 290)

This is a very widespread and variable species with a number of regional forms differing in various characters, including the length of the wings. Two French forms have hitherto been regarded as distinct species: *A. carpentieri* Azam in the Cevennes and *A. kheili* Azam in the French Alps. We have studied the calling songs of both these forms and have found neither of them to differ significantly from that of *A. microptera* (see Figs 737–752). Taken together, however, all these songs are clearly distinct from those of *A. fusca* and *A. tornosi*.

We also consider the morphology of these two forms to fall within the range of variation of *A. microptera*. *A. kheili* is very brachypterous, but no more so than, for example, the Turkish form *A. m. karadagi* Karabağ. We therefore believe, on the basis of both song and morphology, that these forms are better regarded as no more than local variants of *A. microptera*.

Omocestus viridulus (Linnaeus) (p. 304)

This common and very widespread species occurs in the Iberian Peninsula in a form hitherto regarded as a distinct species, *O. kaestneri* (Harz). We have studied the male calling song of this form and found it to be extremely similar to that of typical *O. viridulus* (see p. 305 and Figs 800–803), though tending to be a little shorter. This form is also very similar to typical *O. viridulus* in morphology, though the male has a red patch on the abdomen and shows some slight differences in the supra-anal plate, and the ovipositor shows a variable tendency to develop lateral teeth. There is apparently no overlap in distribution in the Iberian Peninsula, typical *viridulus* occurring only in the vicinity of the Pyrenees, and *kaestneri* occurring elsewhere in the peninsula, especially in the northern half. In view of the close resemblance in both song and morphology, we think this allopatric form is better regarded as being only subspecifically distinct from typical *O. viridulus*, forming the subspecies *O. v. kaestneri* (Harz).

Omocestus antigai (Bolívar) (p. 319)

This species, originally described from Gerona province in Spain, is found in the central and eastern Pyrenees, where it occurs in a number of well isolated populations. Until recently, *O. broelemanni* Azam, described from Val d'Eyne in the French département of Pyrénées-Orientales, was regarded as a distinct species, but Clemente *et al.* (1990) have synonymized it with *O. antigai*. The morphological differences are insignificant and our studies of the male calling and courtship songs at both type localities show that there is also no significant difference in acoustic behaviour. We are therefore in complete agreement with the conclusion of Clemente *et al.* that these two names are synonymous.

Towards the west, in the Spanish provinces of Huesca and Lérida, there is an increase in size and a tendency to develop various small morphological differences; the male calling song remains the same, but the courtship song is a little different in some localities (see p. 326). Clemente *et al.* (1990) regard these western populations as belonging to a distinct species, *O. navasi* Bolívar, described from the Sierra de Guara in Huesca province. Our studies of some 120 specimens from localities widely scattered in the central and eastern Pyrenees, including the analysis of songs recorded from 26 males, suggest that such small differences in morphology and courtship behaviour occur, to a greater or lesser extent, in many small populations over the whole of this region; the extent of the divergence shown by each population no doubt depends on how long it has been isolated. We have found it impossible to draw a clear dividing line between western and eastern forms, on the basis of either morphology or song. We therefore think it preferable, at least for the time being, to regard the populations from all these localities, including the western ones, as being no more than local variants of *O. antigai*, and thus no longer to treat *O. navasi* as a distinct species.

Omocestus minutissimus (Bolívar) (p. 328)

This primarily montane species occurs widely in the eastern half of Spain, but including the Sistema Central. Until recently two local populations were regarded as distinct species, *O. llorenteae* Pascual in the Sierra Nevada and *O. knipperi* Harz, described from a coastal locality in Tarragona, but Clemente *et al.* (1989a) established that these two names were junior synonyms of *O. burri* Uvarov. After studying about 170 specimens from widely scattered localities, and songs recorded from 15 males, we are now convinced that *O. burri* is in turn a junior synonym of *O. minutissimus*, originally described from Escorial in the Sierra de Guadarrama and Cascante in Navarra. The small morphological differences mentioned by Clemente *et al.* (1986, 1990) seem to us to be of a similar order to the variation found over the rest of the range in eastern Spain. The song is highly distinctive and is not significantly different in the Sistema Central from songs recorded in localities widely scattered in the eastern half of the peninsula (see Fig. 930, from a song recorded in the Sierra de Guadarrama, and Figs 931 and 932, from songs recorded in the Sierra Nevada). These recordings have convinced us that the apparent song difference (in the number of echemes in the sequence) between the Sierra de Guadarrama and Sierra Nevada, previously noticed by one of us (Ragge, 1986), is no more than regional variation.

In our experience *O. minutissimus* is a montane species, usually occurring at altitudes above 1000 m, and as high as 2700 m in the Sierra Nevada. The type locality of *O. knipperi* is, however, on low-lying land near the sea.

Stenobothrus stigmaticus (Rambur) and *S. apenninus* Ebner (p. 344, 352)

One of us (Ragge, 1987a) has already remarked on the close similarity in both calling and courtship songs between these two species. There is also some resemblance in morphology and, the two taxa being entirely allopatric, Waeber (1989) regards *S. apenninus* as no more than a subspecies of *S. stigmaticus*. There are, however, some significant morphological differences; for example, apart from the difference in wing development, the male cerci of *S. apenninus* are simply conical, whereas those of *S. stigmaticus* are laterally compressed at the tip. We therefore prefer to retain the specific status of *S. apenninus* until there is stronger evidence that these two taxa represent a single species. It would of course be particularly interesting to determine experimentally whether they will freely interbreed.

Stenobothrus ursulae Nadig (p. 352)

This species was described from a single population in the Piedmont Alps north of Turin (Nadig, 1986). Soon afterwards La Greca (1986) described another new species, *S. nadigi*, from the northern side of the nearby Gran Paradiso massif. In 1990 Nadig collected material of La Greca's species from its type locality and, on the basis of a comparison between these specimens and the type series of *S. ursulae*, together with material from a third locality in this region, concluded that *S. nadigi* was a junior synonym of *S. ursulae* (Nadig, 1991). We have studied the calling song of males from both type localities and can find no significant difference between them (see Fig. 1041, recorded at the type locality of *S. ursulae*, and Fig. 1042, recorded at the type locality of *S. nadigi*). The acoustic behaviour of these populations is thus entirely consistent with Nadig's conclusions based on their morphology.

S. ursulae is quite similar in morphology and calling song to *S. apenninus*. There is, however, some difference in the courtship song and, although these species are probably closely related, we are content to leave them as distinct species.

Chorthippus corsicus (Chopard) (p. 382)

This species is known only from Corsica, where it is widespread in mountains. Not surprisingly, it varies in morphology from one montane locality to another, and until recently it was regarded as four distinct species, the other three being *Omocestus pascuorum* Chopard, *Chorthippus incertus* Chopard and *C. chopardi* Harz. The costal bulge on the anterior margin of the fore wings is not always present, and this has resulted in uncertainty about the generic assignment. Helversen & Helversen (in Harz & Kaltenbach, 1976: 346) and Pfau (1984) have suggested, partly on the basis of the songs, that at least some of these names are synonymous. Our own studies, based on some 70 specimens and the analysis of the songs of 16 males, have convinced us that all these populations belong to one variable species. Since the costal bulge is often present in both sexes, we are confident that the species is best assigned to the genus *Chorthippus*.

Chorthippus cialancensis Nadig (p. 388)

This species was described from the vicinity of the Passo della Cialancia in the Cottian Alps in the Piedmont region of north-western Italy (Nadig, 1986). In the same paper Nadig described another very similar alpine species, *C. sampeyrensis*, from the Colle di Sampeyre, further south in the Cottian Alps. We have studied the male calling songs at both type localities and have found them to be closely similar, while differing markedly from those of most other European species of *Chorthippus* (see Figs 1190–1198). In view of the close resemblance in both song and morphology, we think it preferable to treat these two populations as forms of a single species. Both occur at altitudes above 2000 m, and it seems likely that other small variations in song and morphology will be found in further isolated high-altitude populations of this species in this region of the western Alps.

The composition of each echeme of the calling song of this species shows quite a close resemblance to that of *C. cazurroi* (Bolívar) (Figs 1168–1183), a morphologically similar species known only from the Picos de Europa in northern Spain. However, the echeme-sequence of *C. cazurroi* is much longer than that of *C. cialancensis* and the echeme repetition rate much higher.

The *Chorthippus biguttulus* group (pp. 401–435)

The phonotaxonomy of this group of morphologically similar species has proved to be of exceptional interest. *Chorthippus biguttulus*, the only member of the group to be described by Linnaeus (1758), is one of the commonest grasshoppers of northern and central Europe, and the characteristic male calling song is one of the most familiar summer sounds in much of the European countryside. In the first half of the nineteenth century two further members of the group were described, *C. brunneus* (Thunberg, 1815) and *C. mollis* (Charpentier, 1825), but these species were so similar in morphology to *C. biguttulus* that they were frequently confused with it during most of the following century. Indeed, Fieber (1853*a*) took the view that only one variable species was involved, which he named *C. variabilis*, and this name was widely applied to all three species for many years.

The striking difference in the male calling songs of these species was apparently first observed by Yersin (1852, 1853, 1854*b*), but for many years his discovery went unnoticed. It was not until 1921, when Ramme rediscovered and republished Yersin's observations, that the existence of the three species became firmly established (Ramme, 1921*a*: 91). During the next half-century these species were recorded from localities widely scattered in Europe, including the southern peninsulas, and were believed to be the only members of the group. It was only when the songs began to be studied in southern Europe that it was realized that further members of the group, with quite different songs, occurred in the southern peninsulas. *C. biguttulus* and *C. mollis* were found not to occur south of the Pyrenees and Alps, or south of about latitude 42°S in the Balkan Peninsula. At least six further members of the group, each with a characteristic male calling song, were found to occur in these southern regions.

In western Europe and North Africa the group now comprises seven species: *C. biguttulus*, *C. brunneus*, *C. mollis*, *C. jacobsi*, *C. yersini*, *C. rubratibialis* and *C. marocanus*, the last three of these being originally treated as subspecies of *C. biguttulus*. On the basis of the differences in morphology and, especially, song, we consider that *C. yersini* and *C. marocanus* merit the status of full species (Ragge & Reynolds, 1988). Schmidt (1989) raised *C. rubratibialis* to specific status on the grounds that crossing it with typical *C. biguttulus* produced hybrids of much reduced viability.

The distribution of these species, and the geographical variation that some of them show, are of considerable interest. As can be seen in Fig. 35, typical *C. biguttulus* and *C. mollis* have very similar distributions in western Europe, neither of them occurring in Britain or further south than the Spanish Pyrenees and Italian Alps; *C. mollis* is much more local than *C. biguttulus* and does not extend quite so far north. Both these species have rather different songs in at least parts of the Alps. *C. biguttulus* tends to have a larger number of shorter

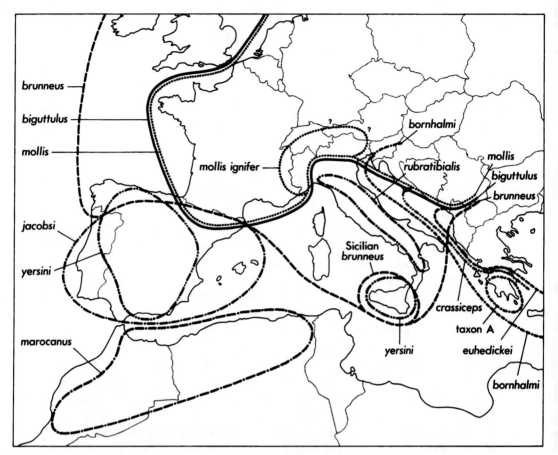

Figure 35 Map showing the approximate distribution of members of the *Chorthippus biguttulus* group in western Europe. 'Taxon A' is an unnamed member of the group. (Modified from Ragge *et al.*, 1990)

echemes in the southern Alps, from Aosta to Carinthia; there is also a tendency towards small differences in colour and morphology. This form is sometimes regarded as a distinct species, *C. eisentrauti* (Ramme, 1931), but the similarity to typical *C. biguttulus* is so close, with an overlap in both song and morphology, that we prefer to regard it as a southern Alpine form of that species; there is no clear geographical separation between the two, so that it is not practicable to treat this form as a subspecies. The song of *C. mollis* tends to be shorter, with a more abrupt ending, in a large area of the Alps, but again especially in the southern parts; there are again small differences in colour and morphology. Studies by one of us on a French population of this Alpine form of *C. mollis*, forming the subject of two recent papers (Ragge, 1981, 1984), suggest that this form may have originated by hybridization between typical *C. mollis* and *C. biguttulus*; the evidence for this, based on morphology and both calling and courtship songs, is given in these two papers. This form is quite well separated geographically from typical *C. mollis* and it is reasonable to treat it as a distinct subspecies, *C. mollis ignifer* Ramme, 1923. These Alpine forms of *C. biguttulus* and *C. mollis* have been more fully discussed in a recent study by Ingrisch (1995), who prefers to treat *C.*

eisentrauti as a distinct species (see also Ingrisch & Bassangova, 1995).

C. brunneus is more widely distributed than *C. biguttulus* and *C. mollis*: it occurs farther north, and is common, as the only member of the group, in the British Isles, Corsica and Sardinia. It also occurs in the Cantabrian Mountains of northern Spain and throughout Italy, including Sicily. In Finland, parts of the southern Alps and Sicily *C. brunneus* produces longer echemes than usual, but the song is still quite distinctive enough for the species to be easily recognized from it.

C. jacobsi is widespread in the Iberian Peninsula and also occurs on Majorca. The only aberrant song we have heard was at a locality in the western Pyrenees (near the Puerto de Artesiaga, Navarra province), where one of us (DRR) recorded songs from two males in which the syllables were less distinct than usual and the echemes were rather short; these males also had an unusually low number of stridulatory pegs. All these characters suggest the possibility of hybridization with *C. brunneus*, which also occurs in this region, but at another locality in Navarra (near the Puerto de Usateguieta) DRR found both species occurring together, each with typical song and morphology, and with no sign of hybridization.

C. yersini is of particular interest. It occurs widely in the uplands of Iberia and Sicily, seldom being found below 1000 m. In Iberia it does not occur north of the R. Ebro and its populations in the main mountain blocks are often well separated from one another. Not surprisingly, some of these populations show differences in male calling song, and there is a strong tendency for body-size to decrease from the northern to the southern part of the peninsula. In most of the range, including Sicily, the calling song is composed mainly of simple trains of about 30–100 syllables, but in the mountains immediately to the south of the R. Ebro, from the Sierra de la Demanda to the mountains of Teruel province, the calling song becomes closer to that of *C. biguttulus*, often consisting of echeme-sequences with the syllables in groups of 2–8. As the Ebro valley separates *C. yersini* from the nearest occurrence of *C. biguttulus*, this resemblance in song in this region suggests that there has been quite recent gene flow between the two. All the other western European populations of *C. yersini* are more widely separated from *C. biguttulus* and show no sign of this trend in the male calling song.

C. rubratibialis occurs throughout the Italian Apennines, from Liguria to Calabria, but not apparently in Sicily. The calling song is quite similar to that of *C. biguttulus*, but Schmidt (1978, 1987a) has demonstrated experimentally that, although hybrid eggs and nymphs can be obtained by crossing these two species, most of them die before reaching the adult stage and the few reaching adulthood die within a few days of the final moult.

C. marocanus is known only from the mountains of Morocco and Algeria. The stridulatory file is shorter in Algeria than in Morocco, especially in the male. It would be interesting to know whether this morphological difference is accompanied by any difference in acoustic behaviour, but the song has so far been studied only in Moroccan males.

The western European members of the *C. biguttulus* group are morphologically very similar to one another and museum specimens are still frequently misidentified. However, in the field they can be easily recognized by the male calling song, which shows striking differences in the members of the group that overlap geographically. Interbreeding experiments have shown that each pair of species tested so far is interfertile, though rather poorly so in the case of *biguttulus* × *rubratibialis*. It seems likely that all these species are quite closely related to one another by fairly recent common ancestry. Resemblances in song and morphology, together with their allopatric distributions, suggest the possibility that *C. jacobsi* and *C. brunneus* are sister species, and that there is also a close relationship between *C. biguttulus*, *C. yersini* and *C. rubratibialis* (Ragge, 1987b; Ragge & Reynolds, 1988; Ragge *et al.*, 1990).

Mason (1991) has recently studied the inter-relationships of the western European members of this group on the basis of mitochondrial DNA, in addition to morphology and song. Unfortunately the mtDNA analysis did not produce a clear result: there was as much variation in mtDNA genotypes within some populations of a single species as there was between populations of different species, and some populations were more similar in mtDNA genotype to heterospecific populations than to other conspecific populations. The author suggests two possible explanations for this variation: that the group has recently arisen from a single polymorphic population or, more probably, that the genotypes have been shared by hybridization between the species. Her quantitative analysis of morphometric characters of the male fore wing and hind leg lends some support to the speculations mentioned in the last paragraph, as does her quantitative analysis of supposedly homologous characters of the male calling songs. Her overall conclusion about the phylogeny of the group, based on all three approaches, is that there was an ancient bifurcation between *C. brunneus*-like and *C. biguttulus*-like ancestral lines. *C. mollis* may also have diverged at about the same time, but her data, based on only one sample of this species, were inadequate to determine this. She considers *C. jacobsi* to be a more recent branch from the *C. brunneus* line, and *C. yersini* and *C. rubratibialis* to be branches from the *C. biguttulus* line. This is in complete accord with our views.

This interesting group has provided a fruitful field of study for many biologists and is likely to continue to do so for many years to come. Further experimental studies on inter-breeding between allopatric pairs or trios, e.g. *brunneus/jacobsi* and *biguttulus/yersini/ marocanus*, would be very welcome, and it would also be most interesting to extend research on all aspects of the group to central and eastern Asia.

Chorthippus binotatus (Charpentier) (p. 435)

This species, known only from France, the Iberian Peninsula and Morocco, is very variable in size, colour and, to some extent, morphology. In the more northerly part of its range it tends to be brachypterous and sometimes generally smaller in mountains, and several of these montane forms have been given subspecific names: *saulcyi* Krauss and *moralesi* Uvarov from the Pyrenees, *algoaldensis* Chopard from the Cevennes and *daimei* Azam from the French Alps. In the southern part of its range it is mainly confined to uplands, where there is no reduction in size or wing-length; indeed, in the Spanish Sistema Central it is particularly large and has been given the subspecific name *dilutus* Ebner, though García *et al.* (1995) have recently taken the view that treating this form as a subspecies is unjustified. We have studied the song in all these regions and have found it to be fairly stable, showing little geographical variation. We think it better to treat all these forms as no more than local variants, not warranting subspecific status.

Chorthippus parallelus (Zetterstedt) and *C. montanus* (Charpentier) (pp. 452, 461)

These species form an interesting pair: they are similar in both morphology and song, and have broadly overlapping distributions. In continental Europe *C. montanus* is much more local than *C. parallelus* and usually occurs in wetter habitats; in Britain *C. montanus* is absent and *C. parallelus* occurs in a wide range of habitats, including wet ones. The difference in calling song is mainly one of tempo, so that the song of a cool *C. parallelus* is rather like that of a warm *C. montanus*; however, when both species are singing at the same temperature the difference is very clear. Jacobs (1953*a*) and Lux (1961) have shown that the females can discriminate between the calling songs of the two species, and Helversen (1979), Helversen & Helversen (1981) and Bauer & Helversen (1987) have shown that this ability is maintained at a range of different temperatures. We agree entirely with the currently prevailing view that *C. parallelus* and *C. montanus* are distinct but quite closely related species.

In the Iberian Peninsula typical *C. parallelus* is replaced by a form differing slightly in morphology and acoustic behaviour. The most notable difference is in the courtship behaviour: typical *C. parallelus* lacks a special courtship song, but the Iberian form has a quite well-developed one (see Figs 1538–1540). One of us (Reynolds, 1980) has taken the view that the Iberian form is best regarded as a subspecies, *C. p. erythropus* Faber, and this view has since been generally accepted. However, Hewitt *et al.* (1987) have shown in laboratory studies on interbreeding that, although female hybrids between these two forms are fully fertile, F_1 male hybrids are sterile. This suggests that the Iberian form is well on the way to achieving the status of a distinct species. The relationship between these forms has recently been fully discussed by Hewitt (1993).

Euchorthippus (p. 463)

All six western European species of this genus have remarkably similar calling songs, so that, as a field character, the song is more useful in recognizing the genus than in distinguishing between the species. Only three of these species overlap with one another in distribution: *E. declivus*, *E. pulvinatus* and *E. chopardi*. To the unaided ear there is no apparent difference between the calling songs of *E. declivus* and *E. chopardi*, though oscillographic analysis reveals that there are some significant differences between them; because of differences in ecological requirements these two species do not, in our experience, occur together in the fairly narrow zone where their distributions overlap. *E. pulvinatus gallicus*, on the other hand, quite often occurs in the same locality with either *E. declivus* or *E. chopardi* and, perhaps significantly, has a noticeably higher echeme repetition rate than either of these species, in addition to further song differences revealed by oscillographic analysis.

The other three western European species do not overlap in range either with each other or with the three species mentioned above and, not surprisingly, do not seem to show any clear diagnostic features in their calling songs. It would be most interesting to know whether the calling songs of the dozen or so Asiatic species of *Euchorthippus* are similar in general pattern to those of the European species.

Using the songs for identification

One of our main aims in producing this book has been to provide a reliable means of identifying the singing Orthoptera of western Europe from their songs. To this end we give descriptions of the songs, illustrate the song-patterns objectively with oscillograms and provide recordings of them on the companion compact discs. The surest way of identifying an orthopteran from its song is to match the song with one of the recordings on the CDs. There are, however, 352 song excerpts on the CDs, amounting to two hours' playing time, and it is not really practicable to play through these excerpts until a matching song is found. The task is lessened by first classifying the song into one of the broad categories set out in Chapter 4 (p. 52), but this in itself is not always easy and some of the categories include a large number of species. We have therefore thought it worth while to provide an identification key based primarily on the songs (Chapter 6). Describing the songs in words is not easy, and the key inevitably suffers from this disadvantage, but we feel it will provide a further aid to the identification of a singing orthopteran in western Europe.

Once he is familiar with their calling songs, the orthopterist is in the happy position of being able to detect and identify all the audibly singing Orthoptera in a locality without needing to see them. He must, however, be aware of the effect of temperature on the tempo of the songs (see p. 45) and of the differences between the daytime and night-time songs of some Tettigoniidae. Species that sing at the same time and in the same place almost always have clearly different songs. The few exceptions to this general rule include those phaneropterine bush-crickets with largely ultrasonic, tick-like songs (e.g. *Leptophyes*), in which the time-lapse between the male tick and the female response tick seems to be the basis of mate-recognition; these songs are in any case too faint to human ears to be audible in the field without the aid of an ultrasound detector.

Chapter 6

Key to the singing Orthoptera of Western Europe, based primarily on their songs

In addition to the song descriptions and oscillograms given in Chapters 7–9, this book provides three further aids to the identification of western European Orthoptera from their songs. First, the recordings of the songs on the companion CDs enable direct comparison to be made with songs heard or recorded in the wild or in captivity. Secondly, the classification of the songs given in Chapter 4 (p. 52) could at least enable a song to be placed in one of the categories defined there, thus limiting the singer to one of the species listed in that category. As a final means of identification we have devised a dichotomous key based primarily on song characters. The key will not be easy to use, but we hope it will nevertheless have some value as a further aid to identification.

The key includes all the 170 western European species of Orthoptera whose songs are described in this book, i.e. all the common ones and most of the rarer ones. The singing western European species not included are mostly local forms, often of doubtful status, whose songs are unknown to us. The key applies only to calling songs produced by isolated males, i.e. the songs most often heard in the field. Diurnal singers are assumed to be singing in warm, sunny weather. For bush-crickets whose songs are too quiet to be heard in outdoor conditions, the key is based on the calling songs produced by captive males in quiet indoor conditions. Sometimes, even in outdoor conditions, such quiet songs can be identified with the aid of an ultrasound detector (provided there is not too much ultrasound from other sources), since the frequency-converted sound will faithfully reproduce the rhythmic pattern of the song.

As far as possible we have used song differences that can be appreciated with the unaided ear, or at most with the help of the second hand (or digital count) of a watch. It should be possible to use the key without resorting to oscillographic analysis, but some of the song characters (especially the faster repetition rates) can be much better appreciated by making a reel-to-reel tape recording of the song and playing it back at a slower speed. We have often used the distribution of the species as an additional aid and, where the song differences are particularly difficult to use, we have supplemented or replaced them by differences in size, structure or colour – although we realize that this is no help to a user who cannot find the insect or is trying to identify a recorded song. When giving morphological characters we have considered the male sex only. Some species can be reached by following either alternative of a couplet, and many species are keyed out several times.

For some couplets the user needs to know how to distinguish bush-crickets from grass-

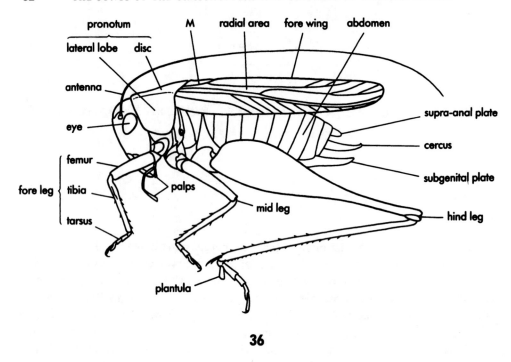

36

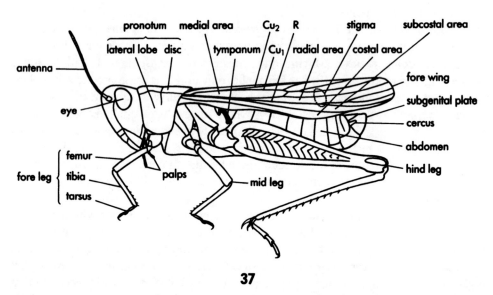

37

Figures 36, 37 Side views of (36) a male bush-cricket and (37) a male grasshopper, illustrating the morphological terms used in the key.

hoppers. In bush-crickets the antennae are like thin threads, tapering to a fine point and usually longer than the body, there is a hearing organ near the base of the fore tibiae and the tarsi have four segments. In grasshoppers the antennae are relatively thick, not usually tapering to a point and usually much shorter than the body, there is no hearing organ on the fore tibiae and the tarsi have three segments. These differences, together with most of the other morphological characters referred to in the key, are illustrated in Figs 36 and 37. The hind wings, when flexed, are folded fanwise beneath the fore wings. The 'knee' is the joint between the femur and tibia. The median and lateral carinae of the pronotum lie along the middle and sides, respectively, of the pronotal disc. The transverse sulcus of the grass-hopper pronotum runs across the disc, dividing it into the prozona and metazona. Grasshopper antennae are described as 'clubbed' when they are conspicuously enlarged at the tip. The titillators of some male bush-crickets are toughened, paired structures lying near the bases of the cerci; they can be difficult to examine in dried specimens – see Ragge (1990: 6).

Terms such as 'echeme' and 'syllable' are defined on pp. 26, 28. If the singing insect is not visible the reader may have to guess how to apply these terms to the song, and so whenever possible we have used self-explanatory terms such as 'song unit' or 'burst of sound'.

After arriving at an identification with the key, the user should always refer for confirmation to the relevant description and oscillograms, and preferably to the recorded song on one of the companion CDs. A few song variants are not included on the CDs; in these cases there is no reference to the CDs in the key, but a description and oscillograms of the song variant will be found where the species is treated in the text.

1	Song with a definite musical pitch, although often shrill or squeak-like (true crickets and mole-crickets 2	
–	Song harsh and unmusical (mostly bush-crickets and grasshoppers) 11	
2(1)	Song a prolonged, continuous, uniform trilling (listen carefully – some intermittent songs may sound continuous when several males are singing together) (mole-crickets) 3	
–	Song broken up into bursts (usually echemes) lasting 5 s at the most (true crickets) 4	
3(2)	Trilling a musical warble with a dominant frequency below 2 kHz. Song coming from the ground in moist places, in the evening or at night .... *Gryllotalpa gryllotalpa* (p. 276; CD 1, track 99, excerpt 1)	
–	Trilling less musical, very shrill and penetrating, with a dominant frequency of at least 3 kHz. Song coming from the ground in dry places, in the evening or at night .... *Gryllotalpa vineae* (p. 278; CD 1, track 99, excerpt 2) [Both these songs could be confused with the churring or reeling songs of *Caprimulgus europaeus* (Nightjar) or warblers of the genus *Locustella* (see pp. 480–484). These birds normally sing from above ground-level, however, and the quality of the sound is different (CD 2, tracks 81–85).]	
4(2)	Each echeme lasting 1–4 s (with a crescendo) and including more than 80 syllables; dominant frequency 6–8 kHz. (Very small, black, ground-dwelling cricket, living in damp places.) .... ***Pteronemobius heydenii*** (p. 274; CD 1, track 97) [This song could be confused with that of *Bufo viridis* (Green Toad) (see p. 478), but the calls of this primarily eastern European amphibian usually last more than 4 s and the dominant frequency is much lower, 1–2 kHz (CD 2, track 79).]	
–	Each echeme usually lasting less than 1 s, with fewer than 40 syllables; dominant frequency below 6 kHz 5	
5(4)	Song consisting, at least partly, of single syllables repeated at variable intervals.	

(Dominant frequency 3–4 kHz. Heard from evening onwards in mountains.) ....
.... ***Eugryllodes pipiens*** (p. 269; CD 1, track 95)
[The short-lived calls of *Alytes obstetricans* (Midwife Toad) (see p. 480) are reminiscent of this cricket's song, but this amphibian is unlikely to occur in the mountain habitats of *E. pipiens* and the dominant frequency of the calls is much lower, under 2 kHz (CD 2, track 80).]

–   Song consisting entirely of echemes of 2 or more syllables 6

6(5)   Song a quiet churring with frequent but rather irregular brief pauses (may sound continuous when several males are singing together). Small brown cricket living in leaf-litter in or near woodland .... ***Nemobius sylvestris*** (p.272; CD 1, track 96)
[This song could be confused with that of *Caprimulgus europaeus* (Nightjar) (see p. 484). The song of this bird does not have frequent pauses, however, and is much louder (CD 2, track 85).]

–   Song louder and more musical, with the echemes more regularly spaced. Not usually heard in woodland 7

7(6)   Each echeme lasting more than a quarter of a second. Delicately built, straw-coloured cricket singing from above ground-level in rough herbage, shrubs or trees...............
.... ***Oecanthus pellucens*** (p. 274; CD 1, track 98)
–   Each echeme usually lasting less than a quarter of a second. Song coming from ground-level 8

8(7)   Each echeme with more than 10 syllables. Small cricket, with fore wings less than 8 mm long .... ***Modicogryllus bordigalensis*** (p. 269; CD 1, track 94)
–   Each echeme with fewer than 10 syllables. Larger crickets, with fore wings more than 8 mm long... 9

9(8)   Brown or straw-coloured cricket, usually found in or near buildings or on rubbish tips
.... ***Acheta domesticus*** (p. 263; CD 1, track 93)
–   Predominantly black crickets, generally found in more natural habitats 10

10(9)   Flightless cricket with vestigial hind wings. Lives in burrows, with male singing from burrow entrance. Sings mainly from May to July. Common and widespread in Europe, range extending northwards to about latitude 55°N ....
.... ***Gryllus campestris*** (p. 259; CD 1, track 91)
–   Fully winged cricket that flies well. Lives in crevices rather than burrows. Sings mainly in mid to late summer. Occurs in the Mediterranean Region, south of latitude 45°N
.... ***Gryllus bimaculatus*** (p. 263; CD 1, track 92)

11(1)   Rattling, whirring or buzzing sounds (crepitation) made by a flying grasshopper ....
.... 12
–   Song produced by an insect while sitting or walking 17

12(11)   Crepitation a loud rattling or whirring sound produced by a large grasshopper with partly reddish hind wings 13
–   Crepitation a buzzing or quieter rattling sound produced by a less bulky grasshopper with no red on its hind wings 14

13(12)   Crepitation a clear rattling sound; flight never very long. Red colour reaching the anterior margin of the hind wings. Occurs widely in Europe ....
.... ***Psophus stridulus*** (p. 282; CD 2, track 1)
–   Crepitation more of a whirring than a clear rattling sound; flight short when disturbed, but males have an undulating display flight, often for distances of 100 m or more, during which crepitation is intermittent. Red colour less extensive, not reaching the

anterior margin of the hind wings. Very local in western Europe ....
.... ***Bryodema tuberculata*** (p. 282; CD 2, track 2)

14(12) Crepitation a loud buzz, often produced while hovering (also sometimes by wing-vibra-
        tion while sitting on the the ground) ....
        .... ***Stenobothrus rubicundulus*** (p. 356; CD 2, track 35)
–   Crepitation a rattling sound produced during short flights 15

15(14) Smaller grasshopper, fore wings shorter than 16 mm ....
        .... ***Stenobothrus cotticus*** (p. 359)
–   Larger grasshoppers, fore wings longer than 16 mm 16

16(15) Fore wings longer than 20 mm. Hind femora partly red .... ***Arcyptera fusca*** (p. 286)
–   Fore wings shorter than 20 mm. Hind femora with no red ....
        .... ***Stauroderus scalaris*** (p. 375)
        [The last three species may be easily distinguished by their calling songs (CD 2, tracks 36, 5
        and 40, respectively).]

17(11) Song a continuous buzzing, clicking, rustling or chuffing, lasting more than half a minute
        and composed of sound units repeated at the rate of at least 5/s (listen carefully –
        some intermittent songs may sound continuous when several males are singing to-
        gether) 18
–   Song in bursts lasting less than half a minute or, if longer, composed of sound units
        repeated at a slower rate than 5/s 46

18(17) Song a continuous, uniform buzz composed of sound units repeated too rapidly to be
        distinguishable (more than 30/s) 19
–   Song not completely uniform or, if uniform, composed of distinguishable sound units,
        repeated less rapidly than 30/s 23

19(18) Buzzing with a definite pitch although very shrill, coming from the ground in dry places,
        in the evening or at night ....
        .... ***Gryllotalpa vineae*** (p. 278; CD 1, track 99, excerpt 2)
–   Buzzing harsh, with no definite pitch, coming from vegetation 20

20(19) Brachypterous bush-crickets, fore wings less than 10 mm long 21
–   Fully winged bush-crickets, fore wings more than 15 mm long 22

21(20) Large bush-cricket, with the fore wings protruding only a short distance from under the
        pronotum. Buzz eventually interrupted by one or more loud clicks
        .... ***Polysarcus denticauda*** (p. 128; CD 1, track 14)
–   Small bush-cricket, with the fore wings covering most of the abdomen. Buzz completely
        uniform .... ***Metrioptera roeselii*** (p. 199; CD 1, track 50)

22(20) Large bush-cricket, fore wings more than 25 mm long. Song produced only in the evening
        and at night .... ***Ruspolia nitidula*** (p. 146; CD 1, track 22)
–   Smaller bush-cricket, fore wings less than 25 mm long. Song produced on warm, sunny
        days.. ***Gampsocleis glabra*** (p. 225; CD 1, track 68)
        [The songs of the last four species could be confused with that of *Cicadetta montana* (see p.
        478), but this day-singing cicada is not likely to be heard from mid-July onwards (CD 2, track
        78).]

23(18) Song composed of echemes repeated at the rate of about 4–7/s and showing a gradual
        crescendo during its first half. Small, brachypterous grasshopper known only from
        high altitudes in the Picos de Europa in northern Spain ....
        .... ***Chorthippus cazurroi*** (p. 382; CD 2, track 43)

–   Song not of this kind, produced by bush-crickets not occurring at high altitudes in the Picos de Europa 24

24(23)   Song composed of two alternating quiet sounds, one with a rustling quality and the other a rapid ticking. (Small, graceful bush-crickets, often occurring in damp places.) 25

–   Song not composed of two alternating sounds 26

25(24)   Fully winged bush-cricket, fore wings extending well beyond the tip of the abdomen. Rustling part of the song composed of trisyllabic echemes .... *Conocephalus discolor* (p. 140; CD 1, track 20, excerpt 2)

–   Brachypterous bush-cricket, fore wings not reaching the tip of the abdomen. Rustling part of the song composed of tetrasyllabic echemes .... *Conocephalus dorsalis* (p. 144; CD 1, track 21)

26(24)   Song a prolonged buzzing, eventually interrupted by one or more loud clicks. (Large, brachypterous, mostly green bush-cricket.) .... *Polysarcus denticauda* (p. 128; CD 1, track 14)

–   Song uniform, either without clicks or composed entirely of clicks 27

27(26)   Song a rapid clicking or ticking 28

–   Song a rustling or chuffing sound, not a series of clicks or ticks 43

28(27)   Clicks repeated more rapidly than 8/s (sometimes more slowly at the start of the song) 29

–   Clicks or ticks repeated less rapidly than 8/s 36

29(28)   Fore wings less than 12 mm long, not reaching the tip of the abdomen 30

–   Fore wings more than 18 mm long, reaching at least to the tip of the abdomen 31

30(29)   Large brown bush-cricket, hind femora more than 22 mm long. Song composed of disyllabic echemes repeated less rapidly than 15/s .... *Platycleis sepium* (p. 178; CD 1, track 37, excerpt 2)

–   Smaller, mostly or at least partly green bush-cricket, hind femora less than 20 mm long. Song composed of trisyllabic echemes repeated more rapidly than 15/s .... *Metrioptera bicolor* (p. 196; CD 1, track 49, excerpt 2)

31(29)   Clicks repeated more rapidly than 17/s 32

–   Clicks repeated less rapidly than 17/s 33

32(31)   Daytime sound. Known only from high mountains in Spain .... *Decticus verrucivorus assiduus* (p. 158)

–   Night-time sound. Not in the Iberian Peninsula .... *Tettigonia cantans* (p. 153; CD 1, track 25, excerpt 2)

33(31)   Fore wings with conspicuous dark spots .... *Decticus verrucivorus verrucivorus* (p. 157; CD 1, track 27)

–   Fore wings without conspicuous dark spots (except at the stridulatory organ) 34

34(33)   Fore wings usually more than 40 mm long, extending well beyond the hind knees ....*Tettigonia viridissima* (p. 148; CD 1, track 23)

–   Fore wings usually less than 40 mm long, hardly extending beyond the hind knees ... 35

35(34)   Song a series of sharp, disyllabic clicks, usually interrupted at irregular intervals by brief pauses of less than 1 s. Occurs in the mountains of central Spain .... *Tettigonia hispanica* (p. 152; CD 1, track 24)

–   Song a series of less sharp, monosyllabic sounds, without interruptions. Occurs widely in Europe, but not in the Iberian Peninsula ....
.... *Tettigonia cantans* (p. 153; CD 1, track 25, excerpt 2)

36(28)   Song loud, easily audible in outdoor conditions. Large bush-crickets, pronotum more than 6 mm long 37

–   Song very quiet, not usually audible in outdoor conditions. Small bush-crickets, pronotum less than 6 mm long 39

37(36)   Fore wings not reaching the tip of the abdomen ....
.... *Platycleis sepium* (p. 178; CD 1, track 37, excerpt 2)

–   Fore wings reaching at least to the tip of the abdomen 38

38(37)   Fore wings extending well beyond the hind knees. Clicks monosyllabic. Common in the Mediterranean Region, but not occurring in the mountains of central Spain ....
.... *Decticus albifrons* (p. 162; CD 1, track 29)

–   Fore wings hardly extending beyond the hind knees. Clicks disyllabic. Occurs in the mountains of central Spain .... *Tettigonia hispanica* (p. 152; CD 1, track 24)
[The song of *Cicada orni* (p. 476) is composed of loud bursts of sound repeated at about the same rate as in the last three species, but these sounds are not click-like (CD 2, track 75).]

39(36)   Song a series of ticks, usually repeated rather irregularly and of uneven loudness. Green bush-cricket .... *Barbitistes fischeri* (p. 122; CD 1, track 9)

–   Song a uniform series of ticking, rasping or lisping sounds. Brown or straw-coloured bush-crickets 40

40(39)   Song a series of ticks repeated at the rate of about 4–6/s. Very small, nymph-like bush-cricket (hind femora less than 13 mm long), known only from Sicily ....
.... *Ctenodecticus siculus* (p. 216; CD 1, track 61)

–   Song almost continuous, composed of rasping or lisping sounds repeated at the rate of about 6–8/s. Larger bush-crickets (hind femora more than 13 mm long) 41

41(40)   Fully winged bush-cricket, fore wings extending beyond the tip of the abdomen ....
.... *Platycleis tessellata* (p. 178; CD 1, track 38)

–   Brachypterous bush-crickets, fore wings not reaching the tip of the abdomen 42

42(41)   Pronotum with a median carina along the whole of its length ....
.... *Platycleis nigrosignata* (p. 183)

–   Pronotum with a median carina only in the metazona .... *Platycleis veyseli* (p. 180)

43(27)   Song a continuous rustling sound, composed of echemes repeated more rapidly than than 10/s. Fully winged bush-cricket, fore wings extending beyond the tip of the abdomen .... *Conocephalus discolor* (p. 140; CD 1, track 20, excerpt 1)

–   Song a chuffing sound, composed of echemes repeated less rapidly than 8/s. Brachypterous bush-crickets, fore wings not reaching the tip of the abdomen 44

44(43)   Song a completely uniform series of echemes, each composed of 3–4 syllables ....
.... *Metrioptera brachyptera* (p. 187; CD 1, track 43)

–   Song a series of echemes, each composed of 5 or more syllables; every so often an echeme is followed by a series of about 2–6 microsyllables 45

45(44)   Occurring only in the Apennine range in Italy ....
.... *Metrioptera caprai* (p. 191; CD 1, track 46)

–   Not occurring in the Apennine range ....
.... *Metrioptera saussuriana* (p. 189; CD 1, track 44)
or *Metrioptera buyssoni* (p. 191; CD 1, track 45)
[These two species seem to have identical songs and are difficult to separate morphologically (see p. 189). *M. buyssoni* is known only from the eastern Pyrenees.]

46(17) Song composed of very short-lived ticks lasting less than 50 ms and repeated less rapidly than 25/s 47

– Song composed of longer sounds or of ticks repeated more rapidly than 25/s 75

47(46) Ticks in well-defined groups or bursts lasting less than half a minute 48

– Ticks in a long ungrouped sequence lasting more than half a minute 62

48(47) Song very quiet, including well-defined echemes of about 3–9 syllables, with a syllable repetition rate of 10–15/s; there are often also isolated syllables and sometimes there is a mixture of quieter and louder syllables or echemes. Small, fully-winged bush-cricket occurring very locally in southern Europe. ....
.... ***Conocephalus conocephalus*** (p. 138; CD 1, track 19)

– Song not of this kind 49

49(48) Each group of ticks clearly divided into 3–5 subgroups, each composed of up to 6 ticks. (Quiet songs from brachypterous green bush-crickets.) 50

– Groups or bursts of ticks simple, without further subdivision, although spacing of ticks is sometimes rather uneven 51

50(49) Subgroups mostly composed of 2–4 ticks ....
.... ***Barbitistes serricauda*** (p. 120; CD 1, track 7)

– Subgroups mostly composed of 5 or 6 ticks ....
.... ***Barbitistes obtusus*** (p. 120; CD 1, track 8)

51(49) Group or burst composed of at least 50 ticks repeated more rapidly than 5/s 52

– Group composed of fewer than 40 ticks repeated less rapidly than 5/s 56

52(51) Very quiet song from a brachypterous bush-cricket; spacing of ticks rather uneven ......
.... ***Barbitistes fischeri*** (p. 122; CD 1, track 9)

– Loud song from a fully winged bush-cricket or grasshopper; spacing of ticks even ....
.... 53

53(52) Song of constant loudness, continuing for more than half a minute but interrupted at irregular intervals by very brief pauses of less than a second. Large green bush-crickets 54

– Song with a gradual crescendo and lasting less than half a minute. Grasshoppers ....
.... 55

54(53) Fore wings usually more than 40 mm long, extending well beyond the hind knees. Widespread in Europe .... ***Tettigonia viridissima*** (p. 148; CD 1, track 23)

– Fore wings usually less than 40 mm long, hardly extending beyond the hind knees. Occurring in the mountains of central Spain ....
.... ***Tettigonia hispanica*** (p. 152; CD 1, track 24)

55(53) Song usually lasting more than 12 s. Sides of the head and thorax green or greenish brown .... ***Omocestus viridulus*** (p. 304; CD 2, track 14)

– Song usually lasting less than 12 s. Sides of the head and thorax dark brown or black
.... ***Omocestus rufipes*** (p. 305; CD 2, track 15)

56(51) Group composed of fewer than 6 ticks. Slender bush-cricket with long legs and wings
.... ***Tylopsis lilifolia*** (p. 114; CD 1, track 3)

– Group composed of at least 6 ticks. Grasshoppers 57

57(56) Ticks produced by flicking a single hind tibia backwards against the distal part of the flexed fore wing. Large grasshopper occurring in wet habitats ....
.... ***Stethophyma grossum*** (p. 284; CD 2, track 3)

–   Ticks produced by simultaneous movements of both flexed hind legs. Grasshoppers oc-
           curring in drier habitats 58

58(57)   Antennae noticeably thickened towards the tip. Palps black or dark brown at the tip .....
           .... *Stenobothrus grammicus* (p. 343; CD 2, track 30)
–   Antennae not noticeably thickened towards the tip. Palps not black or dark brown at the
           tip 59

59(58)   Maxillary palps with the last segment swollen and usually orange-pink or reddish brown
           .... *Stenobothrus bolivarii* (p. 344; CD 2, track 31)
–   Maxillary palps with the last segment normal in shape and colour 60

60(59)   Hind femora less than 9 mm long .... *Dociostaurus jagoi* (p. 300; CD 2, track 12)
–   Hind femora more than 9 mm long 61

61(60)   Body with a conspicuous pale longitudinal stripe along the top of the head, pronotum
           and corresponding part of the flexed fore wings ....
           .... *Ramburiella hispanica* (p. 292; CD 2, track 8)
–   Body without such a stripe or occasionally with one on the head and pronotum only
           .... *Dociostaurus maroccanus* (p. 298; CD 2, track 11)

62(47)   Fully winged bush-crickets 63
–   Brachypterous bush-crickets 66

63(62)   Fore tibial tympanal apertures slit-like 64
–   Fore tibial tympanal apertures open and oval 65

64(63)   Hind femora more than 20 mm long .... *Tylopsis lilifolia* (p. 114)
–   Hind femora less than 15 mm long ....
           .... *Conocephalus conocephalus* (p. 138; CD 1, track 19)

65(63)   Pronotal lateral lobes at least as long as deep ....
           .... *Phaneroptera falcata* (p. 111; CD 1, track 1, excerpt 1)
–   Pronotal lateral lobes distinctly deeper than long ....
           .... *Phaneroptera nana* (p. 111; CD 1, track 2)

66(62)   Ticks repeated more rapidly than 5/s .... *Barbitistes fischeri* (p. 122; CD 1, track 9)
–   Ticks repeated less rapidly than 3/s 67

67(66)   Entirely brown or straw-coloured 68
–   Partly or entirely green 69

68(67)   Hind femora more than 15 mm long ....
           .... *Pachytrachis striolatus* (p. 214; CD 1, track 60)
–   Hind femora less than 15 mm long. (Song inaudible to most people.) ....
           .... *Yersinella raymondii* (p. 211; CD 1, track 58)

69(67)   Pronotum more than 9 mm long ....
           .... *Eupholidoptera chabrieri* (p. 209; CD 1, track 56)
–   Pronotum less than 9 mm long 70

70(69)   Pronotum extending backwards to cover about half the reduced male fore wings.
           (Occurring in western Europe only in Austria and Italy.) 71
–   Pronotum not extending backwards so far; reduced male fore wings largely exposed
           72

71(70)   Pronotum more than 6 mm long .... *Poecilimon ornatus* (p. 135; CD 1, track 16)
–   Pronotum less than 6 mm long .... *Poecilimon jonicus* (p. 135; CD 1, track 17)

72(70)   Body with a white or yellowish lateral stripe from the lower edge of the pronotal lateral
          lobes to the tip of the abdomen 73
–   Body without such a stripe 74

73(72)   Hind femora less than 15 mm long ....
          .... ***Leptophyes albovittata*** (p. 124; CD 1, track 11)
–   Hind femora more than 15 mm long .... ***Leptophyes boscii*** (p. 126; CD 1, track 12)

74(72)   Male subgenital plate truncate at the tip ....
          .... ***Leptophyes punctatissima*** (p. 122; CD 1, track 10)
–   Male subgenital plate emarginate at the tip ....
          .... ***Leptophyes laticauda*** (p. 126; CD 1, track 13)

75(46)   Song lasting more than half a minute, without interruptions of more than 2 s 76
–   Song lasting less than half a minute, separated from other songs by intervals of more
          than 2 s 149

76(75)   Song a virtually continuous wheezing sound (lasting up to a minute or more), although
          composed of distinguishable units repeated at the rate of 2–6/s. Large, brown,
          brachypterous bush-cricket with box-like pronotum, occurring in Spanish mountains
          .... ***Pycnogaster jugicola*** (p. 253; CD 1, track 88)
–   Song not of this kind, in bursts that are clearly separated although sometimes by very
          short intervals 77

77(76)   Song bursts repeated more often than 1/s 78
–   Song bursts repeated less often than 1/s 117

78(77)   Repetition rate of song bursts noticeably irregular 79
–   Repetition rate of song bursts regular 81

79(78)   Quiet song from a small brown cricket living in woodland leaf-litter............................
          .... ***Nemobius sylvestris*** (p. 272; CD 1, track 96)
–   Loud song from large bush-crickets 80

80(79)   Fully winged green bush-cricket occurring only in high mountains south of the Ebro
          valley in Spain. Song composed of polysyllabic echemes...
          .... ***Decticus verrucivorus assiduus*** (p. 158; CD 1, track 28)
–   Brachypterous brown bush-cricket occurring widely in southern Europe but not south
          of the Ebro valley in Spain. Song composed of disyllabic echemes ....
          .... ***Platycleis sepium*** (p. 178; CD 1, track 37, excerpt 1)

81(78)   Song showing a gradual crescendo during at least the first third of its duration. (Grass-
          hoppers) 82
–   Song of constant loudness 85

82(81)   Song a simple succession of syllables, sometimes followed by a quieter 'aftersong' 83
–   Song more complex: a sequence of echemes, each composed of at least two syllables ....
          84

83(82)   Song with an 'aftersong' of short, quieter sounds. Fore tibiae conspicuously swollen. In
          western Europe only in mountains (but not occurring in southern Spain) ....
          .... ***Gomphocerus sibiricus*** (p. 367; CD 2, track 38)
–   Song without an 'aftersong' (except in the Sierra Nevada in southern Spain). Fore tibiae
          normal. Mainly a lowland species except in the southern peninsulas ....
          .... ***Chorthippus vagans*** (p. 395; CD 2, track 49, excerpts 2–4)

84(82)   Song without ticks. Small brachypterous grasshopper (fore wings less than 9 mm long), known only from the Picos de Europa in northern Spain ....
.... ***Chorthippus cazurroi*** (p. 382; CD 2, track 43)
–   Each echeme of the song with a tick. Larger, fully winged grasshopper (fore wings more than 9 mm long), not known from the Picos de Europa...
.... ***Chorthippus apricarius*** (p. 376; CD 2, track 41)

85(81)   Each song unit (which lasts 0·5–1·0 s) ending in a click; every so often there is a louder buzz of variable duration. Large, green, brachypterous bush-cricket occurring in western Europe only in the Pyrenees and French Alps. ....
.... ***Polysarcus scutatus*** (p. 131; CD 1, track 15)
–   Song not of this kind 86

86(85)   Song units lasting more than a third of a second 87
–   Song units lasting less than a third of a second 94

87(86)   Each song unit consisting of a single long syllable 88
–   Song units divided into two or more sub-units, each lasting less than a quarter of a second 89

88(87)   Occurring in France and Catalonia .... ***Uromenus rugosicollis*** (p. 240; CD 1, track 75)
–   Not occurring in these regions ...:................ ***Uromenus elegans*** (p. 242; CD 1, track 76)
                              or ***Uromenus brevicollis*** (p. 242; CD 1, track 77)
         [These two species are very similar in song and morphology (see p. 242).]

89(87)   Sub-units repeated more rapidly than 15/s 90
–   Sub-units repeated less rapidly than 15/s 91

90(89)   Each song unit a sequence of trisyllabic echemes. Mostly or at least partly green bush-cricket occurring widely in Europe but not in southern Spain...................................
.... ***Metrioptera bicolor*** (p. 196; CD 1, track 49, excerpt 1)
–   Each song unit an echeme of diplosyllables. Mostly dark brown or black bush-cricket known only from the Sierra Nevada in southern Spain ....
.... ***Baetica ustulata*** (p. 249; CD 1, track 84)

91(89)   Each echeme composed of at least 8 syllables. Mostly dark brown or black bush-cricket known only from the Sierra Nevada in southern Spain ....
.... ***Baetica ustulata*** (p. 249; CD 1, track 84)
–   Each echeme composed of fewer than 8 syllables. Paler or mostly green bush-crickets, not occurring in the Sierra Nevada 92

92(91)   Occurring only in the central Apennines. (Echemes composed of 5–7 syllables.) ....
.... ***Ephippiger ruffoi*** (p. 235; CD 1, track 72)
–   Not occurring in the central Apennines. (Echemes composed of 2–5 syllables.) 93

93(92)   Syllable repetition rate usually more than 7/s. Occurring in southern France and the Pyrenees (including the Spanish side) ....
.... ***Ephippiger ephippiger*** (p. 227; CD 1, track 69, excerpts 2 and 3)
–   Syllable repetition rate usually less than 7/s. Occurring only in mountains in Spain, mainly south of the Pyrenees .... ***Ephippigerida areolaria*** (p. 238; CD 1, track 73)

94(86)   Each song unit consisting of a single syllable 95
–   Each song unit consisting of an echeme of two or more syllables 103

95(94)   Syllable repetition rate at least 4/s. Fully winged grasshopper ....
.... ***Chorthippus vagans*** (p. 395; CD 2, track 49, excerpt 1)
–   Syllable repetition rate less than 4/s. Brachypterous bush-crickets 96

96(95)   Each syllable lasting more than a tenth of a second 97
–   Each syllable lasting less than a tenth of a second 100

97(96)   Song very quiet and nocturnal, from a small bush-cricket (pronotum less than 6 mm
            long) .... *Isophya kraussii* (p. 118; CD 1, track 6)
–   Song louder, from larger bush-crickets (pronotum more than 6 mm long) 98

98(97)   Song nocturnal, from a grey or brown bush-cricket occurring in central and southern
            Spain and Portugal .... *Thyreonotus bidens* (p. 225; CD 1, track 67)
–   Song mostly diurnal, from mainly green bush-crickets not occurring in these regions
            99

99(98)   Male supra-anal plate fairly clearly separated from the tenth abdominal tergite. Occur-
            ring in much of western and central Europe ....
            .... *Ephippiger ephippiger* (p. 227; CD 1, track 69, excerpt 1)
–   Male supra-anal plate continuous with the tenth abdominal tergite. Known only from
            the Alps .... *Ephippiger terrestris* (p. 231; CD 1, track 70)

100(96)   Mainly brown, grey or straw-coloured bush-crickets, occurring only in central and south-
            ern Spain and Portugal 101
–   Mainly green bush-crickets, not occurring in central and southern Spain or Portugal
            102

101(100)   Pronotum extended backwards, concealing most of the fore wings ....
            .... *Thyreonotus bidens* (p. 225; CD 1, track 67)
–   Pronotum not extended backwards, most of the fore wings exposed ....
            .... *Antaxius spinibrachius* (p. 222; CD 1, track 66)

102(100)   Large bush-cricket, pronotum more than 8 mm long. Song quite loud ....
            .... *Eupholidoptera chabrieri* (p. 209; CD 1, track 56)
–   Small bush-cricket, pronotum less than 6 mm long. Song very quiet ....
            .... *Isophya pyrenaea* (p. 118; CD 1, track 5)

103(94)   Song very quiet, each echeme composed of one macrosyllable followed by several
            microsyllables; these echemes are sometimes interspersed with much longer ones, last-
            ing 1–2 s. Small brown or straw-coloured bush-cricket, in western Europe occurring
            only in the Italian Peninsula and Sardinia. ....
            .... *Platycleis stricta* (p. 185; CD 1, track 42)
–   Song louder, not of this kind; microsyllables absent or occurring only occasionally
            104

104(103)   Each echeme composed of only 2 syllables. Large bush-crickets 105
–   Each echeme composed of at least 3 syllables 107

105(104)   Fore wings less than 12 mm long, not reaching the tip of the abdomen. Brown bush-
            cricket, occurring widely in southern Europe. ....
            .... *Platycleis sepium* (p. 178; CD 1, track 37)
–   Fore wings more than 20 mm long, extending beyond the tip of the abdomen 106

106(105)   Song often interrupted at irregular intervals by brief pauses of less than 1 s. Green bush-
            cricket, occurring in the mountains of central Spain ....
            .... *Tettigonia hispanica* (p. 152; CD 1, track 24)
–   Song without interruptions. Brown bush-cricket, occurring widely in southern Europe
            .... *Platycleis intermedia* (p. 175; CD 1, track 36)

107(104)   Small dark-coloured cricket, often living in crevices in the ground. Song heard in the evening or at night .... ***Modicogryllus bordigalensis*** (p. 269; CD 1, track 94)
[This song could be confused with that of *Cicada orni* (p. 476), but this cicada sings only during the day from shrubs or trees and the quality of the sound is harsher (CD 2, track 75).]

–   Bush-crickets or grasshoppers. Song heard mainly during the day 108

108(107)   Bush-crickets 109
–   Grasshoppers 115

109(108)   Fore wings extending beyond the tip of the abdomen 110
–   Fore wings not reaching the tip of the abdomen 111

110(109)   Each echeme usually composed of 5 or fewer syllables ....
    .... ***Platycleis albopunctata*** (p. 164; CD 1, track 30)
–   Each echeme usually composed of 6 or more syllables ....
    .... ***Platycleis sabulosa*** (p. 168; CD 1, track 31)

111(109)   Each echeme composed of 4 or fewer syllables ....
    .... ***Metrioptera brachyptera*** (p. 187; CD 1, track 43)
–   Each echeme composed of 5 or more syllables 112

112(111)   Occurring only in the Apennine range in Italy ....
    ....***Metrioptera caprai*** (p. 191; CD 1, track 46)
–   Not occurring in the Apennine range 113

113(112)   Occurring only in the Cantabrian Mountains in northern Spain ....
    ....***Metrioptera burriana*** (p. 193; CD 1, track 48)
–   Not occurring in the Cantabrian Mountains 114

114(113)   Male cerci very short, with an inner tooth at the base (usually hidden under the tenth abdominal tergite) .... ***Metrioptera abbreviata*** (p. 193; CD 1, track 47)
–   Male cerci longer, with an inner tooth about halfway along their length ....
    .... ***Metrioptera saussuriana*** (p. 189; CD 1, track 44)
    or ***Metrioptera buyssoni*** (p. 191; CD 1, track 45)
[These two species seem to have identical songs and are difficult to separate morphologically (see p. 189). *M. buyssoni* is known only from the eastern Pyrenees.]

115(108)   Each echeme composed of more than 40 syllables. Pronotal lateral carinae strongly inflexed; hind wings strongly smoky. Known only from the French Cottian Alps. ......
    .... ***Stenobothrus cotticus*** (p. 359; CD 2, track 36)
–   Each echeme composed of fewer than 10 syllables. Pronotal lateral carinae straight or almost so; hind wings transparent 116

116(115)   Each echeme lasting more than a fifth of a second. Fore wings projecting beyond the hind wings (when flexed) .... ***Euchorthippus declivus*** (p. 463; CD 2, track 69)
–   Each echeme lasting less than a fifth of a second. Fore wings not projecting beyond the hind wings (when flexed) ....
    .... ***Euchorthippus pulvinatus*** (subspecies ***gallicus*** and ***elegantulus***)
    (p. 465; CD 2, track 70)

117(77)   Song units lasting more than a third of a second 118
–   Song units lasting less than a third of a second 129

118(117)   Each song unit consisting of a single long syllable 119
–   Each song unit divided into two or more sub-units, each lasting less than a quarter of a second 120

119(118)   Occurring in France and Catalonia ....
.... ***Uromenus rugosicollis*** (p. 240; CD 1, track 75)

–   Not occurring in these regions .... ***Uromenus elegans*** (p. 242; CD 1, track 76)
or ***Uromenus brevicollis*** (p. 242; CD 1, track 77)
[These two species are very similar in song and morphology (see p. 242).]

120(118)   Sub-units repeated more rapidly than 15/s 121

–   Sub-units repeated less rapidly than 15/s 125

121(120)   Fully winged grasshopper .... ***Euchorthippus chopardi*** (p. 469; CD 2, track 71)

–   Brachypterous bush-crickets 122

122(121)   Each song unit a sequence of trisyllabic echemes. Mostly or at least partly green bush-cricket .... ***Metrioptera bicolor*** (p. 196; CD 1, track 49, excerpt 1)

–   Each song unit a simple echeme of syllables. Mostly straw-coloured, brown or black bush-crickets 123

123(122)   Mostly dark brown or black bush-cricket known only from the Sierra Nevada in southern Spain .... ***Baetica ustulata*** (p. 249; CD 1, track 84)

–   Paler bush-crickets not occurring south of the Pyrenees 124

124(123)   Syllable repetition rate more than 20/s. Plantulae (side-flaps) of the hind tarsi as long as the first tarsal segment .... ***Rhacocleis germanica*** (p. 216; CD 1, track 62)

–   Syllable repetition rate less than 20/s. Plantulae of the hind tarsi shorter than the first tarsal segment .... ***Antaxius pedestris*** (p. 220; CD 1, track 64)

125(120)   Each echeme with at least 8 syllables 126

–   Each echeme with fewer than 8 syllables 127

126(125)   Mostly dark brown or black bush-cricket known only from the Sierra Nevada in southern Spain .... ***Baetica ustulata*** (p. 249; CD 1, track 84)

–   Mostly paler, often green bush-cricket known only from the eastern Pyrenees ....
.... ***Uromenus catalaunicus*** (p. 242; CD 1, track 78)

127(125)   Occurring only in the central Apennines. (Echemes with 5–7 syllables.) ....
.... ***Ephippiger ruffoi*** (p. 235; CD 1, track 72)

–   Not occurring in the central Apennines. (Echemes with 2–5 syllables.) 128

128(127)   Syllable repetition rate usually more than 7/s. Occurring in southern France and the Pyrenees (including the Spanish side) ....
.... ***Ephippiger ephippiger*** (p. 227; CD 1, track 69, excerpts 2 and 3)

–   Syllable repetition rate usually less than 7/s. Occurring only in mountains in Spain, mainly south of the Pyrenees .... ***Ephippigerida areolaria*** (p. 238; CD 1, track 73)

129(117)   Each song unit consisting of a single syllable 130

–   Each song unit consisting of two or more syllables 138

130(129)   Fully winged bush-crickets 131

–   Brachypterous bush-crickets 132

131(130)   Pronotal lateral lobes at least as long as deep ....
.... ***Phaneroptera falcata*** (p. 111; CD 1, track 1, excerpt 1)

–   Pronotal lateral lobes distinctly deeper than long ....
.... ***Phaneroptera nana*** (p. 111; CD 1, track 2)

132(130)   Entirely brown or straw-coloured. (Song inaudible to most people.) 133

–   Partly or entirely green 134

133(132)   Each syllable lasting less than 50 ms ....
                          .... ***Yersinella raymondii*** (p. 211; CD 1, track 58)
–   Each syllable lasting more than 100 ms ....
                          .... ***Yersinella beybienkoi*** (p. 214; CD 1, track 59)

134(132)   Song very quiet and nocturnal, from a small bush-cricket (pronotum less than 6 mm long).................................... ***Isophya pyrenaea*** (p. 118; CD 1, track 5)
–   Song louder, from larger bush-crickets (pronotum more than 6 mm long) 135

135(134)   Each syllable lasting less than a tenth of a second... 136
–   Each syllable lasting more than a tenth of a second 137

136(135)   Pronotum smooth. Widespread in southern Europe but absent from the Iberian Penin-sula .... ***Eupholidoptera chabrieri*** (p. 209; CD 1, track 56)
–   Pronotum rugose. In western Europe known only from southern Spain ....
                          .... ***Ephippigerida taeniata*** (p. 238; CD 1, track 74)

137(135)   Male supra-anal plate fairly clearly separated from the tenth abdominal tergite. Occur-ring in much of western and central Europe ....
                          .... ***Ephippiger ephippiger*** (p. 227; CD 1, track 69, excerpt 1)
–   Male supra-anal plate continuous with the tenth abdominal tergite. Known only from the Alps .... ***Ephippiger terrestris*** (p. 231; CD 1, track 70)

138(129)   Bush-crickets 139
–   Grasshoppers 145

139(138)   Fully winged, fore wings extending beyond the tip of the abdomen. In western Europe occurring only in Italy, including Sardinia. ....
                          .... ***Platycleis stricta*** (p. 185; CD 1, track 42)
–   Brachypterous, fore wings not reaching the tip of the abdomen 140

140(139)   Fore wings longer than the pronotum 141
–   Fore wings shorter than the pronotum 142

141(140)   Large brown bush-cricket, hind femora more than 22 mm long. Song composed of loosely grouped disyllabic echemes ....
                          .... ***Platycleis sepium*** (p. 178; CD 1, track 37, excerpt 1)
–   Smaller, mostly or at least partly green bush-cricket, hind femora less than 20 mm long. Song composed of dense sequences of trisyllabic echemes ....
                          .... ***Metrioptera bicolor*** (p. 196; CD 1, track 49, excerpt 1)

142(140)   Pronotum rugose. Occurring only in the Iberian Peninsula 143
–   Pronotum smooth. Not occurring in the Iberian Peninsula 144

143(142)   Each echeme lasting more than 150 ms and composed of at least 5 syllables ....
                          .... ***Uromenus martorellii*** (p. 246; CD 1, track 82)
–   Each echeme lasting less than 150 ms and composed of fewer than 5 syllables ....
                          .... ***Uromenus perezii*** (p. 246; CD 1, track 81)

144(142)   Each echeme lasting more than a fifth of a second and composed of more than 7 syllables
                          .... ***Rhacocleis germanica*** (p. 216; CD 1, track 62)
–   Each echeme lasting less than a fifth of a second and composed of fewer than 7 syllables
                          .... ***Rhacocleis neglecta*** (p. 220; CD 1, track 63)

145(138) Each echeme with a total of at least 10 intrasyllabic gaps. Fore wings projecting beyond the hind wings (when flexed) ....
.... ***Euchorthippus declivus*** (p. 463; CD 2, track 69)

– Each echeme usually with a total of fewer than 10 intrasyllabic gaps. Fore wings not projecting beyond the hind wings (when flexed) 146

146(145) Small, pronotum less than 2·1 mm long. Occurring only on Sardinia ....
.... ***Euchorthippus sardous*** (p. 475; CD 2, track 74)

– Larger, pronotum more than 2·1 mm long. Not occurring on Sardinia 147

147(146) Occurring only on Sicily ....
.... ***Euchorthippus albolineatus siculus*** (p. 471; CD 2, track 72)

– Not occurring on Sicily 148

148(147) Echeme repetition rate usually more than 0·8/s. Syllable repetition rate more than 30/s
.... ***Euchorthippus pulvinatus*** (subspecies ***gallicus*** and ***elegantulus***)
(p. 465; CD 2, track 70)

– Echeme repetition rate usually less than 0·8/s. Syllable repetition rate less than 30/s
.... ***Euchorthippus chopardi*** (p. 469; CD 2, track 71)

149(75) Song consisting of a very quiet nocturnal drumming on a leaf or twig, in sequences of shorter and longer bursts (lasting about 0·1–0·2 s and 0·7–1·3 s, respectively). Pale green bush-cricket living in deciduous trees ....
.... ***Meconema thalassinum*** (p. 136; CD 1, track 18)

– Song not of this kind 150

150(149) Song units very brief, lasting less than half a second, separated from other units by intervals at least as long as the units themselves 151

– Song units longer, lasting at least half a second, or, if shorter, separated from other units by intervals shorter than the units themselves 195

151(150) Each song unit consisting of a single syllable 152
– Each song unit consisting of two or more syllables 170

152(151) Bush-crickets 153
– Grasshoppers 168

153(152) Fully winged, fore wings extending beyond the tip of the abdomen 154
– Brachypterous, fore wings not reaching the tip of the abdomen 158

154(153) Fore tibial tympanal apertures slit-like 155
– Fore tibial tympanal apertures open and oval 157

155(154) Hind femora more than 20 mm long .... ***Tylopsis lilifolia*** (p. 114)
– Hind femora less than 15 mm long 156

156(155) Mostly green with a brown dorsal stripe .... ***Conocephalus conocephalus*** (p. 138)
– Entirely brown or straw-coloured .... ***Platycleis tessellata*** (p. 178)

157(154) Pronotal lateral lobes at least as long as deep ....
.... ***Phaneroptera falcata*** (p. 111; CD 1, track 1, excerpt 1)

– Pronotal lateral lobes distinctly deeper than long ....
.... ***Phaneroptera nana*** (p. 111; CD 1, track 2)

158(153) Occurring only in southern Spain 159
– Not occurring in southern Spain 160

159(158)   Pronotum with well developed lateral carinae. In Europe known only from Cadiz prov-
            ince .... *Ephippigerida taeniata* (p. 238; CD 1, track 74)
–   Pronotum with poorly developed lateral carinae. Occurring widely in Andalusia ....
            .... *Uromenus andalusius* (p. 249; CD 1, track 83)

160(158)   Pronotum smooth 161
–   Pronotum rugose 167

161(160)   Pronotum more than 9 mm long. Mainly green bush-cricket ....
            .... *Eupholidoptera chabrieri* (p. 209; CD 1, track 56)
–   Pronotum less than 8 mm long. Mainly brown or straw-coloured bush-crickets 162

162(161)   Fore wings longer than the pronotum 163
–   Fore wings shorter than the pronotum 164

163(162)   Pronotum with a median carina along the whole of its length ....
            ....*Platycleis nigrosignata* (p. 183; CD 1, track 40)
–   Pronotum with a median carina only in the metazona ....
            .... *Platycleis veyseli* (p. 180; CD 1, track 39)

164(162)   Hind femora more than 15 mm long 165
–   Hind femora less than 15 mm long. (Song inaudible to most people.) 166

165(164)   Fore wings more than 5 mm long .... *Pholidoptera aptera* (p. 205; CD 1, track 52)
–   Fore wings less than 3 mm long .... *Pachytrachis striolatus* (p. 214; CD 1, track 60)

166(164)   Each syllable lasting less than 50 ms ....
            .... *Yersinella raymondii* (p. 211; CD 1, track 58)
–   Each syllable lasting more than 100 ms ....
            .... *Yersinella beybienkoi* (p. 214; CD 1, track 59)

167(160)   Male supra-anal plate fairly clearly separated from the tenth abdominal tergite. Occur-
            ring in much of western and central Europe ....
            .... *Ephippiger ephippiger* (p. 227; CD 1, track 69, excerpt 1)
–   Male supra-anal plate continuous with the tenth abdominal tergite. Known only from
            the Alps .... *Ephippiger terrestris* (p. 231; CD 1, track 70)

168(152)   Maxillary palps with the last segment swollen and usually orange-pink or reddish brown
            .... *Stenobothrus bolivarii* (p. 344; CD 2, track 31)
–   Maxillary palps with the last segment normal in shape and colour 169

169(168)   Hind femora less than 9 mm long .... *Dociostaurus jagoi* (p. 300; CD 2, track 12)
–   Hind femora more than 9 mm long ..... *Ramburiella hispanica* (p. 292; CD 2, track 8)

170(151)   Bush-crickets 171
–   Grasshoppers 180

171(170)   Pronotum smooth 172
–   Pronotum rugose 176

172(171)   Each echeme composed of more than 7 syllables ....
            .... *Rhacocleis germanica* (p. 216; CD 1, track 62)
–   Each echeme composed of fewer than 6 syllables 173

173(172)   Fore wings at most 2 mm long. (Each echeme composed of 3–5 syllables.) ....
            .... *Rhacocleis neglecta* (p. 220; CD 1, track 63)

–   Fore wings more than 2 mm long. (Each echeme usually composed of 3 syllables.) 174

174(173)   Pronotal lateral lobes with a very narrow pale margin ....
                    Pholidoptera griseoaptera (p. 203; CD 1, track 51)

–   Pronotal lateral lobes with a broad pale margin 175

175(174)   Hind femora more than 21 mm long ....
                    Pholidoptera femorata (p. 207; CD 1, track 54)

–   Hind femora less than 21 mm long .... ***Pholidoptera fallax*** (p. 207; CD 1, track 53)

176(171)   Syllables of each echeme uniform, equally loud and repeated less rapidly than 14/s.
Pronotum with lateral carinae poorly developed or absent ....
               Ephippiger ephippiger (p. 227; CD 1, track 69, excerpts 2 and 3)

–   Syllables becoming progressively louder during the course of each echeme and repeated
more rapidly than 14/s. Pronotum with well-developed lateral carinae in the posterior
part 177

177(176)   Each echeme lasting less than a fifth of a second, with all of the syllables lasting less than
20 ms 178

–   Each echeme usually lasting more than a fifth of a second, with at least the last syllable
lasting more than 20 ms 179

178(177)   Last syllable in each echeme longer than the others, lasting at least 10 ms ....
                Uromenus perezii (p. 246; CD 1, track 81)

–   All the syllables in each echeme equally short, lasting less than 5 ms ....
               Uromenus martorellii (p. 246; CD 1, track 82)

179(177)   Each echeme usually composed of fewer than 9 syllables ....
               Uromenus asturiensis (p. 243; CD 1, track 79)

–   Each echeme usually composed of at least 9 syllables ....
               Uromenus stalii (p. 243; CD 1, track 80)

180(170)   Maxillary palps with the last segment swollen and usually orange-pink or reddish brown
               Stenobothrus bolivarii (p. 344; CD 2, track 31)

–   Maxillary palps with the last segment normal in shape and colour 181

181(180)   Syllables repeated less rapidly than 15/s, clearly distinguishable to the unaided ear and
countable. Known only from the Iberian Peninsula and Majorca ....
               Chorthippus jacobsi (p. 420; CD 2, track 56)

–   Syllables repeated more rapidly than 15/s, not or hardly distinguishable to the unaided
ear and not countable 182

182(181)   At least the first half of the echeme-sequence showing a crescendo, successive echemes
becoming gradually louder. Antennae strongly clubbed ....
           Myrmeleotettix maculatus (p. 361; CD 2, track 37, excerpts 1 and 2)

–   Echeme-sequence not showing a gradual crescendo. Antennae not clubbed 183

183(182)   Fore wings reaching only halfway along the abdomen and bulging outwards somewhat.
Bright green grasshopper .... ***Euthystira brachyptera*** (p. 296; CD 2, track 10)

–   Fore wings reaching more than halfway along the abdomen and not bulging outwards
.... 184

184(183)   Syllable repetition rate more than 100/s. Flight with crepitation. Known only from the
French Cottian Alps .... ***Stenobothrus cotticus*** (p. 359; CD 2, track 36)

–   Syllable repetition rate less than 80/s. Flight without crepitation 185

185(184)   Pronotal lateral carinae strongly inflexed 186
–   Pronotal lateral carinae only slightly incurved or almost straight 187

186(185)   Fore wings less than 12 mm long. Known only from central Spain. (Song usually com-
           posed of fewer than 6 echemes.) ....
           .... ***Dociostaurus hispanicus*** (p. 300; CD 2, track 13)
–   Fore wings more than 12 mm long. Not occurring in central Spain. (Song usually com-
           posed of more than 6 echemes.) ....
           .... ***Chorthippus brunneus*** (p. 406; CD 2, track 52, excerpts 1–3)

187(185)   Large, fore wings more than 15 mm long. Green with red or orange hind tibiae and tarsi
           .... ***Chorthippus jucundus*** (p. 441; CD 2, track 62)
–   Smaller, fore wings less than 15 mm long. Variously coloured, hind tibiae not red or
           orange 188

188(187)   Song usually composed of more than 6 echemes, repeated regularly at a rate of at least
           one every 2 s 189
–   Song composed of no more than 6 echemes, usually repeated at a slower rate than one
           every 2 s 194

189(188)   Each echeme with a total of at least 10 intrasyllabic gaps. Fore wings projecting beyond
           the hind wings (when flexed) .... ***Euchorthippus declivus*** (p. 463; CD 2, track 69)
–   Each echeme usually with a total of fewer than 10 intrasyllabic gaps. Fore wings not
           projecting beyond the hind wings (when flexed) 190

190(189)   Small, pronotum less than 2·1 mm long. Occurring only on Sardinia ....
           .... ***Euchorthippus sardous*** (p. 475; CD 2, track 74)
–   Larger, pronotum more than 2·1 mm long. Not occurring on Sardinia 191

191(190)   Occurring only on Sicily ....
           .... ***Euchorthippus albolineatus siculus*** (p. 471; CD 2, track 72)
–   Not occurring on Sicily 192

192(191)   Occurring only on the Balearic Islands ....
           .... ***Euchorthippus angustulus*** (p. 471; CD 2, track 73)
–   Not known from the Balearic Islands 193

193(192)   Echeme repetition rate usually more than 0·8/s. Syllable repetition rate more than 30/s
           .... ***Euchorthippus pulvinatus*** (subspecies ***gallicus*** and ***elegantulus***)
           (p. 465; CD 2, track 70)
–   Echeme repetition rate usually less than 0·8/s. Syllable repetition rate less than 30/s
           .... ***Euchorthippus chopardi*** (p. 469; CD 2, track 71)

194(188)   Fore wings with a small bulge near the base of the anterior (ventral) margin. Each echeme
           composed of fewer than 10 syllables repeated less rapidly than 30/s. Occurring in
           western Europe only in eastern Austria and southern Italy ....
           .... ***Chorthippus dichrous*** (p. 452; CD 2, track 65)
–   Fore wings without such a bulge. Each echeme composed of more than 15 syllables re-
           peated more rapidly than 30/s. Widespread in western Europe, from Scandinavia to
           the Pyrenees and Alps .... ***Chorthippus albomarginatus*** (p. 441; CD 2, track 63)

195(150)   Syllable repetition rate less than 3/s 196
–   Syllable repetition rate more than 3/s 202

196(195)   Sound continuous during the song, or virtually so 197

–   Syllables clearly interrupted by intervals of at least 50 ms. (Bush-crickets) 198

197(196)   Brachypterous bush-cricket more than 25 mm long, occurring only in the Iberian Penin-
            sula ....***Pycnogaster jugicola*** (p. 253; CD 1, track 88)
–   Fully-winged grasshopper less than 25 mm long, widespread in Europe ....
            .... ***Stenobothrus lineatus*** (p. 333; CD 2, track 26)

198(196)   Pronotum saddle-shaped, with lateral carinae only on the posterior part 199
–   Pronotum shaped like the top and two sides of a box, with well-defined lateral carinae
            along the entire length 201

199(198)   Occurring in south-western and southern France and Catalonia ....
            .... ***Uromenus rugosicollis*** (p. 240; CD 1, track 75)
–   Not occurring in these regions 200

200(199)   Cerci long and quite slender. In western Europe known only from Italy (including Sar-
            dinia and Sicily) .... ***Uromenus elegans*** (p. 242; CD 1, track 76)
–   Cerci short and very stout. In western Europe known only from Corsica, Sardinia, Mi-
            norca and the extreme south of Spain ....
            .... ***Uromenus brevicollis*** (p. 242; CD 1, track 77)

201(198)   Occurring only in the Sierra Nevada in southern Spain...
            .... ***Pycnogaster inermis*** (p. 258; CD 1, track 90)
–   Not occurring in the Sierra Nevada ....
            ....***Pycnogaster sanchezgomezi*** (p. 255; CD 1, track 89)

202(195)   Very small, black, ground-dwelling cricket, less than 7 mm long...................................
            .... ***Pteronemobius heydenii*** (p. 274; CD 1, track 97)
–   Larger bush-crickets or grasshoppers, more than 10 mm long 203

203(202)   Bush-crickets 204
–   Grasshoppers 227

204(203)   Fore wings well developed, reaching at least to the tip of the abdomen 205
–   Fore wings much reduced, not reaching the tip of the abdomen 215

205(204)   Mainly green, sometimes with a brown dorsal stripe 206
–   Mainly brown, grey or straw-coloured (pronotum and hind femora occasionally green)
            210

206(205)   Song a dense echeme usually lasting 1–3 s, followed by a loud click. In western Europe
            occurring only in Italy, including Sardinia and Sicily, and Corsica. ....
            .... ***Acrometopa servillea*** (p. 116; CD 1, track 4)
–   Song not of this kind, without a click 207

207(206)   Song loud. Pronotum more than 6 mm long 208
–   Song very quiet. Pronotum less than 5 mm long 209

208(207)   Fore wings less than 30 mm long, hardly extending beyond the hind knees. Widespread
            in western Europe .... ***Tettigonia cantans*** (p. 153; CD 1, track 25, excerpt 1)
–   Fore wings more than 30 mm long, extending well beyond the hind knees. In western
            Europe occurring only in eastern Germany and Austria ....
            .... ***Tettigonia caudata*** (p. 157; CD 1, track 26)

209(207)   Hind femora more than 16 mm long. Widespread in western Europe north of the Medi-
            terranean Region .... ***Phaneroptera falcata*** (p. 111; CD 1, track 1, excerpt 3)

–   Hind femora less than 15 mm long. In western Europe very local in the Mediterranean
           Region, in moist places .... *Conocephalus conocephalus* (p. 138; CD 1, track 19)

210(205)   Song consisting of a series of echemes of fairly uniform duration 211
–   Song a mixture of short (less than 1 s) and much longer (1–5 s) echemes 213

211(210)   Each echeme composed of alternately quieter and louder macrosyllables; microsyllables
           absent. Pronotum less than 5 mm long. In western Europe known only from Germany
           and eastern Austria .... *Platycleis montana* (p. 183; CD 1, track 41)
–   Each echeme composed of uniform macrosyllables and often ending with 2–6
           microsyllables. Pronotum more than 5 mm long. In western Europe occurring only in
           the western Mediterranean Region, from Italy westwards 212

212(211)   Apical part of the titillators long and slender; tenth abdominal tergite with blunt lobes.
           Widespread in the western Mediterranean Region ....
           .... *Platycleis falx* (p. 174; CD 1, track 35)
–   Apical part of the titillators shorter and thicker; tenth abdominal tergite with sharply
           pointed lobes. Known only from the Sierra de Gredos in central Spain ....
           .... *Platycleis iberica* (p. 174; CD 1, track 34)

213(210)   Each short echeme composed of a single macrosyllable followed by several microsyllables.
           Pronotum less than 5 mm long .... *Platycleis stricta* (p. 185; CD 1, track 42)
–   Each short echeme composed of more than one macrosyllable, usually followed by sev-
           eral microsyllables. Pronotum more than 5 mm long 214

214(213)   Syllable repetition rate less than 20/s. Widespread in the Mediterranean Region ....
           .... *Platycleis affinis* (p. 169; CD 1, track 32)
–   Syllable repetition rate more than 20/s. Known only from Italy ....
           .... *Platycleis romana* (p. 172; CD 1, track 33)

215(204)   Each echeme lasting more than 7 s. Pronotum shaped like the top and two sides of a box,
           with well-defined lateral carinae along the entire length. Large brown bush-cricket
           occurring in Spanish mountains. ....
           .... *Pycnogaster jugicola* (p. 253; CD 1, track 88)
–   Each echeme lasting less than 7 s. Pronotum not box-shaped, with the lateral carinae
           absent or in the posterior part of the pronotum only 216

216(215)   Each echeme lasting more than 3 s 217
–   Each echeme lasting less than 3 s 219

217(216)   Pronotum saddle-shaped, with lateral carinae in the posterior part. Occurring only in
           the Iberian Peninsula .... *Platystolus martinezii* (p. 251; CD 1, track 86)
–   Pronotum not saddle-shaped, without lateral carinae. Not occurring in the Iberian Penin-
           sula 218

218(217)   Fore wings less than 4 mm long, with a conspicuous pale spot. Echemes composed of
           diplosyllables .... *Antaxius pedestris* (p. 220; CD 1, track 64)
–   Fore wings more than 4 mm long, without a pale spot. Echemes composed of closing
           hemisyllables only .... *Pholidoptera littoralis* (p. 209; CD 1, track 55)

219(216)   Each echeme composed of a rapid series of brief syllables followed by two or more much
           longer syllables, repeated more slowly. (Iberian Peninsula) 220
–   Each echeme composed of syllables of uniform duration, repeated at a uniform rate
           221

220(219) Each echeme with more than 8 brief syllables and fewer than 4 longer ones. Cerci with
          an inner tooth at the base .... **Callicrania selligera** (p. 251; CD 1, track 85)
–   Each echeme with fewer than 8 brief syllables and more than 4 longer ones. Cerci with
          an inner tooth about halfway along their length ....
          .... **Platystolus faberi** (p. 253; CD 1, track 87)

221(219) Each echeme with fewer than 7 syllables 222
–   Each echeme with more than 7 syllables 223

222(221) Syllable repetition rate usually 4–6/s. Known only from northern Italy, south of the Po
          valley .... **Ephippiger perforatus** (p. 235; CD 1, track 71)
–   Syllable repetition rate usually 6–12/s. Not occurring in Italy south of the Po valley .....
          .... **Ephippiger ephippiger** (p. 227; CD 1, track 69, excerpts 2 and 3)

223(221) Pronotum saddle-shaped, with the posterior part raised over the fore wings ....
          .... **Uromenus catalaunicus** (p. 242; CD 1, track 78)
–   Pronotum not saddle-shaped, the posterior part not raised over the fore wings 224

224(223) Occurring only in the Pyrenees .... **Antaxius hispanicus** (p. 222; CD 1, track 65)
–   Not occurring in the Pyrenees 225

225(224) Hind femora less than 14 mm long .... **Anonconotus alpinus** (p. 211; CD 1, track 57)
–   Hind femora more than 14 mm long 226

226(225) Syllable repetition rate more than 20/s. Plantulae (side-flaps) of the hind tarsi as long as
          the first tarsal segment .... **Rhacocleis germanica** (p. 216; CD 1, track 62)
–   Syllable repetition rate less than 20/s. Plantulae of the hind tarsi shorter than the first
          tarsal segment .... **Antaxius pedestris** (p. 220; CD 1, track 64)

227(203) Song including a loud buzzing sound produced by wing-vibration, either on the ground
          or in flight. Montane grasshopper, in western Europe known only from the Alps and
          higher Apennines. .... **Stenobothrus rubicundulus** (p. 356; CD 2, track 35)
–   Song not including a loud buzzing sound produced by wing-vibration 228

228(227) Song including both clearly separated syllables and dense echemes 229
–   Song not a mixture of clearly separated syllables and dense echemes 231

229(228) Each dense echeme lasting more than a second ....
          .... **Arcyptera fusca** (p. 286; CD 2, track 5)
–   Each dense echeme lasting less than a second 230

230(229) Some or all of the dense echemes followed by a scratchy downstroke hemisyllable lasting
          about half a second. Known only from central and western parts of the Iberian Penin-
          sula .... **Arcyptera tornosi** (p. 288; CD 2, track 6)
–   Song without such long hemisyllables. In western Europe known only from the northern
          half of Spain, the Cevennes, the French Alps and Austria ....
          .... **Arcyptera microptera** (p. 290; CD 2, track 7)

231(228) Song a loud rapid alternation of sibilant buzzing and harder rattling, lasting 10–30 s
          .... **Stauroderus scalaris** (p. 375; CD 2, track 40)
–   Song not of this kind 232

232(231) Song consisting of a sequence of at least 15 echemes repeated at the rate of 2·5–5·5/s
          and each beginning with a distinct tick; the sequence begins very quietly and shows a
          gradual crescendo during at least the first third of its duration 233
–   Song not of this kind, without ticks 234

233(232)   Each echeme composed of a tick followed by two syllables; sequence with 50–150 echemes
.... ***Chorthippus apricarius*** (p. 376; CD 2, track 41)
–   Each echeme composed of a tick followed by at least 8 syllables at the start of the se-
quence and more than 20 towards the end; sequence with 15–80 echemes ....
.... ***Chorthippus mollis*** (p. 413; CD 2, tracks 54 and 55)

234(232)   Song consisting of a sequence of at least 5 echemes repeated at the rate of less than 5/s,
without intervals or with intervals of less than 1 s, and with a crescendo during at
least the early part of the sequence 235
–   Song not of this kind; echemes isolated, or fewer than 5, or repeated at the rate of more
than 8/s, or separated by intervals of at least 1 s and all of similar loudness 241

235(234)   Echemes following one another without intervals 236
–   Echemes separated by intervals of comparable duration to the echemes themselves
.... 239

236(235)   Sequence composed of fewer than 10 echemes (or echeme-sequences sounding like
echemes). Occurring only in the Apennines 237
–   Sequence composed of at least 10 echemes. Occurring only in the Alps 238

237(236)   Each echeme a simple train of syllables, with a completely even, uniform sound. Fore
wings not usually reaching the hind knees. Known only from the Monti Reatini ......
.... ***Chorthippus modestus*** (p. 392; CD 2, track 48)
–   Each 'echeme' actually an echeme-sequence, giving it an uneven sound. Fore wings reach-
ing the hind knees. Widespread in the Apennines ....
.... ***Chorthippus rubratibialis*** (p. 430; CD 2, track 58)

238(236)   Brachypterous, fore wings falling well short of the hind knees. Known only from alti-
tudes above 2000 m in the Italian Cottian Alps ....
.... ***Chorthippus cialancensis*** (p. 388; CD 2, track 45)
–   Fully winged, fore wings reaching at least to the hind knees. Widespread in the Alps ....
.... ***Chorthippus mollis ignifer*** (p. 420; CD 2, track 55)

239(235)   Song lasting more than 8 s. Antennae strongly clubbed ....
.... ***Myrmeleotettix maculatus*** (p. 361; CD 2, track 37)
–   Song usually lasting less than 8 s. Antennae not clubbed 240

240(239)   Echeme repetition rate more than 2·5/s. Not occurring in the Iberian Peninsula ....
.... ***Omocestus petraeus*** (p. 312; CD 2, track 17)
–   Echeme repetition rate less than 2·5/s. Occurring only in the Iberian Peninsula ....
.... ***Omocestus minutissimus*** (p. 328; CD 2, track 23)

241(234)   Song consisting of a sequence (lasting 2–5 s) of 2–4 echemes (or echeme-sequences sound-
ing like echemes) following one another in quick succession without intervals
(Apennines) 242
–   Song not of this kind 243

242(241)   Each echeme a simple train of syllables, with a completely even, uniform sound. Fore
wings not usually reaching the hind knees. Known only from the Monti Reatini ......
.... ***Chorthippus modestus*** (p. 392; CD 2, track 48)
–   Each 'echeme' actually an echeme-sequence, giving it an uneven sound. Fore wings reach-
ing the hind knees. Widespread in the Apennines ....
.... ***Chorthippus rubratibialis*** (p. 430; CD 2, track 58)

243(241)   Song a burst of sound lasting at least 5 s 244

–   Song a burst of sound lasting less than 5 s, or several such bursts 257

244(243)   Syllables repeated less rapidly than 2/s, forming slight fluctuations in the continuous,
           wheezy song .... *Stenobothrus lineatus* (p. 333; CD 2, track 26)
–   Syllables repeated more rapidly than 2/s 245

245(244)   Syllables repeated more rapidly than 40/s, not distinguishable to the unaided ear; song
           an echeme-sequence with an echeme repetition rate of about 4–9/s (heard as a slight
           unevenness in the sound). Antennae strongly clubbed and pale-tipped ....
           .... *Gomphocerippus rufus* (p. 370; CD 2, track 39)
–   Syllables repeated less rapidly than 25/s, clearly distinguishable to the unaided ear; song
           an echeme (occasionally with a quieter aftersong). Antennae not clubbed or, if clubbed,
           not pale-tipped 246

246(245)   Syllables repeated more rapidly than 12/s, forming a rapid ticking sound 247
–   Syllables repeated less rapidly than 12/s, too slow to sound like ticks 248

247(246)   Song usually lasting more than 12 s. Side of the head and thorax green or greenish brown
           .... *Omocestus viridulus* (p. 304; CD 2, track 14)
–   Song usually lasting less than 12 s. Side of the head and thorax dark brown or black
           .... *Omocestus rufipes* (p. 305; CD 2, track 15)

248(246)   Total length less than 13 mm. (Spanish mountains and Corsica) 249
–   Total length more than 13 mm 250

249(248)   Known only from mountains in Spain, usually at altitudes above 2000 m ....
           .... *Omocestus uhagonii* (p. 326; CD 2, track 22)
–   Known only from Corsica .... *Chorthippus corsicus* (p. 382; CD 2, track 42)

250(248)   Main echeme followed by an 'aftersong' of short, quieter sounds 251
–   Song without such an 'aftersong' 252

251(250)   Fore tibiae conspicuously swollen. In western Europe occurring only in mountains, but
           not in southern Spain .... *Gomphocerus sibiricus* (p. 367; CD 2, track 38)
–   Fore tibiae normal. Occurring only in the Sierra Nevada in southern Spain ....
           .... *Chorthippus vagans* (Sierra Nevadan form)
           (p. 399; CD 2, track 49, excerpt 4)

252(250)   Fore wings not reaching the hind knees 253
–   Fore wings reaching at least to the hind knees 254

253(252)   Syllable repetition rate more than 7/s. Occurring in the Pyrenees, Cevennes and French
           Alps .... *Chorthippus binotatus* (p. 435; CD 2, track 60)
–   Syllable repetition rate less than 6/s. Occurring in the Italian and south-eastern Alps
           .... *Chorthippus alticola* (p. 390; CD 2, track 47)

254(252)   Syllable repetition rate usually more than 8/s. Tympanal apertures slit-like, at least 4
           times longer than broad 255
–   Syllable repetition rate usually less than 8/s. Tympanal apertures open and oval, less than
           3 times longer than broad 256

255(254)   Medial area of fore wings enlarged, with ladder-like cross-veins. Occurring only in Italy
           .... *Omocestus uvarovi* (p. 333; CD 2, track 25)
–   Medial area of fore wings not enlarged. Not occurring in Italy ....
           .... *Chorthippus binotatus* (p. 435; CD 2, track 60)

256(254)   Tympanal apertures usually more than twice as long as broad. Known only from Alicante province in south-east Spain ....
.... *Chorthippus reissingeri* (p. 399; CD 2, track 50)

–   Tympanal apertures usually less than twice as long as broad. Widespread in Europe, from Denmark southwards ....
.... *Chorthippus vagans* (p. 395; CD 2, track 49, excerpts 1–3)

257(243)   Song a mixture of louder bursts of sound and quieter 'aftersongs'. Known only from the Iberian and Sicilian uplands ....
.... *Chorthippus yersini* (p. 423; CD 2, track 57, excerpts 1 and 3)

–   Song without quieter 'aftersongs' 258

258(257)   Each echeme of the song composed of two differently-sounding parts produced by the hind legs being moved in different ways 259

–   Each echeme of the song sounding similar throughout its duration (although often showing a crescendo) 260

259(258)   Echemes lasting 2–4 s, produced singly. Rare in western Europe, where it occurs only in the vicinity of the Alps, usually on gravel-banks of rivers and streams ....
.... *Chorthippus pullus* (p. 388; CD 2, track 46)

–   Echemes lasting less than 2 s, produced in a series of indefinite duration. Widespread in Europe from southern Sweden southwards ....
.... *Chorthippus dorsatus* (p. 447; CD 2, track 64)

260(258)   Hind wings not reaching the tips of the fore wings (when flexed). Fore wings not reaching the hind knees, usually less than three times the length of the pronotum ....
.... 261

–   Hind wings reaching the tips of the fore wings (when flexed). Fore wings usually reaching at least to the hind knees, usually more than three times the length of the pronotum
.... 272

261(260)   Hind femora more than 12 mm long. Subgenital plate sharply pointed. Bright metallic green grasshopper, widespread in western Europe from Scandinavia to the Pyrenees and Alps. .... *Chrysochraon dispar* (p. 296; CD 2, track 9)

–   Hind femora less than 12 mm long. Subgenital plate not sharply pointed 262

262(261)   Syllables repeated more rapidly than 13/s, not countable. (Apennines, Pyrenees and Sierra Nevada) 263

–   Syllables repeated less rapidly than 13/s, countable (sometimes with difficulty) 265

263(262)   Occurring only in the Apennines ..... *Stenobothrus apenninus* (p. 352; CD 2, track 33)

–   Not occurring in Italy 264

264(263)   Occurring only in the Pyrenees .... *Omocestus antigai* (p. 319; CD 2, track 20)

–   Occurring only in the Sierra Nevada in southern Spain ....
.... *Omocestus bolivari* (p. 326; CD 2, track 21)

265(262)   Occurring only on Corsica .... *Chorthippus corsicus* (p. 382; CD 2, track 42)

–   Not occurring on Corsica 266

266(265)   Fore wings less than 6·5 mm long. Known only from the Sierra Nevada in southern Spain, mostly at altitudes above 2000 m. ....
.... *Chorthippus nevadensis* (p. 384; CD 2, track 44)

–   Fore wings more than 6·5 mm long 267

267(266) Antennae noticeably thickened towards the tip. Known only from the Italian Alps, in the region between Turin and Aosta, at altitudes above 1000 m. ....
.... *Stenobothrus ursulae* (p. 352; CD 2, track 34)
–   Antennae not noticeably thickened towards the tip 268

268(267) Fore wings less than 8 mm long 269
–   Fore wings more than 8 mm long 270

269(268) Pronotal lateral carinae strongly inflexed. Each downstroke hemisyllable with at least 6 gaps. Known only from the central and southern Spanish mountains, usually at altitudes above 2000 m .... *Omocestus uhagonii* (p. 326; CD 2, track 22)
–   Pronotal lateral carinae weakly inflexed. Each downstroke hemisyllable with fewer than 6 gaps. Common and widespread in Europe ....
.... *Chorthippus parallelus* (p. 452; CD 2, track 66)

270(268) Pronotal lateral carinae strongly inflexed. In Europe known only from France and the Iberian Peninsula .... *Chorthippus binotatus* (p. 435; CD 2, track 60)
–   Pronotal lateral carinae weakly inflexed 271

271(270) Each echeme usually lasting less than 2 s; syllables usually repeated at least as rapidly as 7/s. Cerci less than 0·65 mm long. Common and widespread in Europe ....
.... *Chorthippus parallelus* (p. 452; CD 2, track 66)
–   Each echeme usually lasting more than 2 s; syllables usually repeated less rapidly than 7/s. Cerci more than 0·65 mm long. Widespread in Europe north of the Pyrenees and Alps, but local and restricted to marshy habitats ....
.... *Chorthippus montanus* (p. 461; CD 2, track 68)

272(260) Antennae strongly clubbed and pale-tipped. Song an echeme-sequence with an echeme repetition rate of about 4–9/s (heard as a slight unevenness in the continuous sound)
.... *Gomphocerippus rufus* (p. 370; CD 2, track 39)
–   Antennae not clubbed or pale-tipped. Song not an echeme-sequence or, if an echeme-sequence, with an echeme repetition rate slower than 4/s or faster than 9/s 273

273(272) Song composed of echemes lasting no more than a second, produced singly or separated from other such echemes by intervals of at least half a second 274
–   Song composed of echemes or dense echeme-sequences lasting more than a second
.... 281

274(273) Antennae very thick towards the base, tapering gradually to a pointed tip. Elongate, straw-coloured grasshopper, occurring in Europe only near the Mediterranean coast of Spain, and on Sicily and Lipari ....
.... *Brachycrotaphus tryxalicerus* (p. 286; CD 2, track 4)
–   Antennae of quite uniform thickness, not tapering towards the tip 275

275(274) Syllables clearly distinguishable. (Iberian Peninsula and Majorca) 276
–   Syllables not clearly distinguishable, either repeated too rapidly or merging into one another 277

276(275) Echemes with fewer than 14 syllables, usually repeated less rapidly than 13/s and countable. Tympanal apertures slit-like ....
.... *Chorthippus jacobsi* (p. 420; CD 2, track 56)
–   Echemes with more than 14 syllables, usually repeated more rapidly than 13/s and not countable. Tympanal apertures open, oval. (Not occurring after July.) ....
.... *Chorthippus apicalis* (p. 439; CD 2, track 61)

277(275) Pronotal lateral carinae strongly inflexed 278

–   Pronotal lateral carinae almost straight 280

278(277) Each echeme of uniform loudness, beginning abruptly ....
          .... ***Chorthippus brunneus*** (p. 406; CD 2, track 52, excerpt 3)
–   Each echeme beginning with a crescendo 279

279(278) Occurring only in the Iberian Peninsula south of the Cantabrian Mountains and Ebro
          valley, and in Sicily. Fore wings with a small bulge near the base of the anterior (ven-
          tral) margin .... ***Chorthippus yersini*** (p. 423; CD 2, track 57)
–   Not occurring in these parts of Spain or in Sicily. Fore wings without such a bulge ....
          .... ***Stenobothrus nigromaculatus*** (p. 334; CD 2, track 27)

280(277) Large, green grasshopper with red or orange hind tibiae and tarsi; fore wings more than
          16 mm long .... ***Chorthippus jucundus*** (p. 441; CD 2, track 62)

–   Smaller, variously coloured grasshopper, hind tibiae and tarsi not red or orange; fore wings
          less than 12 mm long .... ***Chorthippus albomarginatus*** (p. 441; CD 2, track 63)

281(273) Occurring only on Corsica .... ***Chorthippus corsicus*** (p. 382; CD 2, track 42)
–   Not occurring on Corsica 282

282(281) Pronotal lateral carinae straight or almost so. Occurring only in the Iberian Peninsula
          ....***Omocestus panteli*** (p. 315; CD 2, track 19)
–   Pronotal lateral carinae inflexed 283

283(282) Syllables clearly distinguishable, usually repeated less rapidly than 40/s 284
–   Syllables not clearly distinguishable, usually repeated more rapidly than 40/s 297

284(283) Syllable repetition rate more than 25/s ....
          .... ***Omocestus haemorrhoidalis*** (p. 309; CD 2, track 16)
–   Syllable repetition rate less than 25/s 285

285(284) Each echeme composed of fewer than 14 syllables 286
–   Each echeme composed of more than 14 syllables 289

286(285) Syllable repetition rate usually more than 8/s. Tympanal apertures slit-like, at least 4
          times longer than broad 287
–   Syllable repetition rate usually less than 8/s. Tympanal apertures open and oval, less than
          3 times longer than broad 288

287(286) Hind femora usually with two conspicuous dark transverse bands; hind tibiae and tarsi
          often red. Transverse sulcus of the pronotum cutting the median carina at about its
          mid-point. Often living on shrubs .... ***Chorthippus binotatus*** (p. 435)
–   Hind femora without two conspicuous dark bands; hind tibiae and tarsi not red. Trans-
          verse sulcus of the pronotum cutting the median carina in front of its mid-point.
          Living near or on the ground in grassy places ....
          .... ***Chorthippus jacobsi*** (p. 420; CD 2, track 56)

288(286) Tympanal apertures usually more than twice as long as broad. Known only from Alicante
          province in south-east Spain ....
          .... ***Chorthippus reissingeri*** (p. 399; CD 2, track 50)
–   Tympanal apertures usually less than twice as long as broad. Widespread in Europe, from
          Denmark southwards ....
          .... ***Chorthippus vagans*** (p. 395; CD 2, track 49, excerpts 1–3)

289(285) Tympanal apertures open and oval, less than 3 times longer than broad 290
–   Tympanal apertures slit-like, at least 4 times longer than broad 292

290(289) Syllable repetition rate more than 11/s. Fore wings more than 13 mm long. Often yellow-
          ish green. (Not occurring after July.) ....
          .... ***Chorthippus apicalis*** (p. 439; CD 2, track 61)
− Syllable repetition rate less than 9/s. Fore wings less than 12 mm long. Never green
          291

291(290) Tympanal apertures usually more than twice as long as broad. Known only from Alicante
          province in south-east Spain .... ***Chorthippus reissingeri*** (p. 399; CD 2, track 50)
− Tympanal apertures usually less than twice as long as broad. Widespread in Europe, from
          Denmark southwards ....
          .... ***Chorthippus vagans*** (p. 395; CD 2, track 49, excerpts 1–3)

292(289) Fore wings with a small bulge near the base of the anterior (ventral) margin. Often living
          on shrubs .... ***Chorthippus binotatus*** (p. 435; CD 2, track 60)
− Fore wings without such a bulge. Living near or on the ground in grassy places 293

293(292) Fore wings more than 10 mm long 294
− Fore wings less than 9 mm long 296

294(293) Syllable repetition rate more than 14/s. Hind femora less than 9 mm long ....
          .... ***Omocestus raymondi*** (p. 315; CD 2, track 18)
− Syllable repetition rate less than 14/s. Hind femora more than 9 mm long 295

295(294) Each echeme composed of more than 35 syllables and usually lasting more than 3 s
          .... ***Stenobothrus fischeri*** (p. 337; CD 2, track 28)
− Each echeme composed of fewer than 35 syllables and usually lasting less than 3 s
          .... ***Stenobothrus festivus*** (p. 340; CD 2, track 29)

296(293) Pronotal lateral carinae gently incurved. Fore wings usually with dark spots in the me-
          dial area. Often partly green. Not occurring in south-east Spain ....
          .... ***Stenobothrus stigmaticus*** (p. 344; CD 2, track 32)
− Pronotal lateral carinae more strongly inflexed. Fore wings without dark spots in the
          medial area. Never green. Occurring only in mountains in south-east Spain ....
          .... ***Omocestus femoralis*** (p. 331; CD 2, track 24)

297(283) Song a single uninterrupted burst of sound, not repeated within 5 s 298
− Song a burst of sound interrupted by hesitations, or a series of two or more bursts of
          sound following one another at intervals of less than 5 s 302

298(297) Occurring only in North Africa .... ***Chorthippus marocanus*** (p. 433; CD 2, track 59)
− Not occurring in North Africa 299

299(298) Song a simple echeme, an even whizzing sound with no substructure 300
− Song an echeme-sequence with an echeme repetition rate of 5–20/s, heard as an un-
          evenness in the sound 301

300(299) Occurring only in the central Italian Apennines .... ***Chorthippus modestus*** (p. 392)
− Occurring only in uplands in Spain .... ***Chorthippus yersini*** (p. 423; CD 2, track 57)

301(299) Occurring only in Spanish uplands south of the Ebro valley ....
          .... ***Chorthippus yersini*** (north-eastern form) (p. 429)
− Common and widespread in Europe, but not occurring south of the Ebro valley in Spain
          .... ***Chorthippus biguttulus*** (p. 401; CD 2, track 51)

302(297) Song a single burst of sound (rarely two) lasting 2–5 s and interrupted by one or more
          very brief hesitations (less than a tenth of a second). (Italian Apennines) 303

–   Song a series of two or more bursts of sound, separated from one another by intervals of at least half a second 304

303(302)   Each burst of sound a simple echeme, an even whizzing sound with no substructure. Known only from the Monti Reatini in the central Apennines .... *Chorthippus modestus* (p. 392; CD 2, track 48)

–   Each burst of sound an echeme-sequence with an echeme repetition rate of 10–20/s, heard as an unevenness in the sound. Widespread in the Apennines .... *Chorthippus rubratibialis* (p. 430; CD 2, track 58)

304(302)   Each burst of sound a simple echeme, an even whizzing sound with no substructure 305

–   Each burst of sound an echeme-sequence with an echeme repetition rate of 5–20/s, heard as an unevenness in the sound 306

305(304)   Fore wings extending beyond the hind knees, with a small bulge near the base of the anterior (ventral) margin. Never green. Occurring only in uplands in Spain .... *Chorthippus yersini* (p. 423; CD 2, track 57)

–   Fore wings hardly reaching the hind knees, without such a bulge. Often mostly or partly green. Quite widespread in Europe, including northern Spain .... *Stenobothrus nigromaculatus* (p. 334; CD 2, track 27)

306(304)   Occurring only in North Africa .... *Chorthippus marocanus* (p. 433; CD 2, track 59)
–   Not occurring in North Africa 307

307(306)   Occurring only in Spanish uplands south of the Ebro valley .... *Chorthippus yersini* (north-eastern form) (p. 429)

–   Common and widespread in Europe, but not occurring south of the Ebro valley in Spain .... *Chorthippus biguttulus* (p. 401; CD 2, track 51)

Chapter 7

Tettigoniidae

Bush-crickets

The European bush-crickets produce shrill, often rather harsh, songs, mostly with a broad frequency spectrum extending well into the ultrasonic range. Few of the songs contain much sound below 10 kHz, and in some species most of the sound produced is ultrasonic, so that even the keenest human ear is able to detect only a small part of the output of a singing male. Terms such as 'faint' and 'quiet', used in the descriptions of the songs, refer of course to the *audible* sound. The apparently faint songs of *Phaneroptera*, *Leptophyes*, *Conocephalus* and *Yersinella*, for example, are mainly ultrasonic; they are no doubt loud enough to the insects themselves, and produce a loud response from an ultrasound detector. A few species (e.g. *Ruspolia nitidula*, *Decticus albifrons*) have resonant songs, i.e. songs with an almost pure dominant frequency. In the European species this frequency is too high-pitched to sound musical, unlike the lower-pitched resonant songs of the Gryllidae and Gryllotalpidae.

Most of the audible sound is usually produced by the closing stroke of the fore wings. The opening stroke is normally quicker than the closing stroke and any sound produced thus tends to be higher in frequency; this in itself is likely to reduce the audible sound content. In a few species (e.g. *Acrometopa servillea*, *Pachytrachis striolatus*, *Antaxius pedestris*, *Baetica ustulata*) the opening stroke is of similar duration to the closing stroke and produces an audible sound of similar intensity. In others (e.g. *Conocephalus discolor*, *C. dorsalis*) the opening stroke, although much quicker than the closing stroke, nevertheless produces a sound of comparable loudness. In yet others (e.g. *Phaneroptera falcata*, *P. nana*, *Uromenus andalusius*, and sometimes *Ephippigerida taeniata*) all the audible sound is produced by the opening stroke.

In most of the western European bush-crickets only the males sing. However, in the Phaneropterinae, Ephippigerinae and Pycnogastrinae the females can also stridulate, sometimes in response to the male calling song and sometimes when handled or threatened by a predator. The stridulatory apparatus is on the fore wings, but is quite different from that of the males (see p. 32).

In addition to producing airborne sound, bush-crickets stridulating on shrubs and trees cause the substrate to vibrate, and it has been shown in some species that such substrate-borne vibration also has communicative value (see especially Keuper *et al.*, 1985). In several ephippigerines both sexes transmit substrate-borne signals by silent tremulation (body-shaking), and in at least one species, *Ephippiger perforatus*, the male tremulation is accompanied by a special low-level stridulation, quite different from the normal calling song. The males of *Meconema* have an entirely different method of producing substrate-borne signals: they vibrate one of the hind legs so that the tarsus is drummed on the substrate.

Phanoptera falcata (Poda)

Gryllus (Tettigonia) falcata Poda, 1761: 52.

Sickle-bearing Bush-cricket; (F) Phanéroptère commun; (D) Gemeine Sichelschrecke; (NL) Sikkelsprinkhaan.

REFERENCES TO SONG. **Oscillogram**: Heller, 1988; Kleukers *et al.*, 1997. **Diagram**: Bellmann, 1985*a*, 1988, 1993*a*; Bellmann & Luquet, 1995; Duijm & Kruseman, 1983. **Wing-movement**: Heller, 1988. **Frequency information**: Froehlich, 1989; Heller, 1988. **Verbal description only**: Broughton, 1972*b*; Burr, 1936; Faber, 1928, 1953*a*; Grassé, 1924; Harz, 1957; Yersin, 1857. **Disc recording**: Bellmann, 1993*c* (CD); Bonnet, 1995 (CD); Odé, 1997 (CD). **Cassette recording**: Bellmann, 1985*b*, 1993*b*.

RECOGNITION (Plate 1: 1). An all-green, lightly-built and rather graceful bush-cricket, with the hind wings extending well beyond the fore wings in both sexes. Distinguished from its relative *P. nana* by the pronotal lateral lobes being at least as long as deep, and from *Tylopsis* by the completely exposed fore tibial tympana.

SONG (Figs 38–46. CD 1, track 1). The faint calling song of the male, produced mainly in the evening and at night, consists of isolated syllables, often separated by irregular intervals but sometimes repeated more regularly at intervals of about 1–4 s. Occasionally, well-defined echemes are also produced (Figs 39, 40), each composed of about 8–20 syllables repeated at the rate of about 6–10/s. Each syllable in such an echeme lasts for about 50–100 ms, but isolated syllables are often shorter than 50 ms. Heller (1988) has shown that *Phaneroptera* is most unusual among Tettigoniidae in producing sound only during the opening strokes of the fore wings, so that all the song components are opening hemisyllables.

DISTRIBUTION. Quite widespread in Europe north of the Mediterranean Region and Black Sea, as far north as the extreme south of the Netherlands, central Germany and southern Poland, but absent from the British Isles. In the south it occurs in northern Portugal, the extreme north of Spain and Italy, and the northern part of the Balkan Peninsula. Eastwards, the range extends through central Asia to China and Japan.

Phaneroptera nana Fieber

Phaneroptera nana Fieber, 1853: 173.

Southern Sickle-bearing Bush-cricket; (F) Phanéroptère méridional; (D) Vierpunktige Sichelschrecke.

REFERENCES TO SONG. **Oscillogram**: Heller, 1988; Schmidt, 1996; Zhantiev & Korsunovskaya, 1986. **Diagram**: Zhantiev & Korsunovskaya, 1986. **Wing-movement** and **frequency information**: Heller, 1988. **Verbal description only**: Bellmann & Luquet, 1995; Harz, 1957. **Disc recording**: Bonnet, 1995 (CD).

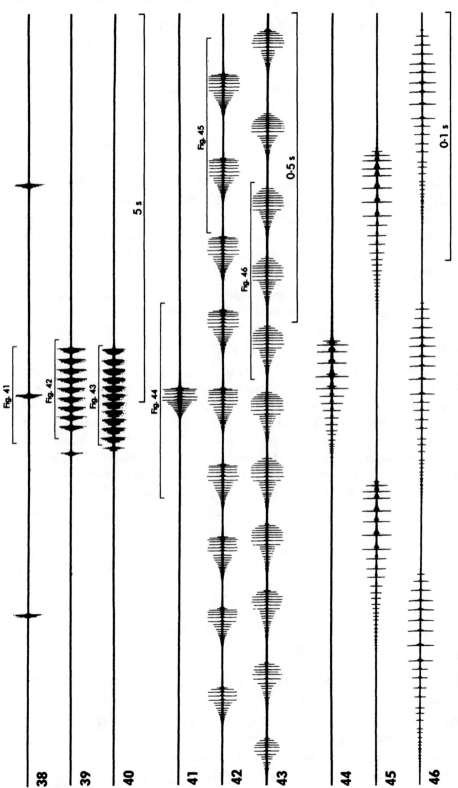

Figures 38–46 Oscillograms at three different speeds of the calling songs of two males of *Phaneroptera falcata.* Figs 38 and 39 show, respectively, isolated syllables and an echeme produced by the same male.

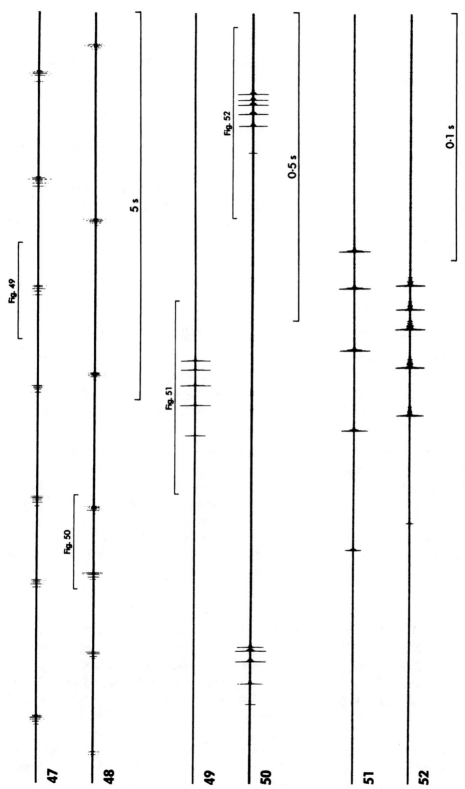

Figures 47–52 Oscillograms at three different speeds of the calling songs of two males of *Phaneroptera nana*.

RECOGNITION. Very similar to *P. falcata*, but can be distinguished from it by the pronotal lateral lobes, which are distinctly deeper than long.

SONG (Figs 47–52. CD 1, track 2). The male calling song, heard mainly at dusk and during the night, is very similar to that of *P. falcata*, consisting of isolated syllables repeated, often quite regularly, at intervals usually within the range 1–3 s. Each syllable lasts for about 50–150 ms and usually consists of 2–7 tooth-impacts. Heller (1988) has shown that, as in *P. falcata*, the sound is produced entirely by the opening stroke of the fore wings.

We have not heard this species produce the clearly-defined echemes that seem to be a normal component of the calling song of *P. falcata*.

DISTRIBUTION. Found throughout the Mediterranean Region, including North Africa, and also occurring in parts of central and northern France, Switzerland and Austria. It is widespread in all the southern peninsulas, and the range extends eastwards through southern Ukraine and Asia Minor to the Caucasus. Absent from the British Isles.

Tylopsis lilifolia (Fabricius)

Locusta lilifolia Fabricius, 1793: 36.

Slender Bush-cricket; (F) Phanéroptère liliacé; (D) Lilienblatt-Sichelschrecke.

REFERENCES TO SONG. **Oscillogram**: Heller, 1988; Schmidt, 1996; Zhantiev & Korsunovskaya, 1986, 1990. **Diagram**: Zhantiev & Korsunovskaya, 1986. **Wing-movement** and **frequency information**: Heller, 1988. **Verbal description only**: Bellmann & Luquet, 1995; Chopard, 1922 (as *thymifolia*); Gerhardt, 1914; Harz, 1957; Ragge, 1974; Yersin, 1857. **Disc recording**: Bonnet, 1995 (CD).

RECOGNITION (Plate 1: 2). Superficially this species is quite similar to *Phaneroptera*, being lightly built and fully winged, with the hind wings extending well beyond the fore wings. *Tylopsis*, however, has slit-like tympanal apertures on the fore tibiae, unlike the exposed, oval tympana of *Phaneroptera*. It is brown, straw-coloured or green in general colour, whereas European adults of *Phaneroptera* are almost always green.

In the field the short echemes of 'tick'-like syllables that typify the day-time song of male *T. lilifolia* are quite characteristic, but when the syllables are more widely spaced and repeated for longer periods (mainly at night) the song could be confused with that of *Phaneroptera nana*.

For a note on the spelling of the specific name of this species see p. 10.

SONG (Figs 53–61. CD 1, track 3). The male calling song usually consists of short echemes of about 2–5 'tick'-like syllables. The syllables in each echeme are separated by intervals of about 300–700 ms and the echemes are repeated at rather irregular and much longer intervals, usually between 10 and 70 s. Each syllable is very short-lived, usually lasting less than 10 ms, and seems to consist of only a few tooth-impacts, usually 3–6 but occasionally up to 10 or more. Sometimes one or more (most often the first) of the syllables in an echeme lasts rather longer, up to 100 ms, with the component sounds more widely spaced.

One Corsican male studied produced (in darkness) long sequences of ungrouped syllables repeated at intervals of about 1–2 s (Figs 55, 58, 61). Each syllable lasted 50–70 ms.

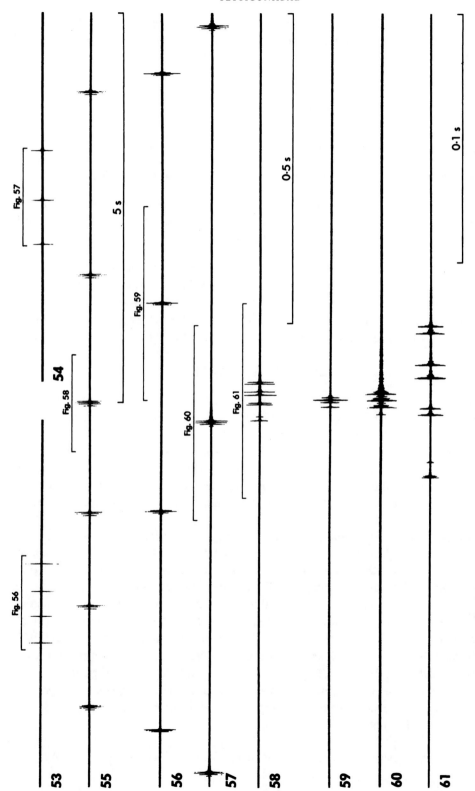

Figures 53–61 Oscillograms at three different speeds of the calling songs of three males of *Tylopsis lilifolia*.

DISTRIBUTION. Widespread and common in the Mediterranean Region, including North Africa and many Mediterranean islands, and extending eastwards into southern Russia and through Turkey into Iran.

Acrometopa servillea Brullé

Phaneroptera servillea Brullé, 1832: 86.

Long-legged Bush-cricket; (F) Phanéroptère corse; (D) Langbeinige Sichelschrecke.

REFERENCES TO SONG. **Oscillogram** and **wing-movement**: Heller, 1988, 1990. **Frequency information**: Heller, 1988. **Verbal description only**: Bellmann & Luquet, 1995; Burr *et al.*, 1923; Krauss, 1879 (as *macropoda*); Willemse, 1979.

RECOGNITION. This large, green, long-legged species may be readily distinguished from other fully winged phaneropterines in western Europe by its relatively broad fore wings. The hind wings of the male protrude beyond the fore wings for a short distance, but those of the female are completely covered by the fore wings when at rest. The tympana on the fore tibiae are covered by auricles, which are rather dilated in the male, less so in the female.

The song, a dense echeme with a final 'click', is highly characteristic.

SONG (Figs 62–67. CD 1, track 4). In warm conditions the male calling song, heard most often in the evening and at night, is a dense echeme of diplosyllables followed immediately by a characteristic 'click'. The echeme usually lasts between 1 s and 3 s, but at very high temperatures can be as short as 0·5 s. The syllables are repeated at the rate of about 30–50/s, usually accelerating slightly and becoming louder during the course of the echeme. In the songs studied the number of syllables in one echeme varied from 20 to 90. The echemes are sometimes repeated at irregular intervals, but can be repeated more regularly for short periods.

Oscillographic analysis shows that there are opening and closing hemisyllables, sometimes of about equal loudness and duration (especially towards the end of the echeme), but more often (at least in the earlier part of the echeme) the opening hemisyllables are slightly louder and longer than the closing ones. Each diplosyllable usually lasts 20–30 ms, but in hot conditions can be a little shorter than this. The main echeme usually ends in a slightly prolonged closing hemisyllable; there is then a very brief interval (about 20–80 ms) followed by a further, 'click'-like, closing hemisyllable, usually lasting 15–50 ms. Heller (1988) has shown that the fore wings are vibrated in a relatively open position during the main echeme and that the final 'click' is produced when they return to a closed position.

Willemse (1979: 136), referring to *Acrometopa* in general, gives the duration of an echeme as 5–10 s. Our own experience suggests that *A. servillea* would produce such long echemes only in very cool conditions.

DISTRIBUTION. Corsica, Sardinia, Italy (including Sicily), Balkan Peninsula and Asia Minor.

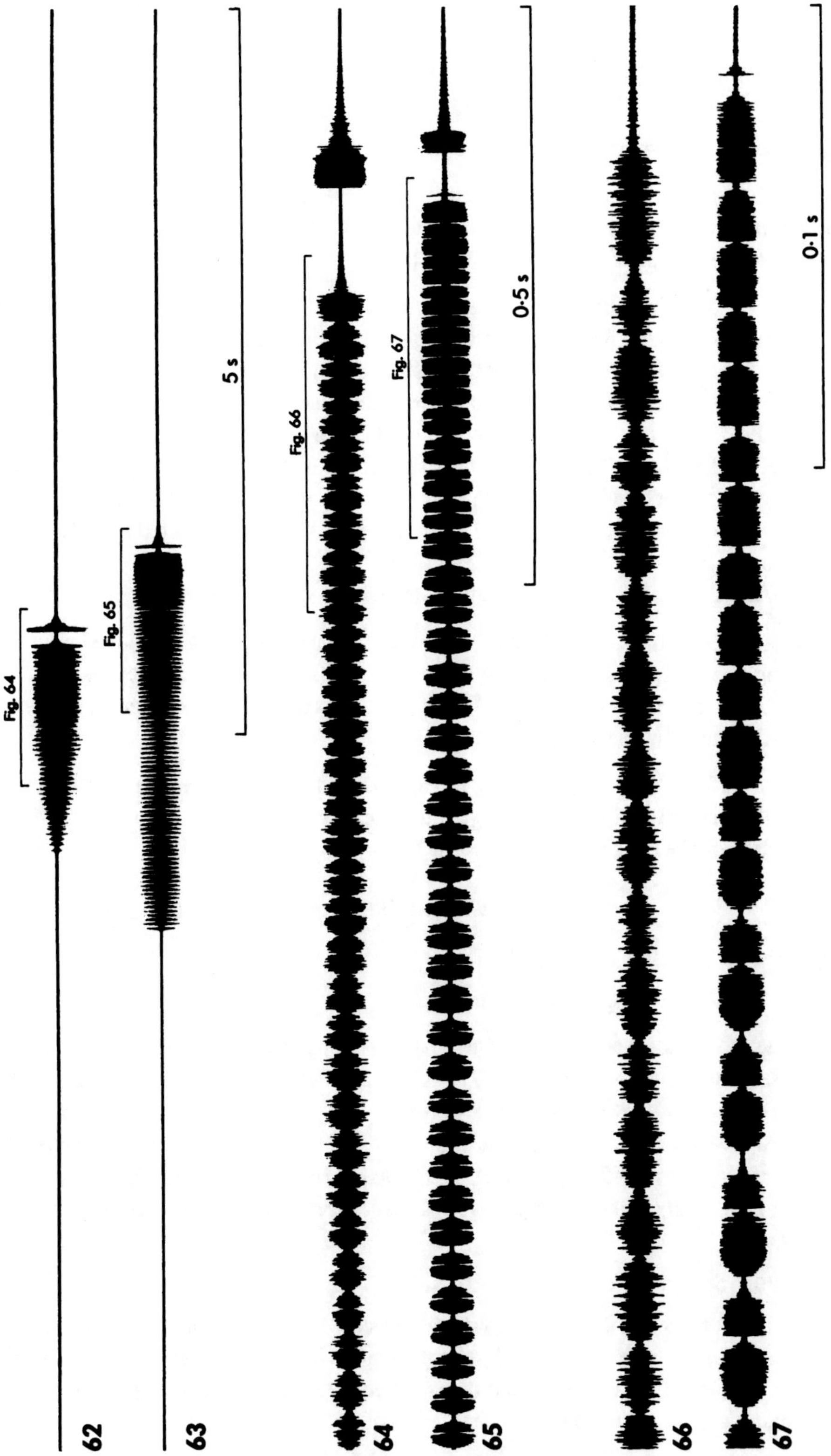

Figures 62–67 Oscillograms at three different speeds of the calling songs of two males of *Acrometopa servillea*.

Isophya pyrenaea (Serville)

Barbitistes pyrenæa Serville, 1838: 481 (but note that this spelling of the specific name appears on p. 766 – see p. 10 of the present work).

Large Speckled Bush-cricket; (F) Barbitiste des Pyrénées.

REFERENCES TO SONG. **Oscillogram**: Heller, 1988. **Diagram**: Duijm & Kruseman, 1983. **Verbal description only**: Bellmann & Luquet, 1995; Béringuier, 1908; Chopard, 1952; Heller, 1990. **Disc recording**: Bonnet, 1995 (CD).

RECOGNITION. The fore wings of *Isophya* are much reduced to a small stridulatory organ in the male and to even smaller flaps, also with a stridulatory function, in the female. *Isophya* may be distinguished from the very similar genus *Barbitistes* by the male subgenital plate, which has no median keel, and the ovipositor, which has a regularly curved ventral margin. *I. pyrenaea* is very similar morphologically to *I. kraussii*, especially in the female. The male subgenital plate is a little broader at the tip than that of *I. kraussii*, but the only reliable morphological distinction is in the stridulatory file: in *I. pyrenaea* there are fewer than 100 teeth and in *I. kraussii* more than 200. Live males may be easily separated by the different calling songs.

For a note on the spelling of the specific name of this species see p. 10.

SONG (Figs 68, 69, 71, 72. CD 1, track 5). The quiet male calling song, produced mainly in the evening and at night, is a series of syllables repeated at the rate of about 0·6–1·5/s. The main part of each syllable lasts about 70–120 ms and is often followed after an interval of about 80–250 ms by a very brief 'after-tick'.

DISTRIBUTION. The Pyrenees (including the Spanish side) and southern France generally; probably also central and eastern Europe but, because of confusion with *I. kraussii*, the eastern extent of the range is at present uncertain.

Isophya kraussii Brunner

Isophya kraussii Brunner, 1878: 65.

Krauss's Bush-cricket; (D) Plumpschrecke.

REFERENCES TO SONG. **Oscillogram**: Grein, 1984 (as *pyrenaea*); Heller, 1988, 1990; Ingrisch, 1991; Schmidt & Baumgarten, 1977 (as *p.*). **Diagram**: Bellmann, 1985a (as *p.*), 1988 (as *p.*), 1993a; Bellmann & Luquet, 1995; Schroth, 1987 (as *p.*). **Wing-movement**: Heller, 1988, 1990. **Frequency information**: Heller, 1988; Froehlich, 1989 (as *p.*). **Verbal description only**: Beier, 1955 (as *p.*); Faber, 1953a (as *p.*); Harz, 1957 (as *p.*). **Disc recording**: Bellmann, 1993c (CD); Grein, 1984 (LP, as *p.*). **Cassette recording**: Bellmann, 1985b (as *p.*), 1993b.

RECOGNITION. See under *I. pyrenaea*.

SONG (Figs 70, 73. CD 1, track 6). The male calling song is similar in general pattern to that of *I. pyrenaea*, but the main part of each syllable is much longer, about 250–330 ms, and the syllables are repeated more rapidly, about 2/s. 'After-ticks' are also usually present, following the main part of the syllable after an interval of about 70–130 ms.

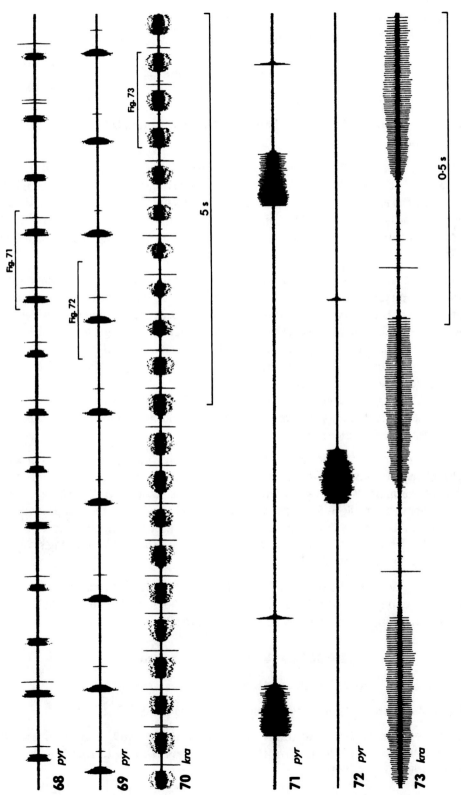

Figures 68–73 Oscillograms at two different speeds of the male calling songs of (68, 69, 71, 72) *Isophya pyrenaea* and (70, 73) *I. kraussii*.

DISTRIBUTION. Southern Germany, Austria and doubtless further east, but the eastern extent of the range as at present uncertain because of confusion with *I. pyrenaea*.

Barbitistes serricauda (Fabricius)

Locusta serricauda Fabricius, 1798: 193.

Saw-tailed Bush-cricket; (F) Barbitiste des bois; (D) Laubholz-Säbelschrecke; (NL) Zaagsprinkhaan.

REFERENCES TO SONG. **Oscillogram**: Heller, 1988; Kleukers *et al.*, 1997. **Diagram**: Bellmann, 1985*a*, 1988, 1993*a*; Bellmann & Luquet, 1995; Duijm & Kruseman, 1983. **Wing-movement**: Heller, 1988. **Frequency information**: Froehlich, 1989; Heller, 1988. **Verbal description only**: Faber, 1928, 1953*a*; Harz, 1957; Heller, 1990. **Disc recording**: Bellmann, 1993*c* (CD); Bonnet, 1995 (CD); Odé, 1997 (CD). **Cassette recording**: Bellmann, 1985*b*, 1993*b*.

RECOGNITION. *Barbitistes* is very similar to *Isophya*, but the male subgenital plate has a median keel and the ventral margin of the ovipositor is straight or almost so for part of its length. Males of *B. serricauda* may be distinguished from *B. obtusus* by their cerci, which are drawn out into a fine tip (thicker and blunter at the tip in *B. obtusus*) and from *B. fischeri* by the subgenital plate, which lacks the very prominent, protruding keel shown by that species. The females of these three species can be distinguished only by difficult and rather unreliable characters.

The male calling song is completely different from that of *B. fischeri*; although broadly similar in pattern to that of *B. obtusus*, it is composed of syllables grouped in twos, threes or fours, rather than fives and sixes.

SONG (Figs 74, 77, 80. CD 1, track 7). The quiet and mainly nocturnal male calling song consists of sequences of 'tick'-like syllables lasting about 3–5 s and separated by intervals of about 4 s or more. Within each sequence the syllables are grouped in a characteristic rhythm. There are usually 3–5 groups, each containing 1–4 syllables; the groups are separated by intervals of about 400–800 ms, and within each group the syllables are separated by intervals of about 60–120 ms. Each 'tick'-like syllable is produced by a closing stroke of the fore wings lasting about 6–10 ms.

DISTRIBUTION. Southern and eastern France, Belgium, Germany and eastwards through central Europe to the Ukraine. The southern limit runs from the Pyrenees (including the Spanish side) through northern Italy and Slovenia to the Black Sea.

Barbitistes obtusus Targioni-Tozzetti

Barbitistes obtusus Targioni-Tozzetti, 1881: 183.

Blunt Bush-cricket; (F) Barbitiste empr.. (D) Südalpen-Säbelschrecke.

REFERENCES TO SONG. **Oscillogram**: Heller, 1988; Schmidt, 1989. **Wing-movement**: Heller, 1988. **Frequency information**: Heller, 1988. **Verbal description only**: Bellmann & Luquet, 1995; Heller, 1990.

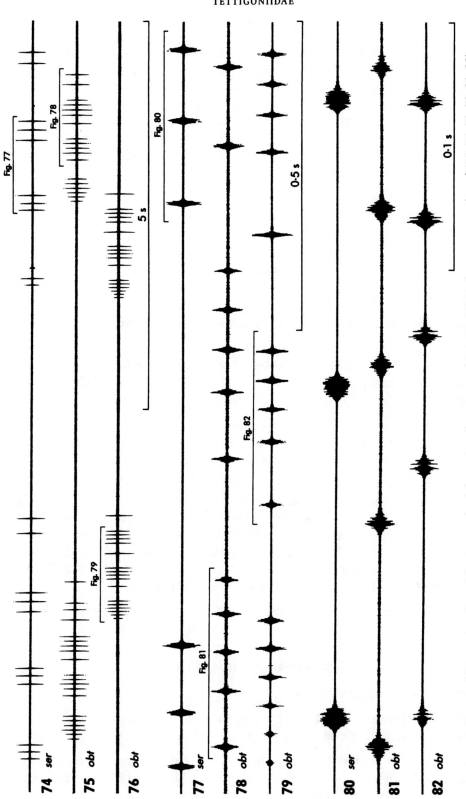

Figures 74–82 Oscillograms at three different speeds of the male calling songs of (74, 77, 80) *Barbitistes serricauda* and (75, 76, 78, 79, 81, 82) *B. obtusus.*

RECOGNITION. See under *B. serricauda* and *B. fischeri*.

SONG (Figs 75, 76, 78, 79, 81, 82. CD 1, track 8). The male calling song is similar to that of *B. serricauda* except for the shorter duration of the sequence, usually 1–2 s, and the way in which the syllables are grouped within each sequence. In *B. obtusus* most of the groups are of five or six syllables, though the last one or two groups in each sequence are often of only one or two. Sometimes the first syllable in one or more of the larger groups is separated from the others by a rather longer interval, so that it could be regarded as an ungrouped syllable. The intervals between groups are of about 140–270 ms and between syllables within a group about 25–120 ms; each syllable is produced by a closing stroke lasting about 6–13 ms.

DISTRIBUTION. Known only from the French, Swiss and Italian Alps, and southwards in Italy as far as the central Apennines.

Barbitistes fischeri (Yersin)

Odontura fischeri Yersin, 1984*a*: 66.

Southern Saw-tailed Bush-cricket; (F) Barbitiste languedocien; (D) Südfranzösische Säbelschrecke.

REFERENCES TO SONG. **Oscillogram**: Heller, 1988. **Frequency information**: Dumortier, 1963*c*. **Verbal description only**: Bellmann & Luquet, 1995; Chopard, 1922 (as *béringuieri*); Heller, 1988. **Disc recording**: Andrieu & Dumortier, 1963 (LP), 1994 (CD); Bonnet, 1995 (CD).

RECOGNITION. The very prominent median keel on the subgenital plate enables males of this species to be easily distinguished from *B. serricauda* and *B. obtusus*.

SONG (Figs 83–88. CD 1, track 9). The quiet male calling song, produced mainly in the evening and at night, consists of long sequences of 'tick'-like syllables, some a little louder than others. Sometimes the louder ones alternate with the softer ones, and often a pair of softer ones alternates with a single louder one (as in Figs 84, 86, 88); sometimes the sequence is less regular. The sequences last for up to 20 s or longer and the overall syllable repetition rate is about 6–8/s; sometimes there is a short sequence of syllables of uniform loudness repeated at about half this rate. Each syllable lasts about 10–15 ms.

DISTRIBUTION. Southern France, the northern half of the Iberian Peninsula and the western Italian Alps.

Leptophyes punctatissima (Bosc)

Locusta punctatissima Bosc, 1792*b*: 45.

Speckled Bush-cricket; (F) Leptophye ponctuée; (D) Punktierte Zartschrecke; (NL) Struiksprinkhaan; (DK) Krumknivgræshoppe; (S) Lövvårtbitare.

REFERENCES TO SONG. **Oscillogram**: Ahlén, 1981; Ahlén & Degn, 1980; Hartley & Robinson, 1976; Heller, 1988; Heller & Willemse, 1989; Kleukers *et al.*, 1997; Robinson *et*

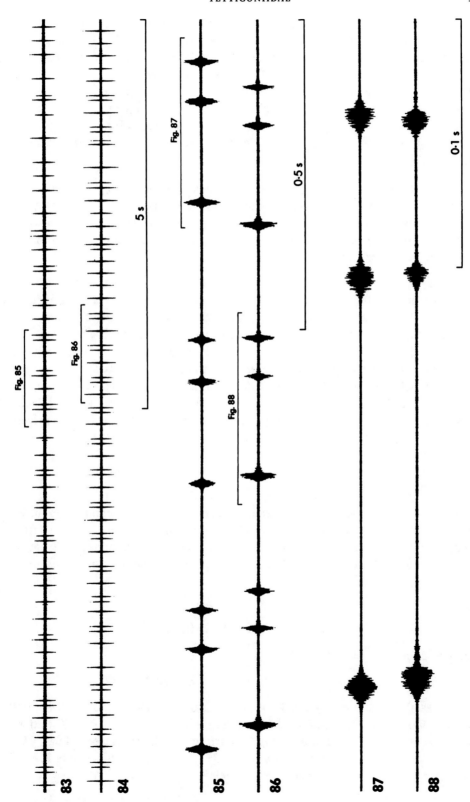

Figures 83–88 Oscillograms at three different speeds of the calling songs of two males of *Barbitistes fischeri*.

al., 1986; Schmidt, 1989; Zimmermann *et al.*, 1989. **Diagram**: Bellmann, 1985*a*, 1988, 1993*a*; Bellmann & Luquet, 1995; Duijm & Kruseman, 1983; Holst, 1970; Ragge, 1965. **Sonagram**: Hartley & Robinson, 1976; Sales & Pye, 1974 (identity uncertain). **Wing-movement**: Heller, 1988; Heller & Willemse, 1989. **Frequency information**: Ahlén, 1981; Ahlén & Degn, 1980; Froehlich, 1989; Heller, 1988; Robinson, 1980; Robinson *et al.*, 1986. **Verbal description only**: Baier, 1930; Broughton, 1952*a*, 1972*b*; Chopard, 1922; Faber, 1953*a*; Harz, 1957; Manzies & Shaw, 1947. **Disc recording**: Andrieu & Dumortier, 1994 (CD); Bellmann, 1993c (CD); Odé, 1997 (CD). **Cassette recording**: Ahlén, 1982; Bellmann, 1985*b*, 1993*b*; Burton & Ragge, 1987.

RECOGNITION (Plate 1: 3). *Leptophyes*, yet another genus of small, brachypterous bush-crickets, has longer antennae than in *Isophya* and *Barbitistes*, usually more than three times the length of the body, and the ovipositor of the female is very finely toothed. *L. punctatissima* is a predominantly green insect, differing from the other species of *Leptophyes* studied by the characters mentioned under those species.

SONG (Figs 89, 90, 92, 93, 95, 96. CD 1, track 10). The male calling song consists of isolated faint 'tick'-like syllables repeated at very variable intervals. Sometimes they are repeated quite regularly every few seconds for long periods (intervals of 1–5 s are common), but at other times the intervals may be much longer and occasionally the syllables may be grouped into short sequences during which the repetition rate may be as high as 2/s; generally the syllables are repeated more rapidly at higher temperatures. The number of tooth-impacts within each syllable also varies greatly: 4–6 is common, but as few as 1 or as many as 8 sometimes occur. Frequently the main group of impacts (which is a closing hemisyllable) is followed after a variable interval (usually between 15 and 100 ms) by another one (rarely two), probably produced by a further closure of the fore wings. A hemisyllable of 4–6 impacts usually lasts between 8 and 20 ms.

The song is produced both during the day and at night, but seldom during the first few daylight hours of the morning.

DISTRIBUTION. This is the most widespread species of *Leptophyes*, occurring from the British Isles and southern Scandinavia through most of Europe to parts of the southern peninsulas and eastwards to the southern part of European Russia. In the British Isles it is very local in Ireland and the extreme south-west of Scotland, and common in southern Britain.

Leptophyes albovittata (Kollar)

Barbitistes albovittata Kollar, 1833: 76.

Striped Bush-cricket; (F) Leptophye rayée; (D) Gestreifte Zartschrecke.

REFERENCES TO SONG. **Oscillogram**: Heller, 1988; Zhantiev & Korsunovskaya, 1986, 1990. **Diagram**: Bellmann, 1985*a*, 1988, 1993*a*; Bellmann & Luquet, 1995; Zhantiev & Korsunovskaya, 1986. **Wing-movement** and **frequency information**: Heller, 1988. **Verbal description only**: Faber, 1953*a*; Harz, 1957. **Disc recording**: Bellmann, 1993c (CD). **Cassette recording**: Bellmann, 1985*b*, 1993*b*.

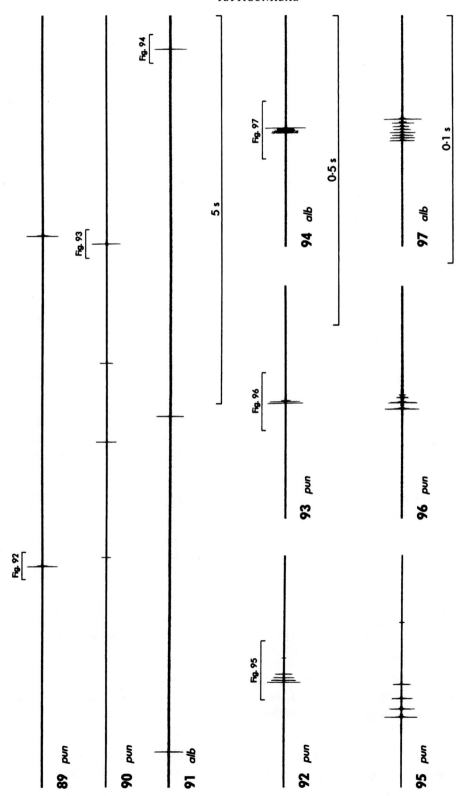

Figures 89–97 Oscillograms at three different speeds of the male calling songs of (89, 90, 92, 93, 95, 96) *Leptophyes punctatissima* and (91, 94, 97) *L. albovittata.*

RECOGNITION. Readily distinguished from *L. punctatissima* and *L. laticauda* by the conspicuous white stripe running along the side of the body from the lower edge of the pronotal lateral lobes to the tip of the abdomen. Conspicuously smaller than *L. boscii* (hind femora less than 15 mm; ovipositor less than 7 mm).

SONG (Figs 91, 94, 97. CD 1, track 11). The male calling song is extremely similar to that of *L. punctatissima* in almost every respect, although the number of tooth-impacts in each hemisyllable tends to be larger (commonly within the range 5–12).

DISTRIBUTION. From Germany (mainly southern and eastern), Austria and the extreme north of Italy eastwards through central Europe, the Balkan Peninsula, Ukraine, central and southern European Russia and Asia Minor to western Kazakhstan.

Leptophyes boscii Fieber

Leptophyes Boscii Fieber, 1853: 260.

Eastern Speckled Bush-cricket; (F) Leptophye sarmate; (D) Boscs Zartschrecke.

REFERENCES TO SONG. **Oscillogram, wing-movement** and **frequency information**: Heller, 1988. **Verbal description only**: Faber, 1953a.

RECOGNITION. A whitish yellow lateral stripe, rather similar to that of *L. albovittata*, distinguishes this species from *L. punctatissima* and *L. laticauda*. It is noticeably larger than *L. albovittata* (hind femora more than 15 mm; ovipositor more than 7 mm).

SONG (Figs 98, 99, 101, 102, 104, 105. CD 1, track 12). Very similar to those of *L. punctatissima* and *L. albovittata*. In the calling songs of the only two males we have studied, the hemisyllables were repeated every 5–10 s, lasted about 20–27 ms and were composed of about 10–17 tooth-impacts.

DISTRIBUTION. From northern Italy, Austria and the Czech Republic eastwards through much of the Balkan Peninsula to the Ukraine.

Leptophyes laticauda (Frivaldsky)

Odontura laticauda Frivaldsky, 1867: 102.

Long-tailed Bush-cricket; (F) Leptophye provençale; (D) Südliche Zartschrecke.

REFERENCES TO SONG. **Oscillogram**: Heller, 1988; Schmidt, 1989. **Wing-movement** and **frequency information**: Heller, 1988.

RECOGNITION. Quite similar to *L. punctatissima* in colour, but males differ from that species in having an emarginate tip to the subgenital plate (truncate in *L. punctatissima*) and females in having a larger ovipositor (more than 9 mm in *L. laticauda*, less than 9 mm in *L. punctatissima*).

SONG (Figs 100, 103, 106. CD 1, track 13). Similar to those of the other *Leptophyes* species studied, but the syllable structure is often more complex, so that there are sometimes up to

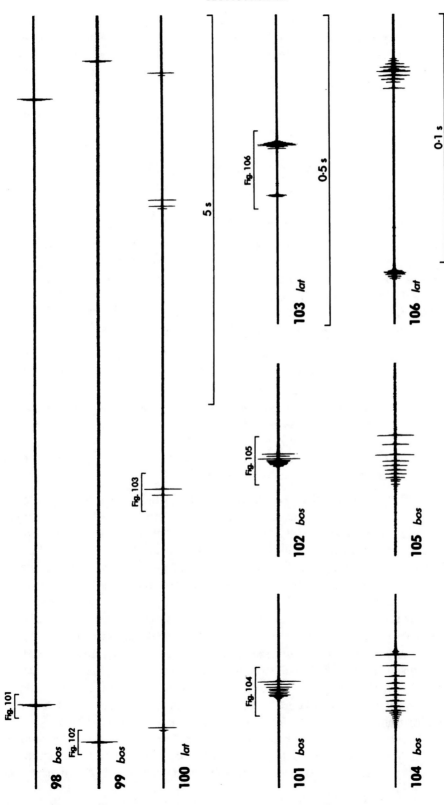

Figures 98–106 Oscillograms at three different speeds of the male calling songs of (98, 99, 101, 102, 104, 105) *Leptophyes boscii* and (100, 103, 106) *L. laticauda.*

three or more groups of tooth-impacts separated by intervals of 15–35 ms; in such cases each group is composed of about 3–10 tooth-impacts and lasts about 3–10 ms. In the males studied the syllables (whether simple or of the more complex kind described above) were repeated at intervals of 1–2 s, although Schmidt (1989) gives the much higher repetition rate of 8/s.

DISTRIBUTION. South-east France, Switzerland, northern Italy and the northern half of the Balkan Peninsula.

Polysarcus denticauda (Charpentier)

Barbitistes denticauda Charpentier, 1825: 99.

Large Saw-tailed Bush-cricket; (F) Barbitiste ventru; (D) Wantschrecke.

REFERENCES TO SONG. **Oscillogram**: Grein, 1984; Heller, 1988, 1990. **Diagram**: Bellmann, 1985*a*, 1988, 1993*a*; Bellmann & Luquet, 1995. **Wing-movement**: Heller, 1988, 1990. **Frequency information**: Heller, 1988. **Musical notation**: Faber, 1953*a*, 1953*b*; Yersin, 1854*b*. **Verbal description only**: Defaut, 1988*b*; Faber, 1928; Harz, 1957. **Disc recording**: Andrieu & Dumortier, 1963 (LP), 1994 (CD); Bellmann, 1993*c* (CD); Bonnet, 1995 (CD); Grein, 1984 (LP). **Cassette recording**: Bellmann, 1985*b*, 1993*b*.

RECOGNITION (Plate 1: 4). The two European species of *Polysarcus* are large, almost wingless insects that could be confused with the superficially similar but quite unrelated genus *Ephippiger*; *Polysarcus*, however, has exposed, oval tympana on the fore tibiae, unlike the slit-like tympanal apertures of *Ephippiger*. *P. denticauda* can be distinguished from *P. scutatus* by its broader vertex (more than twice the width of the first antennal segment in *P. denticauda*, less than twice the width of this segment in *P. scutatus*).

The loud male calling song provides an easy field character for distinguishing this species from *P. scutatus* and any other bush-cricket.

SONG (Figs 107–116. CD 1, track 14). The calling song of the male, produced mainly in warm sunshine, is loud, rather complex and highly characteristic. Typically it consists of a prolonged buzz of rapidly repeated syllables (often produced while the male moves through the vegetation) interrupted at long intervals (often of 1–3 minutes) by a series of loud 'clicks'. The syllable repetition rate is usually within the range 15–35/s for most of the buzzing phase, but for a few seconds before the clicks are produced the rate is increased to about double this value, ranging from 25/s to 70/s or even faster. Oscillographic analysis shows that, during at least the prolonged slower buzz, each syllable is composed of two quite distinct sounds so that it appears to be a diplosyllable; Heller (1988) has shown, however, that both sounds are produced during the closing stroke of the fore wings, the opening stroke being very rapid and silent. This dual structure of the closing hemisyllable often persists during the more rapid buzz that precedes the clicks, but sometimes these more rapidly repeated hemisyllables become almost or completely uniform bursts of sound. The slow hemisyllables usually last about 20–50 ms and the faster ones about 10–30 ms.

The acceleration in syllable repetition rate preceding the clicks usually begins about 5–10 s before the first click, and generally lasts for no more than a second or two until the maximum rate is reached. Sometimes there is only a single click (Fig. 112), but more often

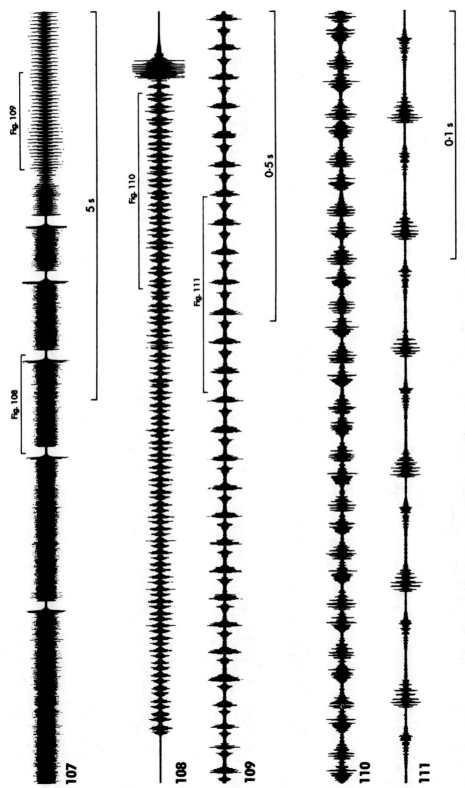

Figures 107–111 Oscillograms at three different speeds of the calling song of a male of *Polysarcus denticauda*.

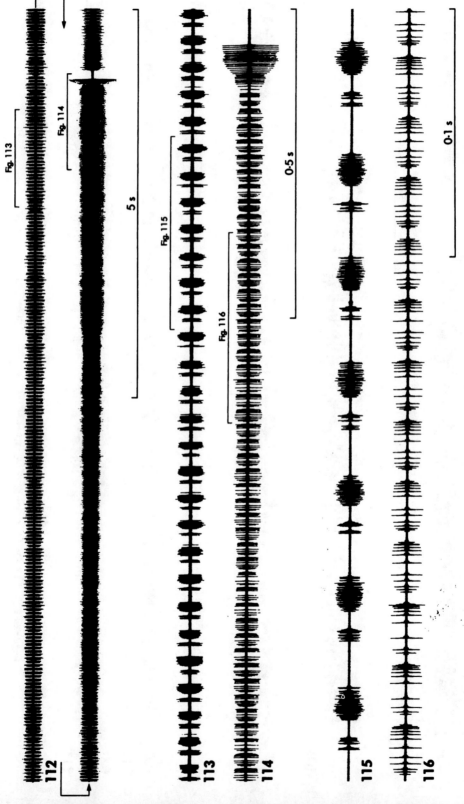

Figures 112–116 Oscillograms at three different speeds of the calling song of another male of *Polysarcus denticauda*. Note that the oscillogram shown in Fig. 112 is split between two lines.

there are a series of clicks (often 2–6, but sometimes as many as 20 or more) with the buzzing continuing between them (Fig. 107). When there are only a few clicks, they are repeated at ever decreasing intervals (the first interval usually being about 1–2 s); each click is followed by a very brief pause (usually within the range 50–150 ms) and then the high-speed buzzing is resumed until the next click. When there are a large number of clicks, the later ones are usually repeated at a fairly constant rate (often about 5–7/s), and the buzzing between them becomes reduced to the slower of the two rates or even slower.

After the last click there is sometimes a pause of several seconds before the cycle begins again with a long period of slow-rate buzzing, but more often there is only a very short interval (less than 200 ms) and sometimes, especially if there are a large number of clicks, the buzzing follows immediately after the last click with no pause whatever. Sometimes there is a short period (usually about 1 s) of rapid buzzing after the last click before the syllable repetition rate is reduced to the slower rate.

Heller (1988) has shown that during the buzzing phase the fore wings are vibrated in a relatively open position, and the clicks are produced by the fore wings being moved to a much more closed position, each click being a closing hemisyllable using a more proximal part of the file.

DISTRIBUTION. Typically an insect of upland meadows, this species occurs in the French, southern German and central European uplands, much of the Alps and Apennines, and eastwards through parts of the Balkan Peninsula to the Carpathians.

Polysarcus scutatus (Brunner)

Orphania scutata Brunner, 1878: 256.

Shielded Bush-cricket; (F) Barbitiste à bouclier.

REFERENCES TO SONG. **Oscillogram** and **frequency information**: Heller, 1988. **Verbal description only**: Chopard, 1952 (from Pussard, 1942); Heller, 1990. **Disc recording**: Bonnet, 1995 (CD).

RECOGNITION. See under *P. denticauda*. In the field the male calling song is again very distinctive.

SONG (Figs 117–126. CD 1, track 15). The calling song of the male, often produced in prolonged bursts of up to a minute or more, is composed of two different kinds of sound, one quieter and one louder. The quieter sound consists of a long sequence of identical echemes following one another in quick succession at the rate of about 1–2/s, and often showing a gradual slight increase in loudness. Each echeme is composed of about 12–30 syllables repeated at the rate of about 35–40/s and ends in a rather longer and louder syllable and a final 'click'. Each syllable in the main part of the echeme lasts about 12–20 ms and the final one, including the click, lasts about 50–80 ms; there is then a brief pause of about 50–100 ms before the next echeme begins. The whole echeme, from the beginning to the click, lasts about 0·5–1·0 s.

The louder sound is an echeme of very variable duration (between 1 and 17 s in the songs studied), composed of syllables similar in duration to the quieter ones but repeated at

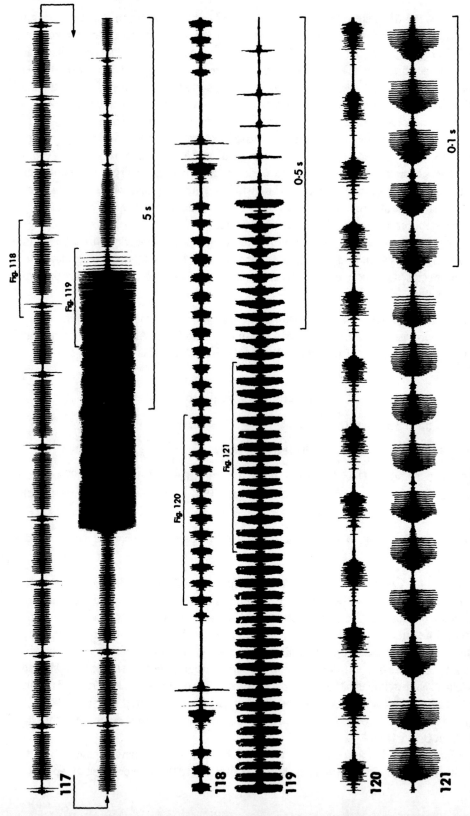

Figures 117–121 Oscillograms at three different speeds of the calling song of a male of *Polysarcus scutatus*. Note that the oscillogram shown in Fig. 117 is split between two lines.

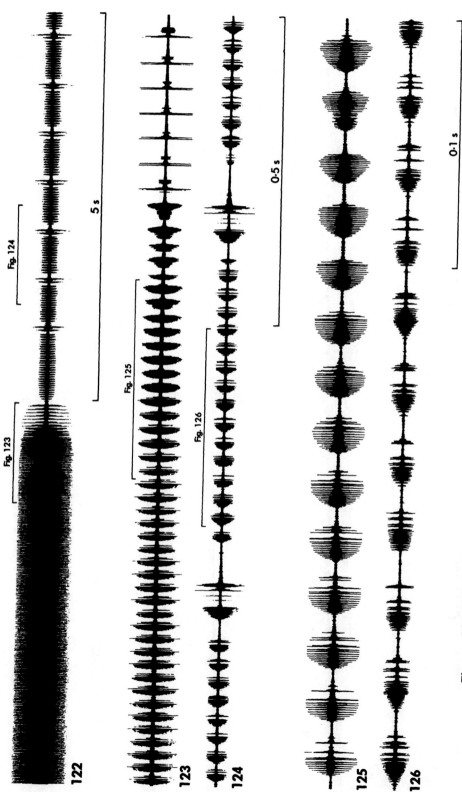

Figures 122–126 Oscillograms at three different speeds of the calling song of another male of *Polysarcus scutatus*.

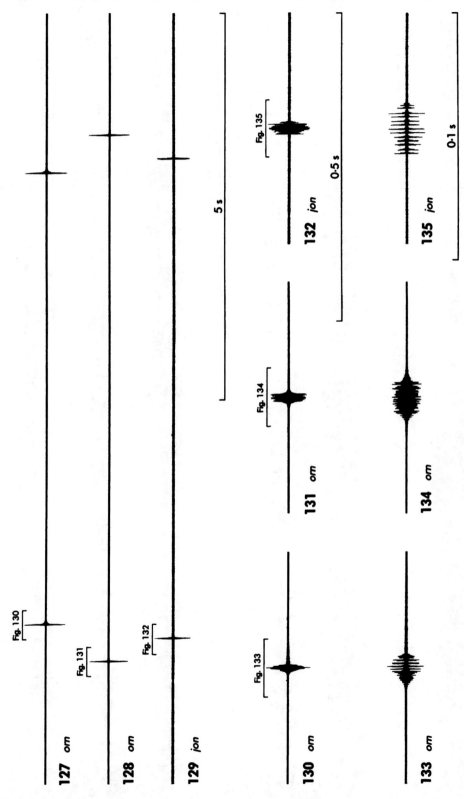

Figures 127–135 Oscillograms at three different speeds of the male calling songs of (127, 128, 130, 131, 133, 134) *Poecilimon ornatus* and (129, 132, 135) *P. jonicus.*

the rate of 45–55/s, and usually ending in a series of about 5–8 microsyllables. This louder echeme is usually preceded and followed by quieter echeme-sequences (as in Fig. 117), but occasionally it is produced quite separately.

As in *P. denticauda* all the syllables seem to be closing hemisyllables, the opening strokes of the fore wings producing little or no sound.

DISTRIBUTION. Found in similar habitats to *P. denticauda* (but much more rarely) in the Pyrenees, French Alps and parts of the Balkan Peminsula.

Poecilimon ornatus (Schmidt)

Ephippigera ornatus Schmidt, 1850: 184.

Ornate Bush-cricket; (D) Südliche Buntschrecke.

REFERENCES TO SONG. **Oscillogram, wing-movement** and **frequency information**: Heller, 1984, 1988; Heller & Helversen, 1986. **Verbal description only**: Heller, 1990; Harz, 1957.

RECOGNITION (Plate 1: 5). *Poecilimon* can be distinguished from the other small, brachypterous Phaneropterinae occurring in western Europe by the pronotum, which is extended backwards to cover about half the male fore wings and almost all of the female fore wings; the transverse sulcus crosses the mid-line about halfway along the pronotum or slightly nearer the front. The antennae are about three times the length of the body and the ovipositor of the female is coarsely toothed at the tip. *P. ornatus* differs from the only other species of *Poecilimon* included in the present study, *P. jonicus*, in being much larger (pronotum more than 6 mm in *P. ornatus*, less than 6 mm in *P. jonicus*).

SONG (Figs 127, 128, 130, 131, 133, 134. CD 1, track 16). The male calling song consists of faint 'tick'-like syllables repeated at very variable intervals (most often every 4–8 s). Each tick is a closing hemisyllable lasting about 20 ms.

DISTRIBUTION. From the Italian Alps and southern Austria through the extreme west of the Balkan Peninsula to Albania and northern Greece.

Poecilimon jonicus (Fieber)

Barbitistes jonicus Fieber, 1853: 175.

Ionian Bush-cricket.

REFERENCES TO SONG. **Oscillogram**: Heller, 1984, 1988; Schmidt, 1989. **Wing-movement** and **frequency information**: Heller, 1984, 1988. **Verbal description only**: Heller, 1990.

RECOGNITION. See under *P. ornatus*.

SONG (Figs 129, 132, 135. CD 1, track 17). The male calling song is very similar to that of *P. ornatus*: single 'tick'-like closing hemisyllables repeated every 4–8 s, or occasionally more or less frequently than this. As in *P. ornatus* each hemisyllable lasts about 20 ms.

DISTRIBUTION. *P. jonicus* occurs in western Europe only as the subspecies *P. j. superbus*

(Fischer), which is found in most of the Italian Peninsula south of the River Po. Other sub-species of *P. jonicus* occur in the southern part of the Balkan Peninsula.

Meconema thalassinum (De Geer)

Locusta thalassina De Geer, 1773: 433.

Oak Bush-cricket; (F) Méconème tambourinaire; (D) Gemeine Eichenschrecke; (NL) Boomsprinkhaan; (DK) Egegræshoppe; (S) Ekvårtbitare.

REFERENCES TO SONG. **Oscillogram**: Heller, 1988; Kleukers *et al.*, 1997. **Diagram**: Bellmann, 1985*a*, 1988, 1993*a*; Bellmann & Luquet, 1995; Duijm & Kruseman, 1983; Ragge, 1965; Sismondo, 1980. **Verbal description only**: Cappe de Baillon, 1921 (as *varium*); Chopard, 1952; Currie, 1953; Faber, 1928 (as *v.*); Gerhardt, 1914 (as *v.*); Harz, 1955, 1957; Holst, 1986; Ragge, 1973. **Disc recording**: Bellmann, 1993*c* (CD); Odé, 1997 (CD). **Cassette recording**: Bellmann, 1985*b*, 1993*b*; Burton & Ragge, 1987.

RECOGNITION (Plate 1: 6). A small, fully winged, pale green bush-cricket living on decidu-ous trees and strictly nocturnal in habits. The male has no obvious stridulatory organ on the fore wings and its cerci are conspicuously long and slender. The female has a relatively long, slightly upcurved ovipositor. The oval tympana on the fore tibiae are completely ex-posed.

SONG (Figs 136–139. CD 1, track 18). The remarkable nocturnal 'song' of this insect is not produced by stridulation but by a rapid drumming of one of the hind legs on the substrate. It is clearly intended for conduction through the substrate and so the airborne sound pro-duced is incidental and very quiet. To produce the sound the male raises both pairs of wings more or less perpendicularly to the body and vibrates one of the hind legs so that its tarsus is drummed on the surface of the leaf, twig or any other substrate that the male happens to be using. The drumming is in short bursts of up to a second or slightly longer and these are almost always in sequences of up to seven bursts, separated from other such sequences by intervals of at least several seconds (3–15 s in the songs studied). Each sequence is usually composed of a combination of shorter and longer bursts, with the shorter ones preceding the longer ones. Each shorter burst usually consists of 3–8 impacts and lasts for about 0·07–0·20 s, while each longer one usually consists of 30–50 impacts and lasts for about 0·7–1·3 s. The impact repetition rate, in both the short and long bursts, is usually 30–40/s at tem-peratures up to 25°C, but can be as much as 60/s or more at higher temperatures. The number of short and long bursts in one sequence is very variable: 2–3 short ones followed by 1 or 2 long ones is a common pattern, but there can be as many as 5 of either kind (though not in the same sequence) and some sequences lack short bursts completely. The intervals between successive bursts last about 0·2–2·0 s.

The quality of the airborne sound depends entirely on the substrate. Drumming on a twig is extremely faint, but if a suitable leaf is used the sound can sometimes be heard from a distance of several metres.

DISTRIBUTION. Very widespread in Europe from the British Isles (where it is rare in Ireland but common in southern Britain) to European Russia, and from southern Scandinavia to

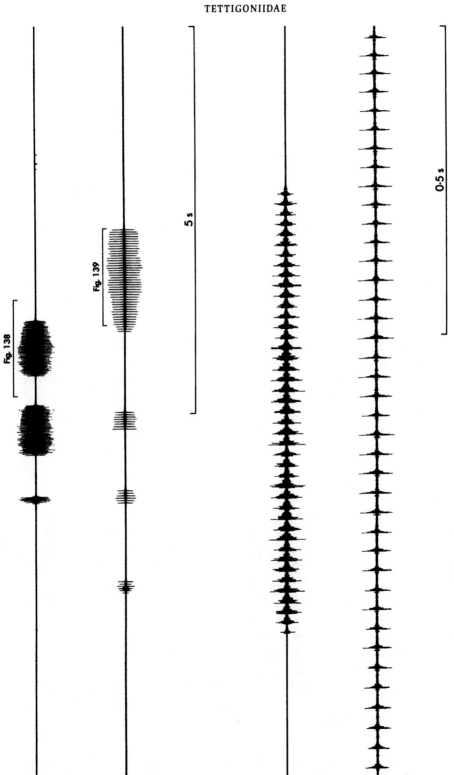

Figures 136–139 Oscillograms at two different speeds of the foot-drumming of two males of *Meconema thalassinum*. The drumming is on a plastic carton in Figs 136 and 138, and on a leaf in Figs 137 and 139.

the northern parts of the Iberian, Italian and Balkan Peninsulas. Introduced into north-eastern USA.

Conocephalus conocephalus (Linnaeus)

Gryllus (Tettigonia) conocephalus Linnaeus, 1767: 696.

Southern Cone-head; (F) Conocéphale africain.

REFERENCES TO SONG. **Oscillogram**: Heller, 1988; Schmidt, 1996. **Sonagram** and **frequency information**: Sales & Pye, 1974.

RECOGNITION. The conical head-shape of *Conocephalus* distinguishes it from all other European Tettigoniidae except *Ruspolia*, which is much bigger (more than 35 mm long, measured to the tips of the flexed fore wings). *C. conocephalus* can be distinguished from *C. discolor* and *C. dorsalis* by the lack of prosternal spines, the smooth-edged ovipositor (finely denticulate in the other two species) and the male cerci, on which the inner tooth is near the base (near the tip in the other two species).

SONG (Figs 140–146. CD 1, track 19). In its simplest form the very faint calling song, produced both by day and at night, is a series of single syllables repeated at rather irregular intervals, but more often it consists of a random mixture of single syllables and echemes. The number of syllables in the echemes is very variable, usually within the range 3–9, but they are repeated within each echeme at a very constant rate, usually within the range 11–14/s. Each syllable is normally a closing hemisyllable lasting about 40–65 ms, whether isolated or forming part of an echeme. In a song including echemes, the brief pauses between successive echemes or isolated syllables are usually of less than a second (commonly 0.2–0.8 s), but there are sometimes longer pauses of up to 10 s or more.

In a common variant of this simple song there is a mixture of quieter and louder syllables; both the quieter and the louder ones may be isolated or in echemes and, apart from a marked contrast in loudness, they are similar in structure and duration. In such songs there is usually a series of quiet echemes or ungrouped syllables followed immediately by a series of loud echemes or ungrouped syllables; there is then an interval of several seconds before another such sequence begins. Sometimes the duration of a sequence of this kind is about equally divided between the quieter and louder parts, but in the most highly developed song studied (shown in Fig. 141) the quieter part of each sequence lasted about 13–15 s and the louder part less than 2 s. In this song, as the oscillograms show, the louder part was heralded by a pair of very quiet syllables, and at this point in the song some of the opening strokes of the fore wings produced brief hemisyllables.

Sometimes soft and loud syllables are produced in pairs, a soft one being followed 1–3 s later by a loud one and then an interval of about 6–8 s before the next pair. Clearly this species has an interesting acoustic repertoire which merits a more comprehensive study.

DISTRIBUTION. A common and widespread African insect which occurs very locally in parts of southern Europe and Asia Minor, usually near the Mediterranean coast.

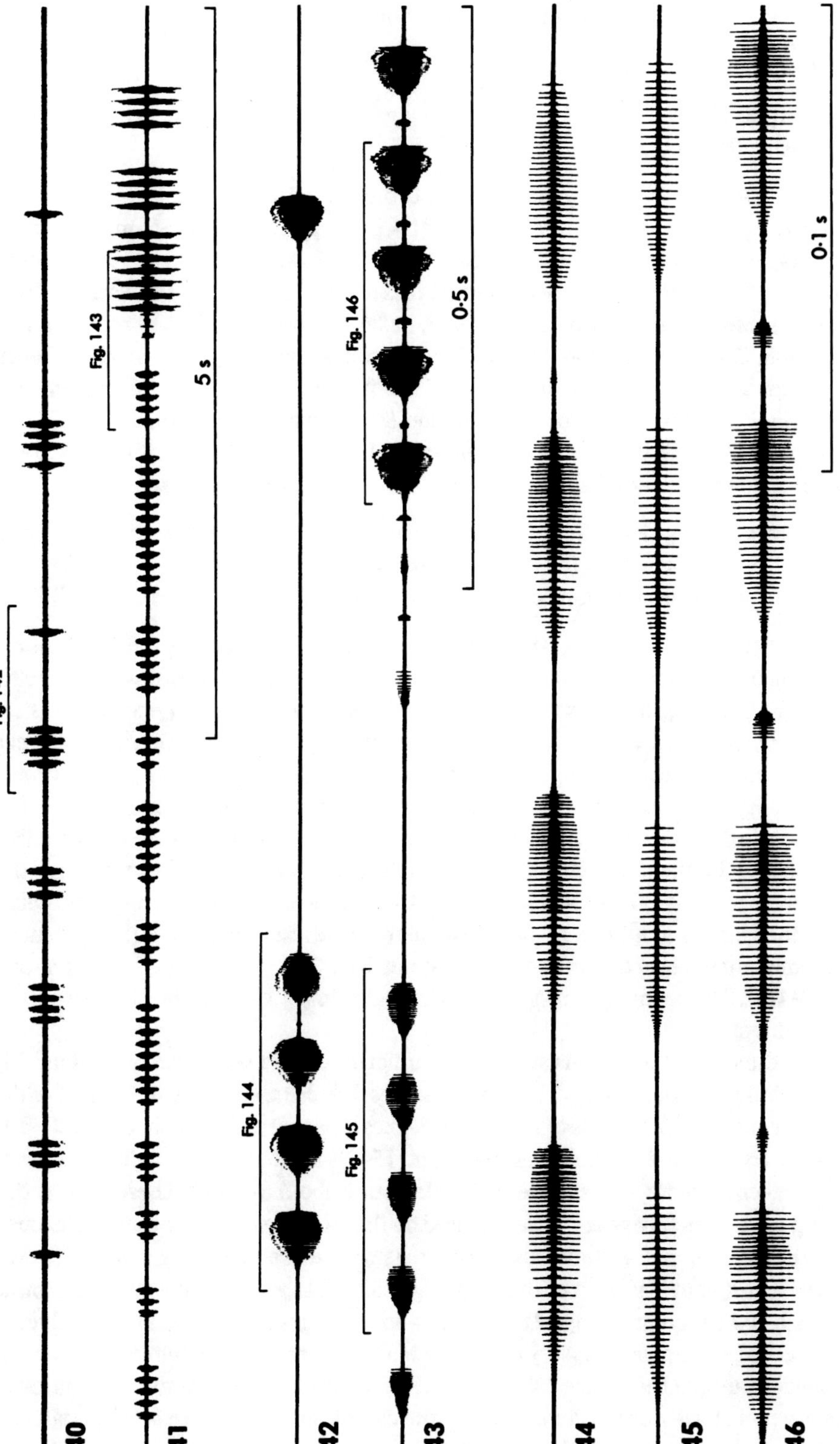

Figures 140–146 Oscillograms at three different speeds of the calling songs of two males of *Conocephalus conocephalus*.

Conocephalus discolor Thunberg

Conocephalus discolor Thunberg, 1815: 275.

Long-winged Cone-head; (F) Conocéphale bigarré; (D) Langflüglige Schwertschrecke; (NL) Zuidelijk Spitskopje.

REFERENCES TO SONG. **Oscillogram**: Bailey & Broughton, 1970; Grein, 1984; Heller, 1988; Kleukers *et al.*, 1997; Schmidt, 1996; Schmidt & Baumgarten, 1977; Schmidt & Schach, 1978. **Diagram**: Bellmann, 1985*a*, 1988, 1993*a*; Bellmann & Luquet, 1995; Duijm & Kruseman, 1983; Ragge, 1965. **Sonagram**: Sales & Pye, 1974. **Wing-movement**: Heller, 1988. **Frequency information**: Bailey, 1967; Bailey & Broughton, 1970; Froehlich, 1989; Heller, 1988; Sales & Pye, 1974. **Verbal description only**: Broughton, 1972*b*; Chopard, 1922 (as *fuscus*); Defaut, 1988*b* (as *f.*); Faber, 1928 (as *f.*); Harz, 1957 (as *f.*); Janssen, 1977; Yersin, 1854*b* (as *f.*). **Disc recording**: Andrieu & Dumortier, 1963 (LP), 1994 (CD, as *f.*); Bellmann, 1993*c* (CD); Bonnet, 1995 (CD); Grein, 1984 (LP); Odé, 1997 (CD); Ragge *et al.*, 1965 (LP). **Cassette recording**: Bellmann, 1985*b*, 1993*b*; Burton & Ragge, 1987.

RECOGNITION. Distinguished from *C. conocephalus* by the characters mentioned under that species and from *C. dorsalis* by the fully developed wings.

For a note on the specific name of this species see p. 10.

SONG (Figs 147–170. CD 1, track 20). The calling song, produced during the day-time, consists of long echeme-sequences lasting up to 4 minutes or more. The sound is faint, with a rustling or sizzling quality. Oscillographic analysis shows the sequences to consist of trisyllabic echemes with a highly characteristic pattern. The first two syllables of each echeme consist of short opening hemisyllables followed by longer closing hemisyllables; the first diplosyllable lasts about 12–21 ms and the second, following after a very short gap of 1–3 ms, is usually very slightly shorter (11–20 ms). There is then a longer gap, usually 3–13 ms, followed by the third syllable, again usually a diplosyllable but lasting longer than the first two, about 17–30 ms. Occasionally the third syllable lacks an opening hemisyllable and may then last for as little as 12 ms. The whole echeme lasts about 45–80 ms and successive echemes are separated by an interval of about 10–20 ms; the echeme repetition rate is about 10–16/s. The opening hemisyllable at the beginning of each echeme is often particularly prominent.

Some of the songs studied consisted only of uniform echeme-sequences of this kind (Figs 147–151), but most included short periods (usually 0·3–2·0 s, rarely up to 6·0 s) of a 'ticking' sound consisting of a series of identical syllables repeated at the rate of about 20–30/s. These were usually diplosyllables lasting about 15–25 ms, but occasionally the opening stroke was silent, leaving a closing hemisyllable lasting about 8–10 ms. These short periods of ticking were sometimes very widely spaced so that they were hardly noticeable, but in about a third of the males studied they occurred much more frequently, sometimes every few seconds during at least part of the song (Figs 152–154). At such times the song sounds superficially like the calling song of *C. dorsalis*, and oscillographic analysis (or speed reduction of a recorded song) is necessary to show that the echemes in the main part of the song are trisyllabic, unlike the tetrasyllabic echemes of *C. dorsalis*. Occasional short bursts of ticking are probably a normal component of the calling song of *C. discolor*, although such bursts are much less prevalent than in *C. dorsalis*.

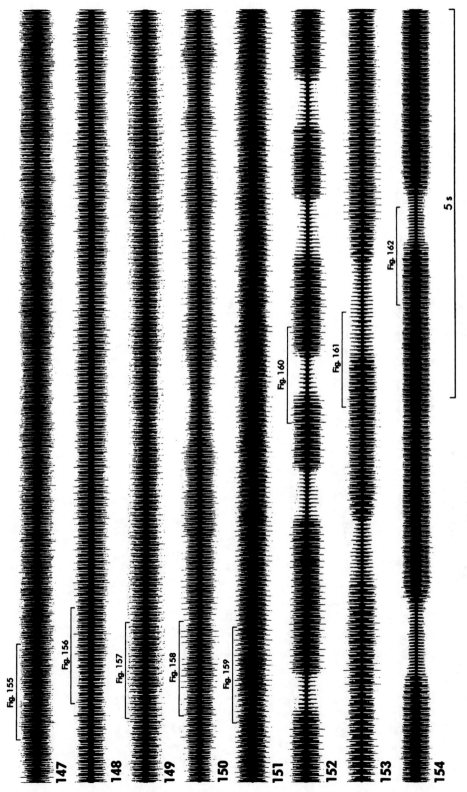

Figures 147–154 Oscillograms of the calling songs of eight males of *Conocephalus discolor*.

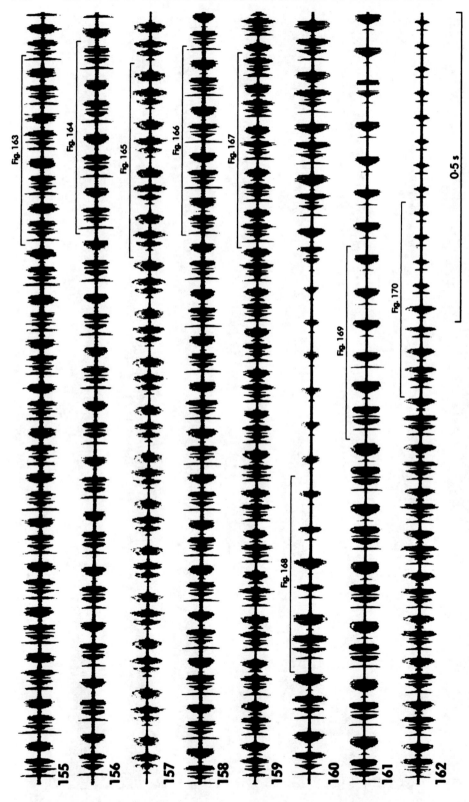

Figures 155–162 Faster oscillograms of the indicated parts of the songs of *Conocephalus discolor* shown in Figs 147–154.

Figures 163–170 Faster oscillograms of the indicated parts of the songs of *Conocephalus discolor* shown in Figs 155–162.

0·1 s

DISTRIBUTION. Widespread in Europe, from northern Germany to the Mediterranean coast and from the extreme south of England to Russia; also known from North Africa and parts of palaearctic Asia.

Conocephalus dorsalis (Latreille)

Locusta dorsalis Latreille, 1804: 133.

Short-winged Cone-head; (F) Conocéphale des Roseaux; (D) Kurzflüglige Schwertschrecke; (NL) Gewoon Spitskopje; (DK) Sivgræshoppe; (S) Sävvårtbitare; (SF) Kaislahepokatti.

REFERENCES TO SONG. **Oscillogram**: Ahlén, 1981; Broughton, 1963a; Grein, 1984; Heller, 1988; Holst, 1970, 1986; Kleukers *et al.*, 1997; Schmidt & Schach, 1978. **Diagram**: Bellmann, 1985a, 1988, 1993a; Bellmann & Luquet, 1995; Duijm & Kruseman, 1983; Holst, 1970; Ragge, 1965; Wallin, 1979. **Sonagram**: Leroy, 1977. **Wing-movement**: Heller, 1988. **Frequency information**: Ahlén, 1981; Broughton, 1952b; Froehlich, 1989; Heller, 1988; Sales & Pye, 1974. **Verbal description only**: Brown, 1955; Broughton, 1972b; Faber, 1928; Weber, 1984. **Disc recording**: Bellmann, 1993c (CD); Bonnet, 1995 (CD); Broughton, 1952b (photographs of disc grooves); Grein, 1984 (LP); Odé, 1997 (CD); Ragge *et al.*, 1965 (LP). **Cassette recording**: Ahlén, 1982; Bellmann, 1985b, 1993b; Burton & Ragge, 1987; Wallin, 1979.

RECOGNITION (Plate 1: 7). Distinguished from *C. conocephalus* and *C. discolor* by being brachypterous (fore wings not reaching the tip of the abdomen; hind wings vestigial). There is a rare macropterous form which can be distinguished from *C. conocephalus* by the characters mentioned under that species, and from *C. discolor* by the more prominent bilobed protuberance from the tenth abdominal tergite of the male and the curved ovipositor of the female (almost straight in *C. discolor*).

SONG (Figs 171–179. CD 1, track 21). The male calling song, produced mainly during the daytime, consists of long bursts (lasting up to 2 minutes or more) of a faint sound made up of two alternating components following each other without pause. One component is an echeme-sequence with a rustling quality, composed of tetrasyllabic echemes repeated at the rate of about 11–17/s; the other is a rapid ticking sound composed of single syllables repeated at the rate of about 15–30/s, with the repetition rate often becoming slower during the period of ticking. Each echeme-sequence usually lasts between 2 and 10s, but longer sequences of up to 30 s or more occasionally occur. The periods of ticking are on average shorter, usually lasting between 1 and 5 s, but there are occasionally longer bursts (up to 20 s or more), especially at the beginning or end of a period of continuous singing.

Oscillographic analysis shows that each echeme of the rustling component of the song consists of four very similar diplosyllables (in contrast to the less uniform trisyllabic echemes of *C. discolor*). The first opening hemisyllable and the last closing hemisyllable of each echeme are usually slightly longer than the remaining ones, so that the first and last syllables, lasting about 12–16 ms, are slightly longer than the intervening ones, each of which lasts about 10–13 ms. Within each echeme the intervals between successive syllables are about 2–3 ms and the whole echeme lasts about 50–65 ms; successive echemes are separated by an interval of about 6–15 ms.

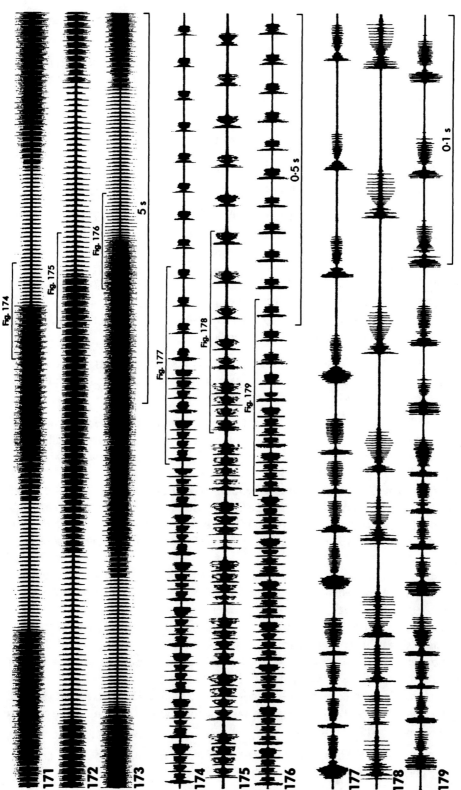

Figures 171–179 Oscillograms at three different speeds of the calling songs of three males of *Conocephalus dorsalis*.

The ticking component of the song is composed of diplosyllables very similar to those forming the ticking component of the song of *C. discolor*.

DISTRIBUTION. Widespread in northern and central Europe, from the southern parts of Scandinavia and the British Isles to the Pyrenees, Alps and northern part of the Balkan Peninsula; also eastwards to European Russia, Asia Minor and central Asia.

Ruspolia nitidula (Scopoli)

Gryllus nitidulus Scopoli, 1786: 62.

Large Cone-head; (F) Conocéphale gracieux; (D) Große Schiefkopfschrecke.

REFERENCES TO SONG. **Oscillogram**: Bailey, 1972; Dubrovin & Zhantiev, 1970; Grein, 1984; Heller, 1988; Schmidt, 1987b, 1996. **Diagram**: Bellmann, 1985a, 1988, 1993a; Bellmann & Luquet, 1995; Duijm & Kruseman, 1983. **Sonagram**: Bailey & Broughton, 1970 (wrongly captioned as being of an African subspecies). **Frequency information**: Bailey, 1967; M.-C. Busnel, 1953; R.-G. Busnel, 1955; Dubrovin & Zhantiev, 1970. **Verbal description only**: Chopard, 1922; Defaut, 1988b; Harz, 1957; Krauss, 1873 (as *mandibularis*). **Disc recording**: Andrieu & Dumortier, 1963 (LP), 1994 (CD); Bellmann, 1993c (CD); Bonnet, 1995 (CD); Grein, 1984 (LP). **Cassette recording**: Bellmann, 1985b, 1993b.

RECOGNITION (Plate 1: 8). The conical head-shape distinguishes this large species from all other European Tettigoniidae except *Conocephalus*, which is much smaller (less than 30 mm long, measured to the tips of the flexed fore wings).

SONG (Figs 180–187. CD 1, track 22). The calling song, produced almost entirely at night, is a long, continuous buzz, often lasting for 10 minutes or more without pause. Oscillographic analysis shows that it consists of a uniform sequence of syllables repeated at the rate of about 70–100/s. The sound is often produced entirely by the closing strokes of the fore wings, but there are sometimes quieter opening hemisyllables; the closing hemisyllables usually last about 5–7 ms. The song is resonant, the audible sound produced being an almost pure tone of about 13–20 kHz (see Figs 186, 187). The frequency usually becomes lower during the course of each hemisyllable, beginning at about 15–20 kHz and ending at about 13–16 kHz.

Bellmann (1985a) describes (p. 90) and illustrates (p. 67) squeaking ('quietschende') sounds occurring at intervals during the song, and these can be heard in his cassette recording (Bellmann, 1985b). Such sounds do not, however, occur in any of our recordings.

The song of this species is superficially similar to those of *Metrioptera roeselii* and *Gampsocleis glabra*, which also produce prolonged buzzes. However, the songs of these two species are non-resonant, comprising a broad band of frequencies, and confusion is in any case unlikely in the field as they sing mainly during the daytime.

Males occasionally produce short echemes lasting only a second or two, but this is almost always in reaction to other males or some other disturbance and cannot be regarded as a normal part of the calling song.

DISTRIBUTION. Much of Europe south of latitude 50°N, parts of North Africa and eastwards into palaearctic Asia. Absent from the British Isles.

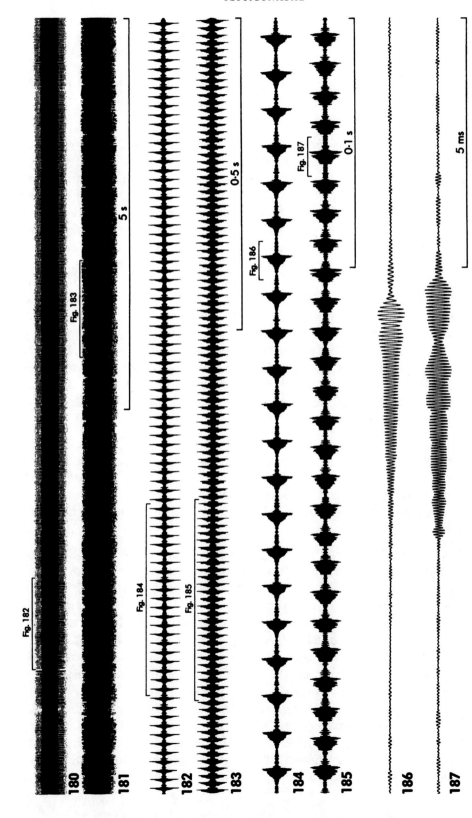

Figures 180–187 Oscillograms at four different speeds of the calling songs of two males of *Ruspolia nitidula*. Figs 186 and 187 show the carrier wave.

Tettigonia viridissima (Linnaeus)

Gryllus (Tettigonia) viridissimus Linnaeus, 1758: 430.

Great Green Bush-cricket; (F) Grande Sauterelle verte; (D) Grünes Heupferd; (NL) Grote Groene Sabelsprinkhaan; (DK) Stor Grøn Løvgræshoppe; (S) Grön Vårtbitare; (N) Grønn Lauvgrashoppe; (SF) Lehtohepokatti.

REFERENCES TO SONG. **Oscillogram**: Ahlén, 1981; Broughton, 1955a; Dubrovin & Zhantiev, 1970; Grein, 1984; Grzeschik, 1969; Heller, 1988; Holst, 1970, 1986; Jatho *et al*. 1994; Keuper *et al*., 1985, 1988a; Kleukers *et al*., 1997; Rheinlaender & Kalmring, 1973; Rheinlaender & Römer, 1980; Schmidt & Baumgarten, 1977; Schmidt & Schach, 1978. **Diagram**: Bellmann, 1985a, 1988, 1993a; Bellmann & Luquet, 1995; Duijm & Kruseman, 1983; Holst, 1970; Ragge, 1965; Wallin, 1979. **Wing-movement**: Heller, 1988. **Frequency information**: Ahlén, 1981; M.-C. Busnel, 1953; R.-G. Busnel, 1955; Dubrovin & Zhantiev, 1970; Froehlich, 1989; Heller, 1988; Jatho *et al*., 1994; Keuper *et al*., 1985, 1988a; Latimer & Broughton, 1984; Sales & Pye, 1974; Rheinlaender & Römer, 1980; Zhantiev & Dubrovin, 1971. **Verbal description only**: Baier, 1930; Beier, 1955; Broughton, 1972b; Burr, 1936; Chopard, 1922; Defaut, 1988b; Faber, 1928; Harz, 1957; Hodge, 1878; Hudson, 1903; Klöti-Hauser, 1921; Laddiman, 1879; Nielsen, 1938, 1972; Nielsen & Dreisig, 1970; Tenant, 1878; Weber, 1984; Yersin, 1854b; Zeuner, 1931. **Disc recording**: Andrieu & Dumortier, 1963 (LP), 1994 (CD); Bellmann, 1993c (CD); Bonnet, 1995 (CD); Broughton, 1952b (photographs of disc grooves); Burton, 1969 (LP, as Great Green Grasshopper); Deroussen, 1993 (CD); Grein, 1984 (LP); Odé, 1997 (CD); Ragge *et al*., 1965 (LP); Roché, 1965 (LP). **Cassette recording**: Ahlén, 1982; Bellmann, 1985b, 1993b; Burton & Ragge, 1987; Wallin, 1979.

RECOGNITION (Plate 1: 9). The European species of *Tettigonia* are medium-sized to large, predominantly green insects, usually living on tall herbs, bushes or trees rather than on or near the ground; the fore wings have no dark spots in the radial area. *T. viridissima* can be distinguished from *T. cantans* and *T. hispanica* by having the fore wings extending well beyond the hind knees, and from *T. caudata* by lacking dark spots on the underside of the hind femora.

SONG (Figs 188–192, 195–199, 202–206. CD 1, track 23). The loud calling song, produced mainly in the late afternoon and after dark, consists of sequences of closely spaced disyllabic echemes. The echeme repetition rate depends on temperature: 10–15/s is common, but in high daytime temperatures the rate may increase to 20/s and on cool nights may be as low as 5/s. Typically the echeme-sequences are interrupted at irregular intervals by very short pauses (Figs 188–190), usually of less than a second and sometimes, at high temperatures, as short as 0·2–0·3 s. Occasionally, long echeme-sequences are produced without even the briefest of interruption (Figs 191, 192); the longest uninterrupted sequence we have recorded lasted 3 minutes 9 seconds.

Oscillographic analysis shows that nearly all the sound is produced by the closing strokes of the fore wings, although quiet opening hemisyllables are sometimes present. The duration of the two closing hemisyllables comprising each echeme again depends on temperature: 15–25 ms for each hemisyllable is common, but in hot conditions they may last 10 ms or even less and on cool nights they may be as long as 50 ms. The second

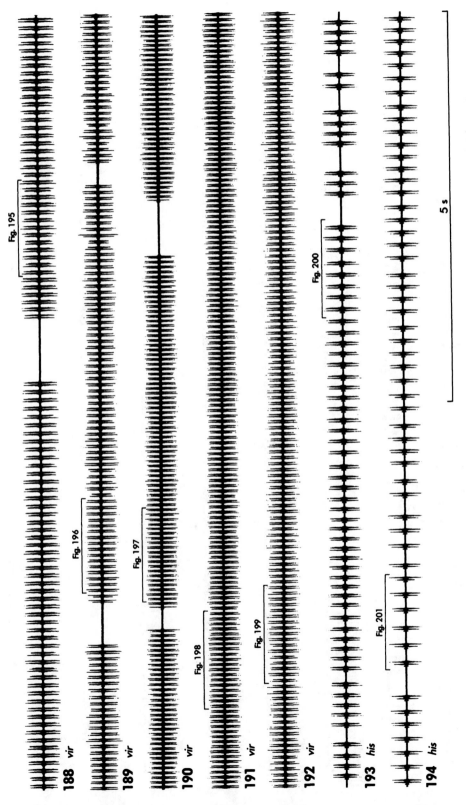

Figures 188–194 Oscillograms of the male calling songs of (188–192) *Tettigonia viridissima* and (193, 194) *T. hispanica*.

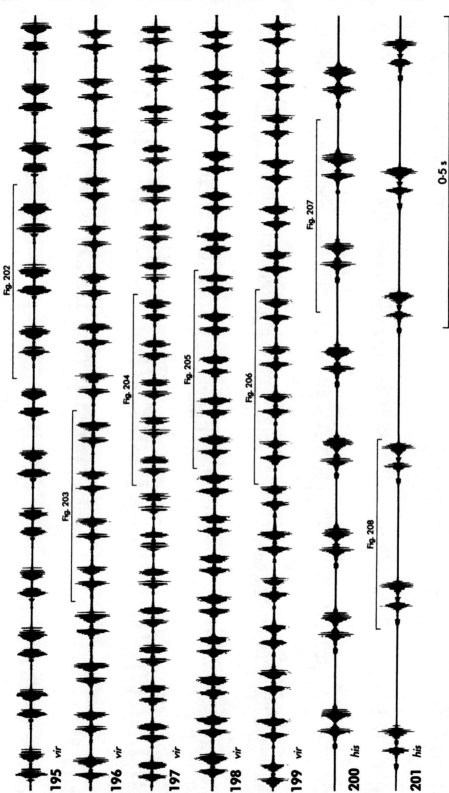

Figures 195–201 Faster oscillograms of the indicated parts of the songs of *Tettigonia viridissima* and *T. hispanica* shown in Figs 188–194.

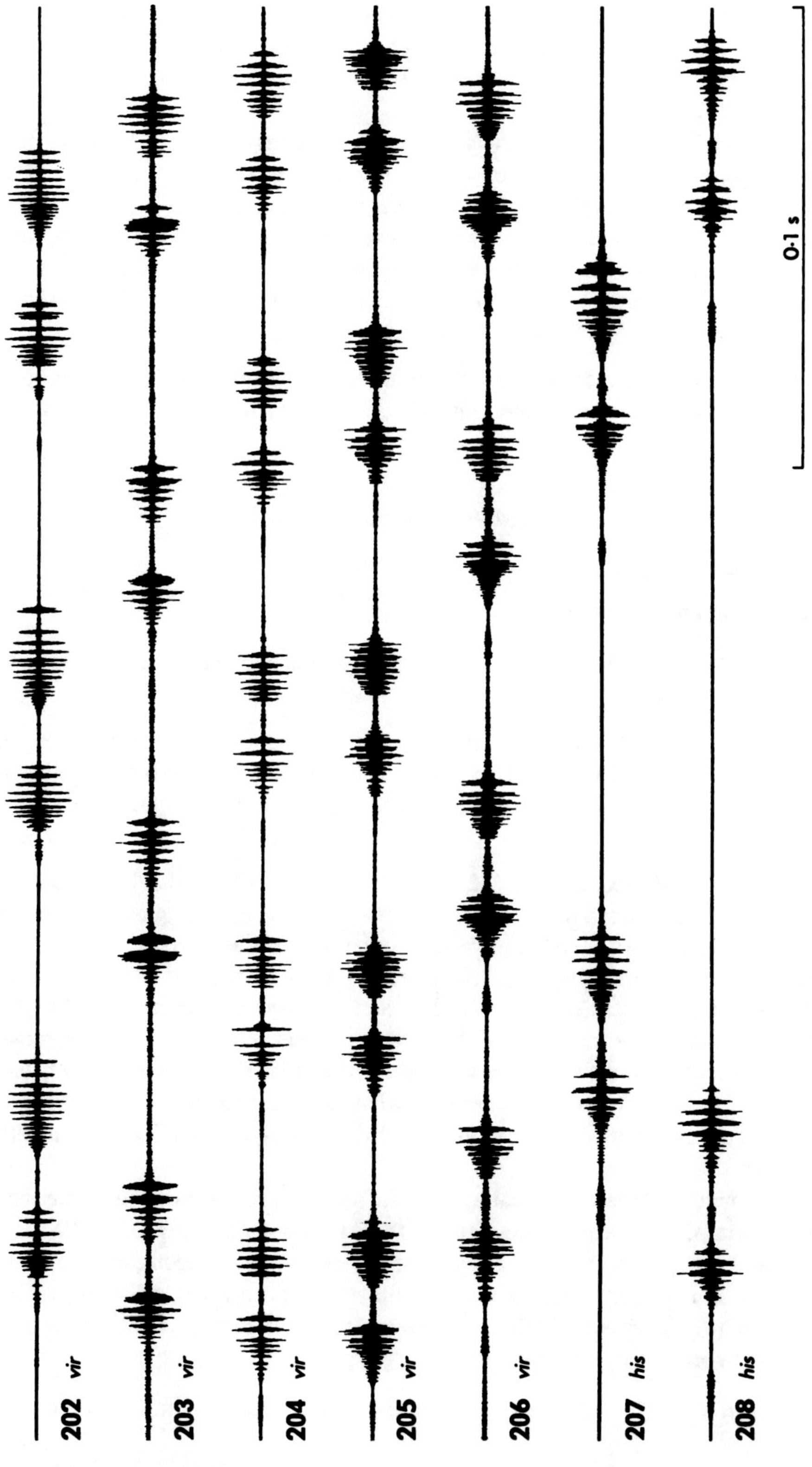

Figures 202–208 Faster oscillograms of the indicated parts of the songs of *Tettigonia viridissima* and *T. hispanica* shown in Figs 195–201.

0.1 s

hemisyllable in each echeme is often slightly longer than the first. The gap between the two hemisyllables varies from about 6 ms (warm conditions) to about 35 ms (cool conditions), and the interval between successive echemes varies from about 20 ms to about 100 ms, again depending on temperature.

Each closing hemisyllable is typically broken up into a very variable number of discrete pulses of sound separated by gaps of varying duration but usually of the order of 1 ms. There are sometimes fewer than 10 of these pulses in one hemisyllable (see for example Fig. 204) and so, at least in such cases, they are unlikely to represent individual tooth-impacts.

The songs of this species and *T. hispanica* could be confused with that of *Platycleis sepium*, but *P. sepium* is brachypterous, thus differing greatly in appearance from *T. viridissima*, and does not overlap in distribution with *T. hispanica*.

DISTRIBUTION. Very widespread in Europe from southern Scandinavia to the southern peninsulas and from southern Britain to Russia. Also occurs in palaearctic Asia and North Africa.

Tettigonia hispanica (Bolívar)

Locusta Hispanica Bolívar, 1893: 24.

Spanish Green Bush-cricket.

REFERENCES TO SONG. **Oscillogram, wing-movement** and **frequency information**: Heller, 1988.

RECOGNITION. The relatively short wings, usually not reaching the hind knees, enable *T. hispanica* to be readily distinguished from *T. viridissima* and *T. caudata*. It is superficially similar to *T. cantans* (see the remarks under that species), but these two species do not overlap in range.

SONG (Figs 193, 194, 200, 201, 207, 208. CD 1, track 24). The loud calling song is very similar in rhythmic pattern to that of *T. viridissima*, consisting of sequences of disyllabic echemes and varying with temperature in a similar way; it is also produced mainly in the afternoon and after dark. However, it has a more sibilant quality and at a given temperature the echeme repetition rate is lower than in *T. viridissima*, varying from 2 to10/s at temperatures ranging from 15 to 30°C. The echeme-sequences are typically interrupted at irregular intervals (usually more frequently than in *T. viridissima*) by very short pauses of about 0·2–1·0 s, but occasionally there are much longer pauses and sometimes the sequences continue for up to a minute or more without interruption, especially at lower temperatures and at night.

Oscillographic analysis shows that most of the sound is produced by the closing strokes of the fore wings, but in all the songs studied there were quiet opening hemisyllables. The duration and structure of the closing hemisyllables and the gap between each pair are similar to those of *T. viridissima*; as in that species, the second closing hemisyllable of each pair is usually slightly longer than the first. The intervals between successive echemes are about 50–300 ms, depending on temperature.

DISTRIBUTION. Primarily a species of the central mountains (Sistema Central) of Spain, where it is widely distributed at altitudes above 600 m. Also believed to occur on Sardinia and in the Sila mountains of southern Italy, but the song has not yet been studied in either of these places.

Tettigonia cantans (Fuessly)

Gryllus cantans Fuessly, 1775: 23.

Upland Green Bush-cricket; (F) Sauterelle cymbalière; (D) Zwitscherschrecke; (NL) Kleine Groene Sabelsprinkhaan; (DK) Syngende Løvgræshoppe; (SF) Idänhepokatti.

REFERENCES TO SONG. **Oscillogram**: Dubrovin, 1977; Dubrovin & Zhantiev, 1970; Grein, 1984; Heller, 1988; Holst, 1970, 1986; Jatho *et al.*, 1994; Kalmring *et al.*, 1985; Keuper & Kühne, 1983; Keuper *et al.*, 1985, 1988*a*; Kleukers *et al.*, 1997; Lottermoser, 1952; Rheinlaender & Kalmring, 1973; Schmidt, 1989; Schmidt & Schach, 1978; Silver *et al.*, 1980. **Diagram**: Bellmann, 1985*a*, 1988, 1993*a*; Bellmann & Luquet, 1995; Duijm & Kruseman, 1983; Holst, 1970. **Sonagram**: Broughton *et al.*, 1975. **Wing-movement**: Heller, 1988. **Frequency information**: Dubrovin, 1977; Dubrovin & Zhantiev, 1970; Dumortier, 1963*c*; Froehlich, 1989; Heller, 1988; Jatho *et al.*, 1994; Keuper & Kühne, 1983; Keuper *et al.*, 1985, 1988*a*; Latimer & Schatral, 1986; Latimer & Sippel, 1987; Lottermoser, 1952; Schatral *et al.*, 1985; Silver *et al.*, 1980; Sippel *et al.*, 1985; Zhantiev & Dubrovin, 1971. **Verbal description only**: Chopard, 1922; Defaut, 1988*b*; Faber, 1928; Harz, 1957; Nielsen, 1938; Nielsen & Dreisig, 1970; Panelius, 1978; Ramme, 1921*a*; Taschenberg, 1871; Weber, 1984; Yersin, 1854*b*; Zeuner, 1931. **Disc recording**: Andrieu & Dumortier, 1963 (LP), 1994 (CD); Bellmann, 1993*c* (CD); Bonnet, 1995 (CD); Deroussen, 1993 (CD); Grein, 1984 (LP); Odé, 1997 (CD). **Cassette recording**: Bellmann, 1985*b*, 1993*b*.

RECOGNITION. *T. cantans* can be easily distinguished from *T. viridissima* and *T. caudata* by its relatively short wings, which hardly extend beyond the hind knees. It is not so easy to distinguish from the equally short-winged *T. hispanica*, but the fore wings of *T. cantans* are broad and smoothly rounded in comparison with those of *T. hispanica*, which are narrower and in which the more distal part of the anterior (ventral) margin is straight; there is in any case no overlap in the ranges of these two species.

SONG (Figs 209–212, 214–217, 219–222. CD 1, track 25). The loud, sibilant calling song is produced both during the day (especially afternoon) and at night, and is much influenced by temperature. In warm, sunny weather the song usually consists of echemes lasting about 1–7 s, with a marked crescendo in the early part of each echeme (Figs 209, 210). As the temperature falls in the late afternoon the echemes become longer, and the nocturnal song (Figs 211, 212) often continues without interruption for several minutes. In warm daytime conditions the syllable repetition rate is usually more than 35/s and can be as high as 55/s in hot sunny weather; on cool nights the rate drops to below 20/s and can be as low as 10/s (rarely 6/s) in cold conditions (*c* 10°C).

Oscillographic analysis shows that the sound is almost always produced only by the closing strokes of the fore wings. In warm conditions the closing hemisyllables last about 12–20 ms, but become longer at lower temperatures, lasting up to 70 ms or more on cool nights. The intervals between successive hemisyllables are also temperature-dependent,

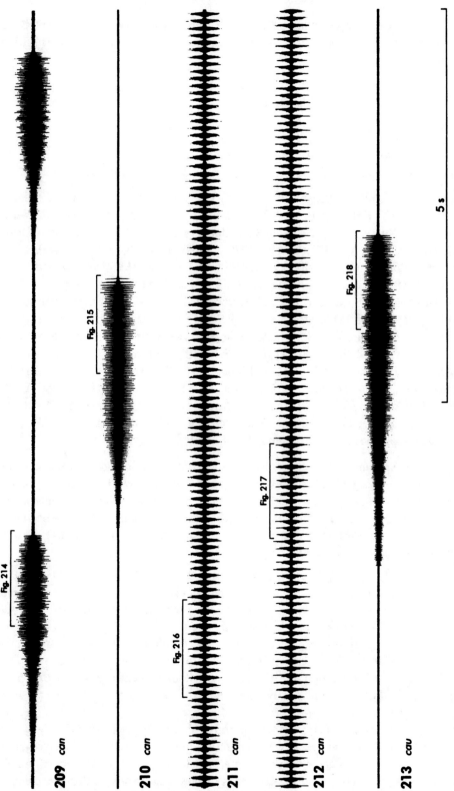

Figures 209–213 Oscillograms of the male calling songs of (209–212) *Tettigonia cantans* and (213) *T. caudata*. The songs shown in Figs 209 and 210 were produced in warm, daytime conditions, and those shown in Figs 211 and 212 were produced in cool conditions at night.

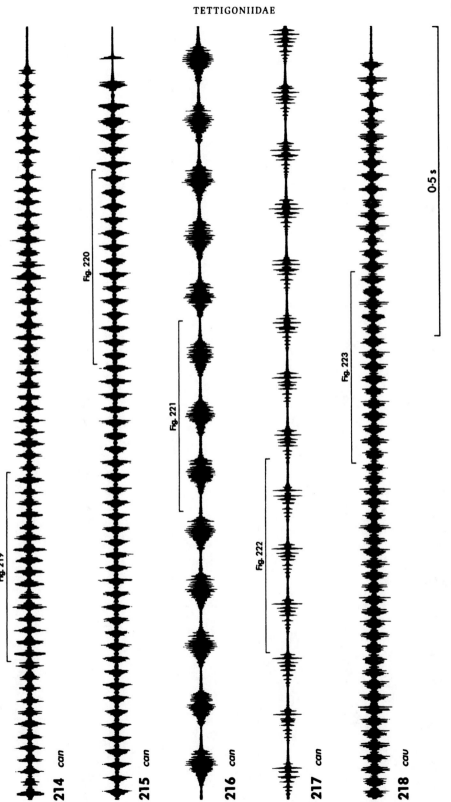

Figures 214–218 Faster oscillograms of the indicated parts of the songs of *Tettigonia cantans* and *T. caudata* shown in Figs 209–213.

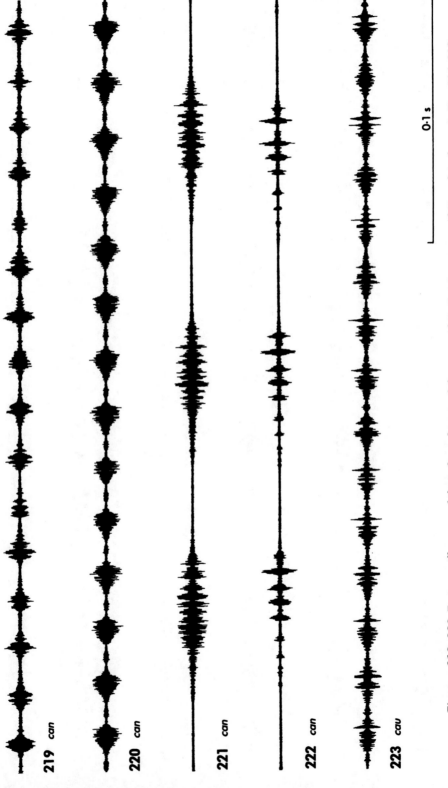

Figures 219–223 Faster oscillograms of the indicated parts of the songs of *Tettigonia cantans* and *T. caudata* shown in Figs 214–218.

ranging from about 5–10 ms at high temperatures to about 30–40 ms in cool conditions.

DISTRIBUTION. *T. cantans* occurs widely in Europe from southern Finland and Denmark to the Pyrenees, Alps and parts of the Apennines. It is absent from the Scandinavian Peninsula and the British Isles, but occurs in the northern parts of the Balkan Peninsula and eastwards through parts of European Russia to Siberia and Manchuria. It is typically found in mountainous areas, but occurs in some lowland localities in the more northerly parts of its range.

Tettigonia caudata (Charpentier)

Locusta caudata Charpentier, 1842: pl. 33.

Eastern Green Bush-cricket; (F) Sauterelle des Grisons; (D) Östliches Heupferd.

REFERENCES TO SONG. **Oscillogram**: Heller, 1988. **Diagram**: Bellmann, 1985*a*, 1988, 1993*a*; Bellmann & Luquet, 1995. **Wing-movement**: Heller, 1988. **Frequency information**: Zhantiev & Dubrovin, 1971. **Verbal description only**: Harz, 1957; Regen, 1903; Taschenberg, 1871; Zeuner, 1931. **Disc recording**: Bellmann, 1993*c* (CD). **Cassette recording**: Bellmann, 1984*b*, 1993*b*.

RECOGNITION. Resembles *T. viridissima* in being fully winged, but may be distinguished from it by the dark spots on the underside of the hind femora.

SONG (Figs 213, 218, 223. CD 1, track 26). The calling song, produced during both day and night, is very similar to the daytime song of *T. cantans*, consisting of loud, sibilant echemes lasting about 1–5 s and with a crescendo in the first half of each echeme. Other characteristics of the song, such as syllable duration and repetition rate, also resemble those of *T. cantans*, and are also dependent on temperature, but the song of *T. caudata*, unlike that of *T. cantans*, never seems to become completely continuous at night.

DISTRIBUTION. Primarily an eastern European and south-west Asian species, occurring in the area studied only in eastern Germany and Austria.

Decticus verrucivorus (Linnaeus)

Gryllus (Tettigonia) verrucivorus Linnaeus, 1758: 429.

Wart-biter; (F) Dectique verrucivore; (D) Warzenbeißer; (NL) Wrattenbijter; (DK) Vortebider; (S) Stor Vårtbitare; (N) Vortebiter; (SF) Niittyhepokatti.

REFERENCES TO SONG. **Oscillogram**: Ahlén, 1981; Dubrovin & Zhantiev, 1970; Grein, 1984; Heller, 1988; Holst, 1970, 1986; Ingrisch *et al.*, 1992; Jatho *et al.*, 1994; Kalmring *et al.*, 1985; Keuper *et al.*, 1985, 1986, 1988*a*; Kleukers *et al.*, 1997; Lottermoser, 1952; Ragge, 1987*b*; Rheinlaender & Kalmring, 1973; Samways, 1989; Schmidt & Schach, 1978; Silver *et al.*, 1980. **Diagram**: Bellmann, 1985*a*, 1988, 1993*a*; Bellmann & Luquet, 1995; Duijm & Kruseman, 1983; Holst, 1970; Ragge, 1965; Wallin, 1979. **Sonagram**: Leroy, 1977. **Wing-movement**: Heller, 1988. **Frequency information**: Ahlén, 1981; M.-C. Busnel, 1953; R.-G. Busnel, 1955; Busnel & Chavasse, 1951; Dubrovin, 1977; Dubrovin & Zhantiev, 1970;

Dumortier, 1963c; Heller, 1988; Jatho *et al.*, 1994; Keuper *et al.*, 1985, 1986, 1988a; Silver *et al.*, 1980; Schatral *et al.*, 1985; Zhantiev & Dubrovin, 1971. **Verbal description only**: Beier, 1955; Broughton, 1972b; Chopard, 1922; Faber, 1928; Harz, 1957; Nielsen, 1938; Weber, 1984; Yersin, 1854b. **Disc recording**: Andrieu & Dumortier, 1963 (LP), 1994 (CD); Bellmann, 1993c (CD); Bonnet, 1995 (CD); Helluin & Ribassin, 1961 (LP); Odé, 1997 (CD); Ragge *et al.*, 1965 (LP). **Cassette recording**: Ahlén, 1982; Bellmann, 1985b, 1993b; Burton & Ragge, 1987; Elliott, 1986; Wallin, 1979.

RECOGNITION (Plate 1: 10). The genus *Decticus* includes two large, heavily-built European species. *D. verrucivorus* can be either green or brown (or a mixture of the two) and the green form could be confused with *Tettigonia*. However, the fore wings of *D. verrucivorus* are almost always conspicuously dark-spotted, with particularly large spots along the radial area; in *Tettigonia* this region of the fore wings is always uniformly green. *D. albifrons* is always brown and can be distinguished from the brown form of *D. verrucivorus* by its longer wings, which extend well beyond the hind knees; the wings seldom reach the hind knees in *D. verrucivorus* and are often much shorter than this, especially in high mountains.

    For notes on the taxonomy of this species see p. 70.

SONG. The calling song is quite uniform in most of western Europe, even where there are morphological differences, e.g. f. *monspeliensis* in southern France (Figs 227, 236, 245) (perhaps now extinct) and f. *aprutianus* in the Apennines (Figs 228, 237, 246). However, *D. v. assiduus* Ingrisch, Willemse & Heller, which occurs in high mountains in central Spain, has a markedly different song. Two separate accounts are therefore given below.

Western Europe generally (Figs 224–228, 233–237, 242–246. CD 1, track 27)
The loud calling song, produced mainly on hot sunny days, consists of a long echeme-sequence, often lasting several (up to 5 or more) minutes, the sound being reminiscent of the loud ticking sound produced by older free-wheeling bicycles. The sequence begins with well-separated echemes, but these are repeated increasingly rapidly until they merge into a dense sequence with an echeme repetition rate of about 8–10/s (rarely slower, down to 6 s).

    Oscillographic analysis shows each echeme to be quite complex, consisting of four syllables, grouped into two pairs. The initial opening of the fore wings at the beginning of the echeme usually produces a distinct hemisyllable, which is sometimes quite loud; the remaining opening hemisyllables are quiet or absent. The first closing hemisyllable is usually quiet and occasionally absent or almost so; the remaining closing hemisyllables are usually well developed. Each pair of closing hemisyllables usually lasts about 12–20 ms (rarely up to 30 ms) with an interval of about 5–10 ms between the two pairs. The second closing hemisyllable of each pair usually follows the first almost immediately, the pause between them being no more than 3 ms at the most. The whole echeme, including the initial opening hemisyllable, usually lasts about 45–80 ms, but can be as short as 30 ms if the initial opening hemisyllable is absent. The interval between successive echemes, once a constant repetition rate has been reached, is usually 50–100 ms, but can be as short as 25 ms or as long as 150 ms.

    Males sometimes produce completely isolated echemes, particularly early in the day, but these are clearly not a fully developed calling song.

D. v. assiduus (Figs 229–232, 238–241, 247–250. CD 1, track 28)
The calling song is also loud and also produced in warm, sunny weather, as well as in cooler

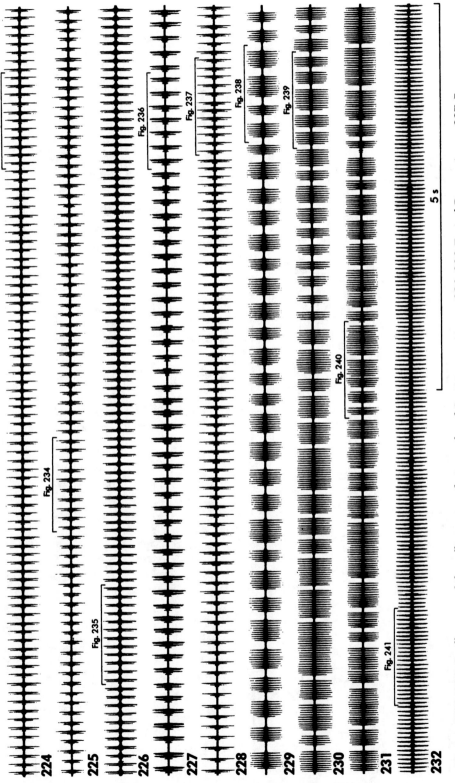

Figures 224–232 Oscillograms of the calling songs of nine males of *Decticus verrucivorus*. **224–226**. Typical *D. verrucivorus*. **227**. Form *monspeliensis* from southern France. **228**. Form *aprutianus* from the Italian Apennines. **229–232**. *D. v. assiduus* from central Spain.

Figures 233–241 Faster oscillograms of the indicated parts of the songs of *Decticus verrucivorus* shown in Figs 224–232.

Figures 242–250 Faster oscillograms of the indicated parts of the songs of *Decticus verrucivorus* shown in Figs 233–241.

0·1 s

242
243
244
245
246
247
248
249
250

conditions. Typically it consists of long sequences of short echemes separated by very brief pauses (Figs 229–231). The sequences begin rather like those of the more typical song of *D. verrucivorus*, with more widely spaced, short echemes of 3–6 syllables. The gaps between the syllables soon close up to about 50–100 ms and the echemes become longer, sometimes consisting of up to 30 or more syllables, though varying greatly and often still including some short ones of 3–5 syllables. A sequence of this kind often lasts for several minutes. Sometimes, in cooler conditions, longer echemes are produced, lasting up to 20 s or more without pause (Fig. 232).

The echemes differ markedly from those of typical *D. verrucivorus* in consisting of simple trains of identical, equally spaced syllables and in varying greatly in the number of syllables they contain. The syllable repetition rate is usually 30–40/s in warm sunny weather, but can be as low as 18/s in cooler conditions. Oscillographic analysis shows that opening hemisyllables are usually very quiet or absent and that the closing hemisyllables usually last about 12–15 ms in warm conditions, but up to 25 ms or more when it is cooler.

DISTRIBUTION. Widespread in Europe and palaearctic Asia, from southern England (where it is very rare) to eastern Russia, and from Scandinavia to the mountains of central Spain, central Italy and the Balkan Peninsula.

Decticus albifrons (Fabricius)

Locusta albifrons Fabricius, 1775: 286.

White-faced Bush-cricket; (F) Dectique à front blanc; (D) Südlicher Warzenbeißer.

REFERENCES TO SONG. **Oscillogram**: Dubrovin & Zhantiev, 1970; Heller, 1988; Jatho *et al.*, 1994; Kalmring *et al.*, 1985; Keuper *et al.*, 1985, 1988*a*; Rheinlaender & Kalmring, 1973; Schmidt, 1996. **Sonagram**: Broughton *et al.*, 1975. **Wing-movement**: Heller, 1988. **Frequency information**: M.-C. Busnel, 1953; R.-G. Busnel, 1955; Busnel & Chavasse, 1951; Dubrovin & Zhantiev, 1970; Heller, 1988; Keuper *et al.*, 1985, 1988*a*; Zhantiev & Dubrovin, 1971. **Verbal description only**: Bellmann, 1993*a*; Bellmann & Luquet, 1995; Chopard, 1952; Dumortier, 1963*c*; Yersin, 1857. **Disc recording**: Bonnet, 1995 (CD); Deroussen, 1993 (CD); Roché, 1965 (LP).

RECOGNITION (Plate 1: 11). See under *D. verrucivorus*.

SONG (Figs 251–257. CD 1, track 29). The loud calling song, a common Mediterranean sound in hot, sunny weather, is a long series of click-like syllables, sometimes repeated fairly regularly at the rate of about 5–7/s or, if less regularly, usually with a mean repetition rate within this range. Depending partly on temperature, the mean repetition rate is occasionally as low as 2/s or as high as 10/s. The interval between successive syllables is usually within the range 0·1–0·4 s, but is sometimes a little under 0·1 s and occasionally as long as 1 s or a little more. The song continues for indefinite periods, often of many minutes.

Oscillographic analysis shows that nearly all (occasionally all) the sound is produced by the closing stroke of the fore wings, each closing hemisyllable lasting about 6–10 ms. The song is resonant, the audible sound produced being an almost pure tone of about 8–9 kHz (see Fig. 257).

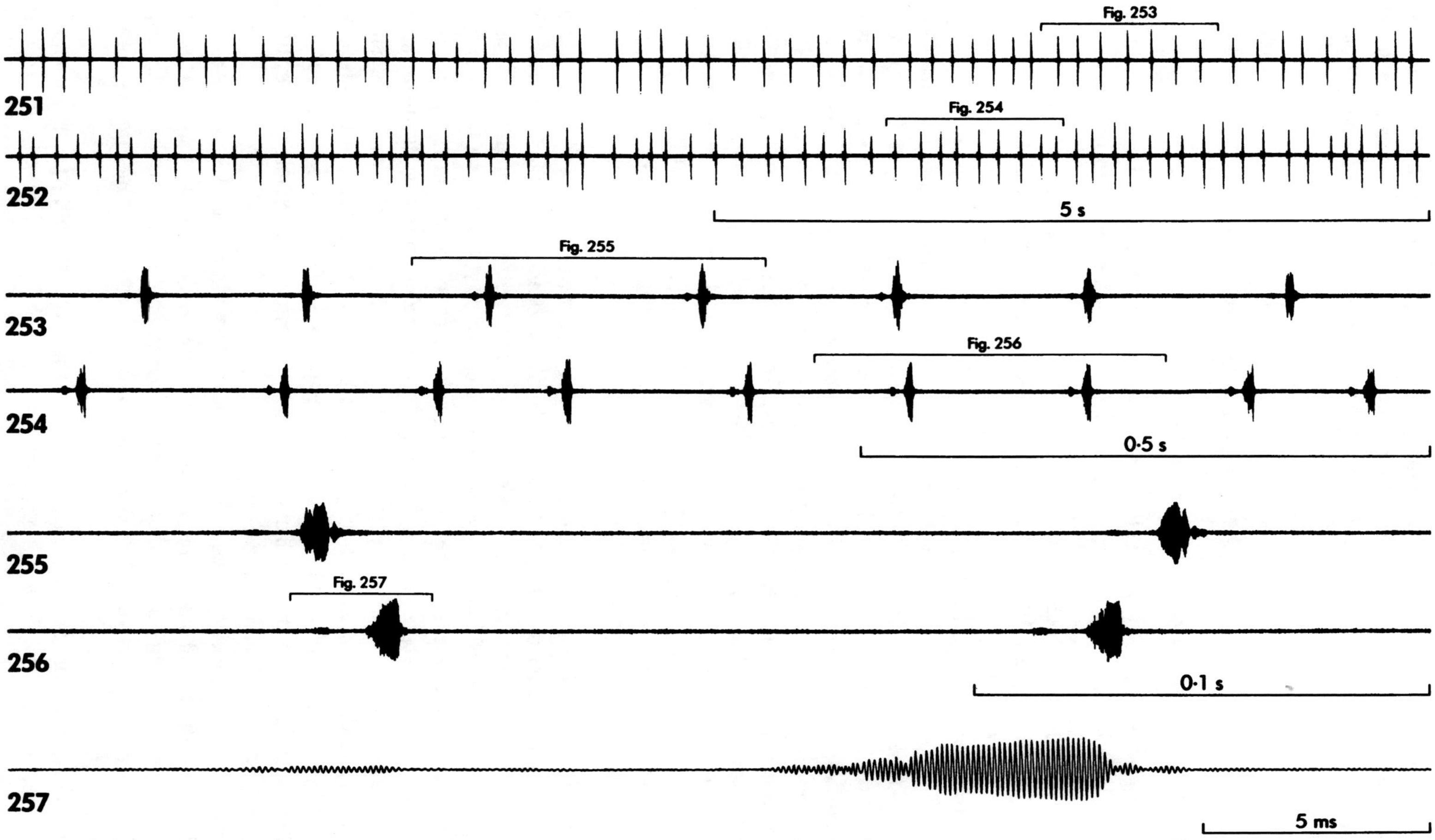

Figures 251–257 Oscillograms at four different speeds of the calling songs of two males of *Decticus albifrons*. Fig. 257 shows the carrier wave.

DISTRIBUTION. The whole of the Mediterranean Region, from Madeira and the Canary Islands to the Caucasus and Iran. The northern limit of the range runs from western France through southern Switzerland and Slovenia to the Black Sea coast of Romania and Bulgaria.

Platycleis albopunctata (Goeze)

Gryllus (Tettigonia) albo-punctatus Goeze, 1778: 89.

Grey Bush-cricket; (F) Decticelle chagrinée; (D) Westliche Beißschrecke; (NL) Duinsabelsprinkhaan; (DK) Sandgræshoppe; (S) Grå Vårtbitare; (SF) Hietahepokatti.

REFERENCES TO SONG. **Oscillogram**: Ahlén, 1981 (as *denticulata*); Broughton, 1965 (as *d.*); Dubrovin & Zhantiev, 1970 (as *intermedia*); Grein, 1984; Heller, 1988; Holst, 1970 (as *d.*), 1986; Jatho *et al.*, 1994; Kleukers *et al.*, 1997; Latimer, 1981*b*, 1981*c*; Latimer & Broughton, 1984; Ragge, 1990; Samways, 1976*b*; Schmidt, 1989 (as *grisea*); Schmidt & Schach, 1978 (as *g.*). **Diagram**: Bellmann, 1985*a*, 1988, 1993*a*; Bellmann & Luquet, 1995; Duijm & Kruseman, 1983; Holst, 1970 (as *d.*); Ragge, 1965 (as *d.*); Samways, 1976*e*; Wallin, 1979 (as *d.*). **Sonagram**: Samways, 1976*b*. **Wing-movement**: Heller, 1988. **Frequency information**: Ahlén, 1981 (as *d.*); Dubrovin & Zhantiev, 1970 (as *i.*); Heller, 1988; Latimer, 1981*a*, 1981*b*, 1981*c*; Latimer & Broughton, 1984; Sales & Pye, 1974 (as *d.*). **Musical notation**: Baier, 1930 (as *g.*); Yersin, 1854*b* (as *g.*). **Verbal description only**: Broughton, 1972*b*; Chopard, 1922; Defaut, 1987; Faber, 1928 (as *g.*); Harz, 1957 (as *d.*); Samways, 1976*d*; Sarra, 1934 (as *g.*); Yersin, 1857 (as *g.*). **Disc recording**: Andrieu & Dumortier, 1994 (CD); Bellmann, 1993*c* (CD); Bonnet, 1995 (CD); Grein, 1984 (LP); Odé, 1997 (CD); Ragge *et al.*, 1965 (LP, as *d.*). **Cassette recording**: Ahlén, 1982 (as *d.*); Bellmann, 1985*b*, 1993*b*; Burton & Ragge, 1987; Wallin, 1979 (as *d.*).

RECOGNITION. *Platycleis*, in the broad sense used here (following Ragge, 1990), is a large Palaearctic genus of about 100 species, of which 18 have been recorded from western Europe. The more common European species can be easily recognized as belonging to *Platycleis* by being fully winged, often quite large, and grey-brown or yellow-brown in colour. The brachypterous species are less easily distinguished from related genera, but they are not as brachypterous as *Pholidoptera* and have a more distinct median carina in the metazona of the pronotum. Partly green, brachypterous bush-crickets, in which the female subgenital plate lacks a median groove but which are otherwise similar to *Platycleis*, are likely to belong to *Metrioptera*, but there is no clear dividing line between these two genera. See Ragge (1990) for a discussion of the classification of these genera and a morphological key to the principal western European species of *Platycleis*.

Females of *P. albopunctata* may be distinguished from most other *Platycleis* species of similar size by the unmodified seventh abdominal sternite, which has no ridges or other prominences; *P. iberica*, in which the female seventh abdominal sternite is only slightly swollen, may be distinguished from *P. albopunctata* by the very broad, subquadrate lobes of the subgenital plate.

Males are much more difficult to recognize, but may be distinguished from the closely similar species *P. sabulosa, P. intermedia, P. affinis* and *P. falx* by the robust titillators, with

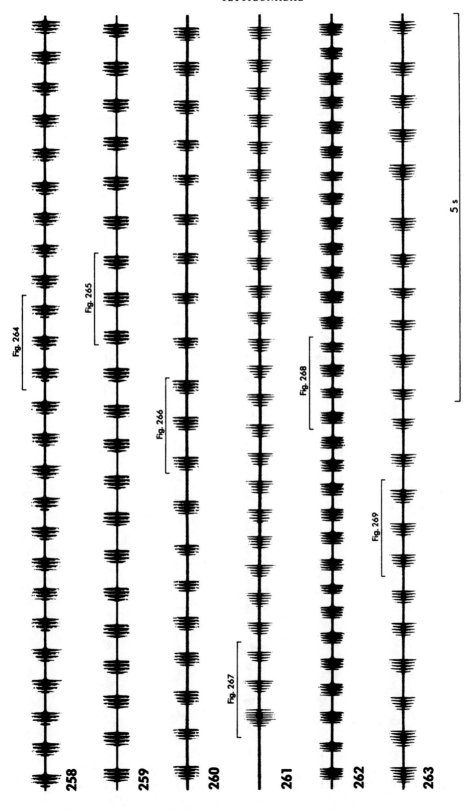

Figures 258–263 Oscillograms of the calling songs of six males of *Platycleis albopunctata*. Figs 258–262 are from males of *P. a. albopunctata* and Fig. 263 is from a male of *P. a. grisea*. (From Ragge, 1990)

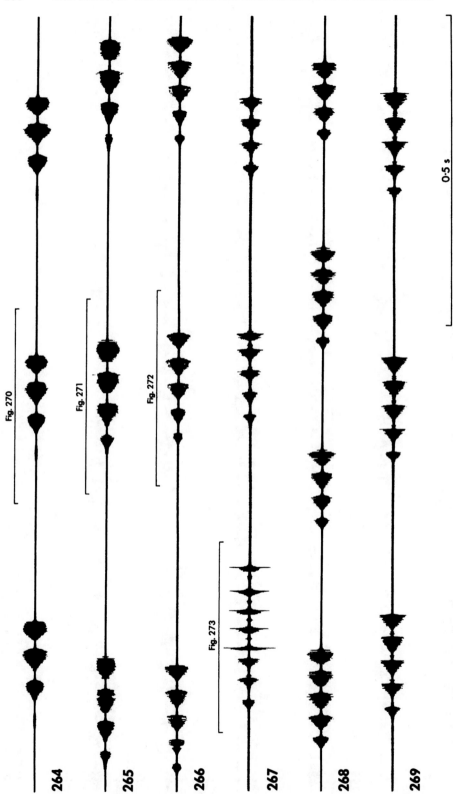

Figures 264–269 Faster oscillograms of the indicated parts of the songs of *Platycleis albopunctata* shown in Figs 258–263. (From Ragge, 1990)

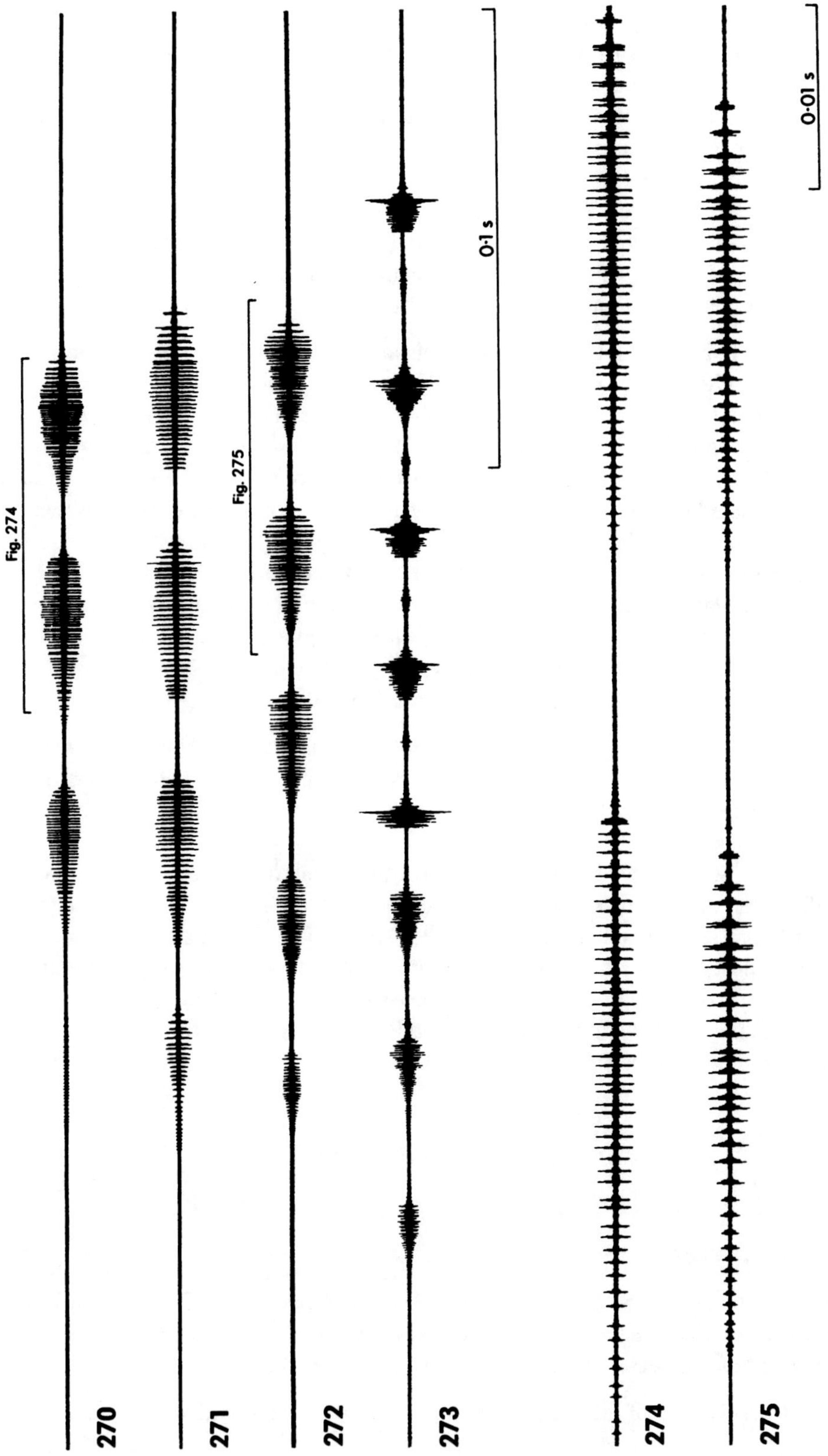

Figures 270–275 Faster oscillograms of the indicated parts of the songs of *Platycleis albopunctata* shown in Figs 264–267. (From Ragge, 1990)

their rather short apical part and broad basal part (see Ragge, 1990). Males of *P. iberica* have quite similar titillators, but can be distinguished from *P. albopunctata* at once in the field by their quite different calling song.

In the field singing males may be recognized by the long sequences of regularly repeated echemes, each usually composed of 3–5 syllables. The echemes of the rather similar calling songs of *P. sabulosa* are normally composed of more than 5 syllables, and those of *P. intermedia* and *P. sepium* are disyllabic. The calling song of *Metrioptera brachyptera* (p. 187) can easily be confused with that of *P. albopunctata* (although at the same temperature the echeme repetition rate is faster), but *M. brachyptera* lives in moister habitats and, being brachypterous, is very different from *P. albopunctata* in appearance.

SONG (Figs 258–275. CD 1, track 30). In warm sunshine the calling song consists of long sequences of echemes repeated fairly regularly at the rate of about 2–4/s and each usually consisting of 3–5 (very rarely 2 and occasionally 6) syllables. Oscillographic analysis shows that opening hemisyllables are often absent and that the closing hemisyllables usually last about 10–30 ms and are repeated within an echeme at the rate of about 25–35/s. The duration of a single echeme of four syllables is about 100–200 ms and the interval between two echemes is about 100–300 ms. In dull weather and at night the echeme repetition rate can drop to less than 1/s and the syllable repetition rate to less than 10/s; in such conditions the closing hemisyllables sometimes last more than 100 ms and a four-syllable echeme more than 600 ms. The first syllable in an echeme is usually quieter than the remaining ones. Microsyllables are usually absent, but occasionally a few are added at the end of an echeme, especially the opening echeme of a sequence (see Figs 261, 267, 273).

DISTRIBUTION. This species occurs very widely in Europe from southern Scandinavia to the southern peninsulas, where it is mainly confined to uplands. It is also recorded from mountains in Morocco. Further song studies are needed to establish the eastern limit of the range of *P. albopunctata*.

The species can be divided broadly into two subspecies on the basis of small morphological differences, mainly the shape of the male titillators and female subgenital plate (see Ragge, 1990); there is no difference in the calling song. The nominate subspecies occurs in the westernmost parts of Europe (including southern Scandinavia, Germany, southern Britain, France and the Iberian Peninsula) and the eastern subspecies, *P. a. grisea* (Fabricius), from Poland, the Czech Republic, Austria and Italy eastwards. In parts of Germany and the Alps transitional forms occur.

Platycleis sabulosa Azam

Platycleis sabulosa Azam, 1901: 157.

Sand Bush-cricket; (F) Decticelle des sables.

REFERENCES TO SONG. **Oscillogram**: Heller, 1988; Latimer & Broughton, 1984; Ragge, 1990; Samways, 1976*b*; Schmidt, 1996. **Diagram**: Samways, 1976*e*. **Sonagram**: Samways, 1976*b*. **Verbal description only**: Defaut, 1987; Samways, 1967*d*.

RECOGNITION. Females of this species are characterized by the seventh abdominal sternite, which has a pair of lateral protuberances (often connected so as to form a transverse ridge). Males are difficult to recognize morphologically, but may be distinguished from *P. albopunctata* by the shape of the titillators and from *P. intermedia* and *P. affinis* by the shape of the tenth abdominal tergite (see Ragge, 1990). Both sexes have a shorter pronotum than *P. falx* (usually less than 6·5 mm in the male, 7·0 mm in the female; usually more than these values in *P. falx*).

In the field males may be distinguished from most other western European members of the genus by their calling song (see under *P. albopunctata*).

SONG (Figs 276–284. CD 1, track 31). The calling song consists of long sequences of echemes repeated fairly regularly at the rate of about 1–3/s and each usually consisting of 6–7 (occasionally 5 and rarely up to 10) syllables. Oscillographic analysis shows that opening hemisyllables are often absent and that the closing hemisyllables usually last about 10–40 ms and are repeated within an echeme at the rate of about 20–40/s. The duration of a single echeme of six syllables is usually about 150–300 ms and the interval between two echemes about 150–500 ms. In dull weather and at night the repetition rates are decreased and the duration of the syllables and echemes increased. The first syllable in an echeme is usually quieter than the remaining ones and the syllables sometimes become steadily louder through most of the echeme. Microsyllables are usually absent.

DISTRIBUTION. The Mediterranean coast of France, the Iberian Peninsula, Morocco and Algeria; also recorded from Italy (including Sicily), the Canary Islands and Israel.

Platycleis affinis Fieber

Platycleis affinis Fieber, 1853: 150.

Tuberous Bush-cricket; (F) Decticelle côtière; (D) Südliche Beißschrecke.

REFERENCES TO SONG. **Oscillogram**: Broughton, 1955b, 1965; Heller, 1988; Latimer & Broughton, 1984; Ragge, 1990; Samways, 1976b, 1976c; Schmidt & Schach, 1978. **Diagram**: Broughton & Lewis, 1979; Samways, 1976e. **Sonagram**: Broughton et al., 1975; Samways, 1976b. **Wing-movement**: Heller, 1988. **Frequency information**: Heller, 1988; Latimer & Broughton, 1984. **Verbal description only**: Bellmann & Luquet, 1995; Defaut, 1987; Samways, 1976d. **Disc recording**: Bonnet, 1995 (CD).

RECOGNITION. Females of this species may be recognized by the seventh abdominal sternite, which has a large median protuberance; there is a similar protuberance in *P. falx*, but in that species the ovipositor is shorter and deeper than in *P. affinis* and the subgenital plate has a paired prominence towards the base. Males may be distinguished from *P. albopunctata* by the shape of the titillators, from *P. sabulosa* and *P. falx* by the shape of the tenth abdominal tergite and (usually) from *P. intermedia* by the conspicuously pale-coloured basal part of vein M in the fore wings (see Ragge, 1990).

In the field singing males may be recognized by the mixture of short and long echemes, mostly ending with microsyllables and with a syllable repetition rate of less than 20/s.

SONG (Figs 285–292. CD 1, track 32). The calling song consists of a mixture of short echemes lasting less than 1 s and usually composed of fewer than 7 macrosyllables, and longer

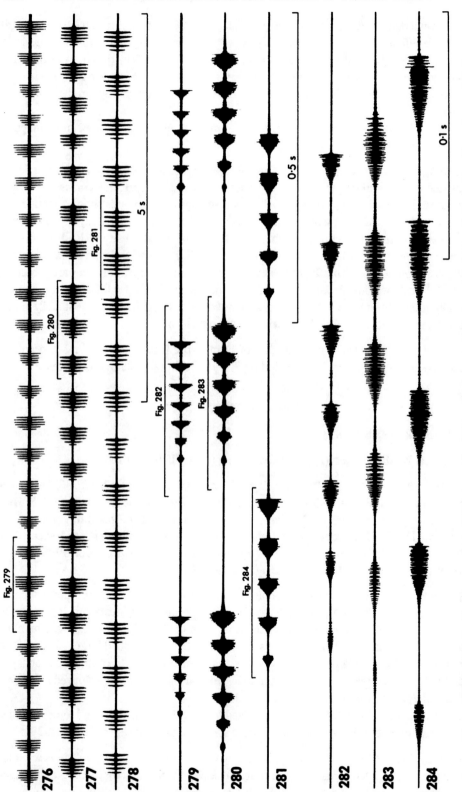

Figures 276–284 Oscillograms at three different speeds of the calling songs of three males of *Platycleis sabulosa*. (From Ragge, 1990)

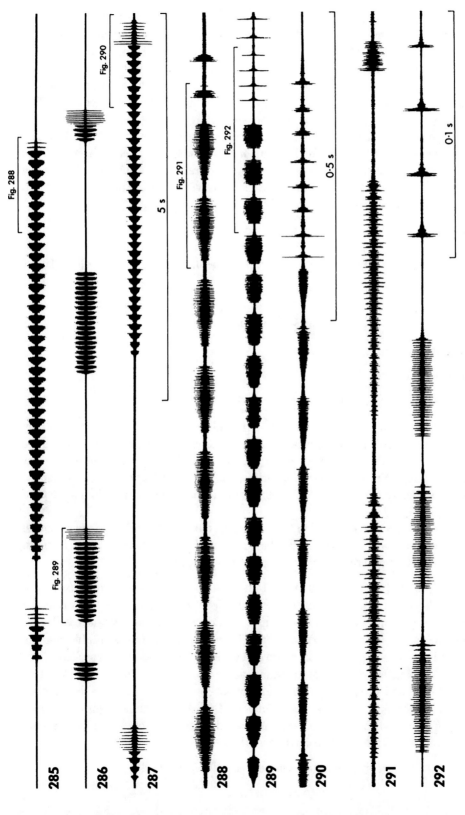

Figures 285–292 Oscillograms at three different speeds of the calling songs of three males of *Platycleis affinis*. Fig. 286 is from a typical daytime song in sunny weather; Figs 285 and 287 are from night-time songs and show the slower rhythm resulting from lower body temperatures. (From Ragge. 1990)

echemes lasting 1–5 s and composed of 8–50 macrosyllables. Each echeme usually ends with a series of 2–9 microsyllables. The echemes are often grouped into one or two short ones followed by a long one, but sometimes they follow one another quite irregularly and occasionally there is a fairly regular sequence of long echemes with few short ones. Oscillographic analysis shows that opening hemisyllables are usually absent and that the closing macrosyllables usually last about 30–110 ms and are repeated at the rate of about 10–20/s. The microsyllables usually last about 3–15 ms and are repeated at the rate of about 17–45/s; the microsyllable sequence at the end of an echeme seldom lasts more than 0·3 s. The echemes often begin quietly and the first one or two syllables are often shorter than the remaining ones. The intervals between echemes vary greatly: long echemes often follow short echemes with an interval of less than a second or with no pause at all; intervals of a few seconds are common and sometimes there are longer intervals of 30 s or more.

DISTRIBUTION. Like *P. intermedia* this species occurs in a large part of the Mediterranean Region, including Morocco and Algeria. Its range extends eastwards to Turkey and southern Asia, and in central Europe it occurs a little further north than *P. intermedia* in Lower Austria and Hungary.

Platycleis romana Ramme

Platycleis romana Ramme, 1927: 142.

Roman Bush-cricket.

REFERENCES TO SONG. **Oscillogram**: Ragge, 1990.

RECOGNITION. Females may be recognized by the subgenital plate, which has unusually broad lobes with a fairly narrow median groove; the seventh abdominal sternite lacks the large protuberance shown by *P. affinis* and *P. falx*, but is swollen in the anterior part. Males may be distinguished from Italian *P. albopunctata*, *P. affinis* and *P. intermedia* by the shape of the titillators, and from *P. sabulosa* and *P. falx* by the shape of the tenth abdominal tergite (see Ragge, 1990).

  Singing males may be recognized in the field by the mixture of short and long echemes, each usually ending with microsyllables. The song could be confused with that of *P. affinis*, but the syllable repetition rate is much higher, more than 20/s.

SONG (Figs 293, 294, 296, 297, 299, 300. CD 1, track 33). The calling song consists of a mixture of short echemes usually lasting less than 0·5 s and composed of fewer than 10 macrosyllables, and longer echemes lasting up to 3 s and composed of 15–80 macrosyllables. Each echeme usually ends with a series of up to 10 microsyllables. Oscillographic analysis shows that quiet opening hemisyllables are often present; the closing macrosyllables last about 20–30 ms and are repeated at the rate of about 25–40/s. The closing microsyllables last about 2–3 ms and are repeated at the rate of about 35–50/s; the microsyllable sequence at the end of an echeme usually lasts less than 0·3 s. The echemes usually begin quietly and the first one or two syllables are often shorter than the remaining ones. The intervals between echemes vary from less than 1 s to more than 4 s.

DISTRIBUTION. Known only from Italy, where it is quite widespread.

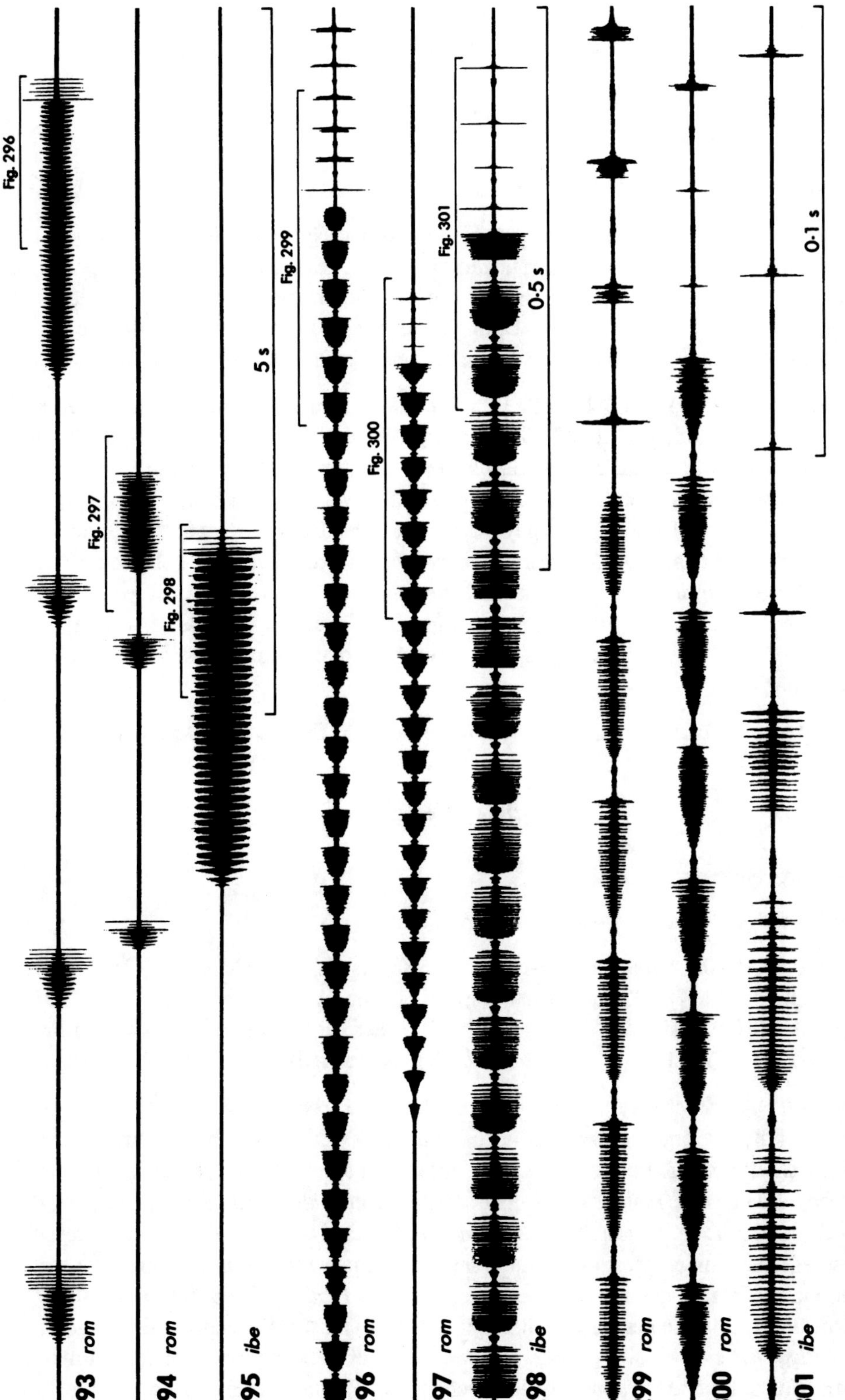

Figures 293–301 Oscillograms at three different speeds of the male calling songs of (293, 294, 296, 297, 299, 300) *Platycleis romana* and (295, 298, 301) *P. iberica*. (The oscillograms of *P. romana* are from Ragge, 1990)

Platycleis iberica Zeuner

Platycleis iberica Zeuner, 1941: 32.

Iberian Bush-cricket.

REFERENCES TO SONG. No published work known to us.

RECOGNITION. Females may be distinguished from other Iberian species of *Platycleis* by the very broad, subquadrate lobes of the subgenital plate. Males are very similar in appearance to *P. albopunctata*, but can be distinguished from it at once in the field by their quite different calling song; *P. falx* has a very similar calling song, but the male tenth abdominal tergite of that species has much less pointed lobes and the apical part of the titillators is more slender.

SONG (Figs 295, 298, 301. CD 1, track 34). The calling song consists of a long sequence of echemes, each usually lasting 2·0–3·5 s, composed of about 35–55 macrosyllables and normally ending with about 2–6 microsyllables. The echemes do not vary greatly in duration within one song and are usually separated by intervals of about 5–15 s. Oscillographic analysis of the song of the single male studied (in complete darkness at an air temperature of 25°C) showed that quiet opening hemisyllables were present in both macro- and microsyllables. The closing macrosyllables lasted about 50–60 ms and were repeated at the rate of about 22–25/s. The microsyllable sequence at the end of an echeme seldom lasted more than 0·2 s. The echemes began quietly, reaching maximum intensity by the third or fourth syllable.

DISTRIBUTION. Known at present only from the Sierra de Gredos in central Spain.

Platycleis falx (Fabricius)

Locusta falx Fabricius, 1775: 286.

Falcate Bush-cricket; (F) Decticelle à serpe.

REFERENCES TO SONG. **Oscillogram**: Heller, 1988; Ragge, 1990; Samways, 1976*b*. **Diagram**: Samways, 1976*e*. **Sonagram**: Samways, 1976*b*. **Wing-movement**: Heller, 1988. **Frequency information**: R.-G. Busnel, 1955; Heller, 1988. **Verbal description only**: Samways, 1976*d*; Yersin, 1857 (as *intermedius*). **Disc recording**: Bonnet, 1995 (CD).

RECOGNITION. Females may be recognized by the paired prominence towards the base of the subgenital plate, and the large median protuberance on the seventh abdominal sternite; there is a similar protuberance in *P. affinis*, but in that species the ovipositor is longer and less deep than in *P. falx*. Males may be distinguished from *P. albopunctata* by the shape of the titillators, and from *P. affinis* and *P. intermedia* by the shape of the tenth abdominal tergite. Both sexes have a longer pronotum than *P. sabulosa* (usually more than 6·5 mm in the male, 7·0 mm in the female; usually less than these values in *P. sabulosa*) (see Ragge, 1990).

   In the field singing males may be distinguished from most other western European species of *Platycleis* by the calling song, which consists of long sequences of fairly uniform echemes (in contrast to the short and long echemes of *P. affinis* and *P. romana*), each lasting more than a second. The calling song of *P. iberica* does not seem to be significantly different, but that species is at present known only from the Sierra de Gredos in central Spain.

SONG (Figs 302–310. CD 1, track 35). The calling song consists of a long sequence of echemes, each usually lasting 1–4 s, composed of about 25–90 macrosyllables and often ending with about 2–5 microsyllables. Within one song the echemes are usually fairly uniform in duration and are usually separated by intervals of about 4–10 s. Oscillographic analysis shows that opening hemisyllables are usually absent and that the closing macrosyllables usually last about 20–85 ms and are repeated at the rate of about 10–30/s. The microsyllables usually last about 1–12 ms and are repeated at the rate of about 20–40/s; the microsyllable sequence at the end of an echeme seldom lasts more than 0·2 s. The echemes usually begin quietly, reaching maximum intensity by the third to seventh syllable, and the first one or two syllables are usually shorter than the remaining ones.

DISTRIBUTION. Found on Madeira and in the western Mediterranean Region from Spain and Morocco to the Italian Peninsula and Tunisia.

Platycleis intermedia (Serville)

Decticus intermedius Serville, 1838: 488.

Intermediate Bush-cricket; (F) Decticelle intermédiaire.

REFERENCES TO SONG. **Oscillogram**: Broughton, 1965 (as *sabulosa*), Elsner & Popov, 1978; Heller, 1988; Latimer & Broughton, 1984; Lewis, 1974; Messina *et al.*, 1980; Ragge, 1990; Samways, 1976*b*, 1976*c*; Schmidt, 1996. **Diagram**: Broughton & Lewis, 1979; Samways, 1976*b*, 1976*e*. **Sonagram**: Broughton *et al.*, 1975; Samways, 1976*b*. **Wing-movement**: Heller, 1988. **Frequency information**: Heller, 1988; Lewis, Seymour & Broughton, 1975; Zhantiev & Dubrovin, 1971. **Verbal description only**: Chopard, 1952; Pinedo, 1985; Samways, 1976*d*. **Disc recording**: Bonnet, 1995 (CD).

RECOGNITION. Females may be recognized by the seventh abdominal sternite, which has two prominences, one near the middle and one near the posterior margin, both variable in shape and size. Males may be distinguished from *P. albopunctata* by the shape of the titillators, from *P. sabulosa* and *P. falx* by the shape of the tenth abdominal tergite and (usually) from *P. affinis* by the inconspicuously coloured basal part of vein M in the fore wings (see Ragge, 1990).

In the field males may be recognized by the mainly nocturnal calling song, which consists of long sequences of disyllabic echemes repeated at the rate of 2–3/s. The echemes of *P. sepium* are also disyllabic, but are repeated more rapidly than 4/s and are often grouped into short sequences of 2–4 echemes.

SONG (Figs 311–316. CD 1, track 36). The calling song, produced mainly in the evening and at night, consists of long sequences of disyllabic echemes repeated fairly regularly at the rate of about 2–3/s. Oscillographic analysis shows that opening hemisyllables are usually absent and that the closing hemisyllables usually last about 40–90 ms, the second of each pair often being a little longer, and sometimes louder, than the first. The duration of each echeme is about 130–180 ms and the interval between two echemes is about 140–250 ms. Microsyllables are usually absent.

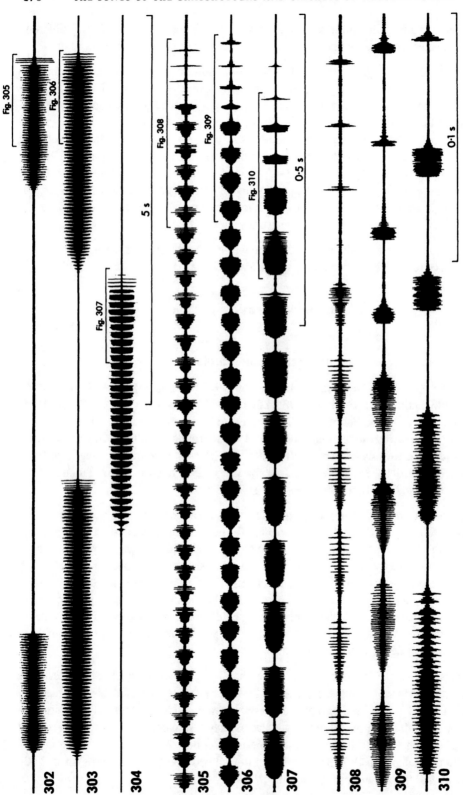

Figures 302–310 Oscillograms at three different speeds of the calling songs of three males of *Platycleis falx*. Figs 302 and 303 are from typical daytime songs in sunny weather; Fig. 304 is from a night-time song and shows the slower rhythm resulting from a lower body temperature. (From Ragge, 1990)

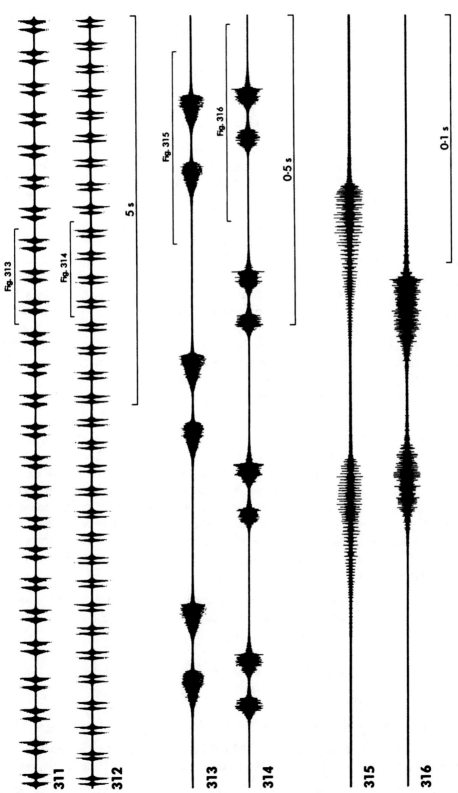

Figures 311–316 Oscillograms at three different speeds of the calling songs of two males of *Platycleis intermedia*. (Figs 311, 313 and 315 are from Ragge, 1990)

DISTRIBUTION. Found in a large part of the Mediterranean Region, from Spain and Morocco to Turkey, and further east into the southern part of temperate Asia.

Platycleis sepium (Yersin)

Decticus sepium Yersin, 1854*a*: 68.

Sepia Bush-cricket; (F) Decticelle échassière; (D) Zaunschrecke.

REFERENCES TO SONG. **Oscillogram**: Dubrovin & Zhantiev, 1970; Heller, 1988; Ragge, 1990; Schmidt, 1989. **Wing-movement** and **frequency information**: Heller, 1988. **Musical notation**: Yersin, 1854*b*. **Verbal description only**: Bellmann & Luquet, 1995; Chopard, 1922. **Disc recording**: Bonnet, 1995 (CD).

RECOGNITION (Plate 1: 12). This species differs from all other western European species of *Platycleis* of comparable size in being brachypterous, the fore wings not reaching the tip of the abdomen. Females are also well characterized by the terminal abdominal sternites: both the sixth and seventh sternites have a bifid prominence and the subgenital plate is uniquely shaped (see Ragge, 1990). *P. sepium* has exceptionally long hind legs, the hind femora exceeding 22 mm; the smaller brachypterous species of *Platycleis* (*veyseli* and *nigrosignata*) have hind femora shorter than 21 mm, as do all the western European species of *Metrioptera*.

In the field males may be recognized by the calling song, which consists of disyllablic echemes repeated at the rate of more than 4/s. The echemes of *P. intermedia* are also disyllabic, but are repeated less frequently than 4/s. The song of *Tettigonia viridissima* can be confused with that of *P. sepium*, but *T. viridissima*, being fully winged, is very different in appearance.

SONG (Figs 317–324. CD 1, track 37). The calling song, heard both during the day and at night, consists of disyllabic echemes repeated at the rate of about 4–15/s (depending on temperature), often grouped into short sequences of 2–4 echemes (Fig. 317, 318) but sometimes in long sequences of indefinite duration (Fig. 319, 320). Oscillographic analysis shows that opening hemisyllables are often present (though always very quiet) and that the closing hemisyllables usually last about 10–50 ms. The duration of each echeme (excluding the first opening hemisyllable, when present) is about 45–100 ms and the intervals between the echemes (within a group or during long sequences) are usually about 60–130 ms. There are no microsyllables.

DISTRIBUTION. Southern Europe from the Mediterranean coast of France to western Georgia, including the Italian and Balkan Peninsulas, Crimea and Asia Minor. From Iberia there are only old records from Catalonia.

Platycleis tessellata (Charpentier)

Locusta tessellata Charpentier, 1825: 121.

Brown-spotted Bush-cricket; (F) Decticelle carroyée; (D) Braunfleckige Beißschrecke.

REFERENCES TO SONG. **Oscillogram**: Grein, 1984; Heller, 1988; Ragge 1990; Schmidt, 1989. **Diagram**: Duijm & Kruseman, 1983. **Wing-movement**: Heller, 1988. **Frequency**

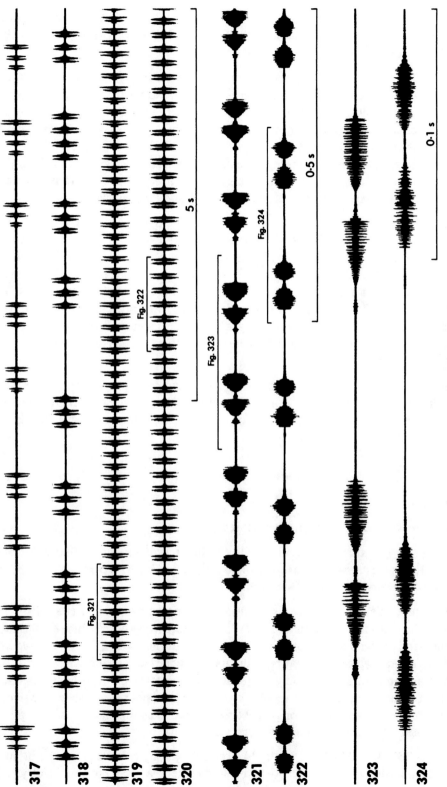

Figures 317–324 Oscillograms at three different speeds of the calling songs of three males of *Platycleis sepium*. Figs 317–320 show the two types of song-pattern produced by this species. Figs 318 and 320 were taken from the same male. (Figs 318, 320, 322 and 324 are from Ragge, 1990)

information: Heller, 1988; Latimer & Broughton, 1984. **Verbal description only**: Bellmann & Luquet, 1995; Chopard, 1922; Rentz, 1963. **Disc recording**: Bonnet, 1995 (CD); Grein, 1984 (LP).

RECOGNITION (Plate 2: 1). Among the small western European species of *Platycleis*, *P. tessellata* may be distinguished from *P. montana* and *P. stricta* by the short, strongly curved ovipositor and modified seventh abdominal sternite of the female (see Ragge,1990), and the male cerci, of which the inner tooth is nearer the tip than the base. Both sexes may be distinguished from *P. veyseli* and *P. nigrosignata* by the fully developed wings.

Live males in captivity may be distinguished from all other fully winged species of *Platycleis* by the calling song. However, the calling song of *P. tessellata* seems to be virtually identical with those of the two brachypterous species *P. veyseli* and *P. nigrosignata*.

SONG (Figs 325–330. CD 1, track 38). The calling song consists of a series of quiet diplosyllables, each lasting about 80–200 ms and composed of an opening hemisyllable lasting about 12–50 ms and a closing hemisyllable lasting about 30–130 ms. The syllables are sometimes repeated regularly at the rate of about 6–8/s for long periods of a minute or more (Fig. 325), but at other times they are repeated much less frequently (about 1–4/s) and less regularly (Fig. 326). Sometimes they are grouped into rather loose echemes of very variable duration. The intervals between syllables vary from about 30–100 ms when they are repeated regularly to about $0 \cdot 1$–$1 \cdot 0$ s or more when the repetition rate is irregular. There are no microsyllables.

DISTRIBUTION. Widespread in the western Mediterranean Region (including North Africa), extending northwards to central France and southern Germany, and eastwards to southern Russia and Asia Minor, but not occurring in Austria or the southern part of the Balkan Peninsula. Introduced into California, USA.

Platycleis veyseli Koçak

Platycleis (Tessellana) veyseli Koçak, 1984: 169.

Veysel's Bush-cricket; (D) Kleine Beißschrecke.

REFERENCES TO SONG. **Oscillogram**: Heller, 1988 (as *vittata*); Ragge, 1990; Schmidt & Schach, 1978 (as *vit.*). **Wing-movement** and **frequency information**: Heller, 1988 (as *vit.*). **Verbal description only**: Harz, 1962 (as *vit.*).

RECOGNITION. This species may be distinguished from most other small species of *Platycleis* occurring in western Europe by its reduced wings, the fore wings not reaching the tip of the abdomen and the hind wings not reaching the tips of the fore wings; from *P. nigrosignata*, which is also brachypterous, it may be distinguished by the lack of a median carina on the prozona of the pronotum. The calling song enables live males in captivity to be distinguished from most western European species of *Platycleis*, but does not seem to show any significant difference from the calling songs of *P. tessellata* and *P. nigrosignata*.

For a note on the specific name of this species see p. 10.

SONG (Figs 331, 333, 335. CD 1, track 39). The calling song consists of a series of quiet diplosyllables, each lasting about 60–80 ms and composed of an opening hemisyllable

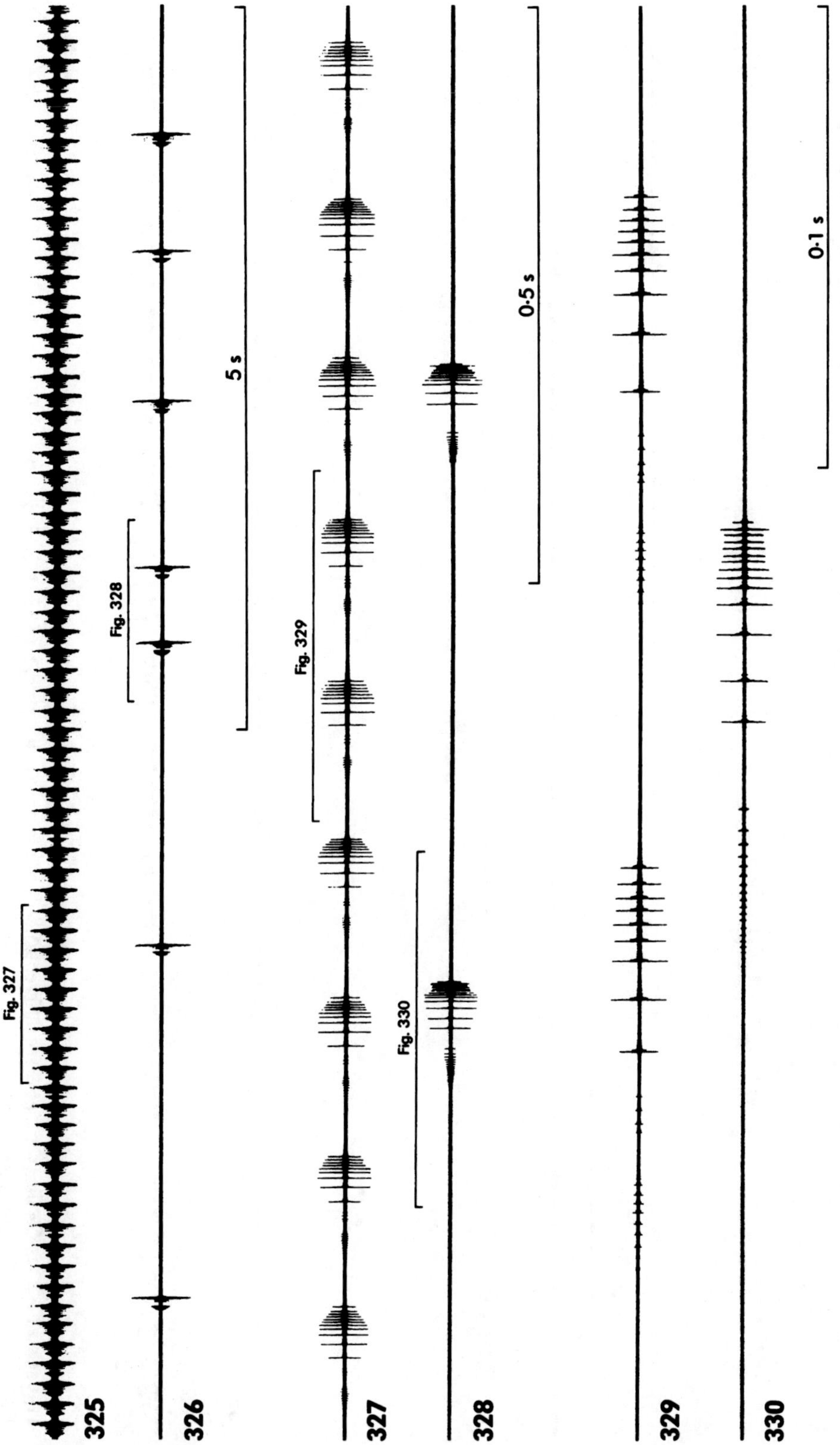

Figures 325–330 Oscillograms at three different speeds of the calling songs of two males of *Platycleis tessellata*, showing the two types of song-pattern produced by this species. (From Ragge, 1990)

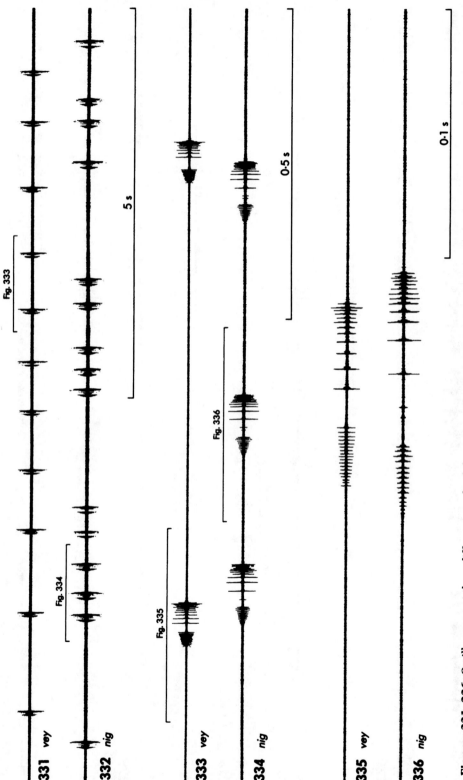

Figures 331–336 Oscillograms at three different speeds of the male calling songs of (331, 333, 335) *Platycleis veyseli* and (332, 334, 336) *P. nigrosignata*. (The oscillograms of *P. veyseli* are from Ragge, 1990)

lasting about 20–25 ms and a closing hemisyllable lasting about 25–35 ms. In the song of the single male studied (in dim light at an air temperature of 26°C) the syllables were grouped into loose echemes of 7–16 syllables, during which the syllable repetition rate averaged about 1/s and the syllables were separated by intervals of about 0·4–1·6 s. It is likely that the study of further males under varying conditions would show that the song varies in the same way as that of *P. tessellata* (p. 180). There are no microsyllables.

DISTRIBUTION. In the area studied *P. veyseli* occurs only in the extreme east of Austria; elsewhere its range extends through south-east Europe (but not the southern part of the Balkan Peninsula) to the southern and central parts of temperate Asia.

Platycleis nigrosignata (Costa)

Decticus (Platycleis) nigrosignatus Costa, 1863: 30.

Black-marked Bush-cricket.

REFERENCES TO SONG. **Oscillogram, wing-movement** and **frequency information**: Heller, 1988.

RECOGNITION. *P. nigrosignata* resembles *P. veyseli* in being small and brachypterous, the fore wings not reaching the tip of the abdomen, but differs from that species in having a median carina along the whole length of the pronotum.

SONG (Figs 332, 334, 336. CD 1, track 40). The calling song is very similar to those of *P. tessellata* and *P. veyseli*, consisting of a series of quiet diplosyllables, often grouped into loose echemes of about 2–8 syllables but sometimes repeated more regularly for long periods. Within a loose echeme the syllables are repeated at the rate of about 2–4/s and are separated by intervals of about 0·1–0·5 s. Oscillographic analysis of the song of the single male studied (in darkness at an air temperature of 22°C) showed that the opening hemisyllables lasted about 30 ms and the closing hemisyllables about 60 ms. There are no microsyllables.

DISTRIBUTION. In the area studied *P. nigrosignata* occurs only in southern Italy; elsewhere its range extends through south-east Europe to Asia Minor.

Platycleis montana (Kollar)

Locusta montana Kollar, 1833: 79.

Steppe Bush-cricket; (F) Decticelle des steppes; (D) Steppen-Beißschrecke.

REFERENCES TO SONG. **Oscillogram**: Heller, 1988; Ragge, 1990. **Diagram**: Bellmann, 1985a, 1988, 1993a; Bellmann & Luquet, 1995. **Cassette recording**: Bellmann, 1985b, 1993b.

RECOGNITION. This species may be distinguished from *P. tessellata, P. veyseli* and *P. nigrosignata* by the long, gently curved ovipositor and unmodified seventh abdominal sternite of the female, and the male cerci, of which the inner tooth is nearer the base than the tip. Females may be distinguished from the related species *P. stricta* by the truncate subgenital plate and males by the shape of the tenth abdominal tergite (see Ragge, 1990).

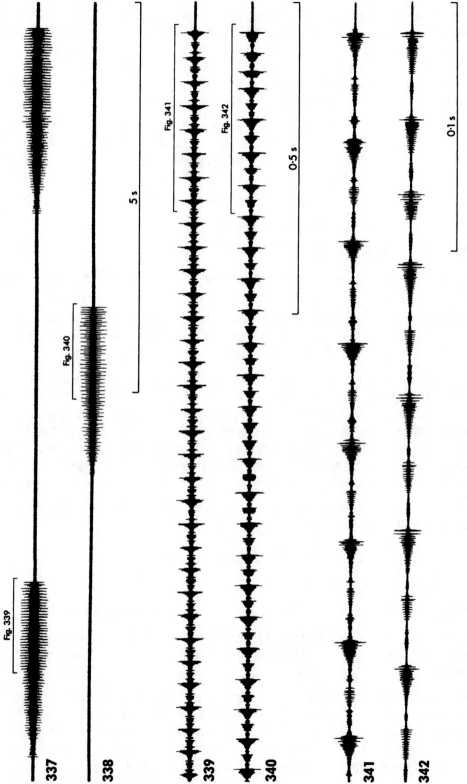

Figures 337–342 Oscillograms at three different speeds of the calling songs of two males of *Platycleis montana*. (From Ragge, 1990)

Live males in captivity may be recognized by the calling song.

SONG (Figs 337–342. CD 1, track 41). The calling song consists of a series of quiet echemes, each lasting about 1–2 s, sometimes repeated regularly at the rate of one every 3–4 s but at other times less frequently and more irregularly. Oscillographic analysis shows each echeme to consist of about 60–140 alternately quieter and louder closing hemisyllables following one another at the rate of about 40–70/s. The quieter closing hemisyllables last about 4–11 ms and the louder ones about 7–17 ms; opening hemisyllables, always quieter than the closing hemisyllables, are often also present. The echemes are usually quieter at the beginning, reaching maximum intensity after about 10–20 syllables. Sometimes the echemes end in a series of loud closing syllables without the interposition of quieter ones (Figs 340, 342), and occasionally there are two quieter closing hemisyllables between two consecutive louder ones. There are no microsyllables.

DISTRIBUTION. This primarily eastern European species occurs in Germany (very locally), eastern Austria, Hungary, Serbia and eastwards to central Asia.

Platycleis stricta (Zeller)

Decticus strictus Zeller, 1849: 116.

Italian Bush-cricket; (D) Südöstliche Beißschrecke.

REFERENCES TO SONG. **Oscillogram**: Heller, 1988; Ragge, 1990.

RECOGNITION. Like *P. montana*, this species may be distinguished from *P. tessellata*, *P. veyseli* and *P. nigrosignata* by the long, gently curved ovipositor and unmodified seventh abdominal sternite of the female, and the male cerci, of which the inner tooth is nearer the base than the tip. Females may be distinguished from *P. montana* by the bilobed subgenital plate, and males by the shape of the tenth abdominal tergite (see Ragge, 1990).

A stationary live male in captivity may be recognized by the calling song.

SONG (Figs 343–347. CD 1, track 42). The fully developed calling song from a stationary male consists of short echemes interspersed with much longer ones and often continues for several minutes. There are about 4–7 short echemes, repeated at intervals of about 0·3–1·2 s, between successive longer echemes; the short echemes last about 50–100 ms and the longer ones about 1–2 s. Oscillographic analysis shows that the short echemes consist of a macrosyllable lasting about 20–30 ms followed by a series of about 3–4 microsyllables, each lasting about 1–5 ms and repeated at the rate of about 50–55/s. The longer echemes follow immediately after a short echeme of this type and consist of a dense sequence of about 90–170 macrosyllables repeated at the rate of about 60–90/s. The syllables in these larger echemes are in groups of three (occasionally four) in which the first one is the quietest and shortest (lasting about 8–12 ms) and the last one the loudest and longest (lasting about 15–20 ms); each group lasts about 40–45 ms and the groups follow one another without any intervening pauses. Opening hemisyllables, always quieter than the closing hemisyllables, are usually present in both the shorter and longer echemes (and are included in the syllable durations given above). There are often small groups of microsyllables interrupting the regular flow of the longer echemes.

While singing males are moving they generally produce only the short echemes.

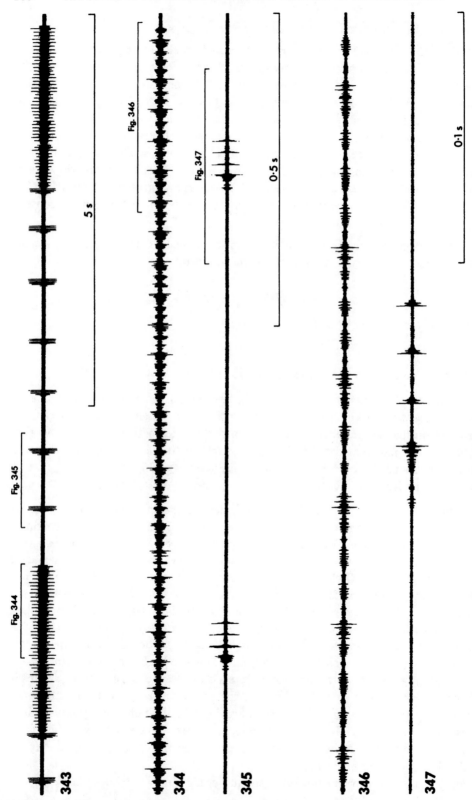

Figures 343–347 Oscillograms at three different speeds of the male calling song of *Platycleis stricta*. (From Ragge, 1990)

DISTRIBUTION. Widespread in Italy and also recorded from Sardinia, the eastern Adriatic coast, Bulgaria and the Ukraine.

Metrioptera brachyptera (Linnaeus)

Gryllus brachypterus Linnaeus, 1761: 237.

Bog Bush-cricket; (F) Decticelle des bruyères; (D) Kurzflüglige Beißschrecke; (NL) Heidesabelsprinkhaan; (DK) Hedegræshoppe; (S) Ljungvårtbitare; (SF) Vihreä Töpökatti.

REFERENCES TO SONG. **Oscillogram**: Ahlén, 1981; Dubrovin & Zhantiev, 1970; Grein, 1984; Grzeschik, 1969; Heller, 1988; Holst, 1970, 1986; Howse *et al.*, 1971; Kleukers *et al.*, 1997; Latimer & Broughton, 1984; Lewis *et al.*, 1971; Rheinlaender & Kalmring, 1973. **Diagram**: Bellmann, 1985*a*, 1988, 1993*a*; Bellmann & Luquet, 1995; Duijm & Kruseman, 1983; Holst, 1970; Lewis *et al.*, 1971; Ragge, 1965; Wallin, 1979; Zhantiev, 1981. **Sonagram**: Broughton *et al.*, 1975. **Wing-movement**: Heller, 1988. **Frequency information**: Ahlén, 1981; Dubrovin, 1977; Dubrovin & Zhantiev, 1970; Dumortier, 1963*c*; Froehlich, 1989; Heller, 1988; Lewis *et al.*, 1971; Sales & Pye, 1974; Zhantiev & Dubrovin, 1974. **Verbal description only**: Beier, 1955; Broughton, 1972*b*; Burr, 1911; Chopard, 1952; Faber, 1928; Harz, 1957; Weber, 1984; Yersin, 1954*b*; Zippelius, 1949. **Disc recording**: Andrieu & Dumortier, 1963 (LP), 1994 (CD); Bellmann, 1993*c* (CD); Bonnet, 1995 (CD); Grein, 1984 (LP); Odé, 1997 (CD); Ragge *et al.*, 1965 (LP). **Cassette recording**: Ahlén, 1982; Bellmann, 1985*b*, 1993*b*; Burton & Ragge, 1987; Wallin, 1979.

RECOGNITION (Plate 2: 2). The ten or so western European species of *Metrioptera* are all brachypterous, though occasional macropterous forms occur in several species. They differ from the smaller brachypterous species of *Platycleis* in lacking conspicuous dark markings on the fore wings and having no median groove on the female subgenital plate, and from *P. sepium* in having shorter hind legs (hind femora less than 21 mm long) and lacking the paired prominences on the sixth and seventh abdominal sternites of the female; there is, however, no clear dividing line between these two genera. *Metrioptera* is less brachypterous than *Pholidoptera* and usually has a more distinct median carina in the metazona of the pronotum.

Females of *M. brachyptera* may be distinguished from all other western European species of *Metrioptera* by the very small median incision on the broadly rounded subgenital plate. Males have well separated, backwardly directed (not diverging, as in *M. saussuriana*) processes on the tenth abdominal tergite, and the inner tooth on the cerci is about halfway along their length (not nearer the apex, as in *M. bicolor* and *M. roeselii*).

In the field the calling song of *M. brachyptera* can be distinguished from that of *M. saussuriana* in being composed of completely uniform echemes, none having microsyllables. It could be confused with the calling song of *Platycleis albopunctata*, but the echeme repetition rate is faster and these two species are very different in appearance.

SONG (Figs 348–356. CD 1, track 43). In warm sunshine the calling song consists of long sequences of echemes repeated regularly at the rate of 4–7/s. Oscillographic analysis shows that each echeme normally consists of four syllables but, as the first syllable is usually short and often very quiet, the echemes are frequently described as being trisyllabic. Opening

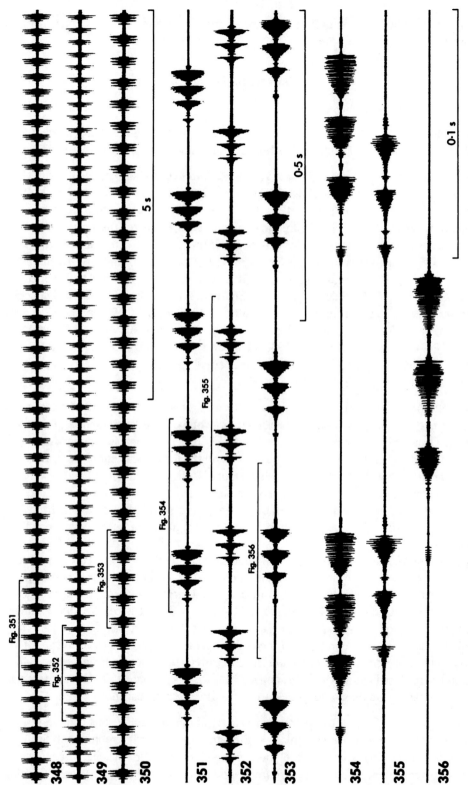

Figures 348–356 Oscillograms at three different speeds of the calling songs of three males of *Metrioptera brachyptera*.

hemisyllables are usually present, though often quiet; the fore wings are held in an open position between echemes, so that there is no opening hemisyllable before the first (quiet) closing syllable of each echeme but, most unusually for a tettigoniid song, there is usually an opening hemisyllable after the last closing hemisyllable. The closing hemisyllables usually last about 15–20 ms and are repeated within an echeme at the rate of about 35–55/s. The duration of a single echeme is about 75–120 ms and the interval between two successive echemes is about 65–110 ms. In dull weather and at night the echeme repetition rate can drop to as low as 2/s and the syllable repetition rate to 10/s; in such conditions the closing hemisyllables (of which the first is often louder than at higher temperatures) sometimes last as long as 90 ms and an echeme 400 ms or more. Under any conditions the echemes are very uniform in structure and there are no microsyllables.

DISTRIBUTION. Widespread in northern and central Europe, from Lapland to the Pyrenees, northern Italy and the northern part of the Balkan Peninsula. The range includes southern Britain and extends eastwards to the far east of temperate Asia.

Metrioptera saussuriana (Frey-Gessner)

Platycleis Saussuriana Frey-Gessner, 1872: 8.

Saussure's Bush-cricket; (F) Decticelle des alpages; (D) Gebirgs-Beißschrecke.

REFERENCES TO SONG. **Oscillogram**: Heller, 1988; Pfau, 1986; Ragge, 1987*b*. **Wing-movement** and **frequency information**: Heller, 1988. **Verbal description only**: Burr, 1911; Chopard, 1952; Defaut, 1981. **Disc recording**: Bonnet, 1995 (CD).

RECOGNITION. Females may be easily distinguished from *M. brachyptera* by the large median incision in the subgenital plate, and males, less easily, by the divergent posterior processes on the tenth abdominal tergite. Separating *M. saussuriana* and *M. buyssoni* is much more difficult, but the lobes of the female subgenital plate are broader and less pointed in *M. saussuriana* than in *M. buyssoni*, and the titillators can be used to separate males (see Defaut, 1981); the head and pronotum are brown in *M. saussuriana*, mostly green in *M. buyssoni*. In the field, careful listening enables the calling song of *M. saussuriana* to be distinguished from that of *M. brachyptera* by the microsyllables added to some of the echemes, but there seems to be no difference between the calling songs of *M. saussuriana* and *M. buyssoni*.

SONG (Figs 357, 358, 360, 361, 363, 364. CD 1, track 44). The calling song consists of long sequences of echemes repeated fairly regularly at the rate of 2–6/s, depending partly on temperature. Every so often a series of about 2–6 microsyllables is added to an echeme; the frequency of the echemes prolonged in this way is very variable, but there are usually between 8 and 25 simple echemes between two prolonged echemes. Oscillographic analysis shows that each simple echeme consists of about 5–10 macrosyllables repeated within the echeme at the rate about 50–90/s and becoming gradually louder during the course of the echeme; each closing macrosyllable lasts about 5–15 ms. The closing microsyllables that sometimes follow an echeme of this kind usually last about 1–3 ms each and are repeated at the rate of about 40–70/s. Quiet opening hemisyllables are sometimes present in both the macro- and microsyllables. A simple echeme usually lasts about 70–150 ms and

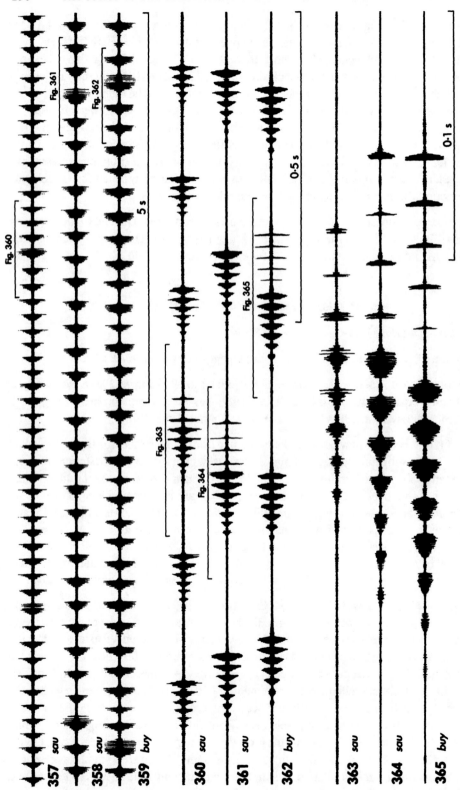

Figures 357–365 Oscillograms at three different speeds of the male calling songs of (357, 358, 360, 361, 363, 364) *Metrioptera saussuriana* and (359, 362, 365) *M. buyssoni*.

one followed by microsyllables about 150–300 ms. In one song studied, produced in dull conditions at an air temperature of 18°C, the simple echemes lasted about 180 ms and the closing macrosyllables about 18 ms; the echemes were repeated at a rather irregular rate averaging less than 1/s and more than half of them were followed by microsyllables.

DISTRIBUTION. An upland species found in the eastern Pyrenees, most of the French uplands and the Alps.

Metrioptera buyssoni (Saulcy)

Platycleis Buyssoni Saulcy, 1887: 190.

Buysson's Bush-cricket; (F) Decticelle albigeoise.

REFERENCES TO SONG. **Oscillogram**: Ragge, 1987*b*. **Verbal description only**: Defaut, 1981.

RECOGNITION. See under *M. saussuriana*. For notes on the taxonomic status of this species see p. 70.

SONG (Figs 359, 362, 365. CD 1, track 45). The calling song seems to be identical with that of *M. saussuriana* and the description given under that species applies in every respect to *M. buyssoni*.

DISTRIBUTION. Known at present only from the eastern half of the Pyrenees.

Metrioptera caprai Baccetti

Metrioptera caprai Baccetti, 1956: 113.

Capra's Bush-cricket.

REFERENCES TO SONG. **Oscillogram**: Ragge, 1987*b*. **Verbal description only**: Pfau, 1986.

RECOGNITION. *M. caprai* closely resembles *M. saussuriana*, showing only small differences in the genitalia. However, as these species are completely allopatric confusion is unlikely in practice.
    For notes on the taxonomic status of *M. caprai* see p. 70.

SONG (Figs 366–371. CD 1, track 46). The calling song consists of long sequences of echemes repeated fairly regularly at the rate of about 3–5/s and each composed of about 7–20 macrosyllables. Every so often a series of about 3–6 microsyllables is added to an echeme; there are usually about 5–15 simple echemes between two echemes prolonged in this way (Pfau, 1986, gives 30–76, based on a recording made by Herbst). Oscillographic analysis shows that the macrosyllables are repeated within an echeme at the rate of about 60–80/s and become gradually louder during at least most of the course of the echeme; each closing macrosyllable lasts about 8–12 ms. The microsyllables that sometimes follow such an echeme usually last about 2–3 ms each and are repeated within a series at the rate of about 65–85/s. Opening hemisyllables are often present (and sometimes quite loud) in both macro- and microsyllables. A simple echeme usually lasts about 100–270 ms and one followed by microsyllables about 170–300 ms.

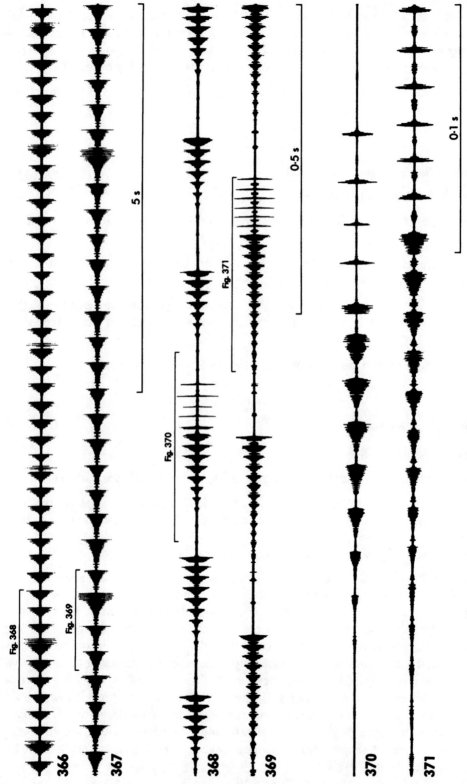

Figures 366–371 Oscillograms at three different speeds of the calling songs of two males of *Metrioptera caprai*.

In comparison to M. *saussuriana*, the number of macrosyllables in each echeme is larger, at least in the type locality (Terminillo in the Monti Reatini), where it averaged about 12 in the three males studied (Figs 366, 368, 370). However, the calling song of a male from the Gran Sasso d'Italia was more similar to *M. saussuriana*, with about 8 macrosyllables per echeme (Figs 367, 369, 371).

DISTRIBUTION. Known only from the Apennine range in Italy.

Metrioptera abbreviata (Serville)

Decticus abbreviatus Serville, 1838: 490.

Basque Bush-cricket; (F) Decticelle aquitaine.

REFERENCES TO SONG. **Oscillogram**: Heller, 1988; Pfau, 1986. **Wing-movement** and **frequency information**: Heller, 1988. **Verbal description only**: Defaut, 1987. **Disc recording**: Andrieu & Dumortier, 1963 (LP), 1994 (CD).

RECOGNITION. Males may be recognized easily by the very short cerci, with the inner tooth right at the base and normally hidden under the tenth abdominal tergite. Females may be recognized by the deeply incised subgenital plate, narrowed slightly just before the incision to form a 'waist'; the female subgenital plate of *M. burriana* is also deeply incised but has no 'waist'.

SONG (Figs 372–380. CD 1, track 47). The calling song is very similar to those of *M. saussuriana* and *M. buyssoni*, consisting of long sequences of echemes repeated fairly regularly at the rate of about 2–4/s and each composed of about 7–9 macrosyllables. The interval between two successive echemes is about 150–350 ms, and tends to be longer in relation to echeme duration than in the other two species (usually about three times the duration of an echeme in *M. abbreviata*, usually less than twice the duration of an echeme in *M. saussuriana* and *M. buyssoni*). Another difference is that, at least in our experience, a completely isolated male of *M. abbreviata*, uninfluenced by the songs of any other males, does not normally produce microsyllables; our recordings of four out of the six males we have studied contain no microsyllables. The other two males, which were quite near to other singing males, produced a series of about 3–7 microsyllables every so often at the end of an echeme; there were 14–81 simple echemes between two echemes prolonged in this way, and such prolonged echemes were, on average, more widely separated than in *M. saussuriana* and *M. buyssoni*. As in these two species, there are sometimes quiet opening hemisyllables in both the macro- and microsyllables produced by *M. abbreviata*.

DISTRIBUTION. Known for certain only from the central and western Pyrenees.

Metrioptera burriana Uvarov

Metrioptera burriana Uvarov, 1935: 75.

Burr's Bush-cricket.

REFERENCES TO SONG. **Oscillogram**: Heller, 1988; Pfau, 1986.

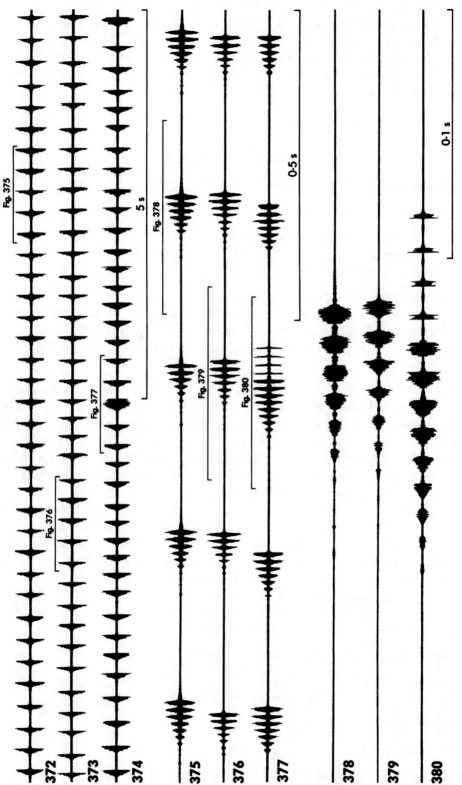

Figures 372–380 Oscillograms at three different speeds of the calling songs of three males of *Metrioptera abbreviata*.

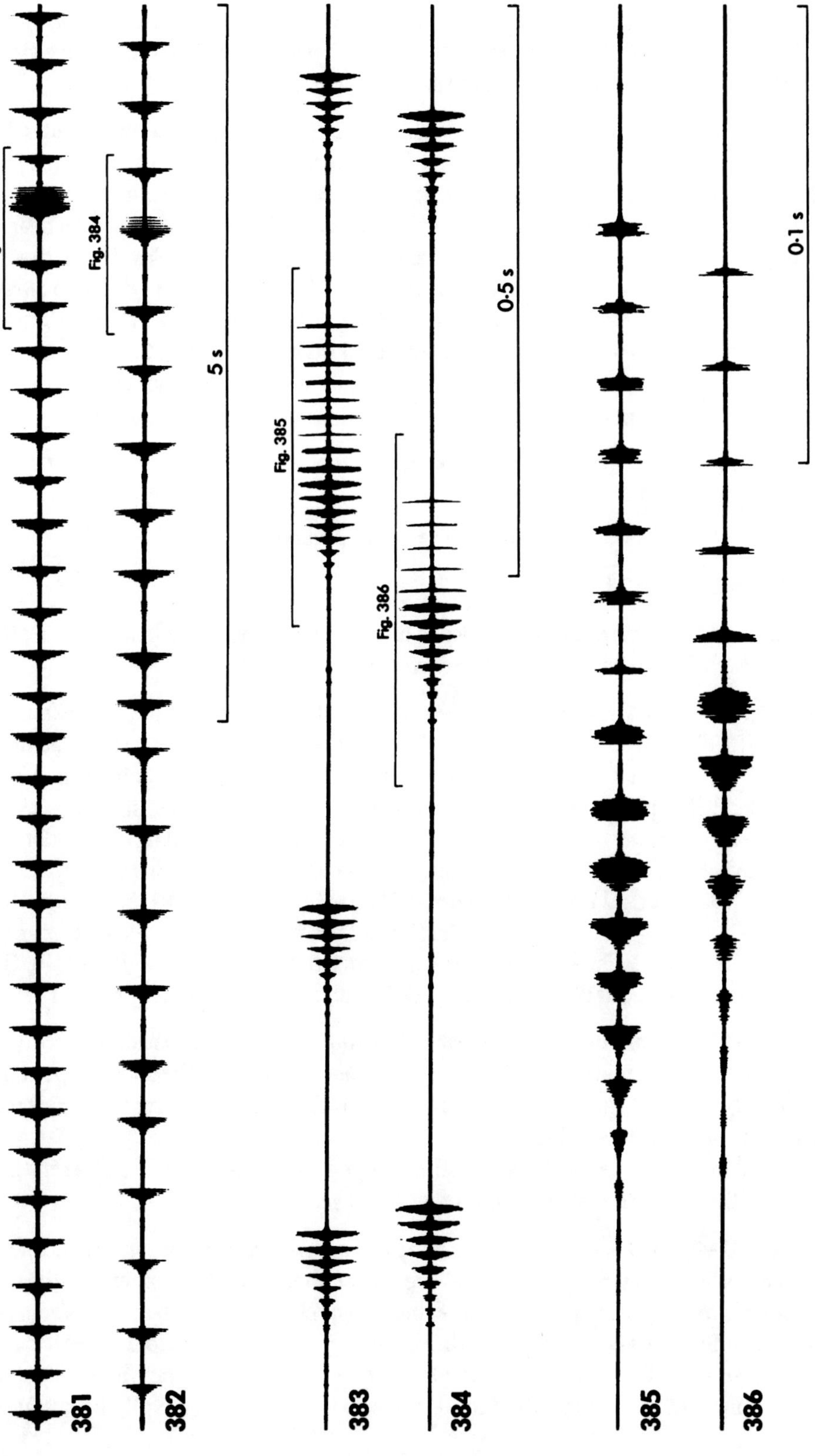

Figures 381–386 Oscillograms at three different speeds of the calling songs of two males of *Metrioptera burriana*.

RECOGNITION. Very similar to *M. abbreviata*, from which males may be distinguished by the much longer cerci, of which the inner tooth, although near the base, is quite easily visible, and females by the lack of a 'waist' on the subgenital plate.

SONG (Figs 381–386. CD 1, track 48). The calling song seems to show no significant difference from that of *M. abbreviata*, consisting of long sequences of echemes repeated fairly regularly at the rate of about 2–4/s and each composed of about 6–10 macrosyllables. As in *M. abbreviata*, our experience suggests that completely isolated males, unable to hear the songs of other conspecific males, are unlikely to produce microsyllables. Males that are near to other singing males produce a series of about 5–7 microsyllables every so often at the end of an echeme; in the four males studied the number of simple echemes between two successive echemes prolonged in this way varied from 47 to over 500.

The sound is often produced solely by the closing strokes of the fore wings, but there are sometimes quiet opening hemisyllables in both the macro- and microsyllables.

DISTRIBUTION. Found only in the vicinity of the Cantabrian Mountains in northern Spain.

Metrioptera bicolor (Philippi)

Locusta bicolor Philippi, 1830: 24.

Two-coloured Bush-cricket; (F) Decticelle bicolore; (D) Zweifarbige Beißschrecke; (NL) Lichtgroene Sabelsprinkhaan; (S) Grön Hedvårtbitare.

REFERENCES TO SONG. **Oscillogram**: Ahlén, 1981; Dubrovin, 1977; Dubrovin & Zhantiev, 1970; Grein, 1984; Heller, 1988; Holst, 1970, 1986; Kleukers *et al.*, 1997; Rheinlaender & Kalmring, 1973; Schmidt & Schach, 1978; Zhantiev, 1981. **Diagram**: Bellmann, 1985a, 1988, 1993a; Bellmann & Luquet, 1995; Duijm & Kruseman, 1983; Holst, 1970; Wallin, 1979. **Wing-movement**: Heller, 1988. **Frequency information**: Ahlén, 1981, Dubrovin, 1977; Dubrovin & Zhantiev, 1970; Dumortier, 1963c; Froehlich, 1989; Heller, 1988; Zhantiev & Dubronin, 1971. **Verbal description only**: Chopard, 1952; Faber, 1928; Fischer, 1850; Gerhardt, 1921; Harz, 1957, 1958a; Zippelius, 1949. **Disc recording**: Andrieu & Dumortier, 1963 (LP), 1994 (CD); Bellmann, 1993c (CD); Bonnet, 1995 (CD); Grein, 1984 (LP); Odé, 1997 (CD). **Cassette recording**: Ahlén, 1982; Bellmann, 1958b, 1993b; Wallin, 1979.

RECOGNITION. *M. bicolor* may be distinguished from all the preceding species of *Metrioptera* by the long cerci of the male, with the inner tooth near the tip, and the short ovipositor of the female. From *M. roeselii* it may be distinguished by the lack of a conspicuous pale margin round the pronotal lateral lobes.

In the field males can be recognized by their calling song, which is quite different from that of any other western European species of *Metrioptera*.

SONG (Figs 387–400. CD 1, track 49). The normal daytime calling song consists of a series of trisyllabic echemes repeated regularly at the rate of about 15–25/s. Sometimes the song is broken up into short and quite regular echeme-sequences lasting about 0·3–1·5 s and separated by brief intervals of about 0·1–0·5 s (Figs 387, 388), but at other times it is continued without interruption for long periods of up to 2 minutes or more (Figs 389, 390); when the song is of the intermittent kind, the echeme-sequences usually begin with a

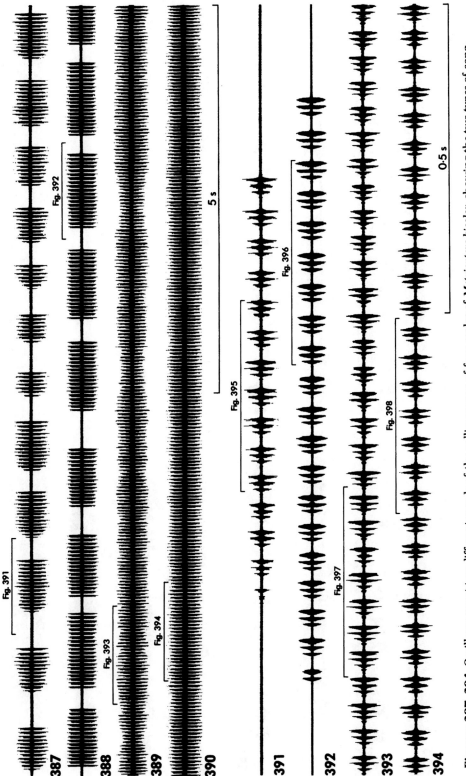

Figures 387–394 Oscillograms at two different speeds of the calling songs of four males of *Metrioptera bicolor*, showing the two types of song-pattern produced by this species.

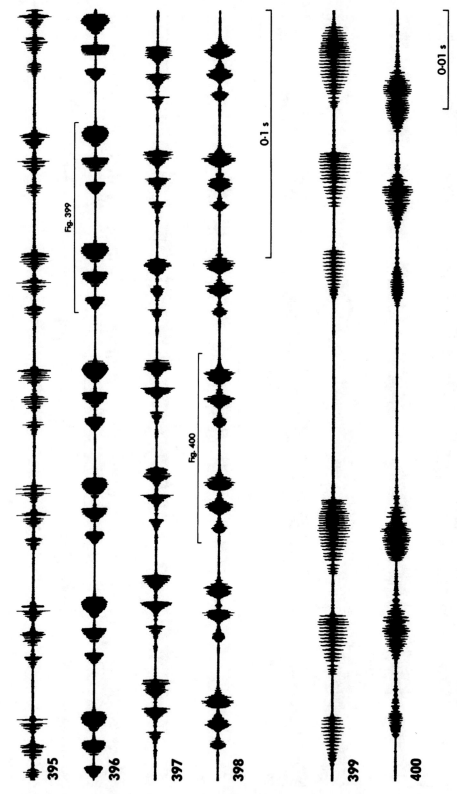

Figures 395–400 Faster oscillograms of the indicated parts of the songs of *Metrioptera bicolor* shown in Figs 391–394.

disyllabic echeme (or occasionally a single isolated syllable), but the remaining echemes are uniformly trisyllabic. Each trisyllabic echeme lasts about 22–32 ms and successive echemes are separated by intervals of about 10–25 ms. The opening strokes of the fore wings are often completely silent, but sometimes there are quiet opening hemisyllables. The last closing hemisyllable in each echeme usually lasts 6–10 ms, but the first is usually quieter and shorter (4–7 ms) and the middle one is often intermediate in duration and occasionally in loudness. Within each echeme the syllable repetition rate is usually between 80 and 120/s. There are no microsyllables.

At night, in much cooler conditions, Ahlén (1981, 1982) recorded a song consisting of regularly repeated echemes of 8–9 syllables. Each echeme lasted about 360–420 ms and each closing hemisyllable about 25–30 ms. The echemes were separated by intervals of about 150–250 ms and the syllable repetition rate within an echeme was about 20/s.

DISTRIBUTION. Widespread in Europe, from the eastern half of France to European Russia and from the extreme south of Sweden to the extreme north of Spain, northern Italy and the northern part of the Balkan Peninsula. Absent from the British Isles and Netherlands.

Metrioptera roeselii (Hagenbach)

Locusta Roeselii Hagenbach, 1822: 39.

Roesel's Bush-cricket; (F) Decticelle bariolée; (D) Roesels Beißschrecke; (NL) Greppelsprinkhaan; (S) Cikadavårtbitare; (SF) Ruskea Töpökatti.

REFERENCES TO SONG. **Oscillogram**: Ahlén, 1981; Broughton, 1965; Dubrovin, 1977; Dubrovin & Zhantiev, 1970; Grein, 1984; Heller, 1988; Holst, 1970, 1976; Kleukers *et al.*, 1997; Latimer & Broughton, 1984; Rheinlaender & Kalmring, 1973. **Diagram**: Bellmann, 1985a, 1988, 1993a; Bellmann & Luquet, 1995; Dubrovin, 1977; Duijm & Kruseman, 1983; Holst, 1970; Ragge, 1965; Wallin, 1979; Zhantiev, 1981. **Sonagram**: Broughton *et al.*, 1975; Sales & Pye, 1974; Samways, 1976a (as *azami*). **Wing-movement**: Heller, 1988. **Frequency information**: Ahlén, 1981; Bailey, 1970; Broughton, 1964; Dubrovin, 1977; Dubrovin & Zhantiev, 1970; Froehlich, 1989; Heller, 1988; Latimer, 1981a, 1981b; Sales & Pye, 1974; Zhantiev & Dubrovin, 1971. **Musical notation**: Yersin, 1854b (as *brevipennis*). **Verbal description only**: Broughton, 1972b; Defaut, 1987, 1988b; Faber, 1928; Haes & Else, 1976; Harz, 1957; Weber, 1984; Zippelius, 1949. **Disc recording**: Andrieu & Dumortier, 1963 (LP), 1994 (CD); Bellmann, 1993c (CD); Bonnet, 1995 (CD); Grein, 1984 (LP); Odé, 1997 (CD); Ragge *et al.*, 1965 (LP). **Cassette recording**: Ahlén, 1982; Bellmann, 1985b, 1993b; Burton & Ragge, 1987; Wallin, 1979.

RECOGNITION (Plate 2: 3). Both sexes of *M. roeselii* can be distinguished from all other western European species of *Metrioptera* by the conspicuous broad pale band round most of the margin of the pronotal lateral lobes.

For notes on the taxonomy of this species see p. 71, and on the spelling of the specific name see p. 10.

SONG (Figs 401–418. CD 1, track 50). The calling song, heard most often in warm, sunny weather, is a continuous penetrating buzz that may last for several minutes with only the briefest occasional hesitation. When a male begins to sing it produces shorter bursts, but

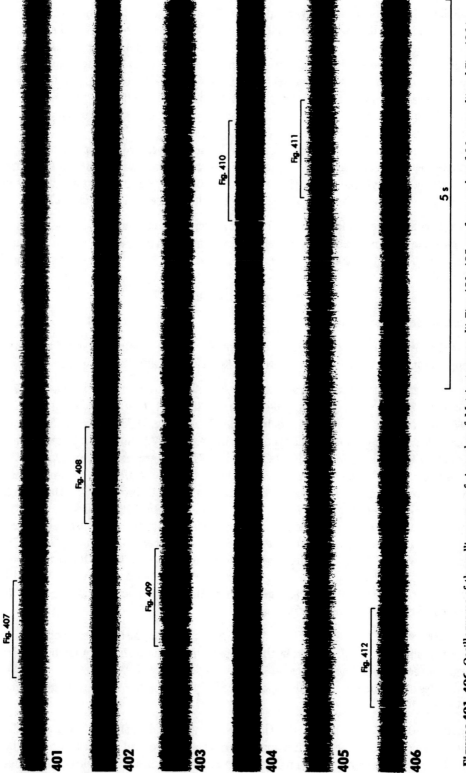

Figures 401–406 Oscillograms of the calling songs of six males of *Metrioptera roeselii*. Figs 401–405 are from males of *M. r. roeselii* and Fig. 406 is from a male of *M. r. fedtschenkoi*. Fig. 405 is from a macropterous male.

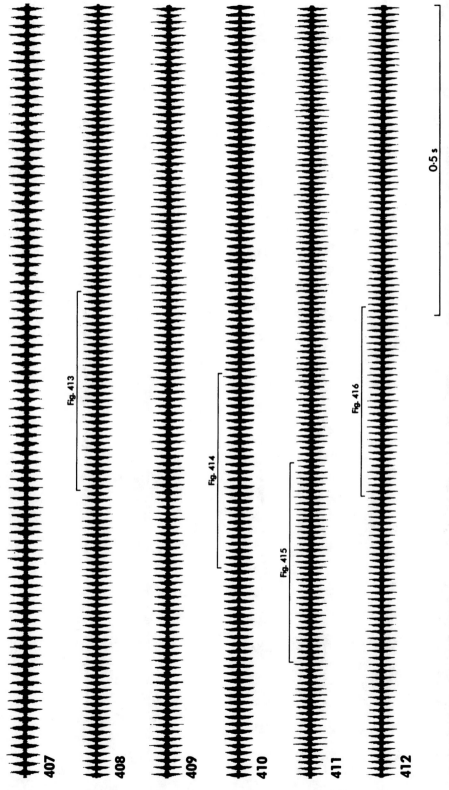

Figures 407–412 Faster oscillograms of the indicated parts of the songs of *Metrioptera roeselii* shown in Figs 401–406. Fig. 411 is from a macropterous male.

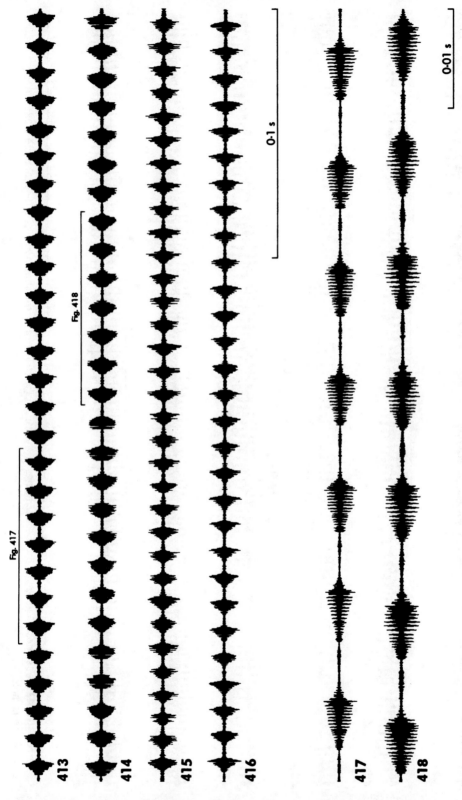

Figures 413–418 Faster oscillograms of the indicated parts of the songs of *Metrioptera roeselii* shown in Figs 407–412. Fig. 415 is from a macropterous male.

these soon lengthen until the sound becomes virtually continuous. The buzz is formed of a dense series of syllables repeated at the rate of about 70–110/s. The sound often consists only of closing hemisyllables, each lasting about 5–10 ms, but there are sometimes quieter opening hemisyllables. The songs of macropterous males (Figs 405, 411, 415) do not seem to differ significantly from those of the usual brachypterous form.

The song is typically a daytime sound, but it may also be heard at night if the temperature is not too low. The syllables of a nocturnal song are repeated much more slowly and can be more readily distinguished by the human ear. Judging by Samways' (1976a) sonagram, which shows a syllable repetition rate of about 40/s, even daytime songs sometimes have much slower wing-movements; this sonagram was based on a song from a captive male inside a car at a temperature of 20°C.

Unlike the otherwise rather similar song of *Ruspolia nitidula* (produced almost entirely at night), the song of *M. roeselii* is non-resonant (see p. 146).

DISTRIBUTION. Widespread in Europe from Ireland (one locality in the south) and southern Britain to Russia and Kazakhstan, and from southern Scandinavia to the Pyrenees, Alps and Balkan Peninsula; eastwards the range extends beyond the Urals to central Asia. Introduced into North America, where it now occurs in south-east Canada and north-east USA. Represented in western Europe almost entirely by the nominate subspecies, *M. r. roeselii*, but a primarily eastern European and Asian subspecies, *M. r. fedtschenkoi* (Saussure), occurs in alpine valleys in northern Italy and a few localities near the Mediterranean coast of France; there appears to be no difference between the calling songs of these two subspecies.

Pholidoptera griseoaptera (De Geer)

Locusta griseoaptera De Geer, 1773: 436.

Dark Bush-cricket; (F) Decticelle cendrée; (D) Gewöhnliche Strauchschrecke; (NL) Bramesprinkhaan; (DK) Buskgræshoppe; (S) Buskvårtbitare; (N) Buskhopper; (SF) Pensashepokatti.

REFERENCES TO SONG. **Oscillogram**: Ahlén, 1981; Dubrovin & Zhantiev, 1970 (as *cinerea*); Grein, 1984; Grzeschik, 1969; Heller, 1988; Holst, 1970, 1986; Jones, 1963, 1964, 1966a, 1966b; Kleukers *et al.*, 1997; Schmidt & Baumgarten, 1977; Schmidt & Schach, 1978. **Diagram**: Bellmann, 1985a, 1988, 1993a; Bellmann & Luquet, 1995; Broughton & Lewis, 1979; Duijm & Kruseman, 1983; Holst, 1970; Ragge, 1965; Wallin, 1979. **Wing-movement**: Heller, 1988. **Frequency information**: Ahlén, 1981; Dubrovin & Zhantiev, 1970 (as *c.*); Heller, 1988; Zhantiev & Dubrovin, 1974 (as *c.*). **Musical notation**: Yersin, 1854b (as *c.*). **Verbal description only**: Baier, 1930 (as *c.*); Broughton, 1972b; Burr, 1911; Chopard, 1922; Defaut, 1987, 1988b; Faber, 1928 (as *c.*); Harz, 1957, Hudson, 1903; Klöti-Hauser, 1921; Taschenberg, 1871 (as *c.*); Weber, 1984. **Disc recording**: Andrieu & Dumortier, 1963 (LP), 1994 (CD); Bellmann, 1993c (CD); Bonnet, 1995 (CD); Grein, 1984 (LP); Odé, 1997 (CD); Ragge *et al.*, 1965 (LP). **Cassette recording**: Ahlén, 1982; Bellmann, 1985b, 1993b; Burton & Ragge, 1987; Wallin, 1979.

RECOGNITION (Plate 2: 4). The five species of *Pholidoptera* occurring in western Europe are more brachypterous than *Metrioptera* and the brachypterous species of *Platycleis*, the fore

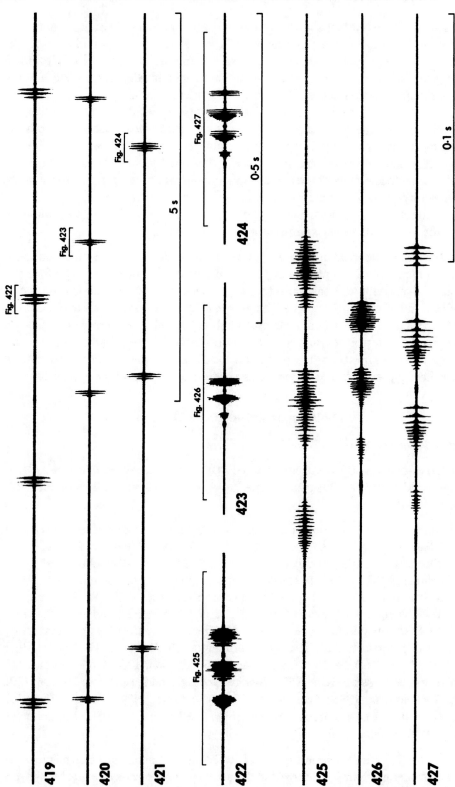

Figures 419–427 Oscillograms at three different speeds of the calling songs of three males of *Pholidoptera griseoaptera*.

wings being reduced to the short, rounded, overlapping flaps of the stridulatory organ in the male, and vestigial, hardly visible, lateral lobes in the female; in addition, the median keel on the metazona of the pronotum is either very indistinct or absent. They live on low vegetation or on the ground and are predominantly brown in colour, with yellow undersides.

P. griseoaptera differs from the other four species in having no more than a very narrow pale margin along the posterior and ventral margins of the pronotal lateral lobes. It is the only species of *Pholidoptera* to occur in the lowlands of northern Europe.

SONG (Figs 419–427. CD 1, track 51). The short, shrill echemes of the calling song, emanating from hedges, nettles and other low vegetation, mainly in the evening and at night, form one of the most familiar summer sounds in the lowlands of central and northern Europe. The echemes are usually repeated rather irregularly (but sometimes quite regularly) at a very variable rate, generally between 10 and 60 per minute. Each echeme usually consists of three (rarely two or four) syllables in quick succession, the first one being usually shorter and often quieter than the rest. The duration of the syllables and echemes is dependent on temperature: at 13°C a closing hemisyllable is likely to last about 20–40 ms and a trisyllabic echeme about 150–200 ms, whereas at 24°C these values can be as little as 10 ms and 60 ms, respectively. There are usually quiet opening hemisyllables, especially at the start of an echeme.

Males within hearing range of one another commonly alternate (and occasionally synchronize) their echemes. Close encounters between males often result in the production of polysyllabic echemes lasting up to 2 s or more and consisting of up to 30 or more syllables.

DISTRIBUTION. Widespread in northern and central Europe, from southern Scandinavia to northern Spain, northern Italy and the northern part of the Balkan Peninsula. In the British Isles it is common in southern Britain and is known from one locality in southern Ireland. Eastwards, the range extends to the Caucasus and Asia Minor.

Pholidoptera aptera (Fabricius)

Locusta aptera Fabricius, 1793: 45.

Alpine Dark Bush-cricket; (F) Decticelle aptère; (D) Alpen-Strauchschrecke.

REFERENCES TO SONG. **Oscillogram**: Grein, 1984; Heller, 1988; Schmidt, 1989. **Diagram**: Bellmann, 1985a, 1988, 1993a; Bellmann & Luquet, 1995. **Wing-movement** and **frequency information**: Heller, 1988. **Musical notation**: Regen, 1908, 1914. **Verbal description only**: Chopard, 1952; Dumortier, 1963d; Harz, 1957; Regen, 1926; Tümpel, 1901. **Disc recording**: Bellmann, 1993c (CD); Grein, 1984 (LP). **Cassette recording**: Bellmann, 1985b, 1993b.

RECOGNITION. Distinguished from the other western European species of *Pholidoptera* by having a broad pale band along the hind margin of the pronotal lateral lobes which is not continued along the ventral margin.

SONG (Figs 428–433. CD 1, track 52). The loud calling song, a familiar nocturnal sound in the Alps and other uplands of central and eastern Europe, consists of sequences of about

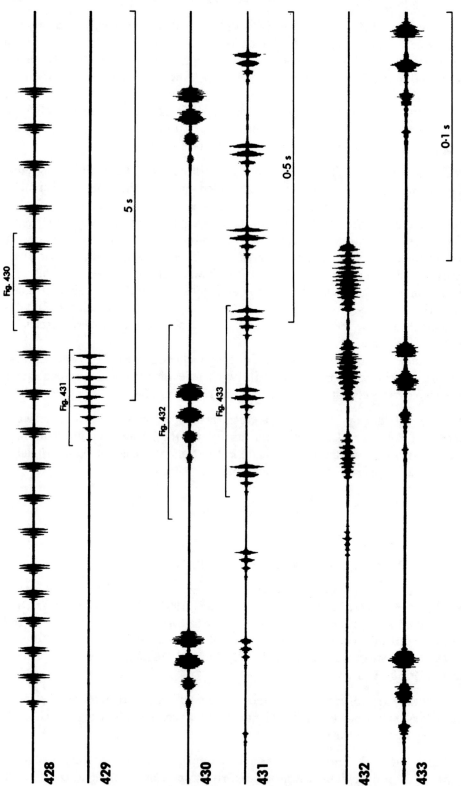

Figures 428–433 Oscillograms at three different speeds of the calling songs of two males of *Pholidoptera aptera*. Figs 428, 430 and 432 are taken from a night-time song and Figs 429, 431, 433 from a daytime one.

5–40 short, sibilant echemes repeated at the rate of about 2–8/s; the repetition rate usually lessens slightly during the course of a sequence. The sequences are repeated at variable intervals (5–10 s is common). When other males are within hearing range they usually respond with similar sequences, so that one often hears choruses of singing males interspersed with periods of silence. The sound is typically heard in the evening and at night, but sometimes also during the day.

Oscillographic analysis shows that the echemes are composed of three principal closing hemisyllables, of which the first is usually quieter and shorter than the other two and is generally preceded by an even quieter and shorter closing hemisyllable. The duration of the syllables and echemes is dependent on temperature: on a cool night a principal closing hemisyllable may last 25–30 ms and an echeme 120–150 ms, whereas in a diurnal song produced in sunshine these values may be reduced to 5–10 ms and 40–50 ms, respectively. There are sometimes very quiet opening hemisyllables.

DISTRIBUTION. From the Alps eastwards through the uplands of central Europe to the Carpathians, and southwards to the Tuscan Apennines and Dinaric Alps.

Pholidoptera fallax (Fischer)

Thamnotrizon fallax Fischer, 1853: 265.

Fischer's Bush-cricket; (F) Decticelle trompeuse; (D) Südliche Strauchschrecke.

REFERENCES TO SONG. **Oscillogram**: Heller, 1988; Schmidt, 1989. **Wing-movement** and **frequency information**: Heller, 1988. **Verbal description only**: Harz, 1962; Yersin, 1857.

RECOGNITION. Distinguished from *P. griseoaptera* and *P. aptera* by the broad pale band along the edge of the pronotal lateral lobes, which is continued along the ventral margin, and from *P. femorata* and *P. littoralis* by the smaller size (hind femora less than 21 mm long in the male, less than 24 mm in the female).

SONG (Figs 434–439. CD 1, track 53). The calling song, produced mainly in the evening and at night, is very similar to that of *P. griseoaptera*, consisting of short, trisyllabic echemes repeated at very variable intervals. The sound is usually produced entirely by the closing strokes of the fore wings.

DISTRIBUTION. From the Pyrenees through southern France to the southern Alps, Italy, the northern part of the Balkan Peninsula and Asia Minor.

Pholidoptera femorata (Fieber)

Pterolepis femoratus Fieber, 1853: 153.

Large Dark Bush-cricket; (F) Decticelle des roselières.

REFERENCES TO SONG. **Oscillogram**: Chopard, 1955; Heller, 1988. **Wing-movement**: Heller, 1988. **Frequency information**: R.-G. Busnel, 1955. **Verbal description only**: Krauss, 1879. **Disc recording**: Bonnet, 1995 (CD).

RECOGNITION. Distinguished from *P. griseoaptera* by the broad pale band along the edge of

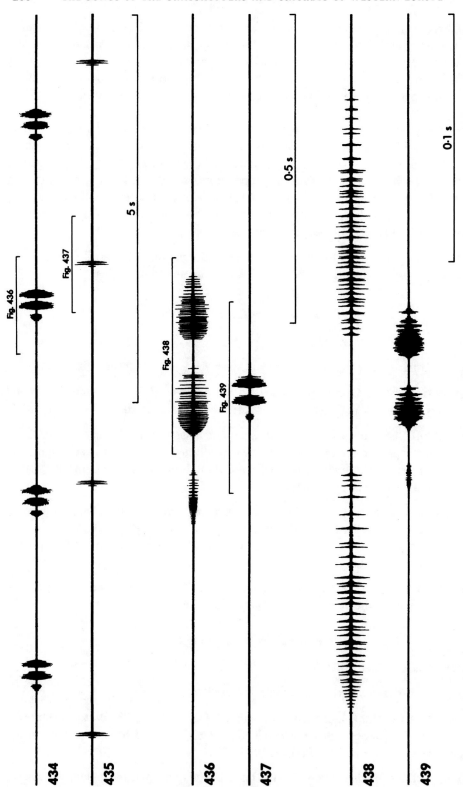

Figures 434–439 Oscillograms at three different speeds of the calling songs of two males of *Pholidoptera fallax*. The song shown in Figs 434, 436 and 438 was produced in much cooler conditions than the song shown in Figs 435, 437 and 439.

the pronotal lateral lobes, and from *P. fallax* by the larger size (hind femora more than 21 mm long in the male, more than 23 mm in the female). From *P. aptera* and *P. littoralis*, males may be distinguished by their very short fore wings, which are half covered by the pronotum and reach no further than the second abdominal tergite, and females by the tubercle on the proximal part of the seventh abdominal sternite.

SONG (Figs 440–442. CD 1, track 54). The calling song is again very similar to that of *P. griseoaptera*, consisting of short, trisyllabic echemes repeated at variable intervals, and produced mainly in the evening and at night. Opening hemisyllables are very quiet or absent.

DISTRIBUTION. Southern France (including Corsica), Italy (including Sardinia and Sicily), much of the Balkan Peninsula and Asia Minor.

Pholidoptera littoralis (Fieber)

Pterolepis littoralis Fieber, 1853: 153.

Littoral Bush-cricket; (D) Küsten-Strauchschrecke.

REFERENCES TO SONG. **Oscillogram**: Heller, 1988; Kalmring *et al.*, 1985; Rheinlaender & Kalmring, 1973. **Wing-movement** and **frequency information**: Heller, 1988. **Verbal description only**: Harz, 1957.

RECOGNITION. Distinguished from *P. griseoaptera* and *P. fallax* by the same characters as *P. femorata* (see above), and from *P. aptera* by the pale band on the pronotal lateral lobes being continued along the ventral margin. Males may be distinguished from *P. femorata* by their larger fore wings, which extend at least to the third abdominal tergite, and females by the lack of a tubercle on the seventh abdominal sternite.

SONG (Figs 443–445. CD 1, track 55). The calling song differs from that of all the other western European species of *Pholidoptera* in consisting of much longer echemes, lasting up to 4 s or more and composed of about 50–70 syllables. Each echeme begins quietly but soon reaches maximum intensity. The syllable repetition rate is about 14–17/s at 21–22°C. Each closing hemisyllable lasts about 20–26 ms; opening hemisyllables are very quiet or absent.

DISTRIBUTION. Italian Alps, Hungary, Rumania and Bulgaria.

Eupholidoptera chabrieri (Charpentier)

Locusta chabrieri Charpentier, 1825: 119.

Chabrier's Bush-cricket; (F) Decticelle splendide; (D) Grüne Strauchschrecke.

REFERENCES TO SONG. **Oscillogram, wing-movement** and **frequency information**: Heller, 1988. **Verbal description only**: Bellmann, 1993a; Bellmann & Luquet, 1995; Krauss, 1879; Tümpel, 1901. **Disc recording**: Bonnet, 1995 (CD).

RECOGNITION (Plate 2: 5). Morphologically quite similar to *Pholidoptera*, but easily recognized by its coloration, usually a vivid green variegated with black and often yellow.

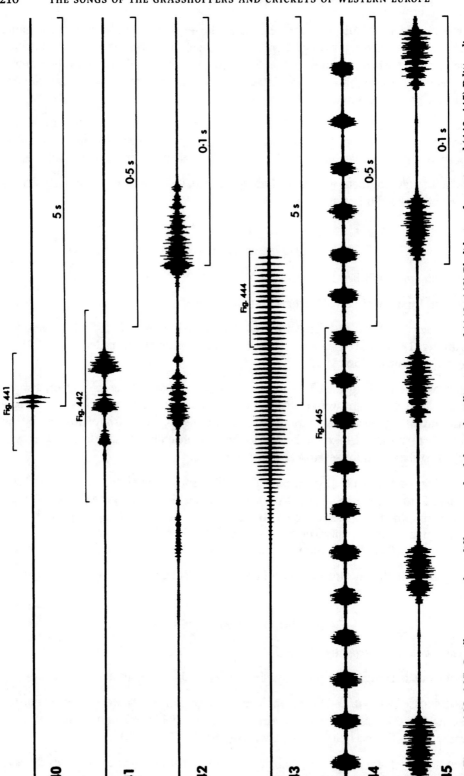

Figures 440–445 Oscillograms at three different speeds of the male calling songs of (440–442) *Pholidoptera femorata* and (443–445) *P. littoralis.*

SONG (Figs 446–451. CD 1, track 56). The calling song, produced both during the day and at night, is a long series of isolated syllables, repeated at intervals of about 0·5–3·0 s. The sound is produced mainly by the closing stroke of the fore wings, which produces a hemisyllable lasting about 10–30 ms in warm conditions but as long as 100 ms or more in cool, dull weather or at night. Quiet opening hemisyllables are often also present.

DISTRIBUTION. Southern France, Switzerland, Italy, southern Austria and the Balkan Peninsula.

Anonconotus alpinus (Yersin)

Pterolepis alpinus Yersin, 1858: 111.

Small Alpine Bush-cricket; (F) Decticelle montagnarde; (D) Alpenschrecke.

REFERENCES TO SONG. **Oscillogram, wing-movement** and **frequency information**: Heller, 1988. **Verbal description only**: Bellmann, 1993*a*; Bellmann & Luquet, 1995; Chopard, 1952; Yersin, 1858. **Disc recording**: Bonnet, 1995 (CD).

RECOGNITION (Plate 2: 6). This small bush-cricket is found only on mountains, usually at altitudes above 1000 m. It is very brachypterous and the reduced fore wings protrude only slightly from under the pronotum, the visible part often being conspicuously pale-coloured. The general colour varies from green to brown, sometimes almost black.

SONG (Figs 452–454. CD 1, track 57). The quiet calling song, typically produced in sunshine, consists of a dense echeme lasting about 1–3 s and composed of about 20–50 diplosyllables repeated at the rate of about 15–35/s. The intervals between successive echemes vary from about 4 s to 30 s or more. Oscillographic analysis shows that the quiet opening hemisyllables last about 5–10 ms and the louder closing hemisyllables about 10–40 s.

DISTRIBUTION. Known only from the Alps, where it is widespread at altitudes up to about 2500 m.

Yersinella raymondii (Yersin)

Pterolepis raymondii Yersin, 1860: 524.

Raymond's Bush-cricket; (F) Decticelle frêle; (D) Kleine Strauchschrecke.

REFERENCES TO SONG. **Oscillogram**: Heller, 1988; ?Schmidt, 1989, 1996 (see below). **Frequency information**: Heller, 1988.

RECOGNITION. *Yersinella* is a brachypterous genus recognizable by its small size, slender build and colouring: pale and straw-coloured above, with a broad, dark brown stripe along each side. The male cerci have no tooth and the ovipositor is upcurved. At present the genus includes only two species, which are so similar to each other that their respective distributions have not yet been determined. The morphological differences between them are detailed by La Greca (1974).

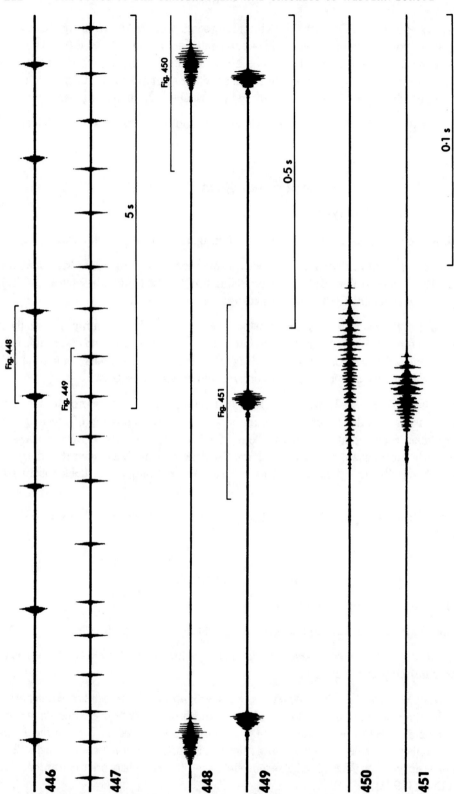

Figures 446–451 Oscillograms at three different speeds of the calling songs of two males of *Eupholidoptera chabrieri*.

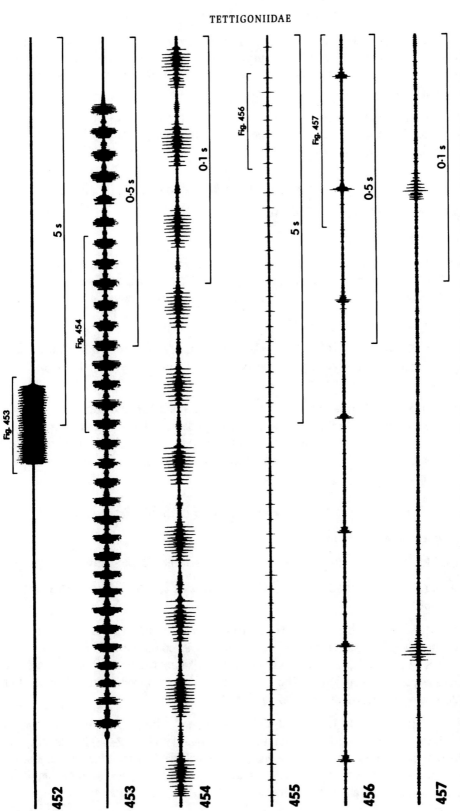

Figures 452–457 Oscillograms at three different speeds of the male calling songs of (452–454) *Anonconotus alpinus* and (455–457) *Ctenodecticus siculus*.

SONG (Figs 458–461. CD 1, track 58). The nocturnal male calling song, almost entirely ultrasonic, consists of brief syllables lasting about 20–30 ms repeated at intervals of 1–3 s or longer. Schmidt (1989, 1996) gives the syllable duration as about 170–200 ms, which we have found to be more typical of *Y. beybienkoi*.

DISTRIBUTION. Originally described from southern France and believed to occur in Switzerland and all the southern European peninsulas, but many of the earlier records may have been based on *Y. beybienkoi*.

Yersinella beybienkoi La Greca

Yersinella beybienkoi La Greca, 1974: 60.

Bei-Bienko's Bush-cricket.

REFERENCES TO SONG. **Oscillogram**: Schmidt's (1989) description and oscillograms of the song of *Y. raymondii* seem to us to be more typical of *Y. beybienkoi*; the song he describes under the name of *Y. beybienkoi* is completely different from the songs we have studied.

RECOGNITION. See under *Y. raymondii*.

SONG (Figs 459, 460, 462, 463. CD 1, track 59). The male calling song is similar to that of *Y. raymondii* except that, in our experience, the syllables are of longer duration, about 120–250 ms, depending on temperature.

DISTRIBUTION. Originally described from the northern Apennines and since found by Nadig (1987) to be widespread in the Apennines from Gran Sasso northwards, and also to occur in parts of the Italian Maritime Alps. We think this species will probably prove to be even more widespread when the earlier records based on the very similar species *Y. raymondii* are carefully checked.

Pachytrachis striolatus (Fieber)

Pachytrachelus striolatus Fieber, 1853: 169.

Striated Bush-cricket; (F) Decticelle striolée; (D) Gestreifte Südschrecke.

REFERENCES TO SONG. **Oscillogram**, **wing-movement** and **frequency information**: Heller, 1988.

RECOGNITION. Another brachypterous brown or straw-coloured bush-cricket occurring in the southern Alps, perhaps most easily recognized by its colour pattern: many dark spots and stripes on the head, pronotum and legs, and a dark longitudinal stripe along each side of the basal part of the abdomen. The male cerci are long, rounded at the tip and without an internal tooth; the ovipositor is long and straight.

SONG (Figs 464–469. CD 1, track 60). The nocturnal calling song consists of single diplosyllables, often grouped together into loose echemes of about 2–6 syllables separated by intervals of 1–2 s (Heller, 1988, gives longer inter-syllable intervals of about 4 s). Oscillographic analysis shows that the opening and closing hemisyllables are quite similar in structure, each lasting about 12–17 ms.

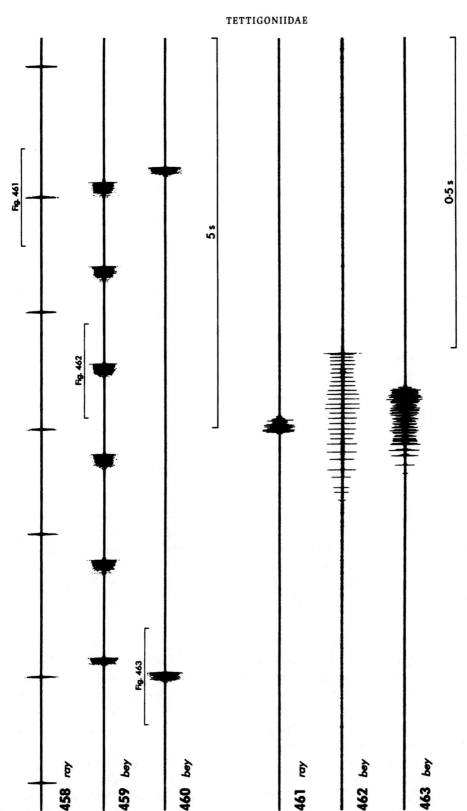

Figures 458–463 Oscillograms at two different speeds of the male calling songs of (458, 461) *Yersinella raymondii* and (459, 460, 462, 463) *Y. beybienkoi*. Note that the oscillograms shown in Figs 458, 460, 461 and 463 were taken from recordings of the frequency-converted sound produced by an ultrasound detector.

DISTRIBUTION. Southern Swiss and Italian Alps, and through Dalmatia to Albania.

Ctenodecticus siculus (Ramme)

Hemictenodecticus siculus Ramme, 1927: 152.

Sicilian Bush-cricket.

REFERENCES TO SONG. No published work known to us.

RECOGNITION. This very brachypterous, brown or straw-coloured, nymph-like species is smaller than any of the other decticine bush-crickets included in this study (hind femora less than 11 mm in the male, less than 13 mm in the female). The fore wings protrude only slightly from under the pronotum and the plantulae (side-flaps) of the hind tarsi are conspicuously long, clearly longer than the first tarsal segment.

SONG (Figs 455–457. CD 1, track 61). The quiet calling song is a simple sequence of 'tick'-like syllables repeated quite regularly for long periods at the rate of about 4–6/s. Each syllable lasts about 7–10 ms and the intervals between successive syllables are about 160–180 ms long.

DISTRIBUTION. Known only from Sicily.

Rhacocleis germanica (Herrich-Schäffer)

Decticus germanica Herrich-Schäffer, 1840: 13.

Mediterranean Bush-cricket; (F) Decticelle orientale; (D) Zierliche Strauchschrecke.

REFERENCES TO SONG. **Oscillogram**: Heller, 1988; Schmidt, 1989, 1996. **Wing-movement and frequency information**: Heller, 1988.

RECOGNITION. *Rhacocleis* is a genus of brachypterous, mottled brown or grey-brown bush-crickets, differing from the otherwise similar genus *Antaxius* in having the plantulae (side-flaps) of the hind tarsi as long as the first tarsal segment (shorter than the first tarsal segment in *Antaxius*). *R. germanica* may be distinguished from *R. neglecta* by the inner tooth of the male cerci, which is long and straight (shorter and bent strongly towards the base of the cercus in *R. neglecta*) and the female subgenital plate, which has a median excision (no excision in *R. neglecta*).

SONG (Figs 470–475. CD 1, track 62). The quiet calling song, produced mainly at night, consists of a sequence of dense echemes, each lasting about 0·2–0·7 s and consisting of about 8–16 diplosyllables repeated at the rate of about 20–55/s. The intervals between the echemes are variable, often between 1 and 3 s but sometimes much longer. Each echeme can be quite uniform in loudness and structure from start to finish, but often there is a crescendo at the beginning or even during the whole of the echeme, and sometimes the first syllable differs slightly from the remaining ones.

Oscillographic analysis shows that there are opening hemisyllables, which may be relatively quiet but are sometimes as loud as, or even louder than, the closing hemisyllables. The opening hemisyllables usually last about 3–15 ms and the closing hemisyllables are

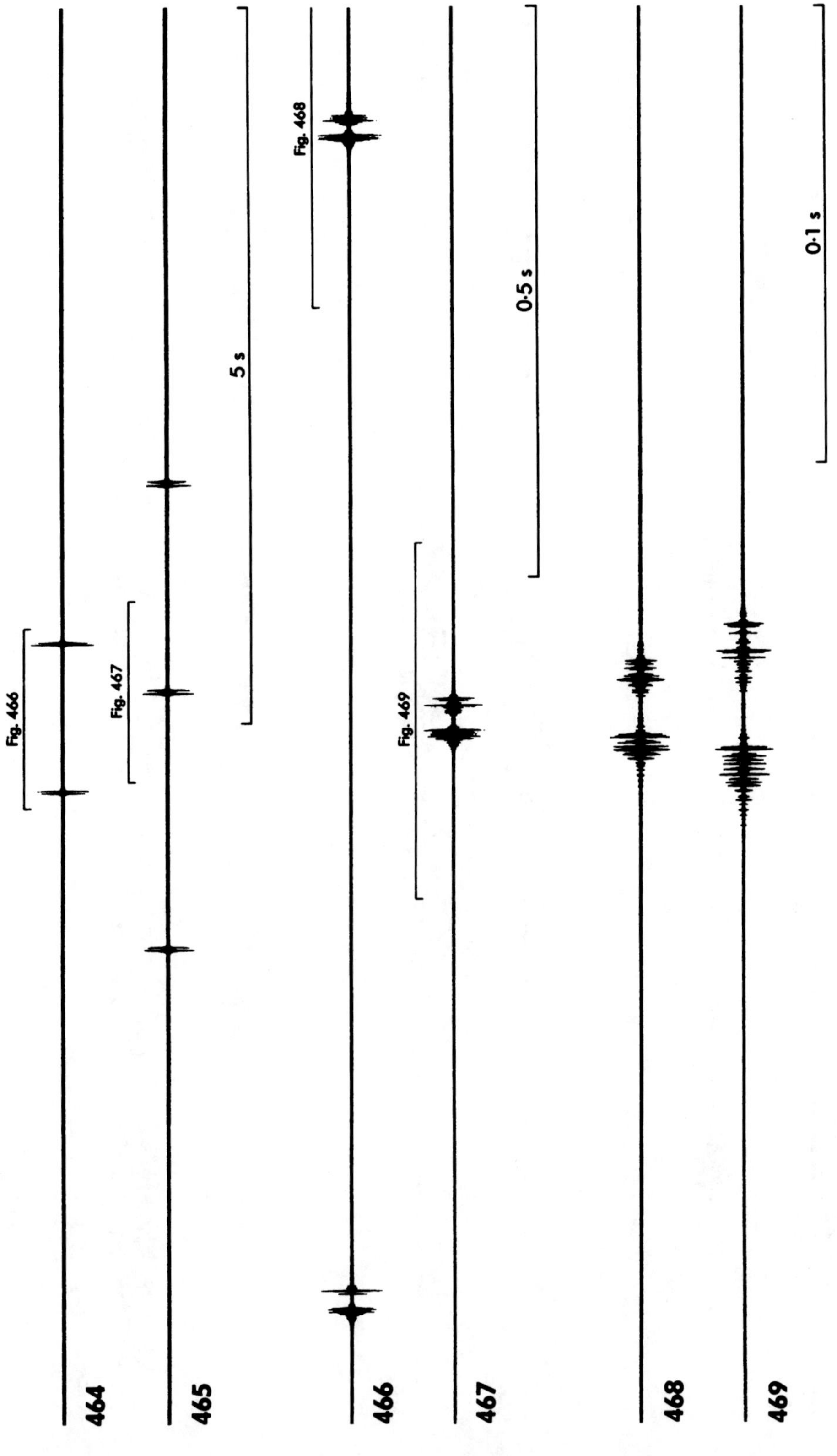

Figures 464–469 Oscillograms at three different speeds of the calling songs of two males of *Pachytrachis striolatus*.

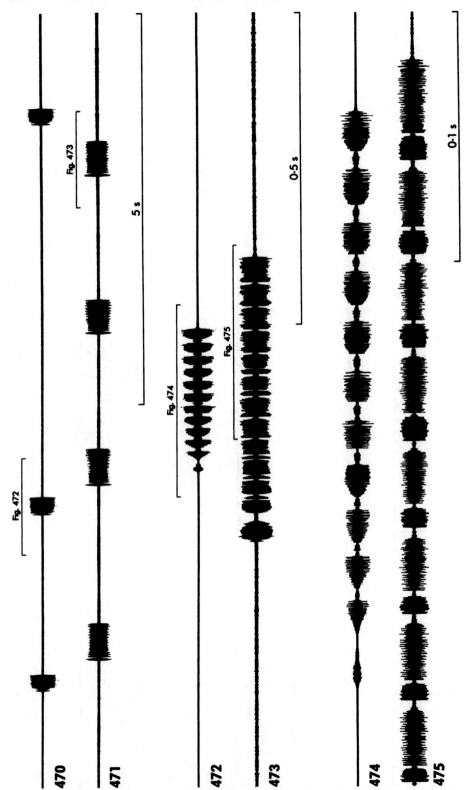

Figures 470–475 Oscillograms at three different speeds of the calling songs of two males of *Rhacocleis germanica.*

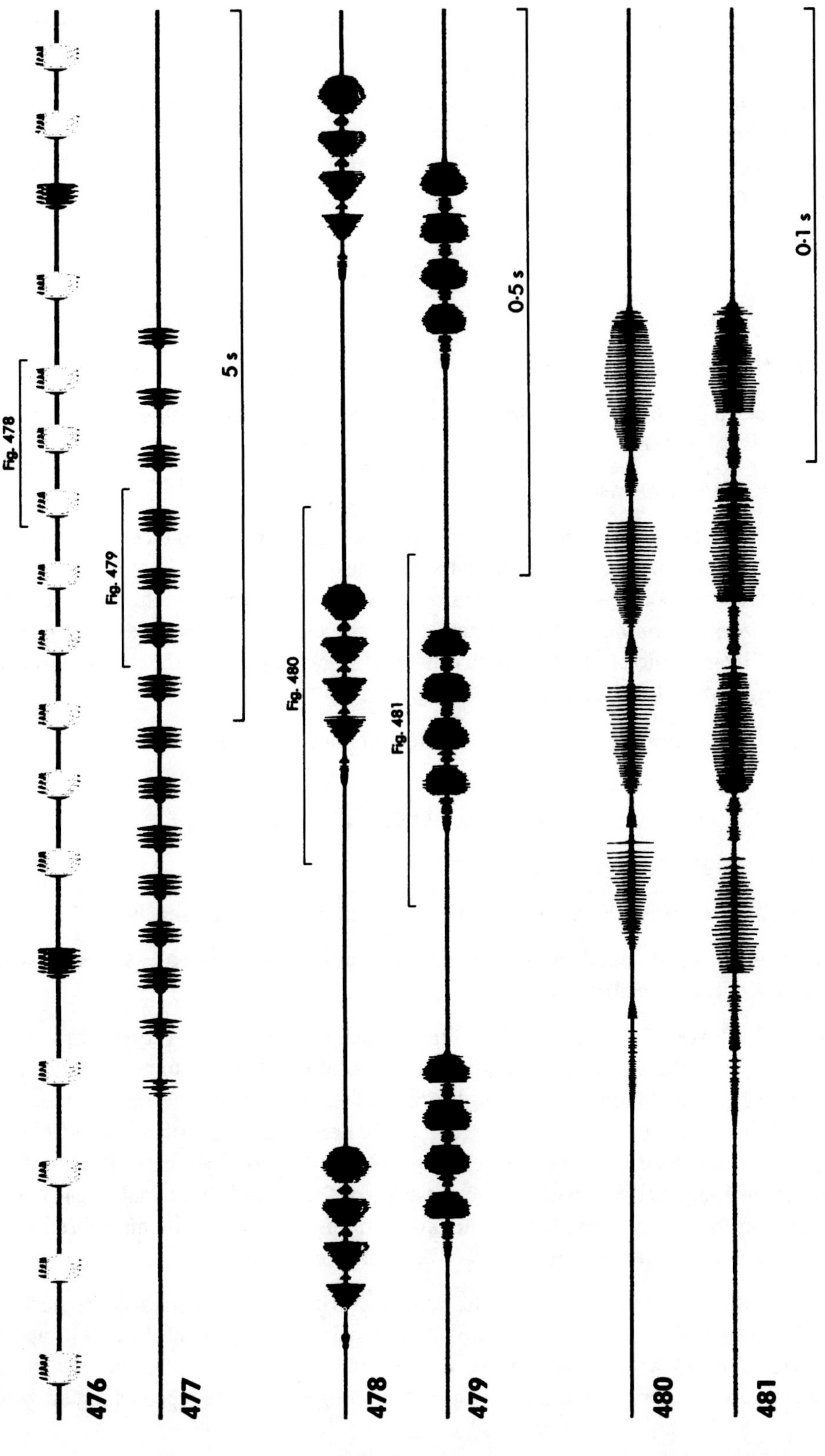

Figures 476–481 Oscillograms at three different speeds of the calling songs of two males of *Rhacocleis neglecta*.

always longer, lasting about 12–30 ms; the syllables tend to become slightly longer during the course of the echeme.

DISTRIBUTION. From near the Mediterranean coast of France through all the countries bordering the north of the Mediterranean Sea to Romania, Moldavia, Bulgaria and Asia Minor; also in Austria and Hungary and on many Mediterranean islands. Absent from the Iberian Peninsula.

Rhacocleis neglecta (Costa)

Pterolepis neglecta Costa, 1863: 27.

Adriatic Bush-cricket.

REFERENCES TO SONG. **Oscillogram**: Heller, 1988; Schmidt, 1996. **Wing-movement** and **frequency information**: Heller, 1988.

RECOGNITION. See *R. germanica*.

SONG (Figs 476–481. CD 1, track 63). The calling song, normally produced at night, is a quiet sequence of tri-, tetra- or pentasyllabic echemes repeated, sometimes rather irregularly, at the rate of about 1–3/s. The sequences vary greatly in duration, often lasting 4–10 s but sometimes 30 s or longer. Oscillographic analysis shows that there are quiet opening syllables and that the first of these in each echeme is much longer than the remaining ones. Each closing hemisyllable lasts about 20–30 ms.

DISTRIBUTION. At present known only from Italy, Corsica, Istria and Dalmatia.

Antaxius pedestris (Fabricius)

Locusta pedestris Fabricius, 1787: 235.

Mottled Bush-cricket; (F) Antaxie marbrée; (D) Atlantische Bergschrecke.

REFERENCES TO SONG. **Oscillogram, wing-movement** and **frequency information**: Heller, 1988. **Disc recording**: Bonnet, 1995 (CD).

RECOGNITION (Plate 2: 7). Like *Rhacocleis*, *Antaxius* includes a number of very brachypterous species, usually mottled brown or grey-brown but occasionally green in colour. They can be distinguished from *Rhacocleis* in having the plantulae (side-flaps) of the hind tarsi shorter than the first tarsal segment (they are as long as the first tarsal segment in *Rhacocleis*).

A. *pedestris* can be distinguished from A. *hispanicus* and A. *spinibrachius* by its short fore wings, which do not usually reach the hind margin of the second abdominal tergite in the male and hardly protrude from under the pronotum in the female. The male fore wings have a conspicuous pale spot, but are not entirely pale-coloured.

SONG (Figs 482–487. CD 1, track 64). The quiet nocturnal song is a dense echeme lasting about 0·5–4·0 s and composed of about 8–60 diplosyllables repeated at the rate of about 15–20/s. Sometimes the echemes are separated by quite long and irregular intervals (commonly from 20 s to more than a minute), but at other times they are repeated regularly in

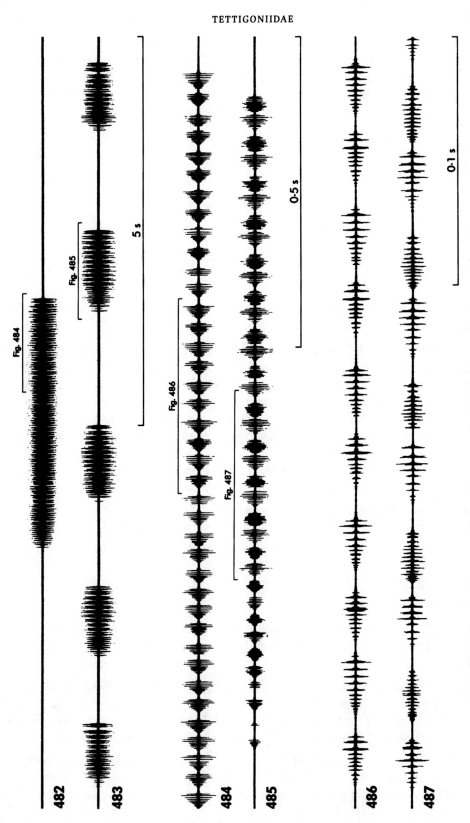

Figures 482–487 Oscillograms at three different speeds of the calling songs of two males of *Antaxius pedestris*.

quite quick succession, as in Fig. 483; occasionally a pair of echemes is produced in even quicker succession, with only the briefest of pauses between them. Oscillographic analysis shows that the opening and closing hemisyllables are remarkably similar in loudness, internal structure and duration, each usually lasting about 18–25 ms. The echemes often begin more quietly, but reach maximum intensity within a few syllables.

DISTRIBUTION. Widely distributed in the Alps, from the extreme west to the Tyrol and southwards as far as the Apuan Alps. Also recorded from the Pyrenees, but these records need confirmation.

Antaxius hispanicus Bolívar

Antaxius hispanicus Bolívar, 1887: 103.

Pyrenean Bush-cricket; (F) Antaxie pyrénéenne.

REFERENCES TO SONG. **Oscillogram** and **wing-movement**: Heller, 1988.

RECOGNITION. Both sexes can be recognized by the exposed parts of the fore wings being entirely yellowish.

SONG (Figs 488–493. CD 1, track 65). The quiet calling song, produced at night, is a dense echeme lasting about 0·5–1·5 s and composed of about 8–25 diplosyllables repeated at the rate of about 15–20/s; as in *A. pedestris*, a pair of echemes is sometimes produced with only a very brief pause between them. The echemes, or pairs of echemes, are separated by very variable intervals, from about 20 s to 2 minutes or more. Oscillographic analysis shows that the opening and closing hemisyllables, unlike those of *A. pedestris*, are usually quite different, the opening ones being composed of a few well-spaced tooth-impacts and the closing ones consisting of a dense sequence of numerous tooth-impacts. As in *A. pedestris*, the echemes usually show a crescendo during the first few syllables.

DISTRIBUTION. Known only from the Pyrenees.

Antaxius spinibrachius (Fischer)

Pterolepis spinibrachius Fischer, 1853: 258.

Spiny-legged Bush-cricket.

REFERENCES TO SONG. **Oscillogram** and **wing-movement**: Heller, 1988. **Frequency information**: Heller, 1988; Latimer & Broughton, 1984. **Verbal description only**: Pantel, 1896.

RECOGNITION. Males may be distinguished from *A. pedestris* and *A. hispanicus* by their fore wings, which are dull-coloured, without a pale spot or any yellowish colouring. The fore wings of the female also lack the yellowish colouring of those of *A. hispanicus* and are much larger than those of *A. pedestris*, reaching well beyond the first abdominal tergite.

SONG (Figs 494–499. CD 1, track 66). The quiet nocturnal calling song is quite different from those of *A. pedestris* and *A. hispanicus* in consisting of a long series of well separated diplosyllables repeated at the rate of about 1·5–3·5/s. Oscillographic analysis shows that

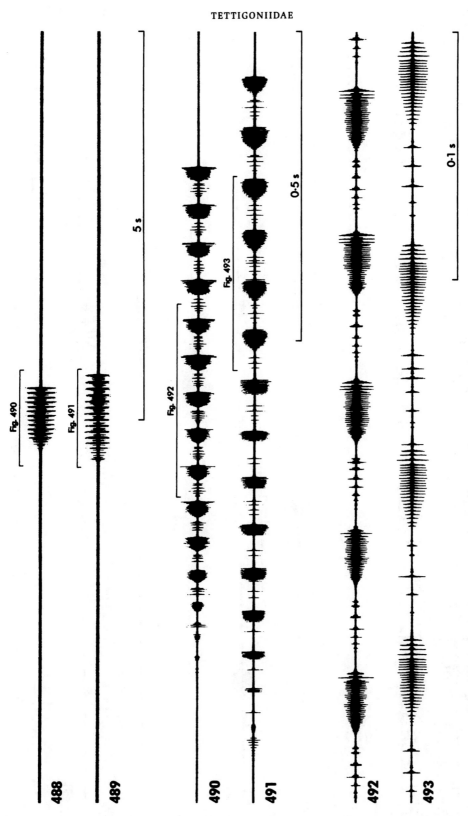

Figures 488–493 Oscillograms at three different speeds of the calling songs of two males of *Antaxius hispanicus*.

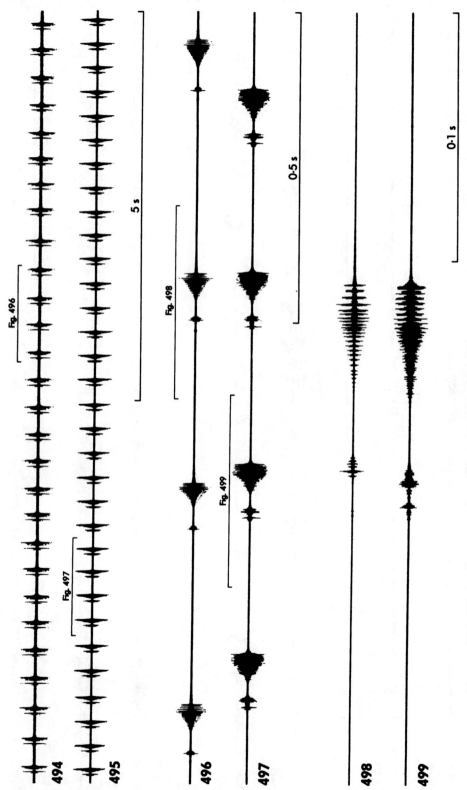

Figures 494–499 Oscillograms at three different speeds of the calling songs of two males of *Antaxius spinibrachius*.

the opening hemisyllables are about half the duration of the closing hemisyllables, which last about 40–65 ms. The interval between two successive syllables is about 0·2–0·4 s.

DISTRIBUTION. Known only from central Spain and Portugal.

Thyreonotus bidens Bolívar

Thyreonotus bidens Bolívar, 1887: 101.

Two-toothed Bush-cricket.

REFERENCES TO SONG. **Oscillogram** and **wing-movement**: Heller, 1988. **Frequency information**: Heller, 1988; Latimer & Broughton, 1984.

RECOGNITION. A moderately large, brachypterous, grey or brown bush-cricket, in which the pronotum is somewhat extended backwards, covering much of the reduced fore wings. There is a conspicuous dark spot on each side of the pronotum, at the point where the posterior margin of each lateral lobe reaches the disc.

SONG (Figs 500–502. CD 1, track 67). The nocturnal calling song is a long series of single syllables repeated at the rate of about 2/s. Most of the sound consists of closing hemisyllables lasting about 100–120 ms, but quieter and very brief opening hemisyllables are usually also present.

DISTRIBUTION. Central and southern Spain and Portugal.

Gampsocleis glabra (Herbst)

Locusta glabra Herbst, 1786: 193.

Heath Bush-cricket; (F) Dectique des brandes; (D) Heideschrecke; (NL) Ratelsprinkhaan.

REFERENCES TO SONG. **Oscillogram**: Heller, 1988; Kleukers *et al.*, 1997; Latimer, 1980; Latimer & Broughton, 1984; Schmidt, 1987*b*; Schmidt & Schach, 1978. **Diagram**: Duijm & Kruseman, 1983. **Sonagram**: Latimer, 1980. **Frequency information**: M.-C. Busnel, 1953; R.-G. Busnel, 1955; Latimer, 1981*a*. **Verbal description only**: Bellmann, 1985*a*, 1988, 1993*a*; Bellmann & Luquet, 1995; Harz, 1957, 1958*b*; Lunau, 1952; Pantel, 1896. **Disc recording**: Bellmann, 1993*c* (CD); Odé, 1997 (CD). **Cassette recording**: Bellmann, 1993*b*.

RECOGNITION. This species resembles a smaller and more slender version of *Decticus verrucivorus*. The general colour is green with conspicuous dark brown spots on the fore wings; the pronotal lateral lobes have a narrow pale margin. The ovipositor is long and slightly downcurved.

SONG (Figs 503–505. CD 1, track 68). The calling song, produced on warm, sunny days, is a penetrating, continuous buzz, often lasting for several minutes. It is composed of units repeated at the rate of about 40–60/s, but the stridulatory movements of the fore wings have not yet been studied and so it is not known at present whether these units are syllables or echemes. Each unit typically consists of a loud sound lasting about 2–3 ms followed by

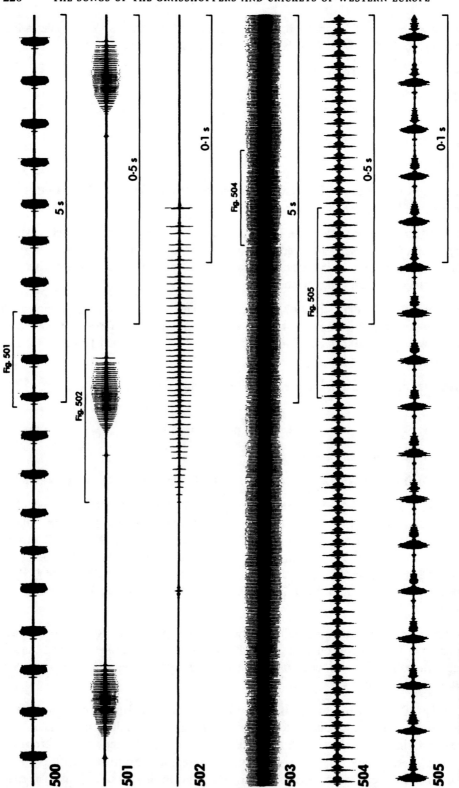

Figures 500–505 Oscillograms at three different speeds of the male calling songs of (500–502) *Thyreonotus bidens* and (503–505) *Gampsocleis glabra*.

one or more quieter sounds. During pauses in the loud buzzing, the male often continues to vibrate the fore wings, producing a much quieter sound in which the units are repeated rather more slowly. Occasionally the song is in shorter bursts of about 3 s, each with a gradual crescendo at the start and preceded by a few seconds of quieter wing-vibration.

DISTRIBUTION. Quite widespread but very local in western Europe, from the Netherlands and Germany (but not the British Isles) in the north to France, northern Spain and the Austrian lowlands in the south. Eastwards, the range extends through eastern Europe to Asia Minor and western Siberia.

Ephippiger ephippiger (Fiebig)

Gryllus ephippiger Fiebig, 1784: 260.

Saddle-backed Bush-cricket; (F) Éphippigère des vignes; (D) Steppen-Sattelschrecke; (NL) Zadelsprinkhaan.

REFERENCES TO SONG. **Oscillogram**: Busnel, 1963 (as *bitterensis* and *cunii*); Busnel, Busnel & Dumortier, 1956; Busnel, Pasquinelly & Dumortier, 1955; Chopard, 1955 (as *b.*); Duijm, 1990; Dumortier, 1956, 1963*a*, 1963*c*; Grein, 1984; Heller, 1988; Jatho *et al.*, 1992, 1994; Kalmring, 1975; Kalmring *et al.*, 1990; Keuper *et al.*, 1985, 1988*a*; Kleukers *et al.*, 1997; Pasquinelly & Busnel, 1955 (as *b.*); Rheinlaender & Kalmring, 1973; Ritchie, 1991, 1992*a*, 1992*b*; Schmidt & Schach, 1978; Stiedl *et al.*, 1991. **Diagram**: Bellmann, 1985*a*, 1988, 1993*a*; Bellmann & Luquet, 1995; Duijm & Kruseman, 1983. **Wing-movement**: Heller, 1988; Pasquinelly & Busnel, 1955 (as *b.*). **Frequency information**: M.-C. Busnel, 1953; R.-G. Busnel, 1955; Busnel & Chavasse, 1951; Dumortier, 1956, 1963*c*; Froehlich, 1989; Heller, 1988; Kalmring, 1975; Keuper *et al.*, 1985, 1988*a*; Leroy, 1966; Stiedl *et al.*, 1991. **Verbal description only**: Bertkau, 1879 (as *vitium*); Burr, 1911 (as *v.*); Defaut, 1987, 1988*b*; Dumortier *et al.*, 1957; Harz, 1957; Meissner, 1917 (as *v.*); Stiedl & Bickmeyer, 1991; Stiedl & Kalmring, 1989. **Disc recording**: Andrieu & Dumortier, 1963 (LP), 1994 (CD); Bellmann, 1993*c* (CD); Bonnet, 1995 (CD); Grein, 1984 (LP); Odé, 1997 (CD); Roché, 1965 (LP). **Cassette recording**: Bellmann, 1985*b*, 1993*b*.

RECOGNITION (Plate 2: 8). The genus *Ephippiger* belongs to a group of very brachypterous European and North African genera in which the pronotum is saddle-shaped, the posterior part being raised to cover much of the rounded flaps that are all that remains of the fore wings. They are rather sluggish insects, crawling rather than jumping. The females can stridulate with the fore wings as in the males, although the stridulatory teeth are arranged in a quite different way (Fig. 10); in some species the females sing in response to the male calling song, but in others the sound seems to be produced purely in defence. *Ephippiger* closely resembles *Ephippigerida*, the males being recognizable mainly by the supra-anal plate, which is usually quadrangular (triangular or rounded in *Ephippigerida*). For notes on the taxonomy of some of the species of *Ephippiger* see pp. 71 and 72.

E. *ephippiger* is the only species of *Ephippiger* to occur in all but the southernmost parts of its range. It is very variable in appearance, but males may always be distinguished from *E. terrestris* by the supra-anal plate, which is fairly clearly separated from the tenth abdominal tergite (continuous with this tergite in *E. terrestris*).

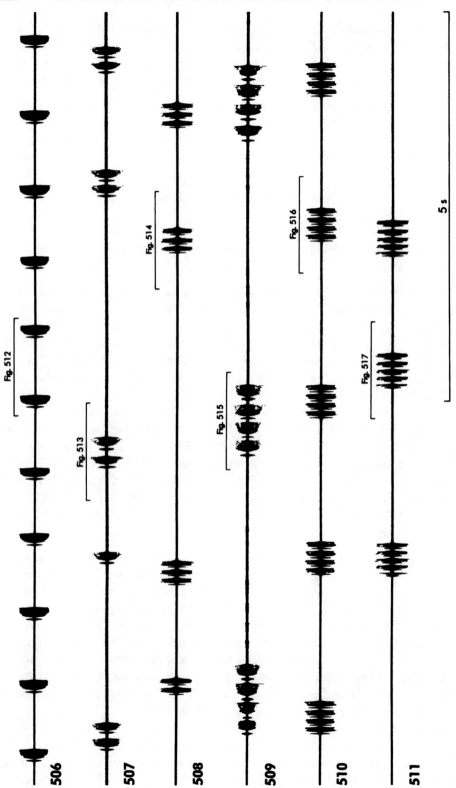

Figures 506–511 Oscillograms of the calling songs of six males of *Ephippiger ephippiger*.

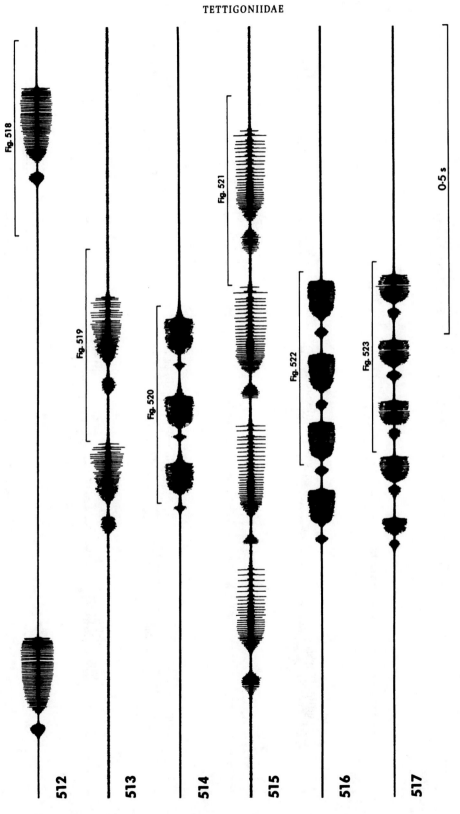

Figures 512–517 Faster oscillograms of the indicated parts of the songs of *Ephippiger ephippiger* shown in Figs 506–511.

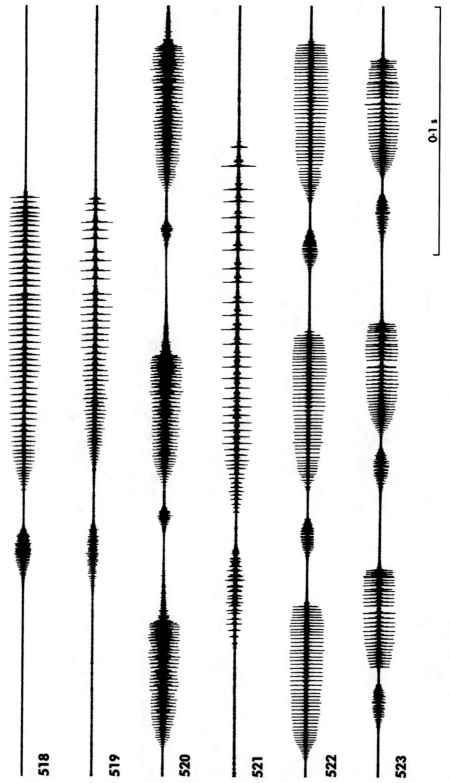

0·1 s

Figures 518–523 Faster oscillograms of the indicated parts of the songs of *Ephippiger ephippiger* shown in Figs 512–517.

518

519

520

521

522

523

SONG (Figs 506–523. CD 1, track 70). Although sometimes produced at night, the loud male calling song is typically a daytime sound, and in many localities singing activity is particularly intense in the morning. The song is composed of diplosyllables, in which the closing hemisyllable usually lasts 3–5 times longer than the opening one and is often louder. Except in parts of southern France and Catalonia, these single syllables are usually isolated from one another by intervals of at least 0·3 s and repeated for long periods more or less regularly at the rate of about 0·5–1·5/s (Fig. 506); sometimes, however, there is an occasional echeme of two (rarely three) diplosyllables with no pause between them, and sometimes such echemes occur more regularly (Fig. 507). As one approaches the Languedoc and Roussillon regions of France, polysyllabic echemes become a normal feature of the song, and in the French départements of Aude and Pyrénées-Orientales (and on the Spanish side of the eastern Pyrenees) one rarely hears an isolated syllable; echemes of 3–5 (occasionally up to 8) syllables are common in this region (Figs 508–511).

Oscillographic analysis shows that the duration of the syllables is partly dependent on temperature and also tends to be shorter in polysyllabic echemes: isolated syllables usually last about 130–200 ms and syllables in echemes about 60–110 ms. The syllable repetition rate within an echeme is usually about 6–12/s, but can be as low as 3–4/s in cool conditions (see Fig. 509, taken from a song recorded at an air temperature of 11°C).

DISTRIBUTION. Much of western Europe from the Netherlands southwards to northern Spain and Italy north of the Po valley. The range extends eastwards through central Europe and the Balkan Peninsula to the southern part of European Russia. Absent from the British Isles.

Ephippiger terrestris (Yersin)

Ephippigera terrestris Yersin, 1854*a*: 63.

Alpine Saddle-backed Bush-cricket; (F) Éphippigère terrestre; (D) Südalpen-Sattelschrecke.

REFERENCES TO SONG. **Oscillogram**: Duijm, 1990; Heller, 1988. **Wing-movement** and **frequency information**: Heller, 1988. **Musical notation**: Yersin, 1854*b*. **Verbal description only**: Chopard, 1952; Dumortier, 1963*c*. **Disc recording**: Bonnet, 1995 (CD).

RECOGNITION. See under *E. ephippiger*.

SONG (Figs 524–538. CD 1, track 70). There seems to be no significant difference between the male calling song of this species and that of *E. ephippiger*, except that *E. terrestris* almost always produces single syllables. Only in one locality (Col St-Martin, near St-Martin-Vésubie in the Alpes-Maritimes) have we heard disyllabic echemes (Figs 527, 532, 537); here they were quite common, although always mixed with single syllables in the same song. In the songs of the 12 males studied, the repetition rate of isolated syllables varied from 0·6/s to 1·7/s and each syllable lasted between 110 and 180 ms.

Receptive females sing in response to the male calling song, producing syllables that are clearly different in pattern from those produced by the male (Figs 528, 533, 538).

DISTRIBUTION. Parts of the French and Italian Alps, and also the Ticino canton of Switzerland.

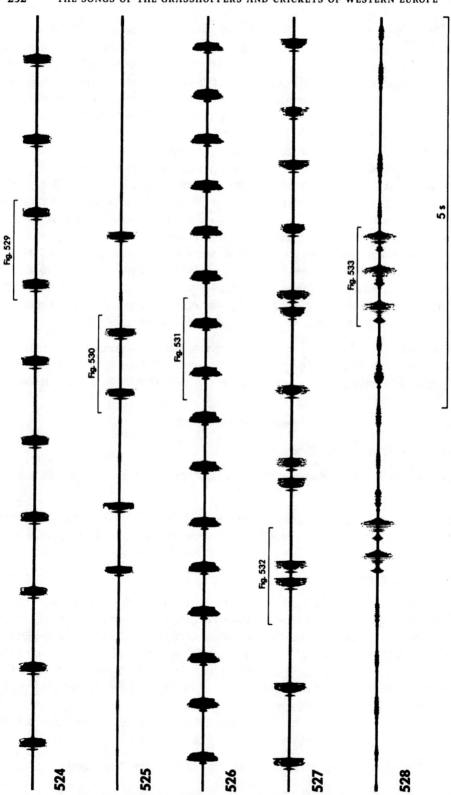

Figures 524–528 Oscillograms of (524–527) the calling songs of four males and (528) the response song of a female of *Ephippiger terrestris*. The quieter background sounds in Fig. 528 are the calling songs of nearby males.

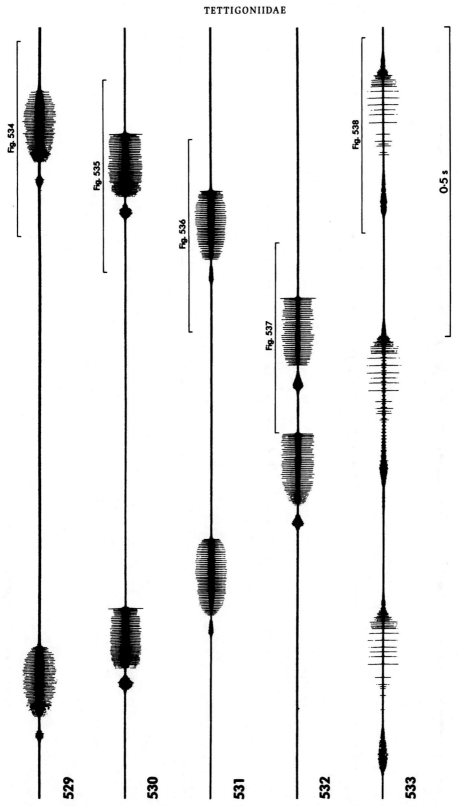

Figures 529–533 Faster oscillograms of the indicated parts of the songs of *Ephippiger terrestris* shown in Figs 524–528.

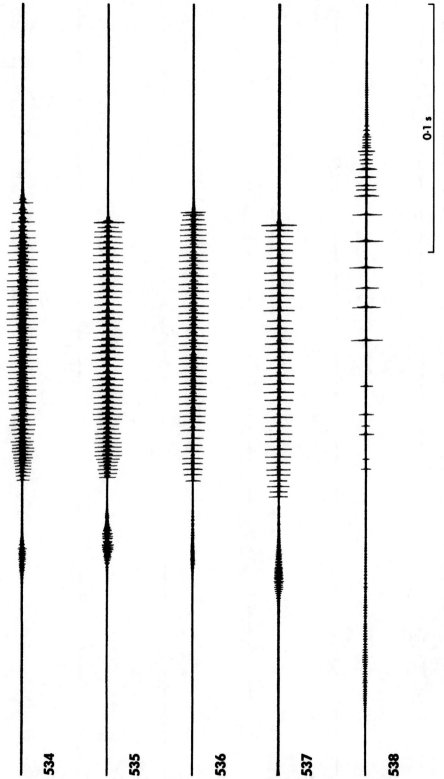

Figures 534–538 Faster oscillograms of the indicated parts of the songs of *Ephippiger terrestris* shown in Figs 529–533.

Ephippiger perforatus (Rossi)

Locusta perforata Rossi, 1790: 267.

North Apennine Bush-cricket.

REFERENCES TO SONG. **Oscillogram**: Heller, 1988; Jatho *et al.*, 1992, 1994; Schmidt, 1989.

RECOGNITION. This species does not overlap in distribution with *E. ephippiger* or *E. terrestris*, but may overlap with *E. ruffoi* in the southern part of its range. Males may be easily distinguished from *E. ruffoi* by their cerci, which have the inner tooth near the base and lack an apical tooth (those of *E. ruffoi* have the inner tooth near the tip, where there is an additional sharp apical tooth), and females by their long ovipositor (longer than 25 mm in *E. perforatus*, shorter than 20 mm in *E. ruffoi*).

SONG (Figs 539–546. CD 1, track 71). The male calling song, heard especially in the morning, consists of echemes of about 4–6 diplosyllables in quick succession. The intervals between the echemes are very variable: they are often quite long, but occasionally the echemes are repeated at intervals of only a few seconds. Oscillographic analysis shows that each syllable lasts about 150–200 ms and consists of a short opening hemisyllable (lasting about 30–50 ms) followed by a longer closing hemisyllable (lasting about 100–150 ms). The syllable repetition rate within an echeme is about 4–6/s.

As in other species of *Ephippiger*, males of *E. perforatus* give short bursts of tremulation from time to time, during which the body is vigorously shaken for about 400–500 ms. However, unlike the males of any other ephippigerine we have studied, those of *E. perforatus* stridulate quietly during each burst of tremulation. The sound is apparently produced by the fore wings, but is much quieter than the normal echemes of the calling song (see Figs 539, 540) and is also structurally different, consisting of an echeme of about 10–15 syllables. Oscillographic analysis suggests that these are diplosyllables composed of relatively few tooth-impacts, and that the opening hemisyllables are longer than the closing ones (Figs 543, 544). The syllable repetition rate is about 20–30/s.

DISTRIBUTION. Known for certain only from northern Italy, from Liguria to Lazio, especially in the Apennines.

Ephippiger ruffoi Galvagni

Ephippiger Ruffoi Galvagni, 1955: 39.

Ruffo's Bush-cricket.

REFERENCES TO SONG. No published work known to us.

RECOGNITION. Like *E. perforatus*, this species does not overlap in range with either *E. ephippiger* or *E. terrestris*. For the distinction between *E. ruffoi* and *E. perforatus*, see under the latter species.

SONG (Figs 547–552. CD 1, track 72). The male calling song, produced during the day, consists of echemes repeated for long periods at intervals of about 0·4–1·1 s and each composed

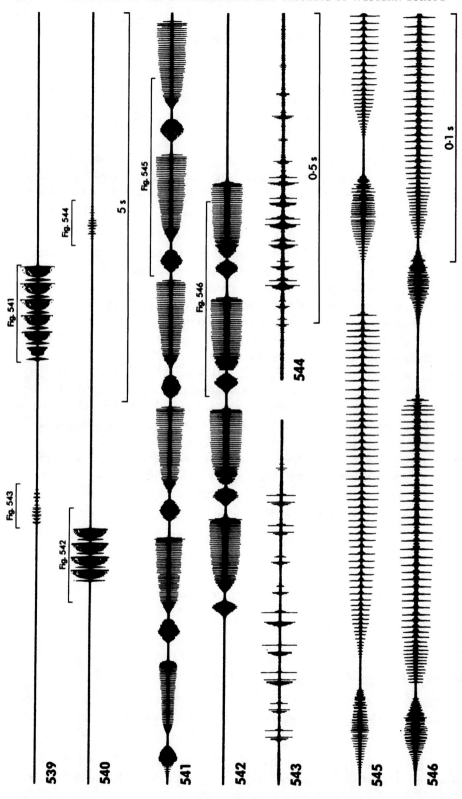

Figures 539–546 Oscillograms at three different speeds of the calling songs of two males of *Ephippiger perforatus*. The quieter sounds shown in Figs 539 and 540 were produced during tremulation (see text).

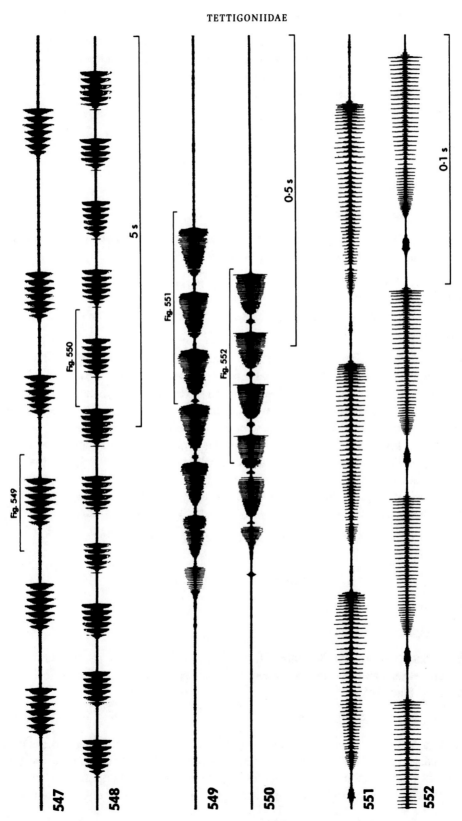

Figures 547–552 Oscillograms at three different speeds of the calling songs of two males of *Ephippiger ruffoi*.

of about 5–7 syllables in quick succession. Oscillographic analysis shows that the syllables usually consist of a short opening hemisyllable followed by a louder closing hemisyllable lasting about 7–12 times as long. Each syllable lasts about 70–110 ms and the syllable repetition rate within an echeme is about 10–12/s.

DISTRIBUTION. Known only from the central Apennines.

Ephippigerida areolaria (Bolívar)

Ephippiger areolarius Bolívar, 1877: 292.

Spanish Mountain Bush-cricket.

REFERENCES TO SONG. **Oscillogram, wing-movement** and **frequency information**: Heller, 1988. **Verbal description only**: Defaut, 1987; 1988*b*.

RECOGNITION. *Ephippigerida* is very similar to *Ephippiger*, but the male supra-anal plate tends to be triangular or rounded rather than quadrangular. *E. areolaria* differs from *E. taeniata* in being much smaller, the hind femora measuring less than 18 mm (more than 20 mm in *E. taeniata*).

SONG (Figs 553, 554, 556, 557, 559, 560. CD 1, track 73). The diurnal male calling song consists of a long series of echemes repeated fairly regularly every 0·5–2·5 s. Each echeme is usually composed of 3–5 diplosyllables, in which the closing hemisyllables are about 10 times longer than the opening ones; the syllable repetition rate within an echeme is about 4–7/s, depending on temperature. Each diplosyllable lasts about 120–250 ms, and a 4–syllable echeme lasts about 0·6–0·9 s; the syllables (and echemes) last longer at lower temperatures. The first opening hemisyllable in an echeme is usually a little longer than the others and is separated from the following closing hemisyllable by a slightly longer gap.

The song is thus quite similar to that of Pyrenean populations of *Ephippiger ephippiger*, but the syllable repetition rate is significantly lower at the same temperature.

DISTRIBUTION. Widespread in mountains in the Iberian Peninsula.

Ephippigerida taeniata (Saussure)

Ephippigera tæniata Saussure, 1898: 238.

Large Striped Bush-cricket.

REFERENCES TO SONG. **Oscillogram**: Grzeschik, 1969; Heller, 1988; Stiedl *et al.*, 1994. **Wing-movement** and **frequency information**: Heller, 1988. **Verbal description only**: Schroeter & Pfau, 1987.

RECOGNITION. See under *E. areolaria*.

SONG (Figs 555, 558, 561. CD 1, track 74). The diurnal male calling song consists of a short syllable repeated for long periods at intervals of about 1–5 s. Oscillographic analysis shows that most (sometimes all) of the sound is produced by the opening stroke of the fore wings, which gives rise to a hemisyllable lasting about 15–50 ms, depending on tempera-

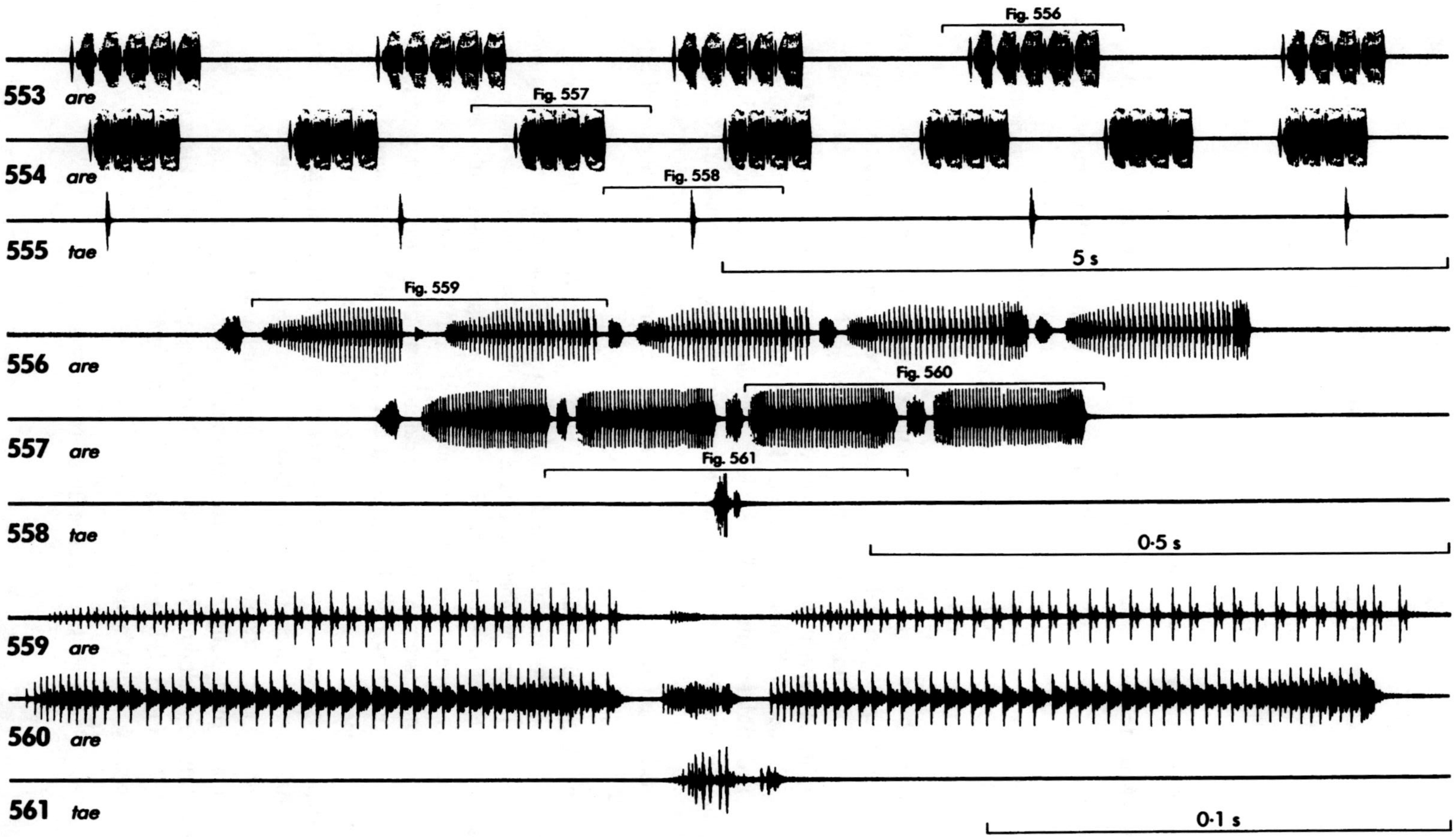

Figures 553–561 Oscillograms at three different speeds of the male calling songs of (553, 554, 556, 557, 559, 560) *Ephippigerida areolaria* and (555, 558, 561) *E. taeniata*.

ture. The closing hemisyllable, when present, is quieter and usually shorter, lasting about 5–30 ms.

DISTRIBUTION. Known only from the extreme south of Spain (Cadiz province), Morocco and Algeria.

Uromenus rugosicollis (Serville)

Ephippiger rugosicollis Serville, 1838: 475.

Rough-backed Bush-cricket; (F) Éphippigère carénée; (D) Kantige Sattelschrecke.

REFERENCES TO SONG. **Oscillogram**: Busnel, Busnel & Dumortier, 1956; Busnel, Dumortier & Busnel, 1956; Dumortier, 1963c; Heller, 1988. **Wing-movement**: Heller, 1988. **Frequency information**: Heller, 1988; Keuper et al., 1988a, 1988b. **Verbal description only**: Chopard, 1922; Defaut, 1984a; Marquet, 1877 (as *durieui*). **Disc recording**: Andrieu & Dumortier, 1963 (LP), 1994 (CD).

RECOGNITION. The genus *Uromenus* includes more than 20 species occurring in the western Mediterranean Region and especially the Iberian Peninsula. They are all brachypterous and similar in appearance to *Ephippiger* and *Ephippigerida*, but differ from both these genera in having quite well developed lateral carinae on the posterior raised part of the pronotum.

Males of *U. rugosicollis* may be recognized by the supra-anal plate, which is fused on to the tenth abdominal tergite and forms a long, rounded, posterior extension of it between the cerci. The females have a fairly short, strongly upcurved ovipositor and a subgenital plate deeply divided into long pointed lobes. In the field the male calling song enables this sex to be easily distinguished from any other species occurring in the same parts of France and Spain.

SONG (Figs 562, 563, 566, 567. CD 1, track 75). *U. rugosicollis* and *U. elegans* are remarkable among the European Tettigoniidae in having very slow stridulatory wing-movements, so that the hemisyllables produced by the closing strokes of the fore wings are normally at least half a second long and often last as long as a second or even longer. They are, in this respect, the equivalent among the bush-crickets of the grasshopper *Stenobothrus lineatus* (see p. 333). The male calling song of *U. rugosicollis*, produced particularly in the evening and during the night, consists of long echemes of closing hemisyllables repeated at the rate of about 0·5–1·1/s. The stridulatory file has very dense teeth and so, even when the wing-movements are at their slowest, the tooth-impact rate is about 100/s; nevertheless these impacts can be appreciated by the human ear and impart a quality to the sound reminiscent of a finger-nail being scraped rapidly along the teeth of a comb. In the songs studied the echemes lasted from 9 s (including 6 syllables) to nearly 2 minutes (including 122 syllables) and were separated by intervals varying from 2 s to 35 s. The intervals between successive syllables varied from 300 to 550 ms. There were no opening hemisyllables.

DISTRIBUTION. South-western and southern France; Catalonia.

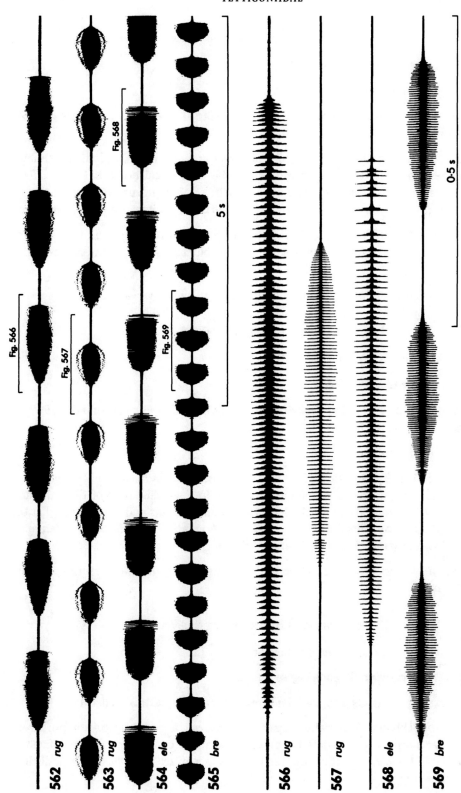

Figures 562–569 Oscillograms at two different speeds of the male calling songs of (562, 563, 566, 567) *Uromenus rugosicollis*, (564, 568) *U. elegans* and (565, 569) *U. brevicollis*.

Uromenus elegans (Fischer)

Ephippigera elegans Fischer, 1853: 219.

Elegant Bush-cricket.

REFERENCES TO SONG. **Oscillogram, wing-movement** and **frequency information**: Heller, 1988.

RECOGNITION. *U. elegans* is the only species of *Uromenus* known from the Italian mainland, Peloponnese and Crete. On Sardinia and Sicily it may be distinguished from *U. brevicollis* by the long and relatively slender male cerci and the strongly upcurved ovipositor; in *U. brevicollis* the male cerci are short and very stout, and the ovipositor is only gently upcurved.

SONG (Figs 564, 568. CD 1, track 76). Extremely similar to *U. rugosicollis*. In the calling song of the only male studied, the closing hemisyllables lasted 1·2 s and were repeated at the rate of 0·8/s.

DISTRIBUTION. The Italian Peninsula, Sardinia, Sicily, Peloponnese and Crete.

Uromenus brevicollis (Fischer)

Ephippigera brevicollis Fischer, 1853: 219.

Short-backed Bush-cricket.

REFERENCES TO SONG. **Verbal description only**: Chopard, 1952; Peyerimhoff, 1908 (as *confusus*); Schroeter & Pfau, 1987.

RECOGNITION. See under *U. elegans*.

SONG (Figs 565, 569. CD 1, track 77). Similar to those of *U. rugosicollis* and *U. elegans*, and also produced mainly in the evening and at night. In the calling song of the only male studied (at 28°C), the closing hemisyllables lasted only 260 ms and were repeated at the rate of 2·3/s; however, Chopard (1952) gives a repetition rate of about 50 per minute, agreeing closely with that of *U. rugosicollis* and *U. elegans*.

DISTRIBUTION. Corsica, Sardinia, Minorca, Sicily, Algeria and the extreme south of Spain, near Algeciras.

Uromenus catalaunicus (Bolívar)

Ephippigera (Steropleurus) Catalaunica Bolívar, 1898b: 125.

Catalan Bush-cricket; (F) Éphippigère catalane.

REFERENCES TO SONG. **Oscillogram** and **frequency information**: Heller, 1988.

RECOGNITION. This and the remaining species of *Uromenus* are not easy to distinguish from one another morphologically, differing mainly in genitalic characters that are difficult to use. Males of *U. catalaunicus* can, however, be easily recognized in the field by their distinctive calling song.

SONG (Figs 570–578. CD 1, track 78). The male calling song, produced during the day, is a long series of well-defined echemes, each consisting of about 9–15 syllables repeated at the rate of about 8–16/s. Each echeme lasts about 0·7–1·5 s and successive echemes are separated by intervals of about 0·7–2·1 s. All these values are fairly constant during any one song. Oscillographic analysis shows that most of the sound consists of closing hemisyllables lasting about 40–100 ms, but there are often quieter and very brief opening hemisyllables. The first closing hemisyllable in an echeme is usually quieter and shorter than the remaining ones.

DISTRIBUTION. Catalonia and the eastern French Pyrenees.

Uromenus asturiensis (Bolívar)

Ephippigera (Steropleurus) Asturiensis Bolívar, 1898*b*: 144.

Asturian Bush-cricket.

REFERENCES TO SONG. **Oscillogram**: Stephen & Hartley, 1991 (as *nobrei*). **Verbal description only**: Hartley, 1990 (as *n.*).

RECOGNITION. See under *U. catalaunicus*. Females may be distinguished from related species by the relatively short ovipositor, usually less than 15 mm long. In the field males may be recognized by their distinctive calling song.

SONG (Figs 579–587. CD 1, track 79). The male calling song, produced during the day, consists of a very brief echeme lasting about 250–350 ms and composed of about 6–8 syllables repeated at the rate of about 25–50/s. The echemes are repeated at very variable intervals, usually between 6 and 25 s. Oscillographic analysis shows that each echeme begins with a series of increasingly loud, short syllables (usually opening hemisyllables, but sometimes with even briefer closing hemisyllables) followed immediately by a longer diplosyllable in which the opening hemisyllable lasts about 25–40 ms and the closing one about 60–120 ms.

Some calling songs consist only of echemes of this kind, but in others some or all of the echemes are followed after a short interval (about 200–500 ms) by a further diplosyllable (Figs 579, 581).

DISTRIBUTION. North-western Iberian Peninsula.

Uromenus stalii (Bolívar)

Ephippiger Stâlii Bolívar, 1877: 284.

Stål's Bush-cricket; (F) Éphippigère du Larzac.

REFERENCES TO SONG. **Oscillogram**: Hartley *et al.*, 1974; Heller, 1988, 1990. **Wing-movement**: Heller, 1988, 1990. **Frequency information**: Heller, 1988; Latimer & Broughton, 1984. **Verbal description only**: Robinson, 1990.

RECOGNITION (Plate 2: 9). See under *U. catalaunicus*. The male calling song is similar to

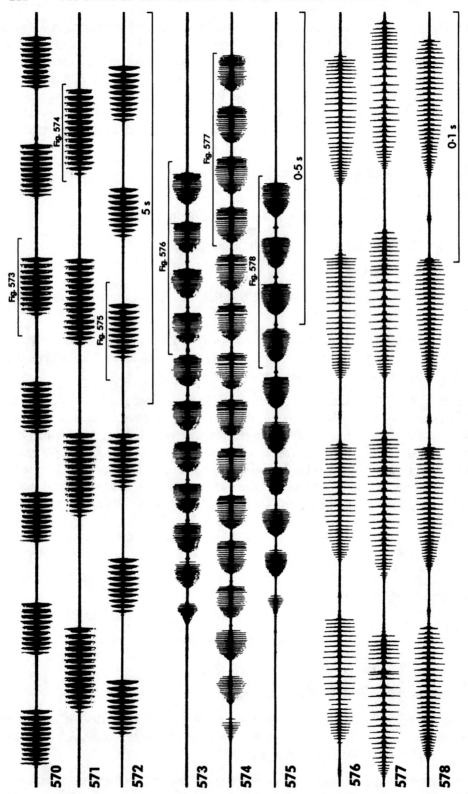

Figures 570–578 Oscillograms at three different speeds of the calling songs of three males of *Uromenus catalaunicus*.

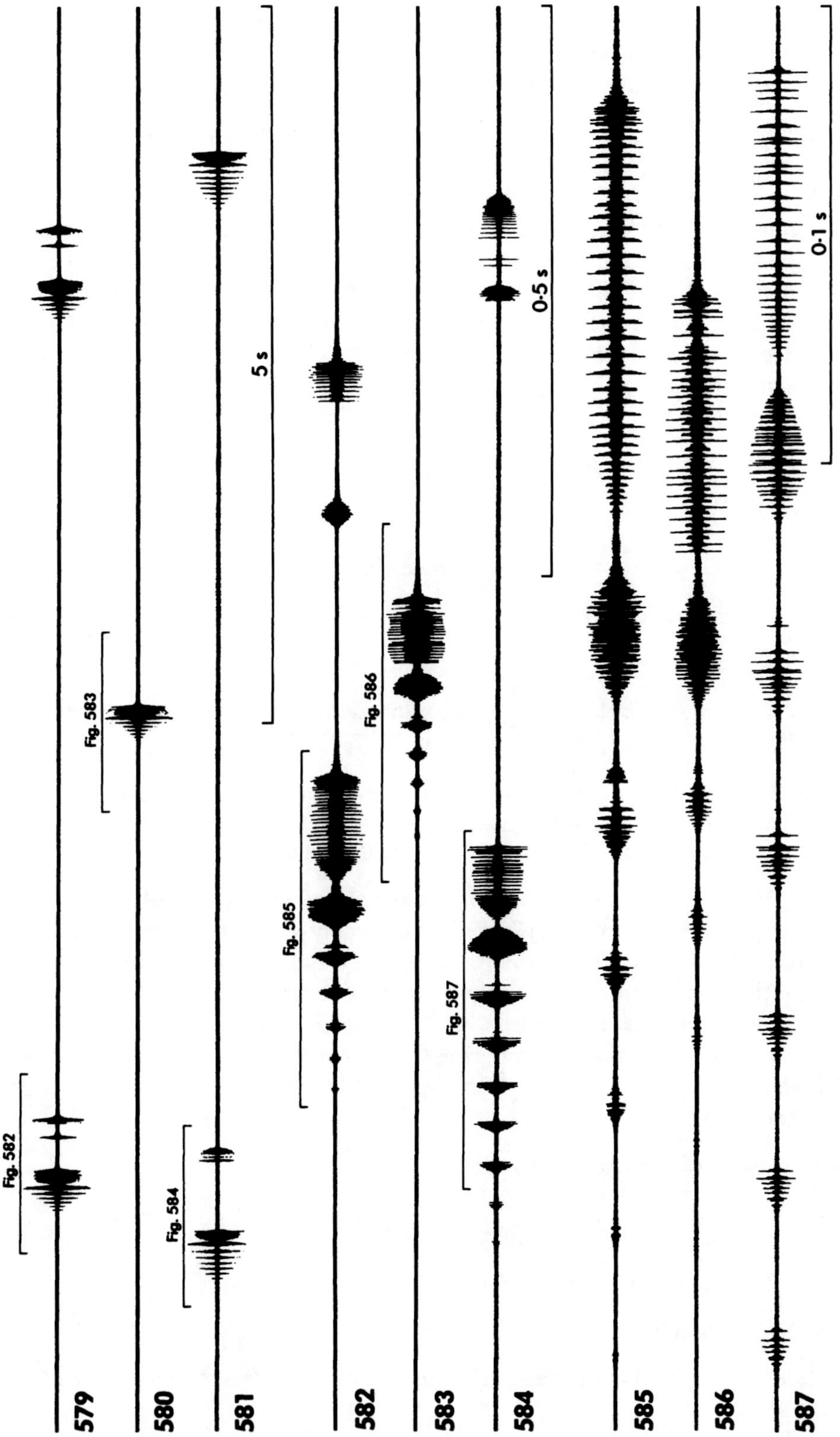

Figures 579–587 Oscillograms at three different speeds of the calling songs of three males of *Uromenus asturiensis*.

that of *U. asturiensis*, but the number of syllables in each echeme is higher, usually more than 8.

For a note on the spelling of the specific name of this species see p. 10.

SONG (Figs 588–596. CD 1, track 80). The male calling song, typically produced during the day, is an echeme lasting about 180–320 ms and composed of about 7–13 syllables repeated at the rate of about 40–50/s. The echemes are usually repeated at intervals of about 5–30 s, but sometimes up to a minute or more. Oscillographic analysis shows that the syllables become gradually louder throughout the echeme, and each is usually composed of a well-developed opening hemisyllable followed by a much shorter and often quieter closing hemisyllable. The final syllable of the echeme is sometimes longer than the others, with a well-developed and more prolonged closing hemisyllable (as in *U. asturiensis*), but some-times similar in structure to the earlier ones. In some songs some of the echemes are followed after a short interval by a further diplosyllable, as in *U. asturiensis*.

DISTRIBUTION. Central and northern Iberian Peninsula, and recorded from Aveyron in southern France.

Uromenus perezii (Bolívar)

Ephippiger Perezii Bolívar, 1877: 282.

Pérez's Bush-cricket.

REFERENCES TO SONG. **Oscillogram, wing-movement** and **frequency information**: Heller, 1988. **Verbal description only**: Defaut, 1987, 1988*b*.

RECOGNITION. See under *U. catalaunicus*. The male calling song, with its very brief echemes of fewer than 5 syllables, is characteristic.

For a note on the spelling of the specific name of this species see p. 10.

SONG (Figs 597–599. CD 1, track 81). The diurnal male calling song consists of a regularly repeated very brief echeme of 2–4 syllables. The echeme repetition rate seems to depend on temperature, being about 1/s in warm sunshine and about one in every 5 s in cool, dull conditions. The sound consists almost entirely of closing hemisyllables. Oscillographic analy-sis shows that the syllables are repeated within an echeme at the rate of about 15–30/s and that they become longer and louder during the course of the echeme, the last closing hemisyllable usually lasting about 30–50 ms. A trisyllabic echeme lasts about 100–120 ms in warm conditions.

DISTRIBUTION. Northern and eastern Spain.

Uromenus martorellii (Bolívar)

Ephippiger (Steropleurus) Martorellii Bolívar, 1878: 444.

Martorell's Bush-cricket.

REFERENCES TO SONG. **Oscillogram, wing-movement** and **frequency information**: Heller, 1988.

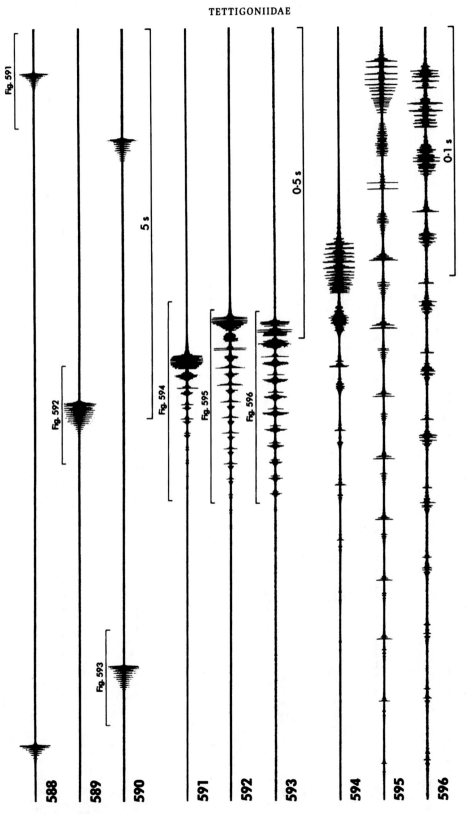

Figures 588–596 Oscillograms at three different speeds of the calling songs of three males of *Uromenus stalii*.

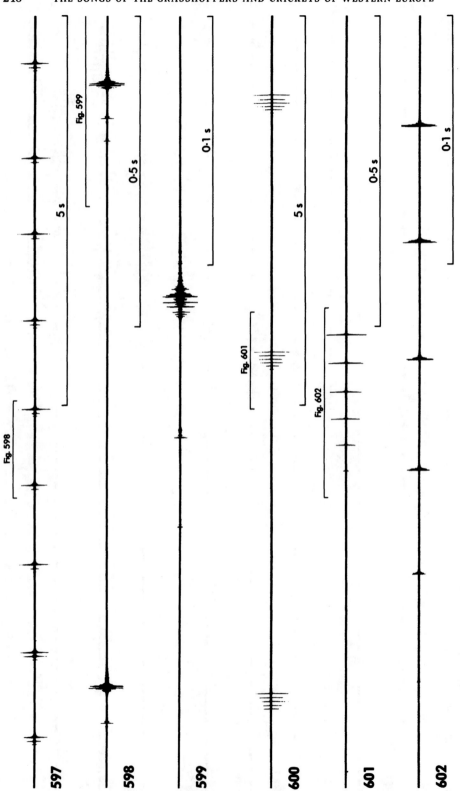

Figures 597–602 Oscillograms at three different speeds of the male calling songs of (597–599) *Uromenus perezii* and (600–602) *U. martorellii.*

RECOGNITION. See under *U. catalaunicus*. The male calling song is similar to that of *U. perezii*, but in the song of the only male studied the number of syllables per echeme was greater and the syllables were all equally short-lived.

For a note on the spelling of the specific name of this species see p. 10.

SONG (Figs 600–602. CD 1, track 82). The calling song of the only male studied (in continuous sunshine) consisted of very brief echemes of 5–6 closing hemisyllables; the echemes were repeated every 3–5 s. Oscillographic analysis showed that the hemisyllables tended to become louder during the course of each echeme, but all were equally brief, lasting only about 2 ms. The syllable repetition rate within an echeme was about 23/s and a 5–syllable echeme lasted about 180 ms. There were no opening hemisyllables. Heller (1988) gives the smaller number of 1–4 syllables per echeme.

DISTRIBUTION. The eastern provinces of Spain, from Barcelona to Almería and Granada.

Uromenus andalusius (Rambur)

Ephippiger Andalusius Rambur, 1838: 49.

Andalusian Bush-cricket.

REFERENCES TO SONG. **Oscillogram, wing-movement** and **frequency information**: Heller, 1988.

RECOGNITION. See under *U. catalaunicus*. The single opening hemisyllables of the male calling song are characteristic.

SONG (Figs 603–605. CD 1, track 83). The calling song of the only male studied consisted of a single opening hemisyllable lasting about 80 ms and repeated at intervals of 3–6 s. The closing strokes of the fore wings were silent or almost so.

DISTRIBUTION. Known only from Andalusia and Murcia in southern Spain.

Baetica ustulata (Rambur)

Ephippiger ustulatus Rambur, 1838: 52.

Sierra Nevadan Bush-cricket.

REFERENCES TO SONG. **Oscillogram, wing-movement** and **frequency information**: Heller, 1988.

RECOGNITION. This bush-cricket, known only from 2500 m upwards in the Spanish Sierra Nevada, is unlikely to be confused with any other species. It differs from its nearest relative in this region, *Uromenus andalusius*, in being predominantly dark brown or black in colour and having no lateral carinae on the pronotum, which almost or entirely covers the reduced fore wings.

SONG (Figs 608–613. CD 1, track 84). The diurnal male calling song consists of a long series of echemes repeated regularly at the rate of about 0·5–1·5/s, depending on temperature.

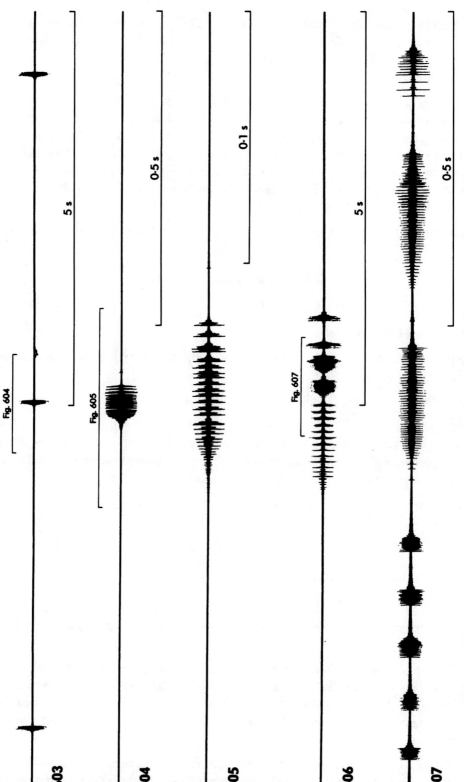

Figures 603–607 Oscillograms of the male calling songs of (603–605) *Uromenus andalusius* and (606, 607) *Callicrania selligera.*

Each echeme is composed of about 7–15 diplosyllables repeated at the rate of about 10–25/s, with a crescendo during the first few syllables. The echemes last about 0·5–1·0 s and the intervals between them are of a similar duration. Oscillographic analysis shows that the opening and closing hemisyllables are of about equal duration (*c* 20–35 ms); they may be quite similar in structure (Fig. 612) or clearly distinct (Fig. 613).

DISTRIBUTION. Known only from the Sierra Nevada in southern Spain, at altitudes of 2500 m upwards.

Callicrania selligera (Charpentier)

Barbitistes selligera Charpentier, 1825: 99.

Lusitanian Bush-cricket.

REFERENCES TO SONG. **Oscillogram**: Heller, 1988; Pfau, 1996.

RECOGNITION. This species, known at present only from the mountains of northern Portugal, is similar in appearance to *Uromenus*, but the well-developed pronotal lateral carinae are toothed. The pronotal disc is quite smooth, especially in the metazona. The ovipositor is long and almost straight.
  For a note on the generic assignment of this species see p. 9.

SONG (Figs 606, 607. CD 1, track 85). The male calling song consists of an echeme lasting about 1·5–2·5 s. At the beginning of the echeme is a series of about 15 very brief syllables (of which at least the earlier ones are diplosyllables) repeated at the rate of about 12–14/s and showing a crescendo during the first few. These brief syllables are followed immediately by 2–3 much longer closing hemisyllables, and these may in turn be followed by some further shorter sounds. The short closing hemisyllables each last up to 30 ms, and the longer ones about 200–250 ms. The echemes are separated by very variable intervals, 3–10 s in the song of the only male studied.

DISTRIBUTION. Known only from mountains in the northern half of Portugal.

Platystolus martinezii (Bolívar)

Ephippigera Martinezii Bolívar, 1873: 222.

Martínez's Bush-cricket.

REFERENCES TO SONG. **Oscillogram**: Heller, 1988; Pfau, 1996; Pfau & Schroeter, 1988.

RECOGNITION. *Platystolus* is generally similar to *Uromenus*, but the male tenth abdominal tergite (although distinct from the supra-anal plate) is prolonged backwards between the cerci. *P. martinezii* differs from *P. faberi* in that the fore wings extend well beyond the hind margin of the pronotum and the inner tooth of the male cerci is at the tip; in *P. faberi* the fore wings are almost entirely covered by the pronotum and the inner tooth of the male cerci is about halfway along their length. In the field the male calling songs of these two species are very distinctive.

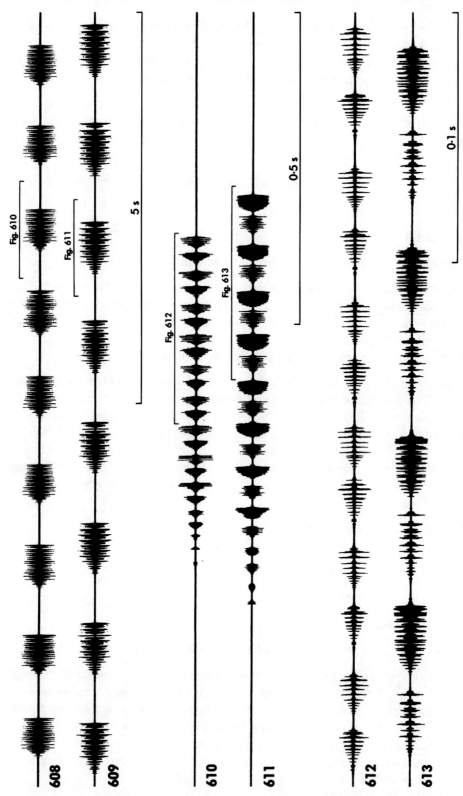

Figures 608–613 Oscillograms at three different speeds of the calling songs of two males of *Baetica ustulata*.

For a note on the spelling of the specific name of this species see p. 10.

SONG (Figs 614–616. CD 1, track 86). The male calling song, produced in the evening and at night, is an echeme lasting about 3–6 s and consisting of about 30–70 closing hemisyllables. The syllable repetition rate gradually increases from about 8/s at the beginning of the echeme to about 10–15/s at the end; each hemisyllable lasts about 30–50 ms. In the song of the only male studied, the echemes were repeated at intervals of 20–50 s.

The acoustic behaviour of the male and female reacting with each other is described in detail by Pfau & Schroeter (1988).

DISTRIBUTION. Iberian Peninsula, especially the central part.

Platystolus faberi Harz

Platystolus faberi Harz, 1975a: 17.

Faber's Bush-cricket.

REFERENCES TO SONG. **Oscillogram**: Pfau, 1996. **Verbal description only**: Schroeter & Pfau, 1987.

RECOGNITION. See under *P. martinezii*.

SONG (Figs 617, 618. CD 1, track 87). In full sunshine the male calling song is an echeme lasting about 1 s and consisting of about 5–7 very brief diplosyllables followed immediately by a similar number of much longer diplosyllables. The series of brief syllables begins quietly, rapidly becoming louder, and the closing hemisyllables increase in duration from about 2–3 ms to about 20 ms during the course of the series; the syllable repetition rate is about 22/s. The longer syllables are repeated at the rate of about 8/s and their closing hemisyllables last about 100–110 ms. Throughout the echeme the opening hemisyllables are comparatively quiet and last only about 10–15 ms. The echemes are repeated at very variable intervals and are sometimes produced in pairs.

Schroeter & Pfau (1987) give the duration of an echeme as 2–5 s at 20°C; this observation was perhaps made in dull weather. The values given above apply to a song produced in continuous sunshine at a shade air temperature of 23°C.

DISTRIBUTION. Known at present only from the Cantabrian Mountains and Sierra de la Demanda in northern Spain.

Pycnogaster jugicola Graells

Pycnogaster jugicola Graells, 1851: 157.

Fat-bellied Bush-cricket.

REFERENCES TO SONG. **Oscillogram**: Heller, 1988; Pfau, 1988. **Frequency information**: Heller, 1988; Latimer & Broughton, 1984. **Verbal description only**: Ebner, 1925 (as *bolivari*); Pantel, 1896 (as *b.*); Pfau & Schroeter, 1983.

RECOGNITION. *Pycnogaster* is another genus of very brachypterous bush-crickets, but it may

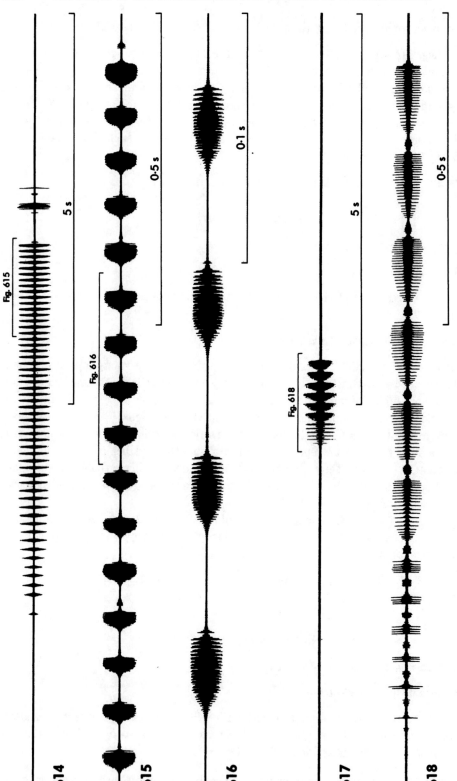

Figures 614–618 Oscillograms of the male calling songs of (614–616) *Platystolus martinezii* and (617, 618) *P. faberi.*

be easily distinguished from the foregoing genera by the pronotum, which is not saddle-shaped but more like the top and two sides of a box, with well-defined lateral carinae along the entire length. *P. jugicola* differs from *P. sanchezgomezi* and *P. inermis* in having the pronotal lateral carinae interrupted only by a groove near the front of the pronotum; in the other two species there is a groove across these carinae about halfway along their length (often in addition to another one nearer the front).

SONG (Figs 619–621, 624–626. CD 1, track 88). The diurnal male calling song is a long, loud echeme of diplosyllables in which the sound is virtually continuous. The duration of the echeme is very variable: in the songs of the eight males studied it varied from 8 s (including 32 syllables) to 61 s (including 280 syllables). The syllable repetition rate, as Pfau (1988) has shown, depends on temperature; in the songs studied it varied from 2·6/s at 20°C to 5·5/s at 32°C. The intervals between the echemes vary greatly, falling within the range 3–45 s in the songs studied. Oscillographic analysis shows that the opening hemisyllables last about 10–20 ms and the closing ones from 150 ms to 400 ms or more, depending on temperature; the opening hemisyllables, although short-lived, are conspicuous sounds and enable the syllable structure of the echemes to be clearly heard.

DISTRIBUTION. Found in mountains in the central, northern and western parts of the Iberian Peninsula.

Pycnogaster sanchezgomezi Bolívar

Pycnogaster Sanchez-Gomezi Bolívar, 1897a: 172.

Sánchez Gómez's Bush-cricket.

REFERENCES TO SONG. **Oscillogram**: Heller, 1988; Pfau, 1988.

RECOGNITION. For the distinction between this species and *P. jugicola* see under that species. Distinguishing between *P. sanchezgomezi* and *P. inermis* is more difficult, but *P. inermis* is restricted to the Sierra Nevada, whereas *P. sanchezgomezi* occurs from the Sierra de Segura northwards.

SONG (Figs 622, 627. CD 1, track 89). The male calling song, produced especially in the evening and at night, is an echeme of closing hemisyllables repeated at the rate of about 1·5–2·5/s, depending on temperature. The whole echeme lasts about 7–20 s. Oscillographic analysis shows that the hemisyllables last about 350–600 ms and the intervals between them about 50 ms. There are no opening hemisyllables.

DISTRIBUTION. Found in the mountains of eastern and south-eastern Spain, from Teruel province in the north to the Sierra de Segura in the south.

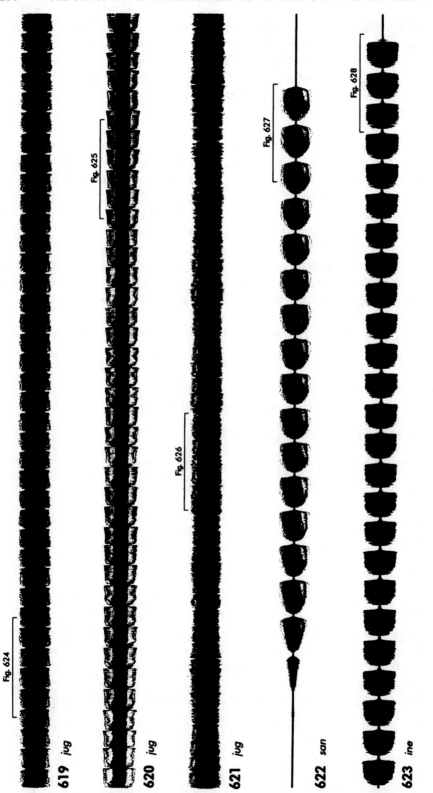

Figures 619–623 Oscillograms of the male calling songs of (619–621) *Pycnogaster jugicola*, (622) *P. sanchezgomezi* and (623) *P. inermis*.

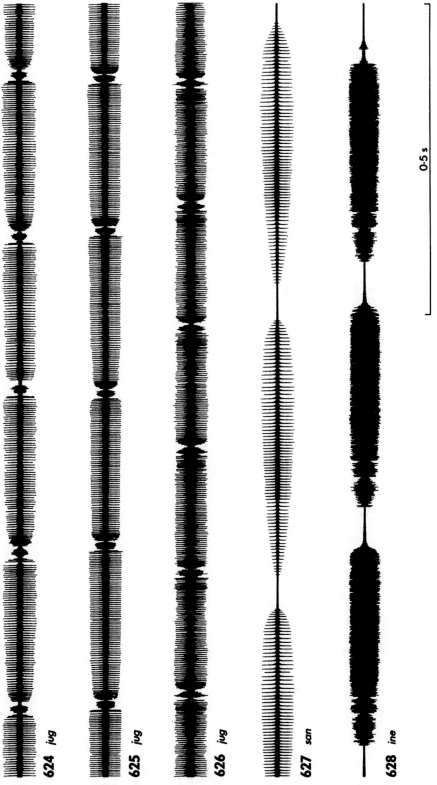

Figures 624–628 Faster oscillograms of the indicated parts of the songs of *Pycnogaster jugicola, P. sanchezgomezi* and *P. inermis* shown in Figs 619–623.

0·5 s

624 *jug*

625 *jug*

626 *jug*

627 *san*

628 *ine*

Pycnogaster inermis (Rambur)

Bradyporus inermis Rambur, 1838: 57.

Unarmed Bush-cricket.

REFERENCES TO SONG. **Oscillogram**: Heller, 1988; Pfau, 1988.

RECOGNITION. See under *P. jugicola* and *P. sanchezgomezi*.

SONG (Figs 623, 628. CD 1, track 90). The male calling song is very similar to that of *P. sanchezgomezi* and is also mainly heard from dusk onwards. It is an echeme lasting up to about 25 s and consisting of syllables repeated at the rate of about 2–3/s, depending on temperature. Oscillographic analysis shows that there are usually no opening hemisyllables and that the closing hemisyllables last about 250–350 ms.

DISTRIBUTION. The Sierra Nevada in southern Spain.

Chapter 8

Gryllidae and Gryllotalpidae

Crickets and mole-crickets

The calling songs of the European true crickets and mole-crickets (unlike those of most bush-crickets) are resonant, with a clearly defined dominant frequency below 8 kHz, usually below 5 kHz, and as a result many of them, especially the lower-pitched ones, sound quite musical. The calling songs of *Gryllus campestris, G. bimaculatus, Acheta domesticus, Modicogryllus bordigalensis* and *Nemobius sylvestris* all have a dominant frequency within the range 3·5–5·5 kHz, and so these species may be said to be broadcasting on the same wave-length, though often using different rhythmic patterns. *Pteronemobius heydenii* sings at the higher frequency of 6·4–7·9 kHz, and the dominant frequencies of the calling songs of *Eugryllodes pipiens* and *Oecanthus pellucens* are at the lower level of 2·5–3·5 kHz. *Gryllotalpa gryllotalpa*, singing at 1·3–1·7 kHz, has the lowest frequency of all the known songs of European Gryllidae and Gryllotalpidae; *G. vineae* sings at frequencies of at least double these values.

The females are normally silent in the true crickets of western Europe (but see p. 33). In the mole-crickets, however, the females can stridulate with their fore wings, though lacking the well developed stridulatory organ of the males. The sound takes the form of short, harsh chirps, quite unlike the male calling songs, and is generally produced when a female encounters another mole-cricket of either sex.

Gryllus campestris Linnaeus

Gryllus (Acheta) campestris Linnaeus, 1758: 428.

Field-cricket; (F) Grillon champêtre; (D) Feldgrille; (NL) Veldkrekel; (DK) Markfårekylling; (S) Fältsyrsan; (SF) Kenttäsirkka; (E) Grillo campestre; (P) Grilo-do-campo; (I) Grillo Canterino.

REFERENCES TO SONG. **Oscillogram**: Bentley, 1977; Bentley & Hoy, 1972; Bentley & Kutsch, 1966; Dambach & Huber, 1974; Dambach & Rausche, 1985; Dambach *et al.*, 1983; Dumortier, 1963*c*; Elsner & Popov, 1978; Faber, 1957*b*; Grein, 1984; Holst, 1970; Huber, 1963, 1970, 1977; Jones & Dambach, 1973; Kämper & Dambach, 1981; Kleukers *et al.*, 1997; Koch *et al.*, 1988; Kutsch, 1969; Kutsch & Otto, 1972; Lottermoser, 1952; Nocke, 1971; Otto, 1971; Popov, 1972, 1975; Popov & Shuvalov, 1974; Popov, Shuvalov, Svetlogorskaya & Markovich, 1974; Schäffner & Koch, 1987; Schmidt & Schach, 1978;

Schmitz *et al.*, 1982; Thorson *et al.*, 1982. **Diagram**: Bellmann, 1985*a*, 1988, 1993*a*; Bellmann & Luquet, 1995; Duijm & Kruseman, 1983; Holst, 1970; Leroy, 1962, 1966; Ragge, 1965. **Sonagram**: Leroy, 1962, 1966, 1977; Popov, 1972. **Wing-movement**: Koch *et al.*, 1988. **Frequency information**: Dumortier, 1963*c*; Lottermoser, 1952; Nocke, 1971; Popov & Shuvalov, 1974; Popov, Shuvalov, Knyazev & Klar-Spasovskaya, 1974; Popov, Shuvalov, Svetlogorskaya & Markovich, 1974; Sales & Pye, 1974. **Musical notation**: Kreidl & Regen, 1905; Yersin, 1854*a*. **Verbal description only**: Alexander, 1967*a*; Baier, 1930; Chopard, 1952; Dathe, 1974; Defaut, 1987, 1988*b*; Elliot & Koch, 1983; Faber, 1928; Harz, 1957; Holst, 1986; Huber, 1983; Hudson, 1903; Popov & Shuvalov, 1977; Regen, 1903, 1913; Rost & Honegger, 1987; Shuvalov & Popov, 1973; Weber, 1984; White, 1789. **Disc record-ing**: Andrieu & Dumortier, 1963 (LP), 1994 (CD); Bellmann, 1993*c* (CD); Bonnet, 1995 (CD); Deroussen, 1993 (CD), 1994 (CD); Helluin & Ribassin, 1961 (LP); Koch & Hawkins, 1969 (LP); Odé, 1997 (CD); Ragge *et al.*, 1965 (LP); Roché, 1965 (LP). **Cassette recording**: Bellmann, 1985*b*, 1993*b*; Burton & Ragge, 1987; Elliot, 1986.

RECOGNITION (Plate 2: 10).   The genus *Gryllus* includes the two large black crickets of Eu-rope. *G. campestris* may be easily distinguished from *G. bimaculatus* by its reduced hind wings, which do not reach the tips of the fore wings; *G. bimaculatus* has fully developed hind wings, reaching well beyond the fore wings, and can fly well.

SONG (Figs 629–642. CD 1, track 91).   From May to July the song of this species is one of the commonest insect sounds of the European countryside, and may be heard at any time of the day or night; in western Europe only a few long-lived adult males remain to sing from August onwards. The calling song, usually delivered at the entrance to a burrow, consists of a long series of loud echemes repeated fairly regularly at the rate of about 3–4/s in warm, sunny conditions, but as slowly as 1/s on a cool night. Each echeme is usually composed of 4 (sometimes 3 and occasionally 5) closing hemisyllables. Oscillographic analysis shows that the first hemisyllable is usually quieter and shorter than the others, and sometimes a crescendo continues throughout the echeme. The later hemisyllables in an echeme each last about 15–20 ms, their duration being unaffected by temperature; the intervals between them, however, depend on temperature, varying from about 10–15 ms in warm, sunny weather (when the syllable repetition rate within an echeme is about 30–40/s) to more than 50 ms on a cool night (repetition rate less than 20/s). (Note that Popov (1972) gives rather longer hemisyllable durations, 20–35 ms, for calling songs in the Ukraine and Azerbaijan.) The duration of a whole 4-syllable echeme varies similarly from about 100 ms in warm conditions to more than double this when it is much cooler. The opening stroke of the fore wings is usually silent, but at lower temperatures there are sometimes quiet open-ing hemisyllables. The frequency of the carrier wave is usually between 4·0 and 5·5 kHz and, as one would expect from the stability of the closing hemisyllable duration, is hardly affected by temperature.

When one male encounters another it often produces longer echemes, composed of up to 8 or more syllables. In the presence of a female the male produces a special courtship song (Figs 638–642), consisting of a quiet purring or rustling sound interspersed with louder ticks and produced with the fore wings held in a lower position over the abdomen than in the call-ing song. The ticks are repeated at the rate of about 2–6/s. Within each tick (Fig. 642) the dom-inant frequency is usually at least three times that of the calling song, usually about 14–17 kHz.

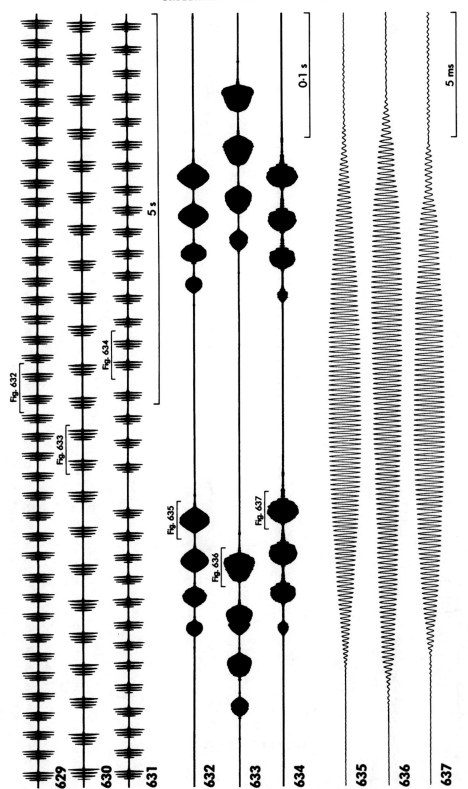

Figures 629–637 Oscillograms at three different speeds of the calling songs of three males of *Gryllus campestris.*

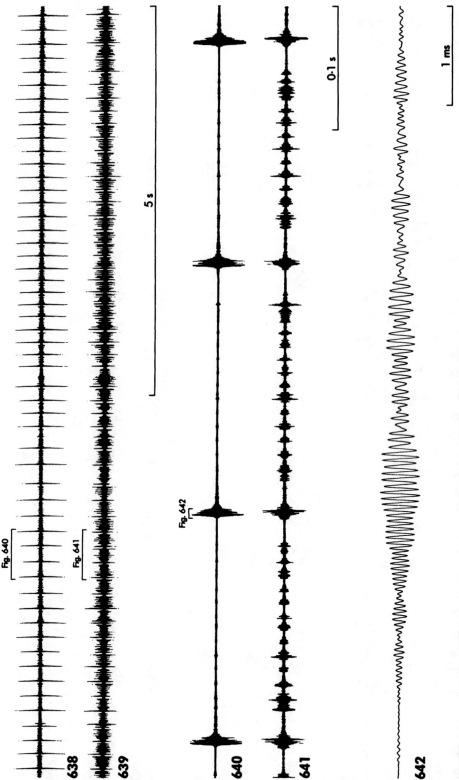

Figures 638–642 Oscillograms at three different speeds of the courtship song of *Gryllus campestris*. Figs 638 and 640 are taken from the unfiltered song, so that the louder high-frequency ticks and quieter rustling sound are shown at their true relative amplitudes; Figs 639 and 641 are taken from the same song excerpts passed through a low-pass filter, so that the rustling sound can be shown at a much higher amplitude.

DISTRIBUTION. Very widespread in Europe, from about latitude 55°N to the Mediterranean Region, including Morocco and Algeria. It occurs very locally in southern England, and the range extends eastwards to western Asia.

Gryllus bimaculatus De Geer

Gryllus bimaculatus De Geer, 1773: 521.

Two-spotted Cricket; (F) Grillon provençal; (D) Mittelmeer-Feldgrille; (NL) Zuidelijke Veldkrekel.

REFERENCES TO SONG. **Oscillogram**: Ahmad & Siddiqui, 1988; Elsner & Popov, 1978; Kutsch, 1969; Libersat *et al.*, 1994; Popov, 1972; Popov & Shuvalov, 1977; Popov, Shuvalov, Knyazev & Klar-Spasovskaya, 1974; Popov, Shuvalov, Svetlogorskaya & Markovich, 1974; Rheinlaender, 1976; Rheinlaender *et al.*, 1976, 1981; Schmidt, 1996; Stephen & Hartley, 1995; Thorson *et al.*, 1982; Zhantiev, 1981; Zhantiev & Dubrovin, 1974. **Diagram**: Leroy, 1962, 1966; Tschuch, 1986; Zhantiev, 1981; Zhantiev & Dubrovin, 1974. **Sonagram**: Leroy, 1962, 1966; Matsuura, 1979; Nagao & Shimozawa, 1987; Otte & Cade, 1984c; Popov, 1972; Simmons, 1988. **Wing-movement**: Koch, 1978, 1980. **Frequency information**: Ahmad & Siddiqui, 1988; Boyd & Lewis, 1983 (as *campestris*); Doherty, 1985; Nocke, 1971; Popov, Shuvalov, Knyazev & Klar-Spasovskaya, 1974; Popov, Shuvalov, Svetlogorskaya & Markovich, 1974; Popov *et al.*, 1975; Rheinlaender *et al.*, 1976, 1981; Stephen & Hartley, 1995; Zhantiev, 1981; Zhantiev & Dubrovin, 1974. **Verbal description only**: Alexander, 1967a; Broughton, 1972b; Chopard, 1922; Dathe, 1974; Janiszewski & Otto, 1989; Kämper & Dambach, 1985; Nielsen & Dreisig, 1970; Tschuch, 1985; Weidemann & Keuper, 1987. **Disc recording**: Bonnet, 1995 (CD).

RECOGNITION. See under *G. campestris*.

SONG (Figs 643–653. CD 1, track 92). The calling song is very similar in almost every respect to that of *G. campestris*, but is heard mainly at night and, in western Europe, later in the year, from July to October. The number of syllables per echeme is usually 3–5, occasionally 2 or 6; it is much less common than in *G. campestris* for the first syllable to be quieter than the remaining ones. In the four males studied the duration of each closing hemisyllable was 20–25 ms.

The courtship song (Figs 649–653) is also similar to that of *G. campestris*, as are the frequency of the carrier wave in the calling song and the frequency within the ticks (Fig. 653) of the courtship song.

DISTRIBUTION. Southern Europe, especially in regions near the Mediterranean coast; southern Asia and Africa. Absent from the British Isles.

Acheta domesticus (Linnaeus)

Gryllus (Acheta) domesticus Linnaeus, 1758: 428.

House-cricket; (F) Grillon domestique; (D) Heimchen; (NL) Huiskrekel; (DK) Husfåre-

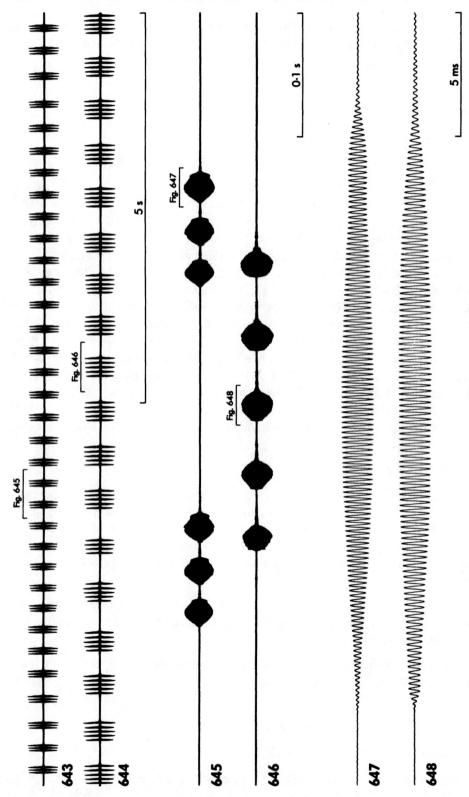

Figures 643–648 Oscillograms at three different speeds of the calling songs of two males of *Gryllus bimaculatus*.

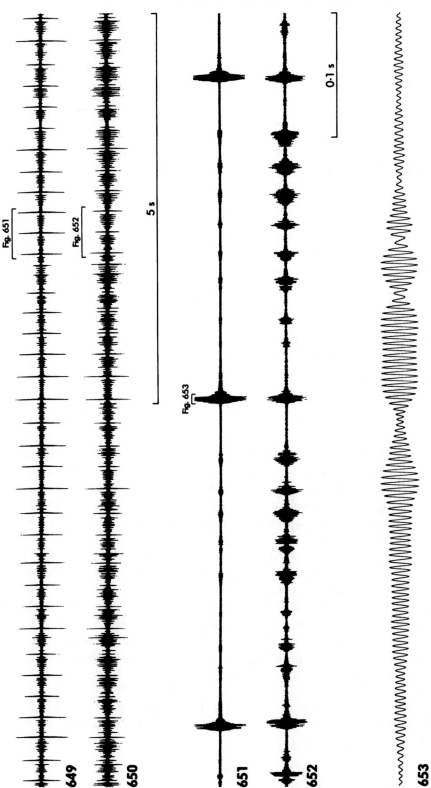

Figures 649–653 Oscillograms at three different speeds of the courtship song of *Gryllus bimaculatus*. Figs 650 and 652 are from the same excerpts as Figs 649 and 651, but filtered so that the quieter rustling component of the song can be shown at a much higher amplitude (see also the legend to Figs 638–642).

kylling; (S) Hussyrsan; (N) Hus-siriss; (SF) Kotisirkka; (E) Grillo doméstico; (P) Grilo doméstico; (I) Grillo del Focolare.

REFERENCES TO SONG. **Oscillogram**: Counter, 1976; Davis, 1968; Ewing & Hoyle, 1965; Faber, 1957b; Grein, 1984; Heiligenberg, 1966, 1969; Holst, 1986; Kleukers *et al.*, 1997; Kutsch, 1969; Loher & Broughton, 1955; Popov, 1971, 1985; Weissman & Rentz, 1977. **Diagram**: Bellmann, 1985a, 1988, 1993a; Bellmann & Luquet, 1995; Duijm & Kruseman, 1983; Holst, 1970; Pierce, 1948; Ragge, 1965; Zhantiev & Dubrovin, 1974. **Sonagram**: Alexander, 1961; Leroy, 1966; Popov, 1971; Sales & Pye, 1974. **Wing-movement**: Davis, 1968. **Frequency information**: M.-C. Busnel, 1953; R.-G. Busnel, 1955; Counter, 1976; Ewing & Hoyle, 1965; Haskell, 1953; Popov, 1985; Sales & Pye, 1974. **Verbal description only**: Broughton, 1972b; Chopard, 1922; Crankshaw, 1979; Faber, 1928; Harz, 1957, 1962; Haskell, 1955a; Khalifa, 1950; Klöti-Hauser, 1921; Nielsen & Dreisig, 1970; Stout *et al.*, 1983; Tschuch, 1985; Weber, 1984; Yersin, 1854b. **Disc recording**: Andrieu & Dumortier, 1963 (LP), 1994 (CD); Bellmann, 1993c (CD); Bonnet, 1995 (CD); Grein, 1984 (LP); Odé, 1997 (CD); Roché, 1965 (LP). **Cassette recording**: Bellmann, 1985b, 1993b; Burton & Ragge, 1957.

RECOGNITION. A brown or straw-coloured, fully winged cricket, normally found only in or near buildings or on rubbish tips. It could be confused with *Modicogryllus bordigalensis*, but that species is smaller (fore wings less than 8 mm long, more than 8 mm long in *A. domesticus*) and often has reduced hind wings.

SONG (Figs 654–665. CD 1, track 93). The calling song, heard mainly in the evening and at night, is a long series of loud echemes repeated fairly regularly at the rate of 1–3/s. Each echeme is commonly composed of 2 or 3 closing hemisyllables, but there are sometimes 4 or more and the number often varies within one song; the first hemisyllable in an echeme is often a little quieter (occasionally much quieter) than the remaining ones. Oscillographic analysis shows that the hemisyllables last about 20–30 ms and the intervals between them are of a similar duration or rather longer. The syllable repetition rate within an echeme is about 15–25/s. Occasionally there are quiet opening hemisyllables. The frequency of the carrier wave is usually between 3·5 and 5·0 kHz.

When males encounter one another they often produce longer echemes, composed of up to 7 or more syllables. In the presence of a female the male produces a special courtship song, consisting of a subdued purring sound punctuated by regularly repeated louder ticks (Figs 660–665). The purring consists of quiet syllables repeated at the rate of about 15–20/s, and the louder ticks are usually repeated at the rate of between 1 and 2/s. The dominant frequency of the quieter syllables (Fig. 664) is usually the same as that of the carrier wave of the calling song, but the frequency within the ticks (Fig. 665) is usually at least four times this, about 15–20 kHz.

DISTRIBUTION. Found throughout Europe (and indeed many other parts of the world) in association with heated buildings or rubbish tips. Fairly common in southern Britain, but now rare elsewhere in the British Isles.

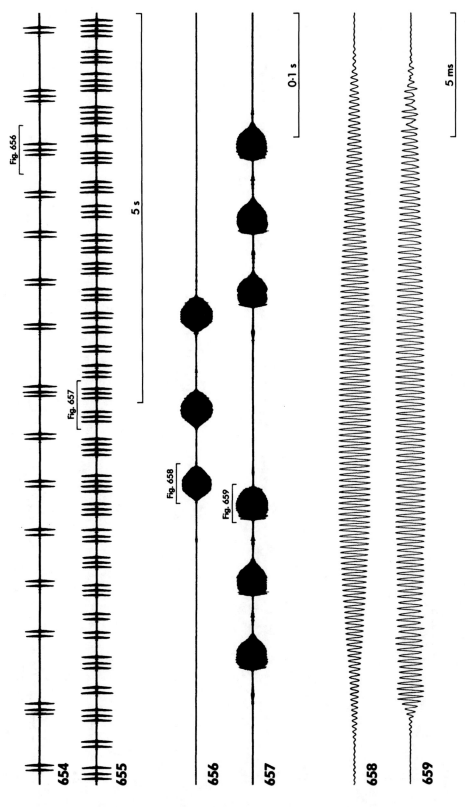

Figures 654–659 Oscillograms at three different speeds of the calling songs of two males of *Acheta domesticus*.

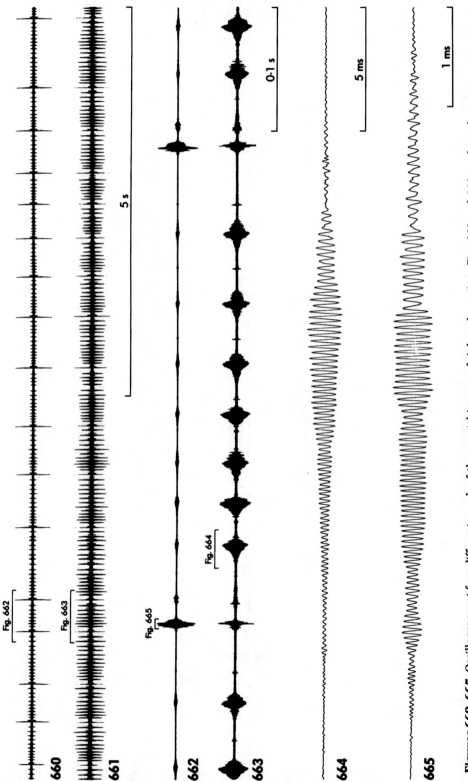

Figures 660–665 Oscillograms at four different speeds of the courtship song of *Acheta domesticus*. Figs 661 and 663 are from the same song excerpts as Figs 660 and 662, but filtered so that the quieter purring component of the song can be shown at a higher amplitude (see also the legend to Figs 638–642).

Modicogryllus bordigalensis (Latreille)

Gryllus bordigalensis Latreille, 1804: 124.

Bordeaux Cricket; (F) Grillon bordelais; (D) Südliche Grille.

REFERENCES TO SONG. **Oscillogram**: Elsner & Popov, 1978; Popov, 1972 (as *Tartarogryllus* sp.), 1985; Popov & Shuvalov, 1977; Popov, Shuvalov, Knyazev & Klar-Spasovskaya, 1974 (as *T.* sp.); Popov, Shuvalov, Svetlogorskaya & Markovich, 1974; Zhantiev & Dubrovin, 1974. **Diagram**: Zhantiev & Dubrovin, 1974. **Sonagram**: Leroy, 1966. **Frequency information**: Popov, 1985; Popov, Shuvalov, Knyazev & Klar-Spasovskaya, 1974; Popov, Shuvalov, Svetlogorskaya & Markovich, 1974; Zhantiev & Dubrovin, 1974. **Verbal description only**: Bellmann, 1993*a*; Bellmann & Luquet, 1995; Chopard, 1922 (as *chinensis*); Defaut, 1987, 1988*b*. **Disc recording**: Bonnet, 1995 (CD); Deroussen, 1994 (CD); Roché, 1965 (LP).

RECOGNITION. Rather like a smaller version of *A. domesticus* (q.v.), but often darker in colour and frequently with reduced hind wings. The calling song is also distinctive.

For a note on the spelling of the specific name of this species see p.11.

SONG (Figs 666–671. CD 1, track 94). The calling song, produced mainly at night, consists of a long series of quite loud echemes repeated at the rate of about 2·5–4·0/s and each composed of about 14–20 syllables. Oscillographic analysis shows that the first few syllables in each echeme are usually quieter than the later ones and are repeated at the rate of about 40–60/s; there is then a sudden crescendo and an increase in repetition rate to about 60–100/s for the remaining syllables. Each echeme lasts about 150–270 ms and each syllable about 5–10 ms. The frequency of the carrier wave is usually about 3·5–5·0 kHz.

DISTRIBUTION. Much of southern Europe, including the whole of the Mediterranean Region and Balkan Peninsula; parts of North Africa and southern palaearctic Asia. Absent from the British Isles.

Eugryllodes pipiens (Dufour)

Grillus [sic] *pipiens* Dufour, 1820: 315.

Mountain-cricket; (F) Grillon testacé; (D) Gelbe Grille.

REFERENCES TO SONG. **Verbal description only**: Bellmann, 1993*a*; Bellmann & Luquet, 1995; Chopard, 1952; Dufour, 1820. **Disc recording**: Bonnet, 1995 (CD); Roché, 1965 (LP).

RECOGNITION. Found in mountainous regions, where it is perhaps most easily recognized by the characteristic nocturnal calling song, consisting, at least partly, of brief isolated syllables. It is straw-coloured with dark brown markings. In the male the fore wings are broadly rounded and scarcely cover the abdomen; the hind wings are reduced to short vestiges. In the female even the fore wings are reduced to short flaps of similar length to the pronotum.

SONG (Figs 672–677. CD 1, track 95). The song, heard only in the evening and at night, consists at least partly of isolated syllables (Figs 672, 674, 676) lasting about 12–30 ms (up

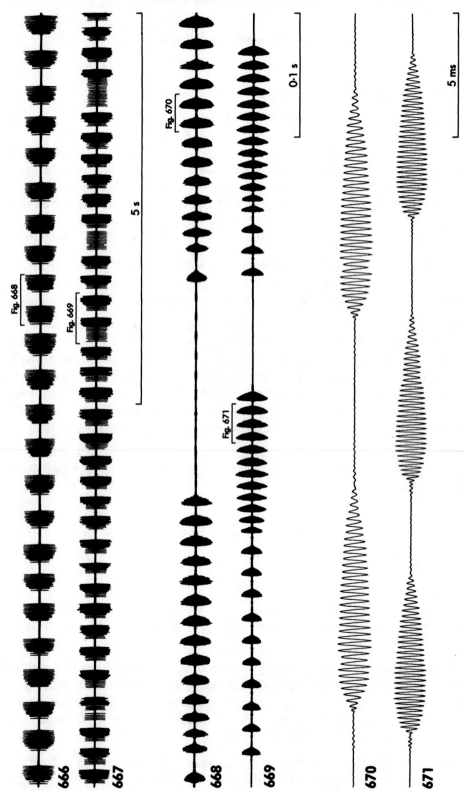

Figures 666–671 Oscillograms at three different speeds of the calling songs of two males of *Modicogryllus bordigalensis*.

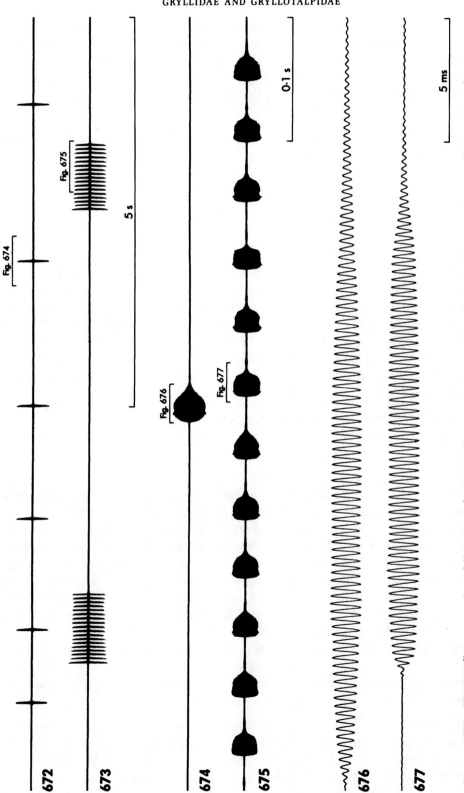

Figures 672–677 Oscillograms at three different speeds of the calling songs of two males of *Eugryllodes pipiens*, showing the two types of song-pattern produced by this species.

to 60 ms on cold nights) and repeated at very variable intervals, usually of between 0·5 and 6·0 s. Sometimes echemes of 2–20 syllables are produced (Figs 673, 675, 677), in which the syllable repetition rate is about 15–30/s; the intervals between the echemes are usually of about 3–7 s. It is not unusual for there to be a mixture of echemes and isolated syllables in the same song. In the eight males studied the frequency of the carrier wave varied from 3·0 to 3·5 kHz and seemed to be unaffected by temperature.

DISTRIBUTION. Mountainous regions of southern France (especially the eastern Pyrenees and French Alps) and the Iberian Peninsula.

Nemobius sylvestris (Bosc)

Acheta sylvestris Bosc, 1792*a*: 44.

Wood-cricket; (F) Grillon des bois; (D) Waldgrille; (NL) Boskrekel.

REFERENCES TO SONG. **Oscillogram**: Faber, 1957*b*; Grein, 1984; Kleukers *et al.*, 1997; Lottermoser, 1952; Schmidt & Baumgarten, 1977. **Diagram**: Bellmann, 1985*a*, 1988, 1993*a*; Bellmann & Luquet, 1995; Duijm & Kruseman, 1983; Fischer, 1853; Ragge, 1965. **Musical notation**: Yersin, 1854*b*. **Verbal description only**: Broughton, 1972*b*; Chopard, 1952; Faber, 1928; Gerhardt, 1921; Girard, 1879; Harz, 1957, Richards, 1952; Weber, 1984. **Disc recording**: Andrieu & Dumortier, 1963 (LP), 1994 (CD); Bellmann, 1993*c* (CD); Bonnet, 1995 (CD); Grein, 1984 (LP); Odé, 1997 (CD); Ragge *et al.*, 1965 (LP). **Cassette recording**: Bellmann, 1985*b*, 1993*b*; Burton & Ragge, 1987.

RECOGNITION (Plate 2: 11). A small dark brown to black cricket with the fore wings extending no more than halfway along the abdomen. It lives in woodland leaf-litter and the calling song is highly distinctive.

SONG (Figs 678–683. CD 1, track 96). The song of this species is one of the commonest sounds of western European woodlands, both during the day and at night, but, being rather quiet, it is seldom noticed. One usually hears a chorus of singing males producing a continuous churring sound rather like a distant nightjar (*Caprimulgus europaeus*) (see p. 484). However, the calling song of each male is broken up into echemes lasting less than a second and separated by intervals of similar duration. The number of syllables in each echeme varies greatly both within and between songs, but is usually within the range 3–30; the duration of the echemes is thus also very variable, usually within the range 70–800 ms. The syllable repetition rate is usually 30–40/s, occasionally up to 50/s. The duration of a syllable is usually between 3 and 10 ms, and the intervals between successive syllables 12–30 ms. In the seven males studied the frequency of the carrier wave varied from 3·8 to 4·6 kHz.

In the presence of a female the male often produces longer echemes, lasting up to 5 s or longer.

DISTRIBUTION. *Nemobius sylvestris* occurs widely in western Europe, from the Low Countries and Germany in the north to the Iberian Peninsula and northern Italy in the south; eastwards, it has been recorded as far as the Ukraine, but not from the Balkan Peninsula. It occurs locally in North Africa and, very locally, in southern England.

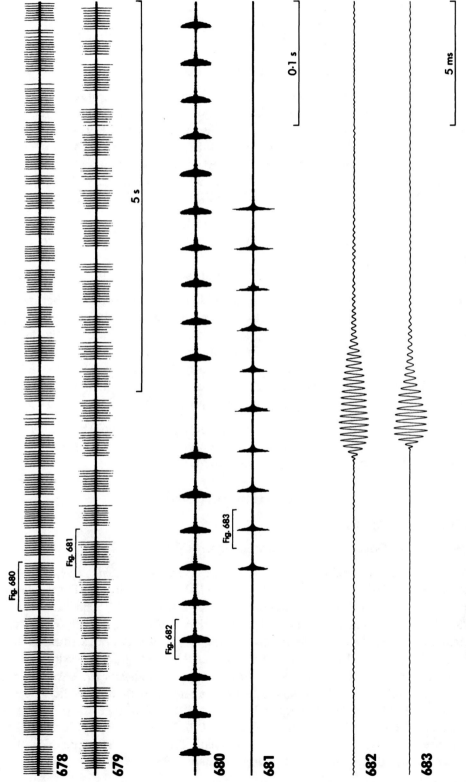

Figures 678–683 Oscillograms at three different speeds of the calling songs of two males of *Nemobius sylvestris*.

Pteronemobius heydenii (Fischer)

Gryllus Heydenii Fischer, 1853: 185.

Marsh-cricket; (F) Grillon des marais; (D) Sumpfgrille.

REFERENCES TO SONG. **Oscillogram**: Grein, 1984; Popov, 1985; Popov, Shuvalov, Knyazev & Klar-Spasovskaya, 1974; Popov, Shuvalov, Svetlogorskaya & Markovich, 1974; Zhantiev & Dubrovin, 1974. **Diagram**: Bellmann, 1985*a* (as *concolor*), 1988, 1993*a*; Bellmann & Luquet, 1995. **Frequency information**: Popov, 1985; Popov, Shuvalov, Knyazev & Klar-Spasovskaya, 1974; Zhantiev & Dubrovin, 1974. **Verbal description only**: Chopard, 1952. **Disc recording**: Andrieu & Dumortier, 1963 (LP), 1994 (CD); Bellmann, 1993*c* (CD); Bonnet, 1995 (CD); Grein, 1984 (LP). **Cassette recording**: Bellmann, 1985*b* (as *c.*), 1993*b*.

RECOGNITION. A very small, dark brown to black cricket living in damp places. The fore wings extend almost to the tip of the abdomen in the male, but are a little shorter in the female. The calling song is unmistakable.

For a note on the specific name of this species and its spelling see p. 11.

SONG (Figs 684–689. CD 1, track 97). The calling song, often heard during the day but especially in the evening and at night, is a long series of well defined echemes, each lasting about 1–4 s. The echemes usually begin quietly, reaching maximum intensity during the first third of their duration. Successive echemes are separated by intervals of about 0·5–2·5 s. Oscillographic analysis shows that the syllable repetition rate within an echeme is about 40–65/s; the syllables last about 9–12 ms and are separated by intervals of about 6–10 ms. In the six males studied the frequency of the carrier wave varied from 6·4 to 7·9 kHz.

DISTRIBUTION. Southern Europe, south of latitude 50°N. It occurs in all the southern peninsulas and its range extends eastwards to parts of central and southern palaearctic Asia; also the High Atlas range in Morocco. Absent from the British Isles.

Oecanthus pellucens (Fischer)

Gryllus pellucens Scopoli, 1763: 109.

Tree-cricket; (F) Grillon d'Italie; (D) Weinhähnchen; (NL) Boomkrekel.

REFERENCES TO SONG. **Oscillogram**: M.-C. Busnel, 1955; Busnel *et al.*, 1954; Chopard, 1955; Dumortier, 1963*c*; Grein, 1984; Kleukers *et al.*, 1997; Pasquinelly & Busnel, 1955; Popov, 1972, 1985; Popov, Shuvalov, Svetlogorskaya & Markovich, 1974; Schmidt & Schach, 1978. **Diagram**: Bellmann, 1985*a*, 1988, 1993*a*; Bellmann & Luquet, 1995; Duijm & Kruseman, 1983; Fischer, 1853. **Sonagram**: Popov, 1972. **Frequency information**: M.-C. Busnel, 1953, 1955; Busnel *et al.*, 1954; Dumortier, 1963*c*; Korsunovskaya, 1978; Pasquinelly & Busnel, 1955; Popov, 1985; Popov, Shuvalov, Svetlogorskaya & Markovich, 1974; Zhantiev & Dubrovin, 1974. **Verbal description only**: Broughton, 1972*b*; Chopard, 1922; Girard, 1879; Harz, 1957; Marquet, 1877; Yersin, 1857. **Disc recording**: Andrieu & Dumortier, 1963 (LP), 1994 (CD); Bellmann, 1993*c* (CD); Bonnet, 1995 (CD); Burton, 1969 (LP); Deroussen, 1993 (CD); Grein, 1984 (LP); Helluin & Ribassin, 1961 (LP); Odé, 1997 (CD); Roché, 1965 (LP). **Cassette recording**: Bellmann, 1985*b*, 1993*b*.

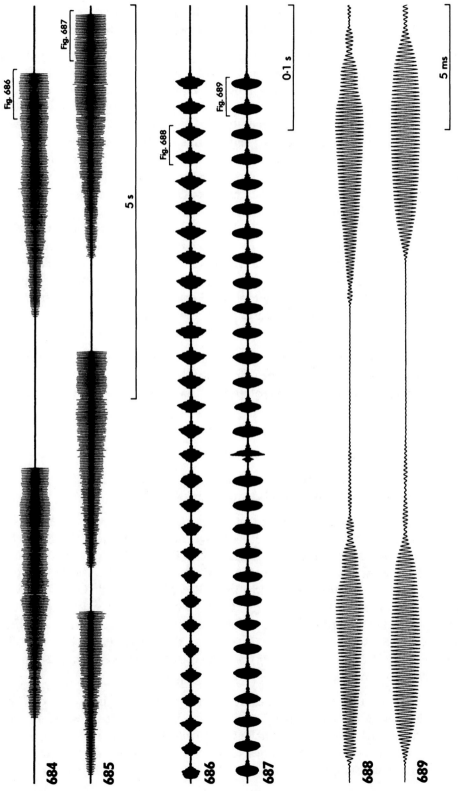

Figures 684–689 Oscillograms at three different speeds of the calling songs of two males of *Pteronemobius heydenii*.

RECOGNITION (Plate 2: 12). A slender, delicately built, straw-coloured cricket, living above ground level on grass, herbs, shrubs or trees. The melodious song is highly characteristic.

SONG (Figs 690–698. CD 1, track 98). The loud calling song, heard only in the evening and at night, is one of the most familiar insect sounds of southern Europe. It consists of regularly repeated echemes rather reminiscent of the warbling of a telephone. The echeme repetition rate depends on temperature, usually being 0·5–1·0/s at temperatures below 20°C, 1·0–1·5/s between 20 and 25°C, and faster than 1·5/s above 25°C. The duration of each echeme is also affected to some extent by temperature, but there is also much individual variation; sometimes the echeme duration varies considerably within one song, but in some songs it is almost constant. The echemes usually last between 300 and 900 ms, but on cool nights they may last more than a second; the intervals between successive echemes are usually about 120–500 ms, but are also sometimes more than a second.

Oscillographic analysis shows that each echeme is usually composed of 15–30 syllables (occasionally up to 40 or more). The syllable repetition rate again depends on temperature, being 20–30/s on cool nights and as high as 40–50/s on very warm ones. Each syllable lasts from about 12 ms to about 22 ms, depending on temperature, and the intervals between successive syllables vary from about 3 ms to about 15 ms. The frequency of the carrier wave also depends to some extent on temperature; it is usually between 2 and 3 kHz, but on warm nights it is often 3·0–3·5 kHz and in one male studied, at 27°C, it was 3·7 kHz.

DISTRIBUTION. Widespread in southern Europe, south of latitude 50°N. It occurs in the whole of the Mediterranean Region, including North Africa, and the range extends eastwards to parts of central and southern palaearctic Asia. Absent from the British Isles.

Gryllotalpa gryllotalpa (Linnaeus)

Gryllus (Acheta) gryllotalpa Linnaeus, 1758: 428.

Mole-cricket; (F) Courtilière commune; (D) Maulwurfsgrille; (NL) Veenmol; (DK) Jordkrebs; (S) Mullvadssyrsan; (N) Jordsiriss; (SF) Maamyyräsirkka.

REFERENCES TO SONG. **Oscillogram**: Bennet-Clark, 1970a, 1970b; Grein, 1984; Kleukers *et al.*, 1997; Zhantiev & Korsunovskaya, 1973. **Diagram**: Bellmann, 1985a, 1988, 1993a; Bellmann & Luquet, 1995; Duijm & Kruseman, 1983; Holst, 1970. **Sonagram**: Bennet-Clark, 1970a. **Frequency information**: Bennet-Clark, 1970a, 1970b; Nickle & Castner, 1984; Popov, 1985. **Verbal description only**: Broughton, 1972b; Chopard, 1952; Dumortier, 1963c; Faber, 1928; Harz, 1957; Holst, 1986; Klöti-Hauser, 1921; Malenotti, 1926 (as *vulgaris*, but identity uncertain); Nickle & Castner, 1984; Ragge, 1965; Regen, 1903 (as *v.*); Weber, 1984; White, 1789; Yersin, 1984b (as *v.*). **Disc recording**: Bellmann, 1993c (CD); Bonnet, 1995 (CD); Grein, 1984 (LP); Odé, 1997 (CD). **Cassette recording**: Bellmann, 1985b, 1993b; Burton & Ragge, 1987.

RECOGNITION (Plate 3: 1). The mole-crickets, with their fore legs highly modified for digging and their bodies covered with fine velvety hair, are very easy to recognize. In some species, including *G. gryllotalpa*, the females can stridulate with their fore wings, though lacking the well-developed stridulatory organ of the males; they produce short harsh sounds, usually when meeting another of their kind.

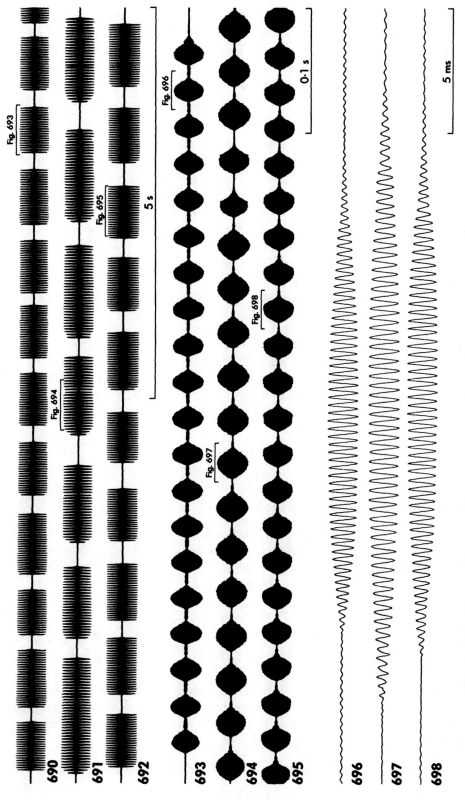

Figures 690–698 Oscillograms at three different speeds of the calling songs of three males of *Oecanthus pellucens*.

In the field *G. gryllotalpa* is most easily distinguished from the only other western European mole-cricket in which the song has been studied, *G. vineae*, by the low dominant frequency of the carrier wave of the calling song: below 2 kHz in *G. gryllotalpa*, usually at least 3 kHz in *G. vineae*. These two species are very similar in appearance, but *G. gryllotalpa* lives in much wetter habitats than *G. vineae* and is the only species of mole-cricket in northern Europe, north of latitude 46°N. Details of the morphological differences are given by Bennet-Clark (1970*b*).

SONG (Figs 699–704. CD 1, track 99, excerpt 1). The male calling song, produced only in the evening and at night during spring and early summer, is a fairly loud, but not shrill, continuous musical warbling. It consists of syllables repeated at a rate dependent on temperature; below 15°C the repetition rate is about 35–45/s, from 20°C to 25°C about 45–55/s, and more than 55/s at temperatures above 25°C. The duration of the syllables is quite constant within one song, but varies greatly from song to song and shows some tendency to diminish at higher temperatures; in the nine males studied the syllable duration ranged from 5 ms to 18 ms. The frequency of the carrier wave seems to be unaffected by temperature; it varied from 1·3 kHz to 1·7 kHz in the males studied.

The song is produced from just below the surface of the ground in a specially made singing burrow with several entrances.

DISTRIBUTION. Widespread in Europe, reaching as far north as southern Sweden and occurring very locally in southern Britain. It occurs widely in France, Germany and much of eastern Europe, but its southern limit cannot be determined without further studies on the song in southern Europe. A song recorded near Burgos, for which no specimen is available, suggests that *G. gryllotalpa* occurs at least as far south as northern Spain.

Gryllotalpa vineae Bennet-Clark

Gryllotalpa vineae Bennet-Clark, 1970*b*: 130.

Vineyard Mole-cricket; (F) Courtilière des vignes.

REFERENCES TO SONG. **Oscillogram**: Bennet-Clark, 1970*a*, 1970*b*, 1976. **Sonagram**: Bennet-Clark, 1970*a*. **Frequency information**: Bennet-Clark, 1970*a*, 1970*b*; Busnel, 1953 (as *gryllotalpa*); Busnel & Chavasse, 1951 (as *g.*). **Disc recording**: Andrieu & Dumortier, 1994 (CD); Bonnet, 1995 (CD); Roché, 1965 (LP, as *g.*).

RECOGNITION. See under *G. gryllotalpa*.

SONG (Figs 705–710. CD 1, track 99, excerpt 2). The nocturnal male calling song is a very loud, shrill, continuous trill, audible on a still night from a distance of at least 500 m. It is composed of syllables lasting about 6–12 ms and repeated at the rate of about 35–90/s, depending partly on temperature. In the six males studied the frequency of the carrier wave varied from 3·0 to 5·0 kHz. As in *G. gryllotalpa* the song is produced from a specially made singing burrow, but the burrow is more regularly shaped than in *G. gryllotalpa*, with a pair of short horn-shaped passages leading to two entrance holes.

DISTRIBUTION. So far known only from southern France and the Iberian Peninsula.

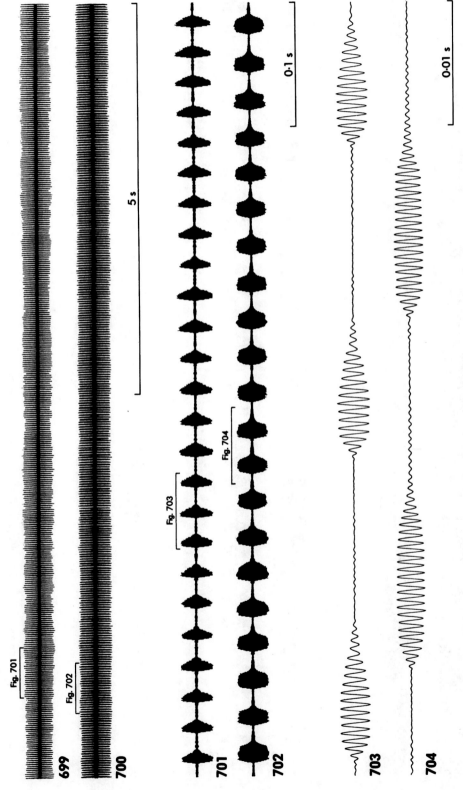

Figures 699–704 Oscillograms at three different speeds of the calling songs of two males of *Gryllotalpa gryllotalpa*.

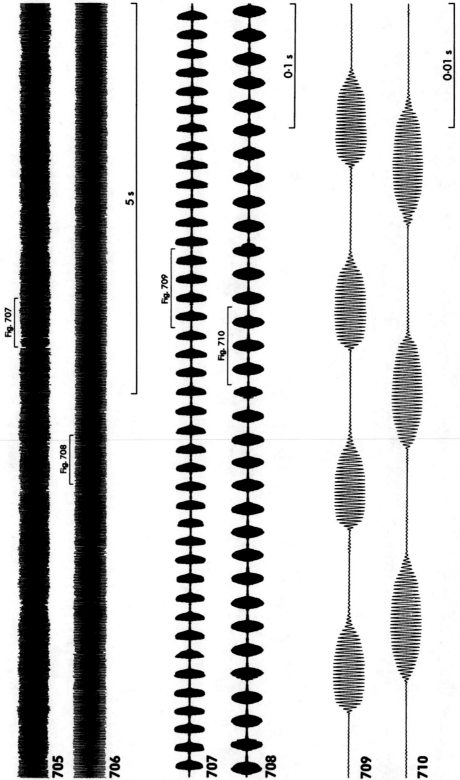

Figures 705–710 Oscillograms at three different speeds of the calling songs of two males of *Gryllotalpa vineae*.

Chapter 9

Acrididae

Grasshoppers

The songs of gomphocerine grasshoppers, produced by rubbing the hind femora against the fore wings, are unmusical buzzing or hissing sounds. They have a broad frequency spectrum, but are generally less high-pitched, and thus easier to hear, than the songs of the Tettigoniidae. In addition to the calling song, males of many species produce a special courtship song when next to a female, and often various other sounds associated with other kinds of behaviour. Our descriptions are mainly of the calling songs, but we have described some of the more notable courtship songs and occasionally other sounds that can be diagnostically useful. It is usually only males that are heard singing in the field, but females that are receptive to mating also sing, sometimes spontaneously but especially in response to singing males, and the sounds they produce are usually similar in pattern to the male calling song. In the wild receptive females are usually mated quickly, after which they stop singing until they have laid at least one batch of eggs.

Most of the sound is usually produced by the downstrokes of the hind legs, as can be observed quite easily in songs produced by relatively slow leg-movements, such as the calling songs of *Chorthippus vagans* and *C. montanus*. In *Stenobothrus lineatus*, which has the slowest stridulatory leg-movements known in the Gomphocerinae, the upstrokes and downstrokes produce sounds of similar intensity.

Some grasshoppers can make a rattling, whirring or buzzing sound in flight, known as crepitation. Examples among the European Gomphocerinae are *Stenobothrus cotticus*, *Stauroderus scalaris*, males of *Arcyptera fusca* and especially the hovering display flight of males of *Stenobothrus rubicundulus*. Crepitation is particularly common in the Locustinae, especially in North America. In Europe the most notable examples are *Psophus stridulus* and *Bryodema tuberculata* (especially the remarkable display flight of the males). Although many European Locustinae can also stridulate by rubbing the hind femora against the fore wings, the sounds produced are usually rather nondescript and of little or no diagnostic value; we have therefore not described them in this book.

In *Stethophyma grossum* the male calling song takes the unusual form of a series of loud 'ticks' produced by flicking one of the hind tibiae backwards against the fore wings. This form of calling song is unique among the European grasshoppers, though such hind tibial flicking sometimes forms part of a courtship song (e.g. that of *Omocestus viridulus*).

Grasshopper nymphs sometimes move their hind legs in a way that matches the stridulatory movements of adults of the same species, but usually no sound is produced.

Psophus stridulus (Linnaeus)

Gryllus (Locusta) stridulus Linnaeus, 1758: 432.

Rattle Grasshopper; (F) Oedipode stridulante; (D) Rotflüglige Schnarrschrecke; (NL) Klappersprinkhaan; (DK) Trommende Græshoppe; (S) Trumgräshoppa; (N) Klapregrashoppe; (SF) Palosirkka.

REFERENCES TO SONG. **Oscillogram**: Kleukers *et al.*, 1997. **Diagram**: Duijm & Kruseman, 1983; Luquet, 1978. **Verbal description only**: Bellmann, 1985*a*, 1988, 1993*a*; Bellmann & Luquet, 1995; Chopard, 1922; Faber, 1928, 1936, 1953*a*; Harz, 1957; Jacobs, 1950*b*, 1953*a*; Karny, 1908; Poulton, 1896; Prochnow, 1907; Regen, 1902; 1903; Stäger, 1930; Wallin, 1979. **Disc recording**: Bellmann, 1993*c* (CD); Odé, 1997 (CD). **Cassette recording**: Bellmann, 1985*b*, 1993*b*; Wallin, 1979.

RECOGNITION. *P. stridulus* belongs to the Locustinae, a group sometimes referred to as the band-winged grasshoppers, in which the hind wings are often brightly coloured and sometimes also have a dark band near the posterior margin. Unlike the Gomphocerinae, these insects do not have well-developed calling songs and lack a stridulatory peg-row on the hind femur. They do, however, have an extra peg- or tubercle-bearing vein (the medial intercalary) on the fore wing, and the males sometimes make quiet sounds, usually of little diagnostic value, by scraping a ridge on the inner side of the hind femur against this vein. Some of them can crepitate, i.e. make a loud rattling or whirring sound with the hind wings when in flight.

P. *stridulus* may be recognized by the very prominent median carina on the pronotum and the bright red hind wings, with dark brown tips. The males are otherwise very dark, sometimes almost black, in colour, the females rather paler. In *Bryodema tuberculata* the pronotum has a much weaker median carina and the red colouring of the hind wings is much less extensive, not reaching the anterior margin.

CREPITATION (Figs 711, 712. CD 2, track 1). The males usually make a loud rattling sound when in flight, the crepitation rate being about 30–35/s. The females have rather reduced wings and are hardly capable of flight; they can nevertheless crepitate during flight-assisted jumps. Both sexes also occasionally crepitate while sitting on the ground. The flight crepitation of the males seems to be facultative: in cool or dull weather they often make silent flights.

DISTRIBUTION. Widely distributed in Europe, from southern Scandinavia to the northern parts of the southern peninsulas. The range extends eastwards through central Asia to eastern China and Korea. Absent from the British Isles.

Bryodema tuberculata (Fabricius)

Gryllus tuberculatus Fabricius, 1775: 290.

Speckled Grasshopper; (F) Oedipode des torrents; (D) Gefleckte Schnarrschrecke; (DK) Hedeskratte; (S) Rosenvingad Gräshoppa; (SF) Ruususiipisirkka.

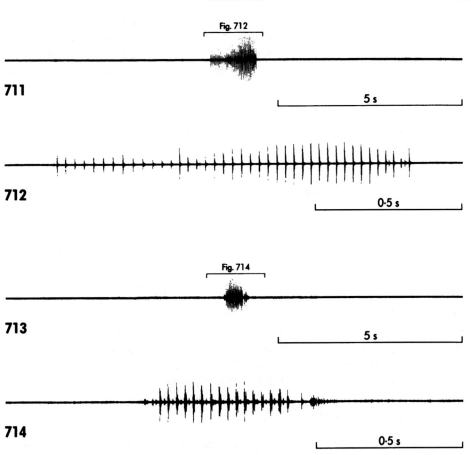

Figures 711–714 Oscillograms of the crepitation during short flights of (711, 712) *Psophus stridulus* and (713, 714) *Bryodema tuberculata*.

REFERENCES TO SONG. **Oscillogram**: Xi *et al.*, 1992. **Verbal description only**: Bellmann, 1985*a*, 1988, 1993*a*; Bellmann & Luquet, 1995; Bornhalm, 1991; Dumortier, 1963*b*; Faber, 1928, 1936, 1953*a*; Graber, 1873; Harz, 1957; Holst, 1986; Jacobs, 1950*b*, 1953*a*; Wallin, 1979. **Disc recording**: Bellmann, 1993*c* (CD); Grein, 1984 (LP). **Cassette recording**: Bellmann, 1985*b*, 1993*b*; Wallin, 1979.

RECOGNITION. See under *Psophus stridulus*. In the field the spectacular flight display of the male is unique in western Europe.

CREPITATION (Figs 713, 714. CD 2, track 2). The male flight display of this species is the nearest equivalent in western Europe to the comparable, but often more impressive, flight displays performed by the males of the much richer locustine fauna of North America. Displaying males of *B. tuberculata* fly for long distances (sometimes several hundred metres), usually at heights of several metres above the ground. The flight is undulating, like that of many small birds, and consists of alternating crepitation, during which the wings are beating vigorously, and silent flight, during which the male appears to glide. Oscillographic analysis of a recording taken from a male in mid-display showed that the bursts of crepitation lasted about 0·4–0·5 s and the intervals between them about 0·2–0·3 s; the crepitation rate was about 55/s. The crepitation is a whirring rather than a rattling sound, the individual wing-beat sounds being much less distinct than in *Psophus stridulus*.

When disturbed both sexes often crepitate during short flights (Figs 713, 714). Recordings taken from three males and one female showed that the crepitation rate during such flights varied between 30 and 40/s.

DISTRIBUTION. Very local in western Europe: now known only from Öland in Sweden and a few localities in Bavaria, eastern Switzerland and the Tyrol. Farther east it occurs in the Ukraine, European Russia and temperate Asia as far as Mongolia and China.

Stethophyma grossum (Linnaeus)

Gryllus (Locusta) grossus Linnaeus, 1758: 433.

Large Marsh Grasshopper; (F) Criquet ensanglanté; (D) Sumpfschrecke; (NL) Moerassprinkhaan; (DK) Sumpgræshoppe; (S) Kärrgräshoppa; (SF) Suosirkka.

REFERENCES TO SONG. **Oscillogram**: Grein, 1984; Kleukers *et al.*, 1997; Schmidt & Baumgarten, 1977. **Diagram**: Bellmann, 1985a, 1988, 1993a; Bellmann & Luquet, 1995; Duijm & Kruseman, 1983; Holst, 1970; Ragge, 1965; Wallin, 1970. **Frequency information**: Meyer & Elsner, 1996. **Verbal description only**: Broughton, 1972b; Faber, 1928, 1953a; Graber, 1873; Harz, 1957, 1962; Holst, 1986; Jacobs, 1950b, 1953a; Taschenberg, 1871; Weber, 1984; Willemse, 1943. **Disc recording**: Bellmann, 1993c (CD); Grein, 1984 (LP); Odé, 1997 (CD); Ragge *et al.*, 1965 (LP). **Cassette recording**: Bellmann, 1985b, 1993b; Burton & Ragge, 1987; Elliott, 1986; Wallin, 1979.

RECOGNITION (Plate 3: 2). The large size and colouring of this species, together with its occurrence on wet bogs, make it easy to recognize. The general colour is greenish yellow, green or greenish brown (rarely reddish purple), with a conspicuous yellowish stripe along the anterior (ventral) margin of the fore wings; the underside of the hind femora is red (occasionally yellow) and the hind tibiae are yellow with black spines. It has a medial intercalary vein on the fore wings and so is usually placed in the Locustinae, but the general appearance is more suggestive of a gomphocerine. In the field the male calling song is highly characteristic.

Males are superficially similar in appearance to the larger species of *Arcyptera*, but the members of that genus have red hind tibiae and are found in drier habitats in mountains.

For a note on the generic assignment of this species see p. 11.

SONG (Figs 715–720. CD 2, track 3). The male calling song is unique among the European grasshoppers. It consists of a series of quite loud 'ticks' produced by flicking one of the hind tibiae backwards against the distal part of the flexed fore wing. The number of ticks is very variable: it is usually between 5 and 10, but can be fewer and is occasionally more than 15. The repetition rate of the ticks also varies, even within one series, but is usually between 1 and 3/s. The tibiae are usually used one at a time and there is no regular alternation between left and right; occasionally both tibiae are used simultaneously. Sometimes ticks of this kind are produced, by both sexes, as a response to disturbance.

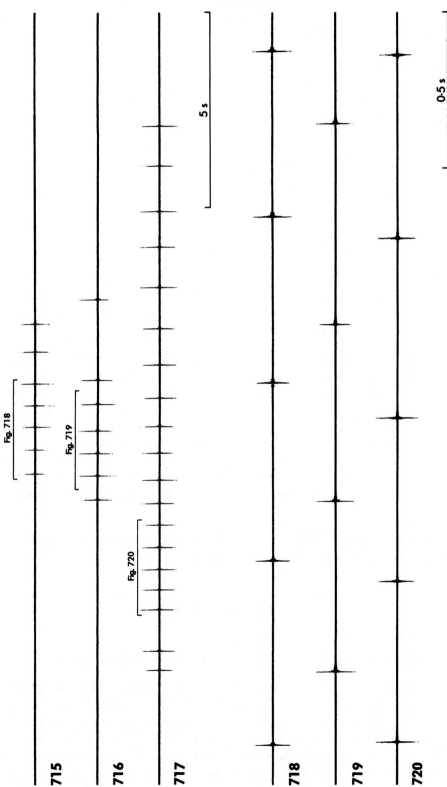

Figures 715–720 Oscillograms at two different speeds of the songs of three males of *Stethophyma grossum*. Figs 715 and 716 show typical calling songs and Fig. 717 shows a longer song produced in the presence of a female.

In the presence of a female the male often produces longer sequences (Fig. 717), sometimes including as many as 20 ticks.

DISTRIBUTION. Very widespread in Europe in suitably wet habitats from the Pyrenees and Alps northwards, occurring as far north as Lapland and locally in southern England and western Ireland. In the southern peninsulas it is confined to a few localities in the northernmost parts. Eastwards, the range extends to Kazakhstan and Siberia.

Brachycrotaphus tryxalicerus (Fischer)

Opomala tryxalicera Fischer, 1853: 305.

Savanna Grasshopper.

REFERENCES TO SONG. **Oscillogram**: Green (1995).

RECOGNITION. A small but elongate, straw-coloured grasshopper found in Europe only near the Mediterranean coast of Spain, and on Sicily and Lipari. The antennae are very thick towards the base, tapering gradually to a pointed tip, and the costal area of the male fore wings is expanded, with ladder-like cross-veins.

SONG (Figs 721–726. CD 2, track 4). The male calling song is a rather quiet, brief echeme usually lasting between 0·5 and 1·0 s and composed of about 5–15 syllables repeated at the rate of about 8–15/s. Sometimes a male produces mainly short echemes of about 0·5 s interspersed with an occasional longer echeme of about 1·0 s. The intervals between successive echemes vary greatly, but are usually of more than 5 s. Oscillographic analysis shows that the first one or two syllables in an echeme are louder and often longer and more complex than the remaining ones. Each syllable is usually a double sound, but the leg-movements have not been studied and so the exact way in which the sounds are produced is not yet known; it is not even certain that these constituent sounds are syllables, as the leg-movements are very rapid.

DISTRIBUTION. Widespread in tropical Africa; known in Europe only from near the Mediterranean coast of Spain, and on Sicily and Lipari.

Arcyptera fusca (Pallas)

Gryllus (Locusta) fuscus Pallas, 1773: 727.

Large Banded Grasshopper; (F) Arcyptère bariolée; (D) Große Höckerschrecke.

REFERENCES TO SONG. **Oscillogram**: Bukhvalova, 1993a; Bukhvalova & Zhantiev, 1993; Faber, 1957a; García *et al.*, 1987; Grein, 1984; Xi *et al.*, 1992. **Diagram**: Bellmann, 1985a, 1988, 1993a; Bellmann & Luquet, 1995; Jacobs, 1950b, 1953a. **Sonagram**: Sales & Pye, 1974. **Frequency information**: Dumortier, 1963c; Sales & Pye, 1974. **Musical notation**: Yersin, 1854b (as *variegatum*). **Verbal description only**: Chopard, 1952; Dumortier, 1963d; Faber, 1928; 1953a; Harz, 1957; Karny, 1908; Poulton, 1896. **Disc recording**: Andrieu &

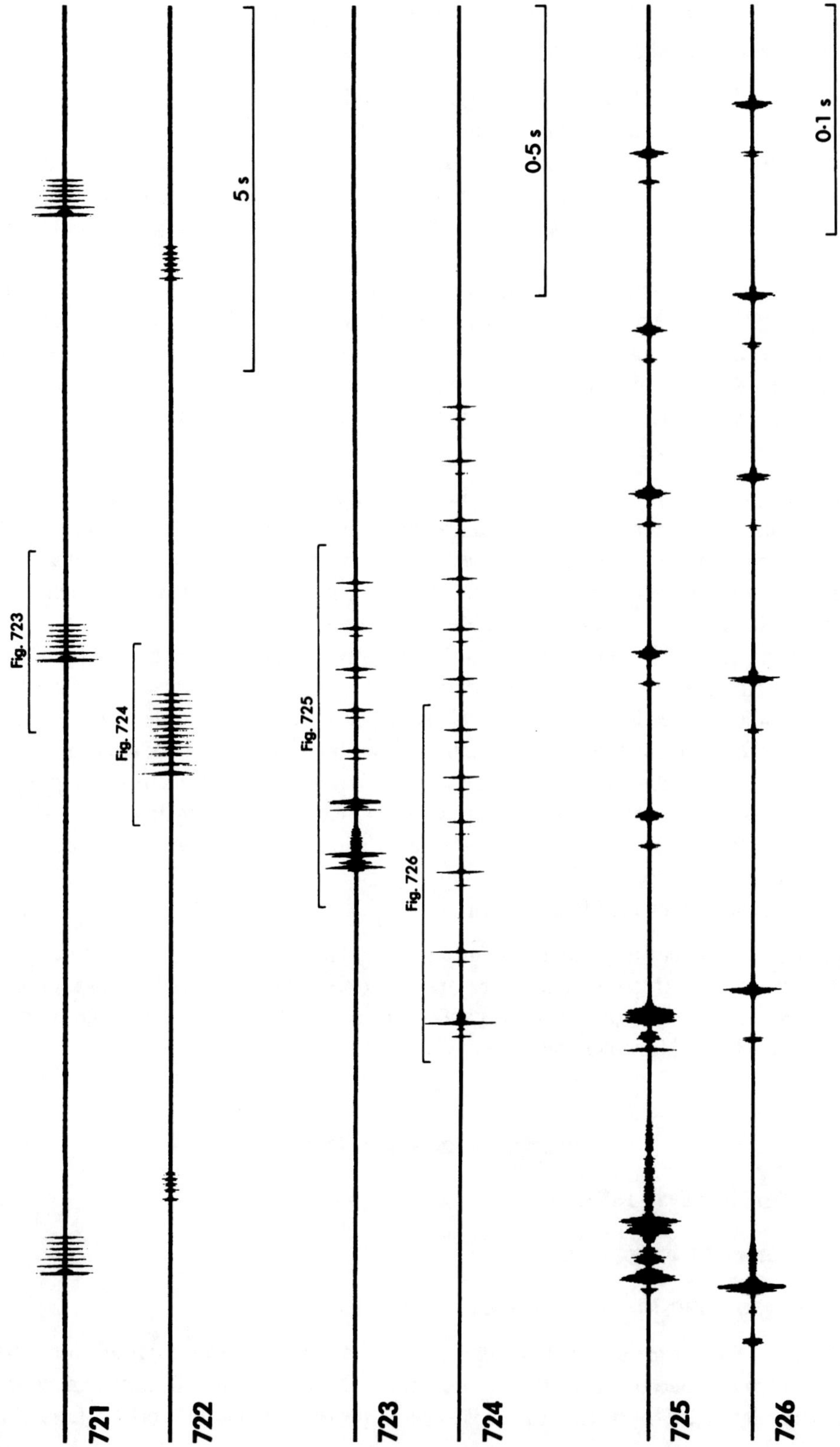

Figures 721–726 Oscillograms at three different speeds of the calling songs of two males of *Brachycrotaphus tryxalicerus*.

Dumortier, 1994 (CD); Bellmann, 1993c (CD); Bonnet, 1995 (CD); Grein, 1984 (LP). **Cassette recording**: Bellmann, 1985b, 1993b.

RECOGNITION. (Plate 3: 3). The genus *Arcyptera* includes several European species, mainly found in mountains and most easily recognized by their striking colour-pattern, especially the conspicuous dark and pale bands on the hind femora, which are red or yellow ventrally, and the red hind tibiae. The general colour is yellowish to olive green or brown, and they are sturdily built insects. The wings are usually shortened in the female, and sometimes so in the male.

A. *fusca* is a large insect, in which the males have fore wings longer than 20 mm and the pronotal lateral carinae are straight or almost so in the prozona. The wings are shortened in the female, not reaching the hind knees; the hind wings are strongly darkened in both sexes. In the field the loud male calling song is unmistakable.

Males could be confused with *Stethophyma grossum*, but that species has yellow hind tibiae and is found only in wet habitats.

SONG (Figs 727–730. CD 2, track 5). The loud and highly characteristic calling song of the male consists of a combination of isolated syllables with a dense echeme, forming a sequence lasting about 1·5–4·0 s. A typical sequence consists of one or two clearly separated syllables, each lasting about 100–300 ms, then a very short, explosive syllable lasting no more than about 60 ms followed immediately by a dense echeme lasting about 0·9–2·5 s; this is in turn followed by one or two (occasionally three or four) separate syllables of the same kind as the opening ones. Sometimes two or more sequences of this kind follow one another in quick succession, but they are also often separated by intervals of at least several seconds. Occasionally the isolated syllables are omitted either before or after the dense echeme, and such syllables are quite often produced in complete isolation, unaccompanied by the dense echeme. Oscillographic analysis shows that the isolated syllables are broken up into about 10–20 distinct, evenly spaced sounds lasting no more than 10 ms and separated by momentary gaps of similar duration.

Males of A. *fusca* often crepitate in flight.

DISTRIBUTION. Typically a species of mountains, found in the Pyrenees (French and Spanish), Cevennes, Alps (including the Jura and the uplands of southern Germany) and mountains of eastern Europe, including parts of the Balkan Peninsula. The range extends eastwards to Siberia, Mongolia and China.

Arcyptera tornosi Bolívar

Arcyptera Tornosi Bolívar, 1884: civ.

Iberian Banded Grasshopper.

REFERENCES TO SONG. No published work known to us.

RECOGNITION. Similar to A. *fusca* but smaller (male fore wings shorter than 20 mm); the pronotal lateral carinae are distinctly bent inwards in the prozona and the hind femora are yellow ventrally (red in the other western European species of *Arcyptera*). As in A. *fusca*, the

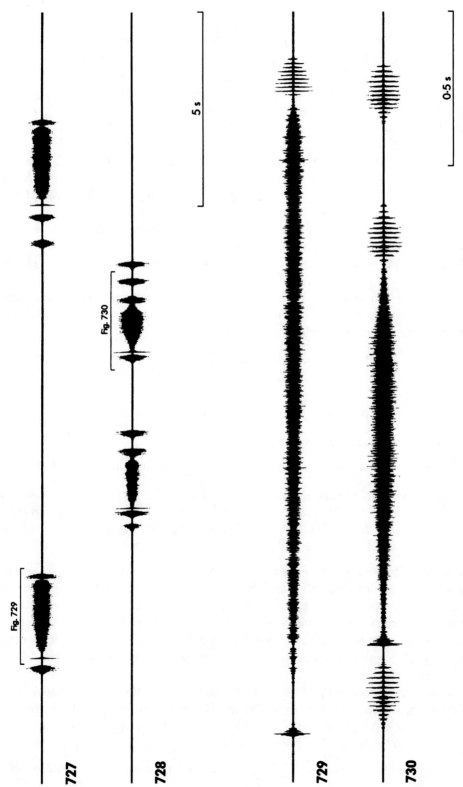

Figures 727–730 Oscillograms at two different speeds of the calling songs of two males of *Arcyptera fusca*.

wings are shortened in the female and the hind wings are darkened in both sexes.

SONG (Figs 731–736. CD 2, track 6). The loud male calling song consists of sequences including separate syllables of two quite different kinds, together with a dense echeme. In the mountains of central Spain, a fully developed sequence consists of a series of very brief 'ticks', repeated at the rate of about 4/s, followed by a dense echeme lasting about 150–550 ms and, after a pause of similar duration, a single, rather 'scratchy' syllable, again of similar duration, produced by a relatively slow downstroke of the hind femora (Figs 732, 733, 735, 736). Commonly, 3–4 of these sequences follow one another in fairly quick succession, separated by intervals of about 1–2 s; it is also common for the first sequence to have only 1–2 ticks (occasionally none) before the dense echeme, while the remaining sequences have about 4–10. During the course of a series of ticks, which tends to show a crescendo, the early ones appear to be single syllables, but the later ones, which last rather longer, seem to be composed of more than one syllable. Oscillographic analysis shows that the slow downstroke hemisyllable following the dense echeme is broken up into discrete sounds (each composed of about 1–4 tooth-impacts) separated by gaps of about 1–3 ms.

In the calling song of a male studied from near Coimbra, Portugal, there was usually only one tick (never more than four) before a dense echeme, and often none at all before the later dense echemes in a series (Figs 731, 734).

DISTRIBUTION. Found only in Portugal and in the central mountains of Spain.

Arcyptera microptera (Fischer de Waldheim)

Oedipoda microptera Fischer de Waldheim, 1833: 384.

Small Banded Grasshopper; (F) Arcyptère savoyarde; (D) Kleine Höckerschrecke.

REFERENCES TO SONG. **Oscillogram**: Bukhvalova & Zhantiev, 1993; García *et al.*, 1987; Grein, 1984; Xi *et al.*, 1992. **Diagram**: Duijm & Kruseman, 1983. **Verbal description only**: Bellmann, 1993a; Bellmann & Luquet, 1995; Faber, 1953a; Harz, 1957; Jacobs, 1953a. **Disc recording**: Bonnet, 1995 (CD, as *carpentieri* and *kheili*); Grein, 1984 (LP).

RECOGNITION. This widespread species varies greatly in different parts of its range. The wings are usually well developed in both sexes, but several local forms are brachypterous, the fore wings sometimes not reaching beyond the fifth abdominal tergite in the male and no further than the second in the female; the hind wings in such forms are vestigial. When the hind wings are sufficiently well developed to show it, they are transparent or at the most darkened a little towards the tip, in contrast to the strongly darkened hind wings of *A. fusca* and *A. tornosi*.

A. carpentieri (occurring in the Causses of the Cevennes in southern France) and *A. kheili* (a brachypterous form occurring in the French Alps) do not differ significantly from *A. microptera* either in morphology or in song, and we consider them to be no more than local forms of this species (see p. 72 and Figs 737–752).

SONG (Figs 737–752. CD 2, track 7). The male calling song is usually composed of both

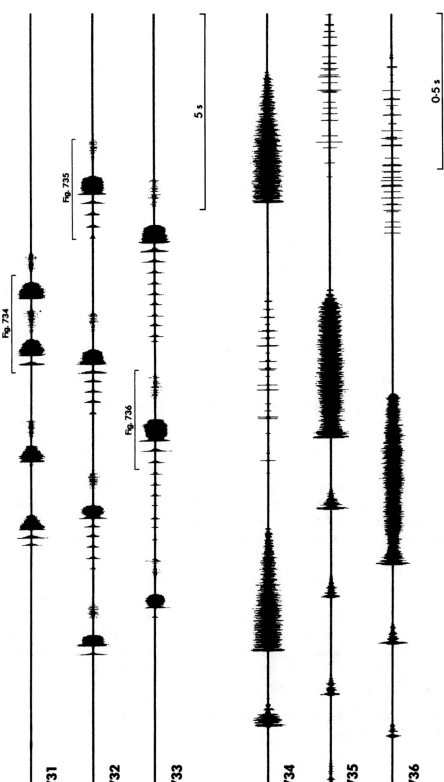

Figures 731–736 Oscillograms at two different speeds of the calling songs of two males of *Arcyptera tornosi*.

short and longer sequences, each containing at least one dense echeme and one or more separate syllables. One common pattern is a short sequence of 1–3 short dense echemes followed after an interval of about 1–3 s by a series of 'ticks' leading to a longer dense echeme (occasionally accompanied by a shorter one). The short echeme usually contains only about 3–5 syllables and the longer one about 8–12. The number of ticks preceding the longer echeme is very variable: it is commonly between 5 and 15, but can be fewer than this or more than 20; there is often a single tick immediately before the shorter echeme. Sometimes there are 2–3 (rarely 4) of the longer dense echemes, each preceded by a series of ticks; sometimes there is more than one sequence of short echemes before a longer one, and occasionally the song ends with such a sequence. The song shows quite a strong similarity to that of *A. tornosi*, but always lacks the slow downstroke hemisyllable so characteristic of that species.

DISTRIBUTION. In western Europe now known only from the northern half of Spain, the Cevennes, the French Alps and Austria; farther east the range extends through eastern Europe to Kazakhstan, Siberia and China.

Ramburiella hispanica (Rambur)

Gryllus Hispanicus Rambur, 1838: 88.

Striped Grasshopper; (F) Criquet des Ibères; (D) Spanischer Grashüpfer.

REFERENCES TO SONG. **Verbal description only**: Defaut, 1987, 1988*b*; Yersin, 1857.

RECOGNITION. This species is most easily recognized by its colour-pattern. There is a conspicuous pale longitudinal stripe along the top of the head, pronotum and corresponding part of the flexed fore wings; these parts are otherwise dark greyish brown. In addition, there are three conspicuous dark bands on the otherwise pale inner side of each hind femur. There is an intercalary vein (without stridulatory pegs) in the medial area of the female fore wings, and a rather poorly developed one in the male. The difference in size between the two sexes is unusually great. In the field the male calling song is very characteristic.

SONG (Figs 753–758. CD 2, track 8). The male calling song is a loose echeme of about 4–20 (sometimes up to 30 or more) very brief diplosyllables, each lasting about 60–160 ms. The syllable repetition rate averages about 0·5–2·0/s, but the syllables are often irregularly spaced. Sometimes the echeme begins quietly, with a crescendo during the first few syllables. Oscillographic analysis shows that each syllable consists of a very rapid upstroke hemisyllable lasting about 8–20 ms, followed immediately by a slower downstroke hemisyllable lasting about 40–100 ms.

DISTRIBUTION. Southern France (especially near the Mediterranean coast), the Iberian Peninsula and North Africa.

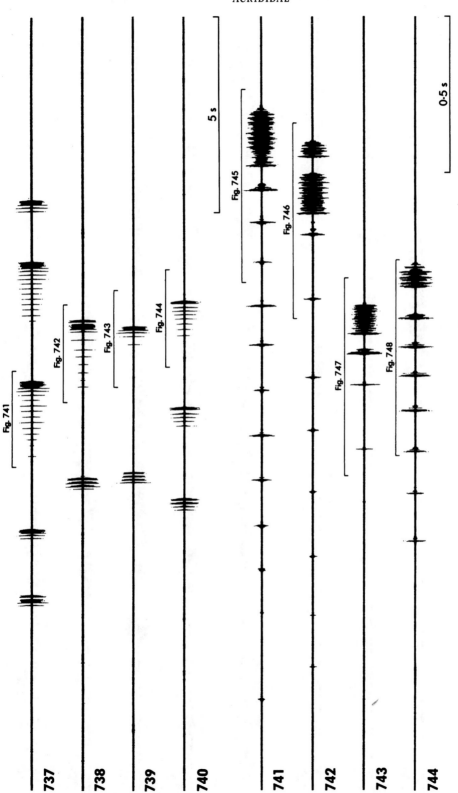

Figures 737–744 Oscillograms at two different speeds of the calling songs of four males of *Arcyptera microptera*. Figs 737 and 738 are from typical Spanish males; Figs 739 and 740 are from males of the forms occurring in the Cevennes and French Alps, respectively. Fig. 738 is from a male with only one hind leg.

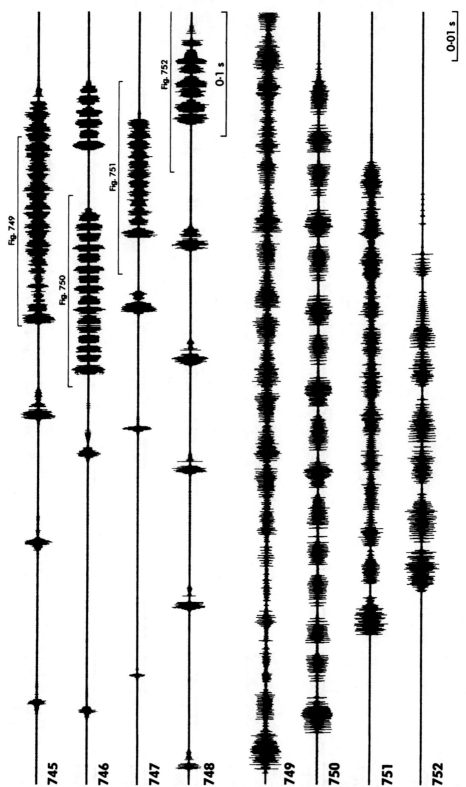

Figures 745–752 Faster oscillograms of the indicated parts of the songs of *Arcyptera microptera* shown in Figs 741–744. Figs 746 and 750 are from a male with only one hind leg.

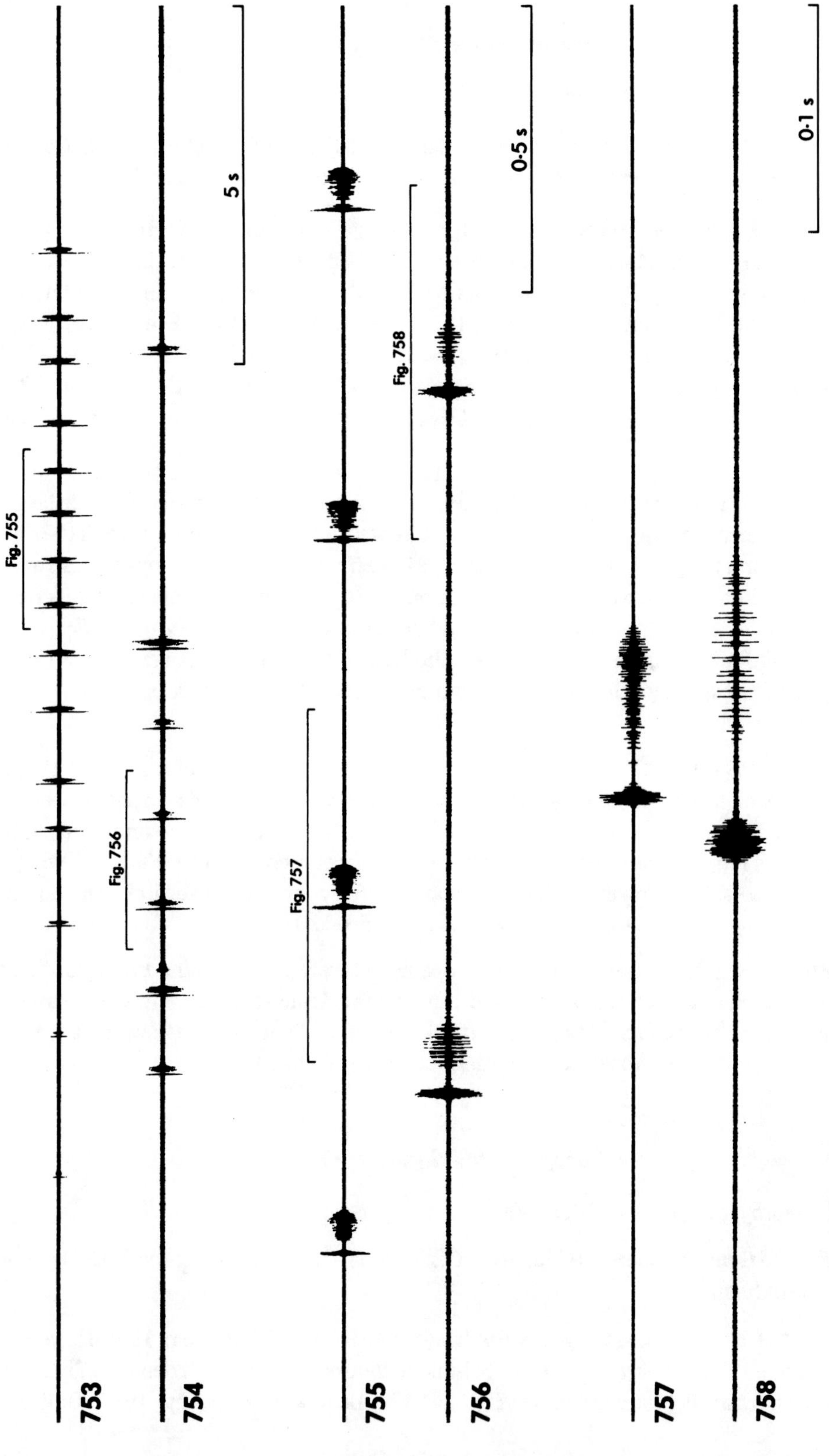

Figures 753–758 Oscillograms at three different speeds of the calling songs of two males of *Ramburiella hispanica.*

Chrysochraon dispar (Germar)

Podisma dispar Germar, 1836: pl. 7.

Large Gold Grasshopper; (F) Criquet des clairières; (D) Große Goldschrecke; (NL) Gouden Sprinkhaan; (DK) Guldgræshoppe; (S) Guldgräshoppa; (SF) Kultaheinäsirkka.

REFERENCES TO SONG. **Oscillogram**: Bukhvalova & Zhantiev, 1993; Grein, 1984; Kleukers *et al.*, 1997; Schmidt & Baumgarten, 1977; Vedenina, 1990; Vedenina & Zhantiev, 1990; Zhantiev, 1981. **Diagram**: Bellmann, 1985a, 1988, 1993a; Bellmann & Luquet, 1995; Duijm & Kruseman, 1983; Jacobs, 1950b, 1953a; Wallin, 1979. **Frequency information**: Meyer & Elsner, 1996. **Verbal description only**: Defaut, 1979; Faber, 1928, 1953a; Renner & Kremer, 1980. **Disc recording**: Bellmann, 1993c (CD); Bonnet, 1995 (CD); Grein, 1984 (LP); Odé, 1997 (CD). **Cassette recording**: Bellmann, 1985b, 1993b; Wallin, 1979.

RECOGNITION. In practice males of this species are most easily recognized by their bright lustrous green colouring. The fore wings are quite well developed, but do not reach the hind knees; the hind wings are vestigial. The females have much shorter fore wings and are quite different in colour: brown (rarely greenish or purplish), with the hind tibiae and underside of the hind femora reddish. Both sexes could be confused with *Chorthippus parallelus*, but can be distinguished from that species by the lack of foveolae on the vertex. See under *Euthystira brachyptera* for the differences from that species. Both sexes occasionally occur in a fully winged form.

SONG (Figs 759–767. CD 2, track 9). The male calling song consists of a series of simple echemes, each lasting about 0·5–2·0 s (longer in cool conditions) and composed of about 12–20 syllables repeated at the rate of about 10–20/s (slower in cool conditions). The echemes are repeated at variable intervals, usually of between 3 and 6 s. Each echeme begins quietly, reaching maximum intensity about halfway through its duration. During the louder part of the echeme each syllable lasts about 50–120 ms.

DISTRIBUTION. Widespread in western Europe from Sweden and Finland to the southern slopes of the Pyrenees and Alps. Eastwards, the range extends through central and eastern Europe (including the more northerly parts of the Balkan Peninsula) to Kazakhstan, Siberia and the far east of temperate Asia. Absent from the British Isles.

Euthystira brachyptera (Ocskay)

Gryllus brachypterus Ocskay, 1826: 409.

Small Gold Grasshopper; (F) Criquet des Genévriers; (D) Kleine Goldschrecke; (NL) Kleine Goudsprinkhaan.

REFERENCES TO SONG. **Oscillogram**: Bukhvalova & Zhantiev, 1993; Grein, 1984; Helversen & Helversen, 1994; Kleukers *et al.*, 1997; Schmidt & Baumgarten, 1977; Vedenina & Zhantiev, 1990. **Diagram**: Bellmann, 1985a, 1988, 1993a; Bellmann & Luquet, 1995; Duijm &

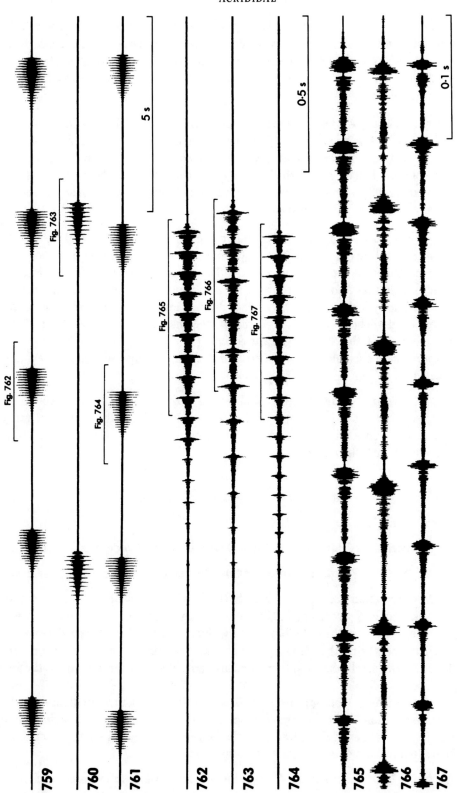

Figures 759–767 Oscillograms at three different speeds of the calling songs of three males of *Chrysochraon dispar*.

Kruseman, 1983; Jacobs, 1950b, 1953a. **Leg-movement**: Helversen & Helversen, 1994. **Musical notation**: Yersin, 1854b. **Verbal description only**: Defaut, 1979; Faber, 1928, 1953a; Harz, 1957; Luquet, 1978; Renner, 1952; Yersin, 1854b. **Disc recording**: Bellmann, 1993c (CD); Bonnet, 1995 (CD); Grein, 1984 (LP); Odé, 1997 (CD). **Cassette recording**: Bellmann, 1985b, 1993b.

RECOGNITION (Plate 3: 4). Like *Chrysochraon dispar*, this species can be readily recognized by its colouring: both sexes are bright green with a metallic lustre. The males have shorter fore wings than in *C. dispar*, reaching only about halfway along the abdomen and bulging outwards somewhat; in the female the fore wings are reduced to small lateral flaps, usually pink in colour, and the ovipositor valves are exceptionally long. As in *C. dispar*, both sexes occasionally occur in a fully winged form.

SONG (Figs 768–776. CD 2, track 10). The male calling song is a very brief and rather quiet echeme, lasting only about 140–300 ms in warm conditions and composed of about 4–7 syllables repeated at the rate of about 20–30/s. The echemes are sometimes separated by long intervals, but may also be in groups of a few echemes separated by short intervals of about 2–4 s. The syllables often show a clear division into quieter upstroke and louder downstroke hemisyllables (Figs 774, 775), the two together lasting about 30–40 ms. There is often a crescendo at the beginning of the echeme.

In the presence of a female the male often produces a much longer sequence of regularly spaced echemes.

DISTRIBUTION. From the Low Countries and central Germany in the north, through the French uplands to the southern slopes of the Pyrenees and Alps in the south. The range extends eastwards through much of central and eastern Europe to Kazakhstan and Siberia. Absent from the British Isles.

Dociostaurus maroccanus (Thunberg)

Gryllus maroccanus Thunberg, 1815: 244.

Moroccan Locust; (F) Criquet marocain; (D) Marokkanische Wanderheuschrecke.

REFERENCES TO SONG. **Frequency information**: Busnel, 1953; Busnel & Chavasse, 1951.

RECOGNITION. The genus *Dociostaurus* has a characteristic colour-pattern. The pronotal lateral carinae are not very prominent, but their position is marked by conspicuous pale stripes, sharply angled inwards at the anterior end of the pronotum and then sharply diverging in the posterior part, so that they tend to form a cross. The hind femora have two or three conspicuous dark dorsal spots, and there are dark spots along the medial area of the fore wings.

D. maroccanus may be distinguished from *D. jagoi* by its much larger size (length of the hind femur more than 9 mm in the male, more than 13 mm in the female; less than these values in *D. jagoi*) and from *D. hispanicus* by the longer fore wings, extending beyond the hind knees (not reaching the hind knees in *D. hispanicus*).

SONG (Figs 777–784. CD 2, track 11). The stridulatory leg-movements of *Dociostaurus* have

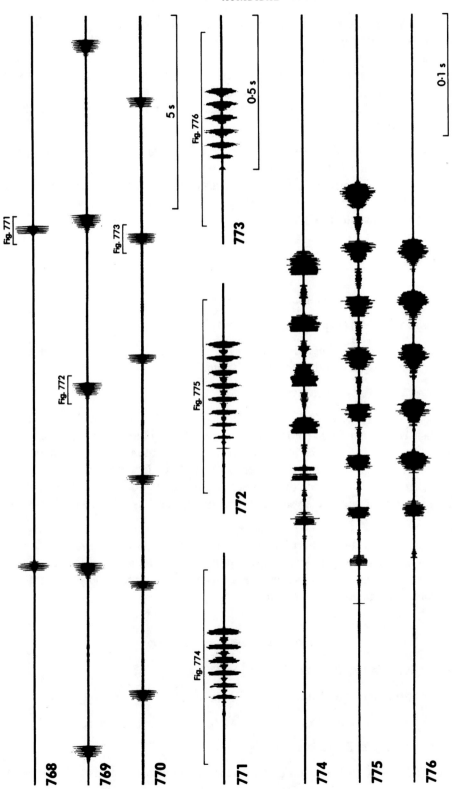

Figures 768–776 Oscillograms at three different speeds of the calling songs of three males of *Euthystira brachyptera*. Figs 769, 772 and 775 are from a macropterous male.

not yet been studied in detail, and so the terminology used in the song descriptions for the three species included here is based solely on careful observation and oscillographic analysis; it may not be entirely correct.

The male calling song of *D. maroccanus* consists of an echeme of about 6–12 very brief, 'tick'-like syllables, each lasting about 30–40 ms. Oscillographic analysis shows that each syllable is broken up into about 4–6 short-lived and well-separated sounds. The syllable repetition rate is about 2/s.

DISTRIBUTION. The whole of the Mediterranean Region, including the southern European peninsulas, North Africa and many Mediterranean islands. In France it occurs mainly in the départements bordering the Mediterranean Sea, and the northern limit runs eastwards along the south side of the Alps and then through Hungary, Romania and the Ukraine. The range includes Somalia and extends eastwards to Kazakhstan and central Asia.

Dociostaurus jagoi Soltani

Dociostaurus (Kazakia) jagoi Soltani, 1978: 26.

Jago's Grasshopper; (F) Criquet de Jago.

REFERENCES TO SONG. **Oscillogram**: Blondheim, 1990; García *et al.*, 1994. **Sonogram**: Blondheim, 1990. **Verbal description only**: Defaut, 1987, 1988*b*.

RECOGNITION. See under *D. maroccanus*.

SONG (Figs 785–787, 789–791, 793–795, 797–799. CD 2, track 12). The male calling song is superficially very similar to that of *D. maroccanus*, consisting of an echeme of about 6–14 brief, 'tick'-like syllables, repeated at the rate of about 1·5–3·0/s and each composed of about 4–7 well-separated sounds. However, each syllable lasts twice as long as in *D. maroccanus*, about 50–100 ms, and the ticks are quieter, as one would expect from a much smaller insect. Oscillographic analysis shows that there is usually a crescendo during the course of each syllable.

DISTRIBUTION. Most of the Mediterranean Region from Portugal to the Levant, including North Africa, although not yet recorded from Italy. In France it has been recorded from a few localities in the west, as well as from the départements bordering the Mediterranean Sea. Eastwards, the range extends to Iran.

Dociostaurus hispanicus (Bolívar)

Stauronotus brevicollis var. *Hispanicus* Bolívar, 1898*a*: 14.

Iberian Cross-backed Grasshopper.

REFERENCES TO SONG. No published work known to us.

RECOGNITION. See under *D. maroccanus*.

SONG (Figs 788, 792, 796. CD 2, track 13). The male calling song consists of a series of

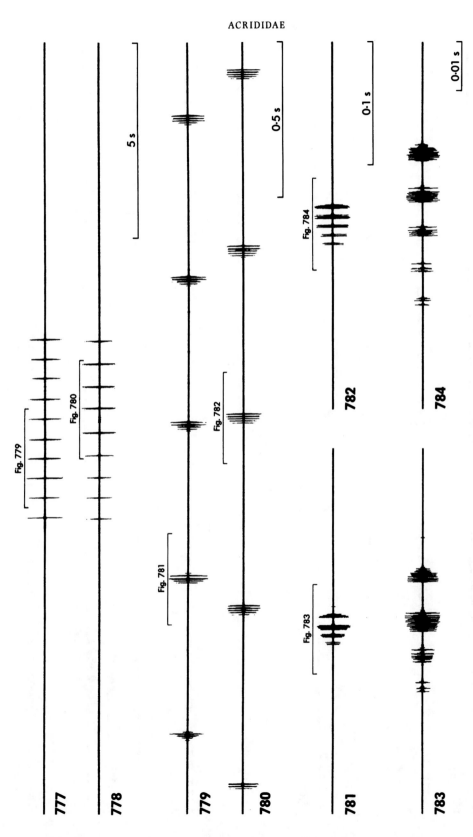

Figures 777–784 Oscillograms at four different speeds of the calling songs of two males of *Dociostaurus maroccanus*.

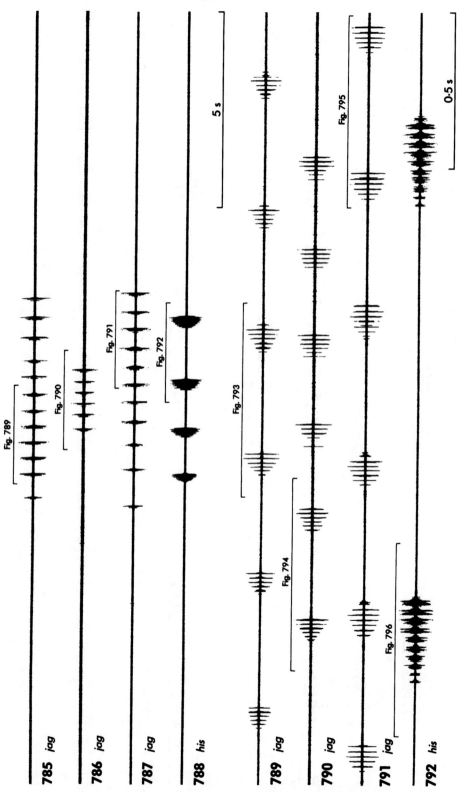

Figures 785–792 Oscillograms at two different speeds of the male calling songs of (785–787, 789–791) *Dociostaurus jagoi* and (788, 792) *D. hispanicus.*

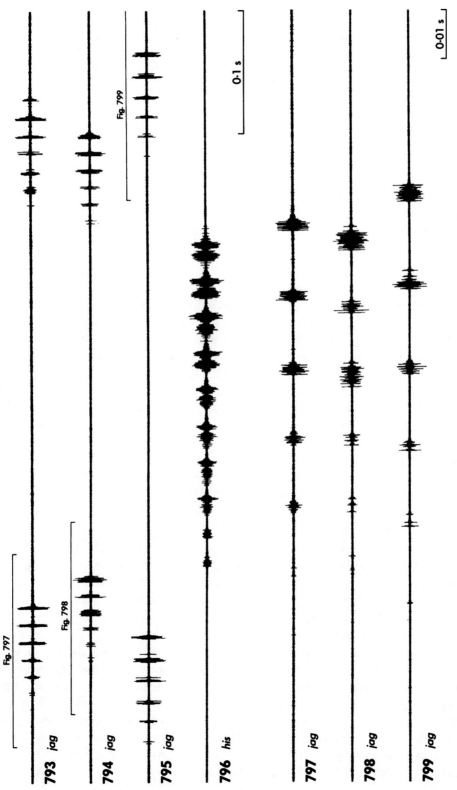

Figures 793–799 Faster oscillograms of the indicated parts of the songs of *Dociostaurus jagoi* and *D. hispanicus* shown in Figs 789–792.

about 2–6 echemes repeated at the rate of about 0·7–2·0/s, each lasting about 150–300 ms and composed of about 6–12 syllables repeated at the rate of about 30–40/s. Oscillographic analysis shows that each echeme begins quietly, reaching maximum intensity towards the end. The later syllables in each echeme have a double structure, but whether these subdivisions correspond to upstrokes and downstrokes of the hind legs is at present unknown.

DISTRIBUTION. Known only from Spain.

Omocestus viridulus (Linnaeus)

Gryllus (Locusta) viridulus Linnaeus, 1758: 433.

Common Green Grasshopper; (F) Criquet verdelet; (D) Bunter Grashüpfer; (NL) Wekkertje; (DK) Lynggræshoppe; (S) Grön Ängsgräshoppa; (N) Grønn Markgrashoppe; (SF) Niittyheinäsirkka.

REFERENCES TO SONG. **Oscillogram**: Clemente, 1987; Eiríksson, 1992, 1993; Elsner, 1974*a*, 1975, 1983*a*; Elsner & Popov, 1978; Grein, 1984; Haskell, 1957, 1958, 1961; Hedwig, 1986*a*, 1986*b*, 1990; Hedwig & Elsner, 1981, 1985; Hedwig & Meyer, 1994; Heinrich & Elsner, 1997; Holst, 1970, 1986; Kleukers *et al.*, 1997; Kutsch, 1976; Kutsch & Schiolten, 1979; Loher & Broughton, 1955; Ragge, 1986; Waeber, 1989; Zhantiev, 1981. **Diagram**: Bellmann, 1985*a*, 1988, 1993*a*; Bellmann & Luquet, 1995; Duijm & Kruseman, 1983; Haskell, 1957; Holst, 1970; Jacobs, 1950*b*, 1953*a*; Ragge, 1965; Wallin, 1979. **Leg-movement**: Elsner, 1974*a*, 1975, 1983*a*; Elsner & Popov, 1978; Hedwig, 1986*a*, 1986*b*, 1990; Hedwig & Elsner, 1981, 1985; Hedwig *et al.*, 1990; Hedwig & Meyer, 1994; Heinrich & Elsner, 1997; Waeber, 1989. **Frequency information**: Haskell, 1957, 1958; Meyer & Elsner, 1996. **Musical notation**: Yersin, 1854*b*. **Verbal description only**: Beier, 1956; Defaut, 1987, 1988*b*; Elsner & Hirth, 1978; Faber, 1928, 1953*a*; Haskell, 1955*a*; Harz, 1957; Skovmand & Pedersen, 1978, 1983; Weber, 1984; Weih, 1951. **Disc recording**: Bellmann, 1993*c* (CD); Bonnet, 1995 (CD); Grein, 1984 (LP); Odé, 1997 (CD); Ragge *et al.*, 1965 (LP). **Cassette recording**: Bellmann, 1985*b*, 1993*b*; Burton & Ragge, 1987; Wallin, 1979.

RECOGNITION (Plate 3: 5). The 20 or so western European species of *Omocestus* are rather nondescript grasshoppers, recognizable as belonging to this genus only by negative characters. They lack the bulge on the anterior (ventral) margin of the fore wings of *Chorthippus*, the lateral teeth on the ovipositor of *Stenobothrus* and the clubbed antennae of *Gomphocerus*, *Gomphocerippus* and *Myrmeleotettix*. The medial area of the fore wings is not enlarged, except in *O. uvarovi*. See Ragge (1986) for further discussion and a morphological key to the principal western European species of *Omocestus*.

Over most of its range *O. viridulus* may be distinguished from the rather similar species *O. rufipes* and *O. haemorrhoidalis* by the lack of any red colouring on the male abdomen and the conspicuously longer ovipositor. In Spain (away from the Pyrenees), where the male abdomen is red-tipped (subspecies *kaestneri* Harz), males of *O. viridulus* may usually be distinguished from *O. rufipes* by being green above (males of *O. rufipes* are brown above) and occurring in moister habitats; *O. haemorrhoidalis* does not occur in Spain except in the vicinity of the Pyrenees. In the field males of *O. viridulus* may be recognized by their loud calling song, which usually lasts for more than 12 s.

For a discussion of the taxonomy of this species in the Iberian Peninsula, see p. 72.

SONG (Figs 800–815. CD 2, track 14). The male calling song is an echeme usually lasting about 12–25 s and consisting of syllables repeated at the rate of about 15–20/s. (In some Spanish localities the calling song of *O. v. kaestneri* is sometimes a little shorter but usually within the range 10–14 s.) The echeme begins quietly (the first few leg-movements producing no audible sound) and gradually increases in loudness until maximum intensity is reached after a few seconds; the echeme then continues at a constant intensity until reaching an abrupt end. The syllable repetition rate is highest at the beginning of the echeme (usually 18–20/s), gradually lessening towards the end (when it is usually 15–18/s). The song is a conspicuous summer sound in much of the European countryside, louder than the songs of most other common grasshoppers.

In the presence of a female the male produces longer echemes, usually lasting more than 30 s and occasionally more than a minute; one of the hind legs is usually moved through a noticeably wider angle than the other and produces most of the sound. After a series of these echemes with short pauses (about 10–15 s) between them, there is a quite different and much quieter echeme lasting about 3–5 s and composed of syllables repeated at the rate of about 12–16/s (Fig. 804). This is normally followed by a series of loud syllables and an attempt to copulate with the female. If this attempt is unsuccessful, the male will usually produce a series of sharp 'ticks' (Fig. 805) by kicking backwards with the hind tibiae (as in the calling song of *Stethophyma grossum*) before beginning another sequence of courtship echemes. Usually the two hind tibiae are kicked back simultaneously, but sometimes one at a time, changing haphazardly (rarely regularly) from one side to the other. The number of ticks is very variable but is usually between 5 and 15, and the repetition rate is rather irregular, generally about 1–2/s.

DISTRIBUTION. Found in fairly moist habitats throughout Europe except north of the Arctic Circle and in much of the southern peninsulas; widespread in the British Isles. Eastwards, the range extends to Siberia and Mongolia.

Omocestus rufipes (Zetterstedt)

Gryllus rufipes Zetterstedt, 1821: 90.

Woodland Grasshopper; (F) Criquet noir-ébène; (D) Buntbäuchiger Grashüpfer; (NL) Negertje; (S) Rödgumpgräshoppa.

REFERENCES TO SONG. **Oscillogram**: Broughton, 1955a (as *ventralis*); Grein, 1984 (as *v.*); Holst, 1986 (as *v.*); Kleukers *et al.*, 1997; Loher & Broughton, 1955 (as *v.*); Ragge, 1986; Schmidt, 1989 (as *v.*); Schmidt & Baumgarten, 1977 (as *v.*); Waeber, 1989 (as *v.*). **Diagram**: Bellmann, 1985a (as *v.*), 1988, 1993a (as *v.*); Bellmann & Luquet, 1995; Duijm & Kruseman, 1983 (as *v.*); Holst, 1970 (as *v.*); Jacobs, 1950b, 1953a; Luquet, 1978 (as *v.*); Ragge, 1965; Wallin, 1979 (as *v.*). **Leg-movement**: Waeber, 1989. **Frequency information**: R.-G. Busnel, 1955 (as *v.*); Loher & Broughton, 1955 (as *v.*). **Musical notation**: Yersin, 1854b. **Verbal description only**: Beier, 1956; Chopard, 1922 (as *v.*); Defaut, 1987 (as *v.*), 1988b (as *v.*); Faber, 1953a (as *v.*); Harz, 1957 (as *v.*); Weber, 1984 (as *v.*); Weih, 1951. **Disc recording**: Andrieu & Dumortier, 1963 (LP), 1994 (CD, as *v.*); Bellmann, 1993c (CD, as *v.*); Bonnet,

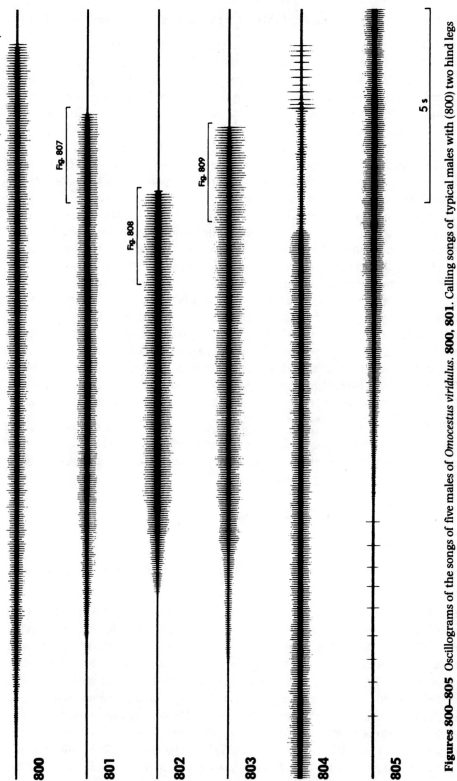

Figures 800–805 Oscillograms of the songs of five males of *Omocestus viridulus*. **800, 801**. Calling songs of typical males with (800) two hind legs and (801) one hind leg. **802, 803**. Calling songs of males of *O. v. kaestneri* from Spain. **804, 805**. Parts of the courtship song showing (804) the concluding part of a main echeme followed by the quieter echeme and loud syllables that precede an attempt to copulate, and (805) the 'ticks' produced before the beginning of a main echeme. (Figs 800, 801, 804 and 805 are from Ragge, 1986)

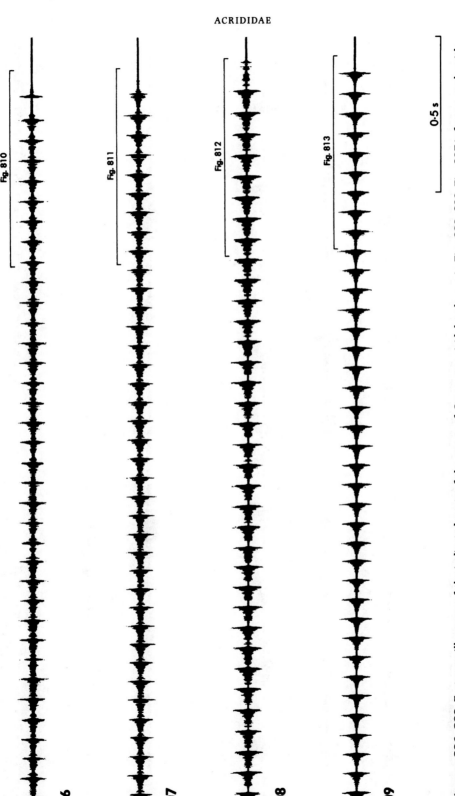

Figures 806–809 Faster oscillograms of the indicated parts of the songs of *Omocestus viridulus* shown in Figs 800–803. Fig. 807 is from a male with only one hind leg. (Figs 806 and 807 are from Ragge, 1986)

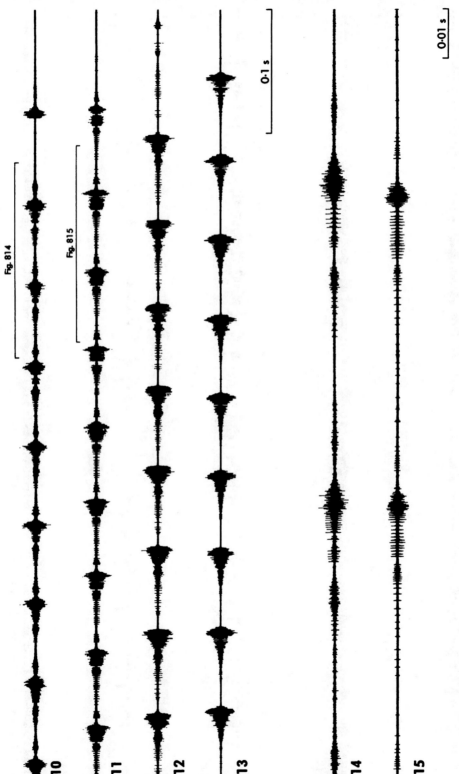

Figures 810–815 Faster oscillograms of the indicated parts of the songs of *Omocestus viridulus* shown in Figs 806–809. Figs 811 and 815 are from a male with only one hind leg. (Figs 810, 811, 814 and 815 are from Ragge, 1986)

1995 (CD); Grein, 1984 (LP, as v.); Odé, 1997 (CD); Ragge *et al.*, 1965 (LP). **Cassette recording**: Bellmann, 1985b (as v.), 1993b (as v.); Burton & Ragge, 1987; Wallin, 1979 (as v.).

RECOGNITION. The red colouring on the underside of the abdomen usually enables both sexes of this species to be distinguished from its relatives. The distal part of the hind wings is much more strongly darkened than in *O. haemorrhoidalis*, in which the abdomen is occasionally reddish on the underside. See under *O. viridulus* for the distinction between *O. rufipes* and *O. v. kaestneri*.

In the field the male calling song is quite characteristic. Although similar in basic pattern to those of *O. viridulus* and *O. haemorrhoidalis*, it differs in duration, typically lasting about half as long as that of *O. viridulus* and about three times as long as that of *O. haemorrhoidalis*; there is, however, some overlap in duration with the shorter calling songs of *O. v. kaestneri*.

For a note on the specific name of this species see p. 11.

SONG (Figs 816–827. CD 2, track 15). The male calling song is an echeme typically lasting 5–10 s and consisting of syllables repeated at the rate of about 13–23/s. The echeme begins quietly and gradually increases in loudness until reaching maximum intensity towards the abrupt end, thus resembling the first half of the calling song of typical *O. viridulus*. As in *O. viridulus* the syllable repetition rate is highest at the beginning of the echeme (usually 17–23/s), gradually lessening towards the end (when it is usually 13–17/s). In the southern parts of the range of this species in western Europe, and especially in Italy, the calling song is often a little more prolonged, lasting up to 15 s and occasionally longer.

In the presence of a female the male produces a courtship song quite similar to that of *O. viridulus*. There is first a series of echemes similar to those of the calling song but usually rather longer; this is followed by a quite different and quieter echeme lasting about 5–10 s and composed of syllables repeated at the rate of about 10–15/s (Fig. 819). There are then several loud syllables followed by an attempt to copulate with the female. The 'ticks' produced by backward kicks of the hind tibiae during the courtship song of *O. viridulus* seem never to occur in that of *O. rufipes*.

DISTRIBUTION. This species can tolerate drier conditions than *O. viridulus* and often occurs in more shaded habitats. Its distribution is rather more southerly: it hardly occurs north of latitude 60°N, and the range extends southwards to central Spain, southern Italy and Greece. Eastwards, it occurs in Asia Minor, Kazakhstan and southern Siberia. Local in southern England.

Omocestus haemorrhoidalis (Charpentier)

Gryllus haemorrhoidalis Charpentier, 1825: 165.

Orange-tipped Grasshopper; (F) Criquet rouge-queue; (D) Rotleibiger Grashüpfer; (NL) Bruin Schavertje; (S) Alvargräshoppa.

REFERENCES TO SONG. **Oscillogram**: Bukhvalova & Zhantiev, 1993; Clemente, 1987; Grein, 1984; Holst, 1986; Kleukers *et al.*, 1997; Ragge, 1986; Schmidt & Schach, 1978; Vedenina & Zhantiev, 1990; Waeber, 1989. **Diagram**: Bellmann, 1985a, 1988, 1993a; Bellmann & Luquet, 1995; Duijm & Kruseman, 1983; Jacobs, 1950b, 1953a; Wallin, 1979. **Leg-movement**: Waeber, 1989. **Frequency information**: Meyer & Elsner, 1996. **Verbal description**

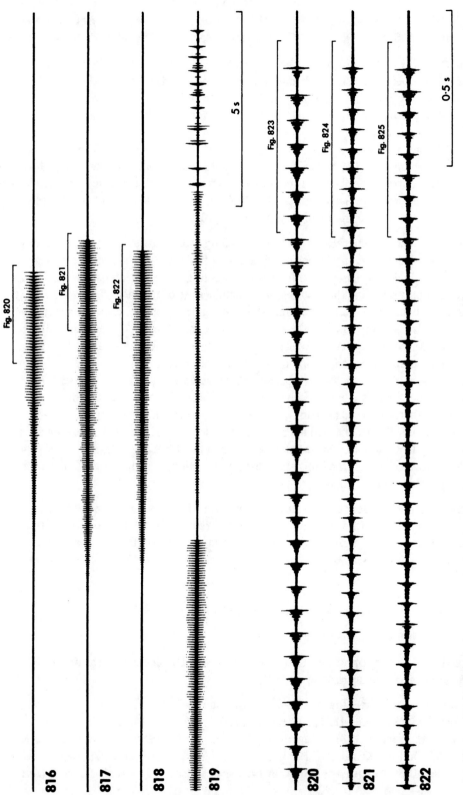

Figures 816–822 Oscillograms of the songs of four males of *Omocestus rufipes*. **816–818.** Calling songs of males with (816, 817) two hind legs and (818) one hind leg. **819.** Part of the courtship song showing the concluding part of a main echeme followed by the quieter echeme and loud syllables that precede an attempt to copulate. **820–822.** Faster oscillograms of the indicated parts of the songs shown in Figs 816–818. (From Ragge, 1986)

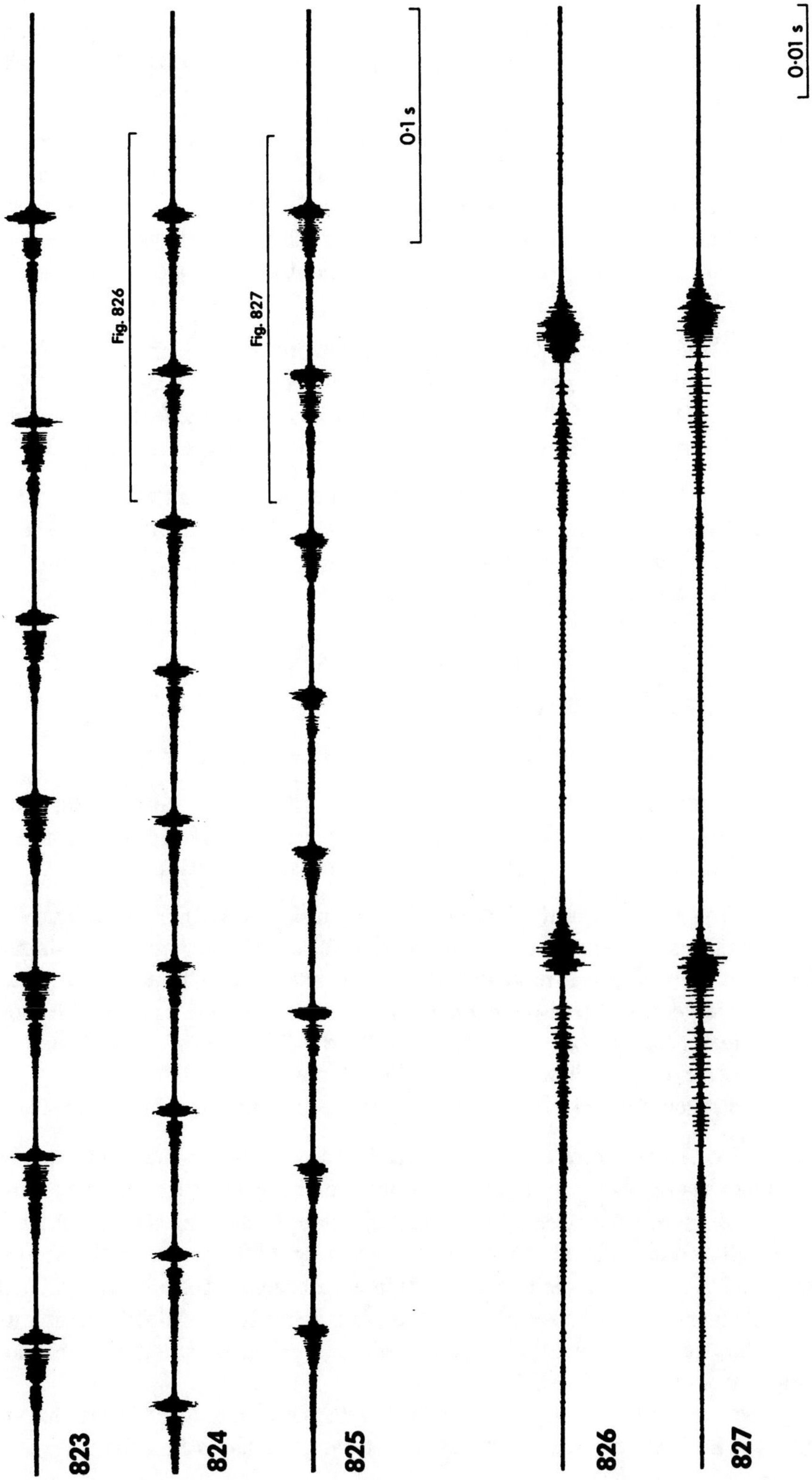

Figures 823–827 Faster oscillograms of the indicated parts of the songs of *Omocestus rufipes* shown in Figs 820–822. Figs 825 and 827 are from a male with only one hind leg. (From Ragge, 1986)

only: Faber, 1953*a*; Harz, 1957; Poulton, 1896; Weber, 1984. **Disc recording**: Bellmann, 1993*c* (CD); Bonnet, 1995 (CD); Grein, 1984 (LP); Odé, 1997 (CD). **Cassette recording**: Bellmann, 1985*b*, 1993*b*; Wallin, 1979.

RECOGNITION. For the distinction between this species and *O. petraeus*, see under that species. Both sexes may be distinguished from *O. viridulus*, *O. rufipes* and *O. raymondi* by the almost transparent hind wings (strongly darkened in the distal part in those three species). In the field males can be recognized quite easily by their calling song, in which the echemes are shorter than those of *O. viridulus* and *O. rufipes*, but much longer than those of *O. raymondi*.

SONG (Figs 828–836. CD 2, track 16). The male calling song is an echeme lasting about 2–4 s and consisting of syllables repeated at the rate of about 25–40/s (about double the rate of *O. viridulus* and *O. rufipes*). The echeme begins quietly but soon reaches maximum intensity. The syllable repetition rate gradually lessens during the course of the echeme.

DISTRIBUTION. Widespread in western Europe, from the southernmost parts of Sweden to the southern slopes of the Pyrenees and central Italy; absent from the British Isles. Farther east it occurs widely in central and eastern Europe (including much of the Balkan Peninsula), Kazakhstan, Siberia, Mongolia and Korea.

Omocestus petraeus (Brisout)

Acridium petræum Brisout, 1856: cxiv.

Rock Grasshopper; (F) Criquet des grouettes; (D) Felsgrashüpfer.

REFERENCES TO SONG. **Oscillogram**: Bukhvalova & Zhantiev, 1993; Ragge, 1986; Waeber, 1989. **Diagram**: Luquet, 1978. **Leg-movement**: Waeber, 1989. **Verbal description only**: Bellmann, 1993*a*; Bellmann & Luquet, 1995; Faber, 1953*a*; Harz, 1957.

RECOGNITION. This species lacks the darkened coloration of the distal part of the hind wings shown by *O. viridulus*, *O. rufipes* and *O. raymondi*, and is also noticeably smaller. Distinguishing it from *O. haemorrhoidalis* is more difficult, but *O. petraeus* is again noticeably smaller, and the head is larger in comparison to the pronotum and more convex above with shorter foveolae. The males lack the red colouring shown by the distal part of the abdomen in *O. haemorrhoidalis*, showing a yellowish colouring in this region instead.

The highly distinctive calling song enables males to be recognized easily in the field.

SONG (Figs 837–845. CD 2, track 17). The male calling song consists of a sequence of about 10–20 echemes lasting about 2–5 s. The sequence begins quietly, reaching maximum intensity after about 6 echemes; even at its maximum intensity the song is rather quiet. The echeme repetition rate is usually about 3–5/s, becoming a little slower as the sequence progresses, and the syllable repetition rate within each echeme about 80–110/s. Each of the louder echemes lasts for about 150 ms and contains about 10–15 syllables. A male from Istria in Croatia studied by Waeber (1989) produced longer echeme-sequences (lasting about 10 s) composed of more echemes (about 30).

In the presence of a female the male produces echemes of a quite different kind (Figs 839, 842, 845). Each echeme begins with a relatively long syllable, lasting about 20–50 ms,

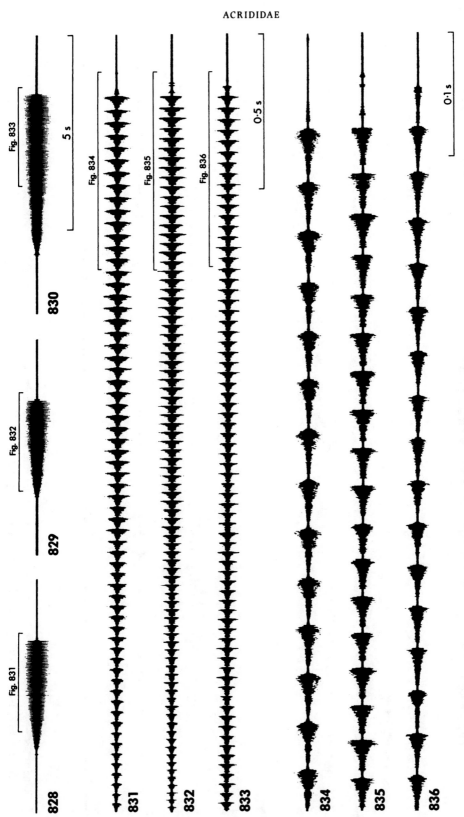

Figures 828–836 Oscillograms at three different speeds of the calling songs of three males of *Omocestus haemorrhoidalis*. (From Ragge, 1986)

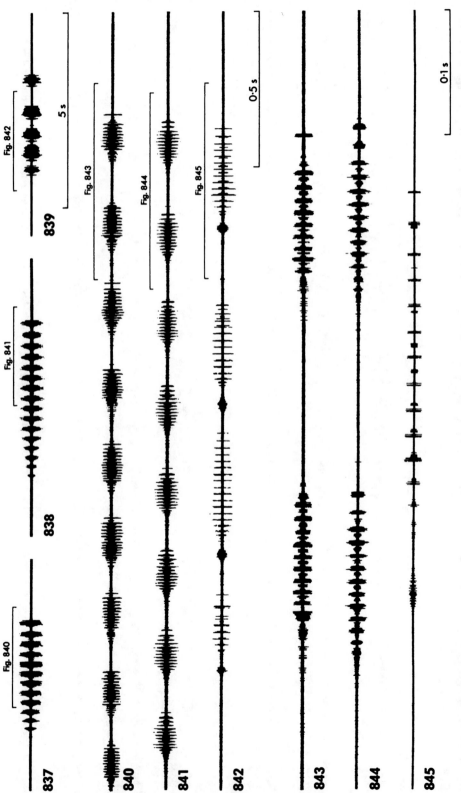

Figures 837–845 Oscillograms at three different speeds of the songs of two males of *Omocestus petraeus*. 837, 838, 840, 841, 843, 844. Calling songs. 839, 842, 845. Courtship song of the same male as Figs 837, 840, 843. (From Ragge 1986)

and this is immediately followed by a series of very short sounds, each lasting less than 5 ms, repeated at the rate of about 50/s. These echemes are produced either singly or, more often, in groups of about 2–5. Faber (1953a) describes the courting male producing echeme-sequences of the calling song type but of longer duration, consisting of 20–34 echemes. It is possible that the special courtship echemes described above are normally an adjunct to echeme-sequences of this kind; the only courting male we have observed produced only the sounds we have described while with the female.

DISTRIBUTION. In western Europe known only from France (especially the southern half), Italy and eastern Austria. Farther east it occurs from the Balkan Peninsula and Asia Minor to Kazakhstan and southern Siberia.

Omocestus raymondi (Yersin)

Stenobothrus Raymondi Yersin, 1863: 289.

Raymond's Grasshopper; Criquet des garrigues; (D) Südfranzösischer Grashüpfer.

REFERENCES TO SONG. **Oscillogram**: Bukhvalova & Zhantiev, 1993; Ragge, 1986; Waeber, 1989. **Leg-movement**: Waeber, 1989. **Verbal description only**: Bellmann, 1993a; Bellmann & Luquet, 1995; Chopard, 1922; Defaut, 1987, 1988b.

RECOGNITION. This all-brown species is most likely to be confused with *O. haemorrhoidalis*, but can be distinguished from it in both sexes by the strongly darkened distal part of the hind wings (almost transparent in *O. haemorrhoidalis*). In the field males may be distinguished from those of *O. haemorrhoidalis* by the very short echemes and slower syllable repetition rate of the calling song.

SONG (Figs 846–854. CD 2, track 18). The male calling song is an echeme lasting about 1·0–1·5 s and consisting of about 18–25 syllables repeated at the rate of about 15–20/s. Each echeme usually begins quietly, rapidly increasing in intensity. Oscillographic analysis shows that the downstroke hemisyllables have a characteristic pattern of gaps, which often become obscured towards the end of the echeme (Figs 849–851). The echemes are often produced singly and repeated at irregular intervals (10–15 s is typical), but sometimes they are in groups of 2–4 with much shorter intervals between them (often 2–5 s) (Figs 847, 848).

DISTRIBUTION. Known only from southern France, the Iberian Peninsula, north-western Italy and North Africa. In the southern part of its range this species has two generations per year, adults of the first generation being most numerous in June and those of the second in September.

Omocestus panteli (Bolívar)

Stenobothrus Panteli Bolívar, 1887: 95.

Pantel's Grasshopper.

REFERENCES TO SONG. **Oscillogram**: Ragge, 1986; Waeber, 1989. **Leg-movement**: Waeber, 1989. **Verbal description only**: Defaut, 1987, 1988b.

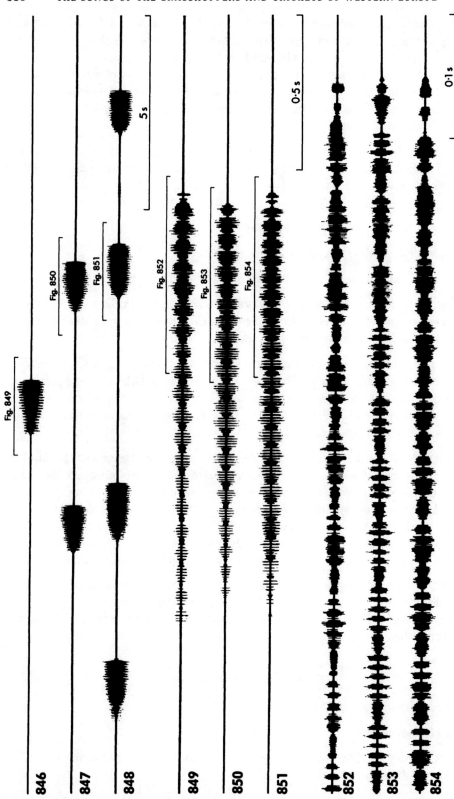

Figures 846–854 Oscillograms at three different speeds of the calling songs of three males of *Omocestus raymondi.* (From Ragge, 1986)

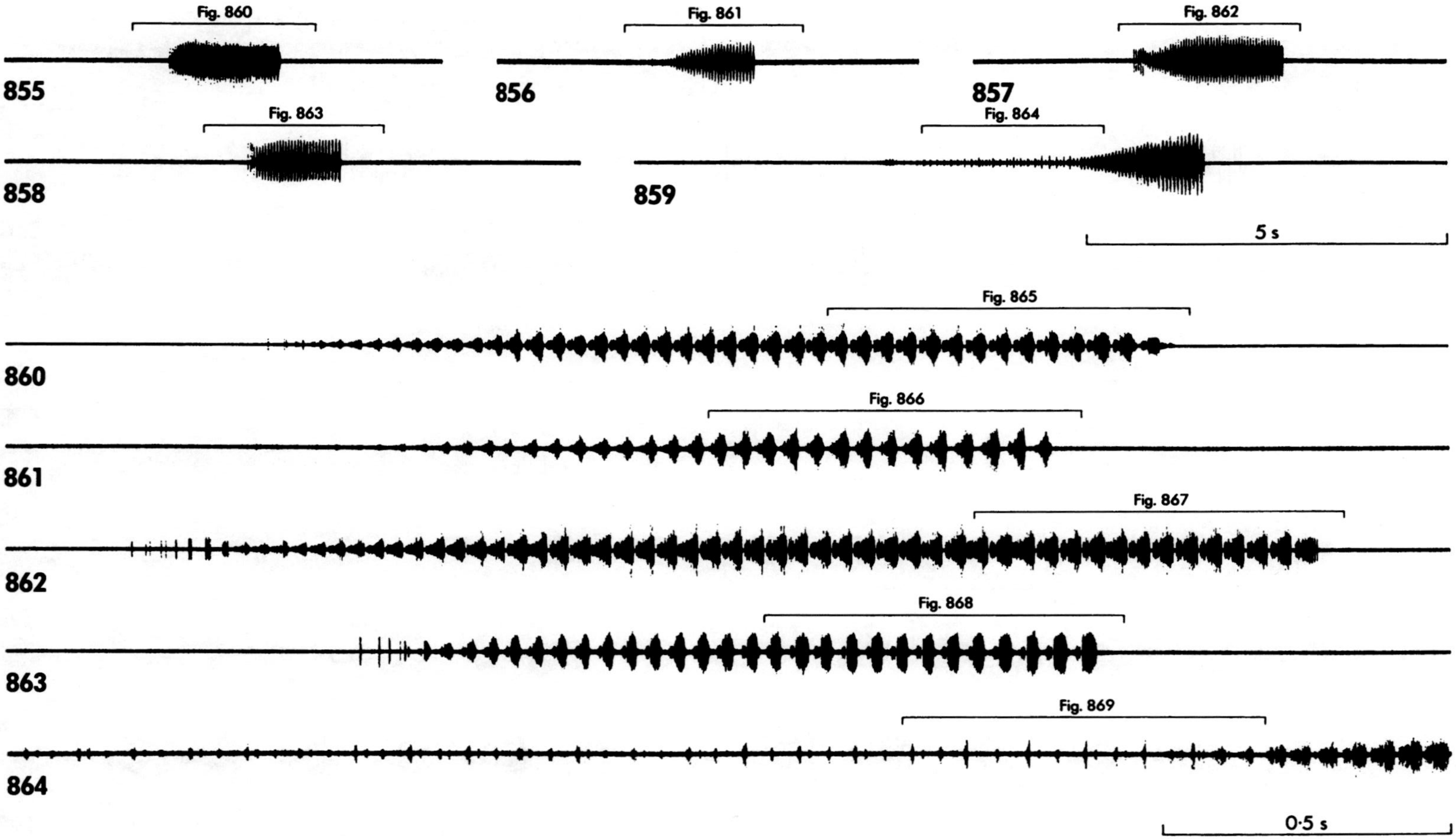

Figures 855–864 Oscillograms of the songs of five males of *Omocestus panteli*. **855–858**. Calling songs of males with (855–857) two hind legs and (858) one hind leg. **859**. Courtship song. **860–864**. Faster oscillograms of the indicated parts of the songs shown in Figs 855–859. (From Ragge, 1986)

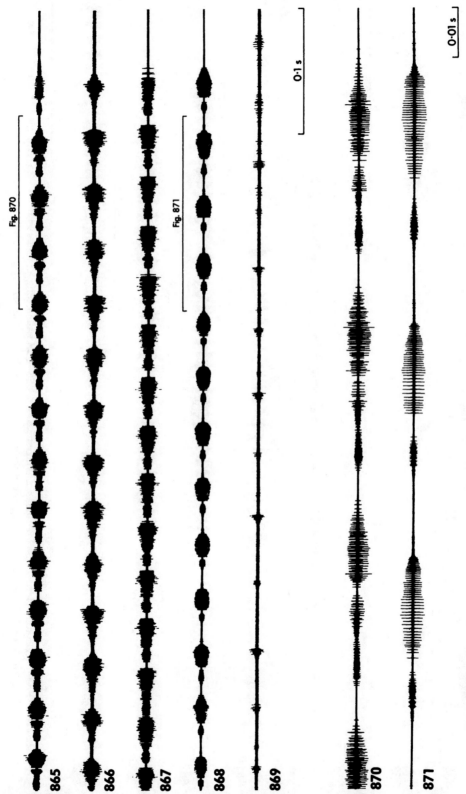

Figures 865–871 Faster oscillograms of the indicated parts of the songs of *Omocestus panteli* shown in Figs 860–864. Figs 868 and 871 are from a male with only one hind leg. (From Ragge, 1986)

RECOGNITION. *O. panteli* is confined to the Iberian Peninsula, where it is most likely to be confused with *O. viridulus*; it is, however, much smaller than that species and has a much shorter ovipositor. There is also a superficial resemblance to *Stenobothrus stigmaticus*, which is similar in size and also widespread in the Iberian Peninsula, but the pronotal lateral carinae are straighter in *O. panteli*, the male cerci are conical throughout (laterally compressed at the tip in *S. stigmaticus*) and the female lacks the lateral teeth on the ovipositor shown by all species of *Stenobothrus*. There is a large form (f. *meridionalis* Bolívar) found in the south of the peninsula, especially in mountains; this form is even more like *viridulus*, but confusion is unlikely as that species does not occur so far south.

SONG (Figs 855–871. CD 2, track 19). The calling song is an echeme lasting 1–2 s and consisting of about 30–55 syllables repeated at the rate of about 18–30/s. The echeme begins quietly (sometimes after a few louder syllables) but soon reaches maximum intensity. The syllable repetition rate gradually lessens during the course of the echeme.

   In the presence of a female the male first produces a different kind of echeme lasting about 2–3 s, and this is immediately followed by an echeme similar in duration and syllable repetition rate to that of the calling song but with an increase in intensity continuing through the whole echeme (Figs 859, 864, 869).

DISTRIBUTION. Found only in the Iberian Peninsula, where it is widespread and very common.

Omocestus antigai (Bolívar)

Stenobothrus (Omocestus) Antigai Bolívar, 1897b: 232.

Pyrenean Grasshopper; (F) Sténobothre catalan.

REFERENCES TO SONG. **Oscillogram**: Clemente, 1987 (as *navasi*); Ragge, 1986 (as *broelemanni*); Reynolds, 1986 (as *b.*); Waeber, 1989. **Leg-movement**: Waeber, 1989. **Verbal description only**: Clemente *et al.*, 1990 (as *n.*).

RECOGNITION. This Pyrenean species is brachypterous, the fore wings not usually reaching the hind knees in the male and reduced to short lobes (about 1·5 times the length of the pronotum) in the female; in both sexes the hind wings fall short of the fore wings by quite a large gap. There is thus no risk of confusion with any of the fully winged species of *Omocestus*. Confusion is also unlikely with any of the brachypterous species occurring farther south in Spain, such as *O. uhagonii* and *O. minutissimus*, which are much smaller (hind femora less than 9·2 mm in the male, less than 11·0 mm in the female).

   For a discussion of the taxonomic problems presented by this species, see p. 72.

SONG (Figs 872–911. CD 2, track 20). The male calling song is a series of echemes, each lasting about 1·5–3·0 s and consisting of about 20–45 syllables repeated at the rate of about 15–25/s. Each echeme begins quietly, reaching maximum intensity after about 1 s. Oscillographic analysis shows that each downstroke hemisyllable is interrupted by gaps, especially during the early part of the echeme. The echemes are sometimes produced singly, but are more often repeated fairly regularly, about every 3–10 s, in a series of indefinite duration.

   In the presence of a female, the male often produces a series of rather longer echemes,

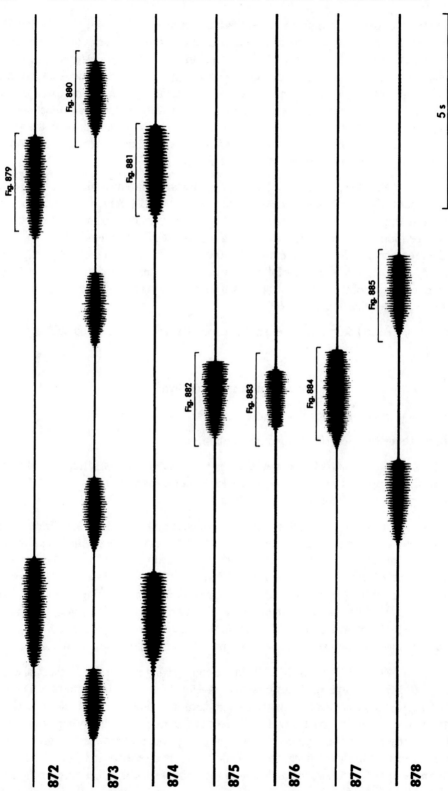

Figures 872–878 Oscillograms of the calling songs of seven males of *Omocestus antigai*. (Figs 872 and 873 are from Ragge, 1986)

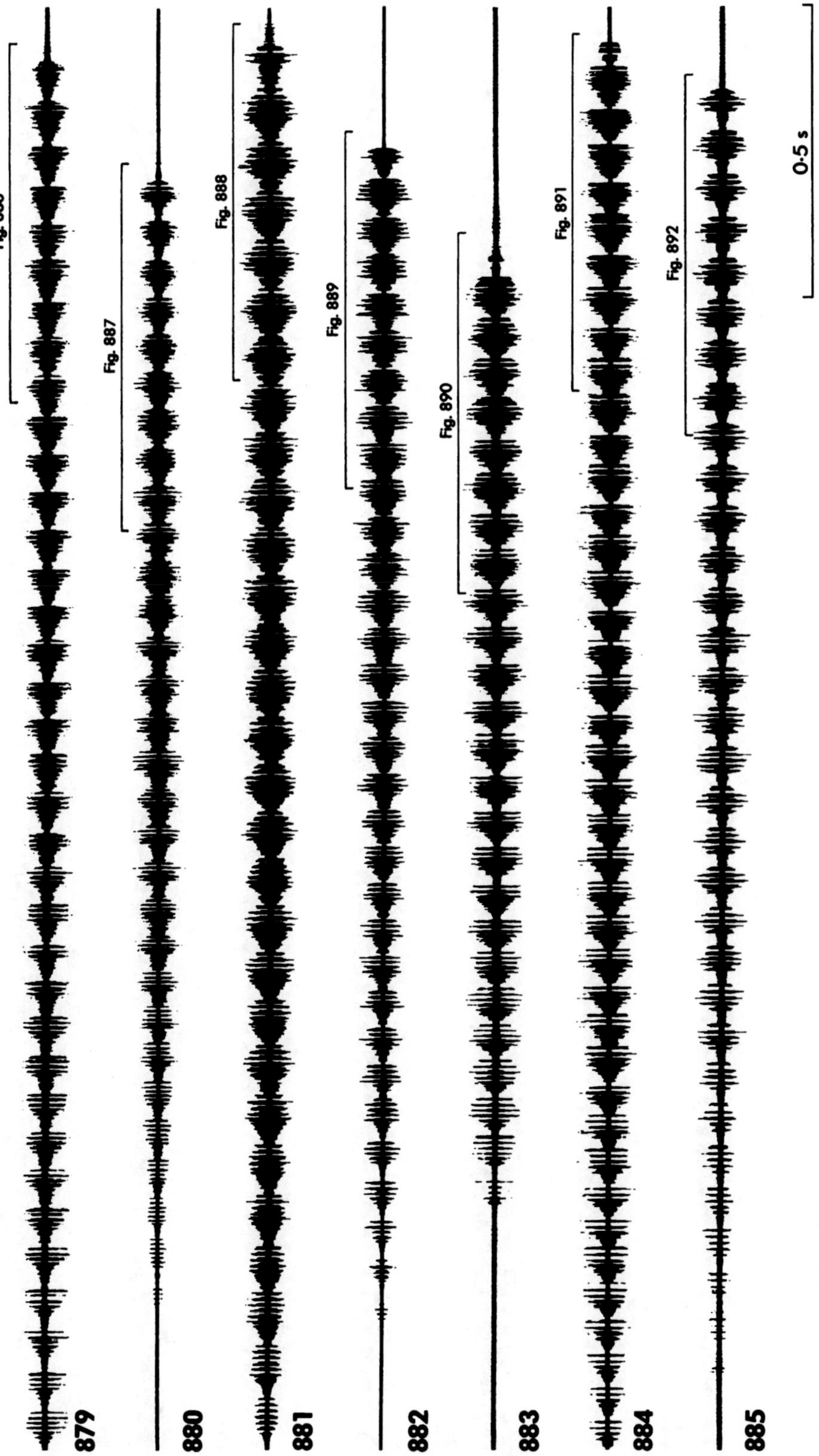

Figures 879–885 Faster oscillograms of the indicated parts of the songs of *Omocestus antigai* shown in Figs 872–878. (Figs 879 and 880 are from Ragge, 1986)

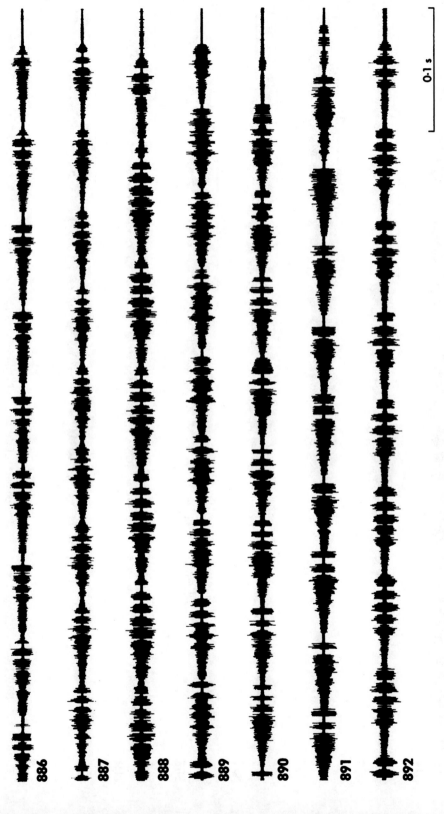

Figures 886–892 Faster oscillograms of the indicated parts of the songs of *Omocestus antigai* shown in Figs 879–885. (Figs 886 and 887 are from Ragge, 1986)

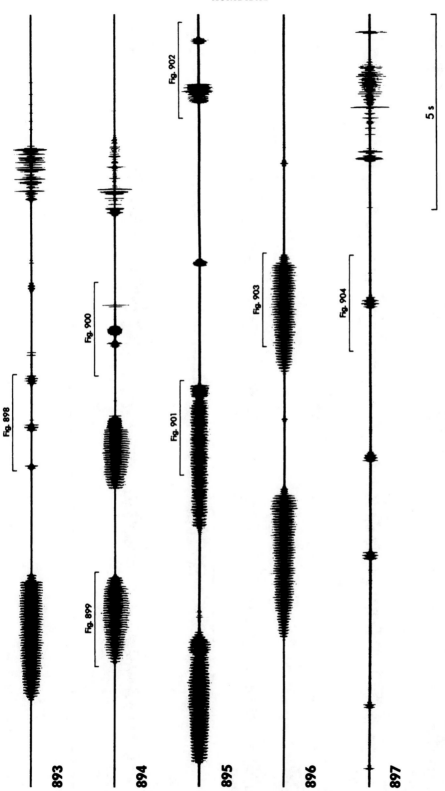

Figures 893–897 Oscillograms of parts of the courtship songs of five males of *Omocestus antigai*. Figs 893, 894 and 897 show the concluding parts of courtship sequences, each ending in the loud syllables that precede an attempt to copulate. (Fig. 893 is from Ragge, 1986)

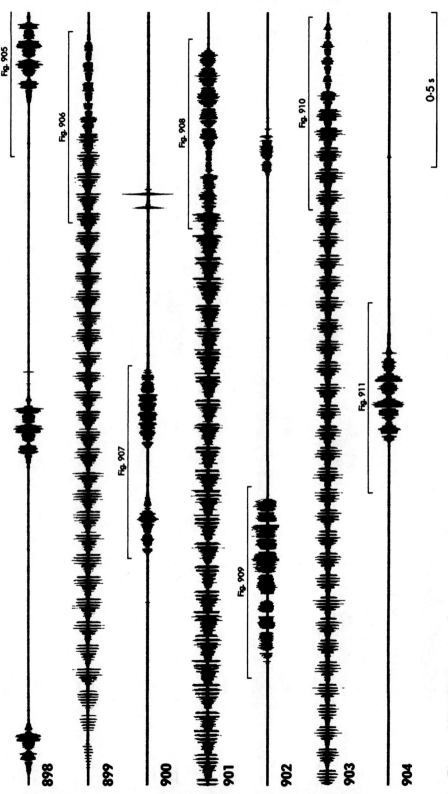

Figures 898–904 Faster oscillograms of the indicated parts of the songs of *Omocestus antigai* shown in Figs 893–897. (Fig. 898 is from Ragge, 1986)

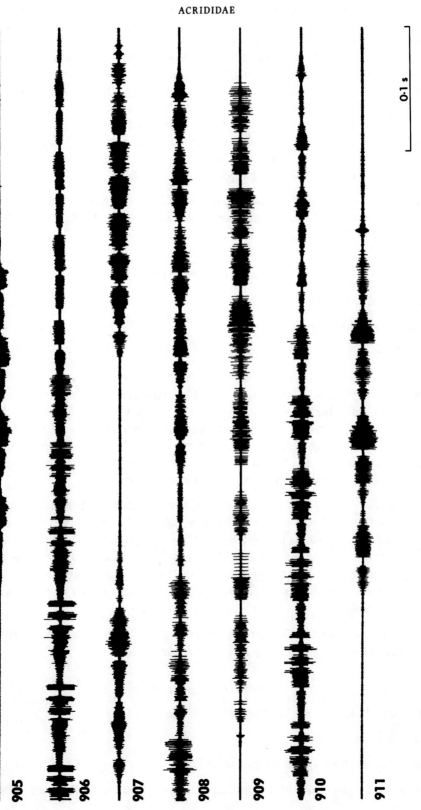

Figures 905–911 Faster oscillograms of the indicated parts of the songs of *Omocestus antigai* shown in Figs 898–904. (Fig. 905 is from Ragge, 1986)

0·1 s

each lasting up to 4 s or occasionally longer (Figs 893–897). These echemes often end in several syllables of a different kind: they are usually either quieter or louder than the immediately preceding normal syllables, they do not have gaps in the downstroke hemisyllables and they are usually produced by only one hind leg (the leg used alternating from one echeme to another). In one locality (Torre de Tamurcia, Lérida) courting males produced the whole echeme with only one hind leg, successive echemes being produced by the left and right legs alternately (Figs 895, 901, 909). When an attempt at copulation is imminent, groups of syllables like those at the end of courtship echemes are often produced separately, without any syllables of the calling song type (Figs 897, 904, 911).

DISTRIBUTION. Known only from the Pyrenees, mainly on the Spanish side from Huesca province eastwards, but also from the French side at a few localities in the département of Pyrénées-Orientales.

Omocestus bolivari Chopard

Omocestus bolivari Chopard, 1939: 172.

Bolívar's Grasshopper.

REFERENCES TO SONG. **Oscillogram**: Ragge, 1986; Waeber, 1989.

RECOGNITION. This species, known in Europe only from the higher parts of the Sierra Nevada in southern Spain, is very similar to *O. uhagonii* and *O. minutissimus*. Males may be distinguished from these species by their very short hind wings, which are less than half the length of the fore wings, and the females by their short fore wings, which are usually less than 1·3 times longer than the pronotum. In the field the isolated echemes of the calling song enable males to be easily distinguished from *O. minutissimus*; from *O. uhagonii* they may be distinguished by the much faster syllable repetition rate.

SONG (Figs 912–920. CD 2, track 21). The calling song is an echeme lasting about 0·5–2·0 s and consisting of about 10–30 syllables repeated at the rate of 14–16/s. The echeme begins quietly, soon reaching maximum intensity. Oscillographic analysis shows that there are gaps in each downstroke hemisyllable and that they occur throughout the echeme; there are, however, only 2–4 gaps per syllable (Figs 915–920), many fewer than in *O. uhagonii*. As in *O. uhagonii* the echemes are repeated at irregular intervals, varying from a few seconds to over a minute.

DISTRIBUTION. Known only from the higher parts of the Sierra Nevada in southern Spain and High Atlas in Morocco, usually at altitudes above 1500 m.

Omocestus uhagonii (Bolívar)

Gomphocerus (Stenobothrus) Uhagonii Bolívar, 1876: 324.

Uhagon's Grasshopper.

REFERENCES TO SONG. **Oscillogram**: Ragge, 1986.

RECOGNITION. Males of this brachypterous montane species can be distinguished from the

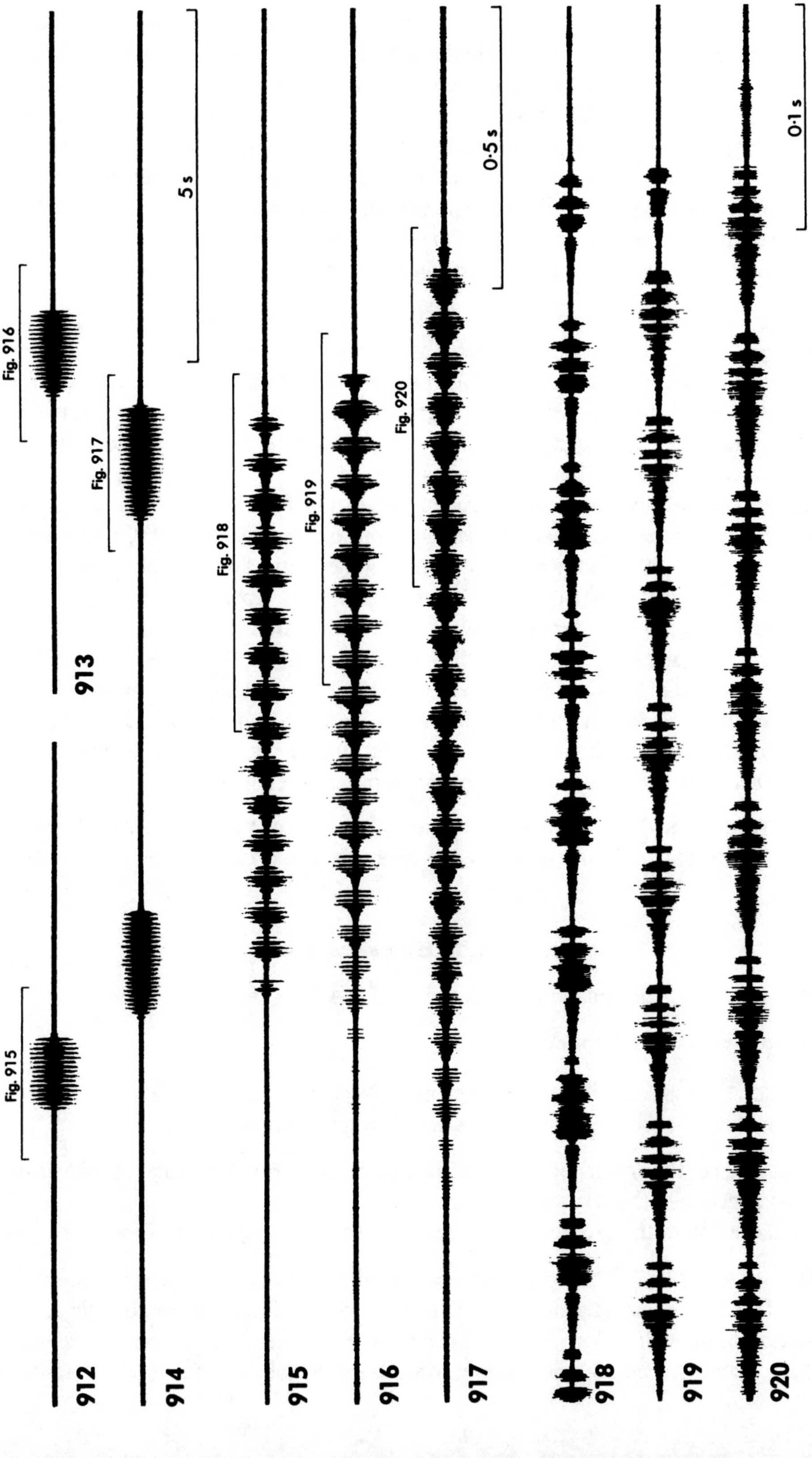

Figures 912–920 Oscillograms at three different speeds of the calling songs of three males of *Omocestus bolivari*. (From Ragge, 1986)

rather similar species *O. bolivari* and *O. minutissimus* by the cerci, which are laterally compressed towards the tip (simply conical in the other two species). Females can be separated from *O. bolivari* by the length of the fore wings, which are 1·4–1·8 times longer than the pronotum (0·9–1·3 in *O. bolivari*), and from *O. minutissimus* by the much less strongly sigmoid ventral profile of the lower valves of the ovipositor (see Ragge, 1986, figs 4, 5).

In the field the male calling song enables this species to be easily distinguished from *O. minutissimus*, which produces echeme-sequences of a quite different kind; from *O. bolivari* it may be distinguished by the much slower syllable repetition rate.

For a note on the spelling of the specific name of this species see p. 10.

SONG (Figs 921–929. CD 2, track 22).   The male calling song is an echeme lasting 1–2 s and consisting of about 10–15 syllables repeated at the rate of 6–7/s. Each echeme begins quietly, rapidly increasing in intensity. Oscillographic analysis shows that each downstroke hemisyllable has a large number of gaps (commonly as many as 8) and that this pattern of gaps is maintained until the end of the echeme (Figs 924, 925, 927, 928). The echemes are repeated at irregular intervals, varying from a few seconds to over a minute.

In the presence of a female the male produces a series of slightly longer echemes (lasting 2–3 s) at much more regular intervals (usually about 6–8 in 40 s) (Fig. 923). The echemes consist of about 15–20 syllables and at the end of some of them (usually about half) there is a group of 3–4 syllables of a different kind (Figs 926, 929), repeated more rapidly (at the rate of about 16/s). In the intervals between the echemes are quieter 'ticking' sounds repeated fairly regularly at the rate of about 3–4/s. The number of echemes in the series is variable but can be more than 20. The series of echemes is followed by a variable period (often more than a minute) in which the quiet ticking continues and which ends in a number of sequences of a different kind of sound, still quiet, and then an attempt at copulation. If this is unsuccessful the cycle begins again with another series of echemes.

DISTRIBUTION. Known only from Spain, in the higher parts of the Sierra de Guadarrama, Sierra de Gredos and (much more locally) Sierra Nevada, usually at altitudes above 2000 m.

Omocestus minutissimus (Bolívar)

Gomphocerus (Omocestus) minutissimus Bolívar, 1878: 424.

Small Mountain Grasshopper.

REFERENCES TO SONG. **Oscillogram**: Clemente, 1987 (as *burri*); Clemente *et al.*, 1989a (as *b.*); Ragge, 1986; Waeber, 1989. **Leg-movement**: Waeber, 1989.

RECOGNITION. For the distinction between this species and the rather similar *O. bolivari* and *O. uhagonii*, see under those species.

For a discussion of the taxonomic problems presented by *O. minutissimus*, see p. 73.

SONG (Figs 930–938. CD 2, track 23).   The rather quiet male calling song is a sequence of about 5–15 echemes lasting about 2–8 s. The sequence usually begins very quietly, reaching maximum intensity after about 2–6 echemes. The echeme repetition rate is about 1·5–2·5/s, usually becoming a little slower towards the end of the sequence. Each echeme

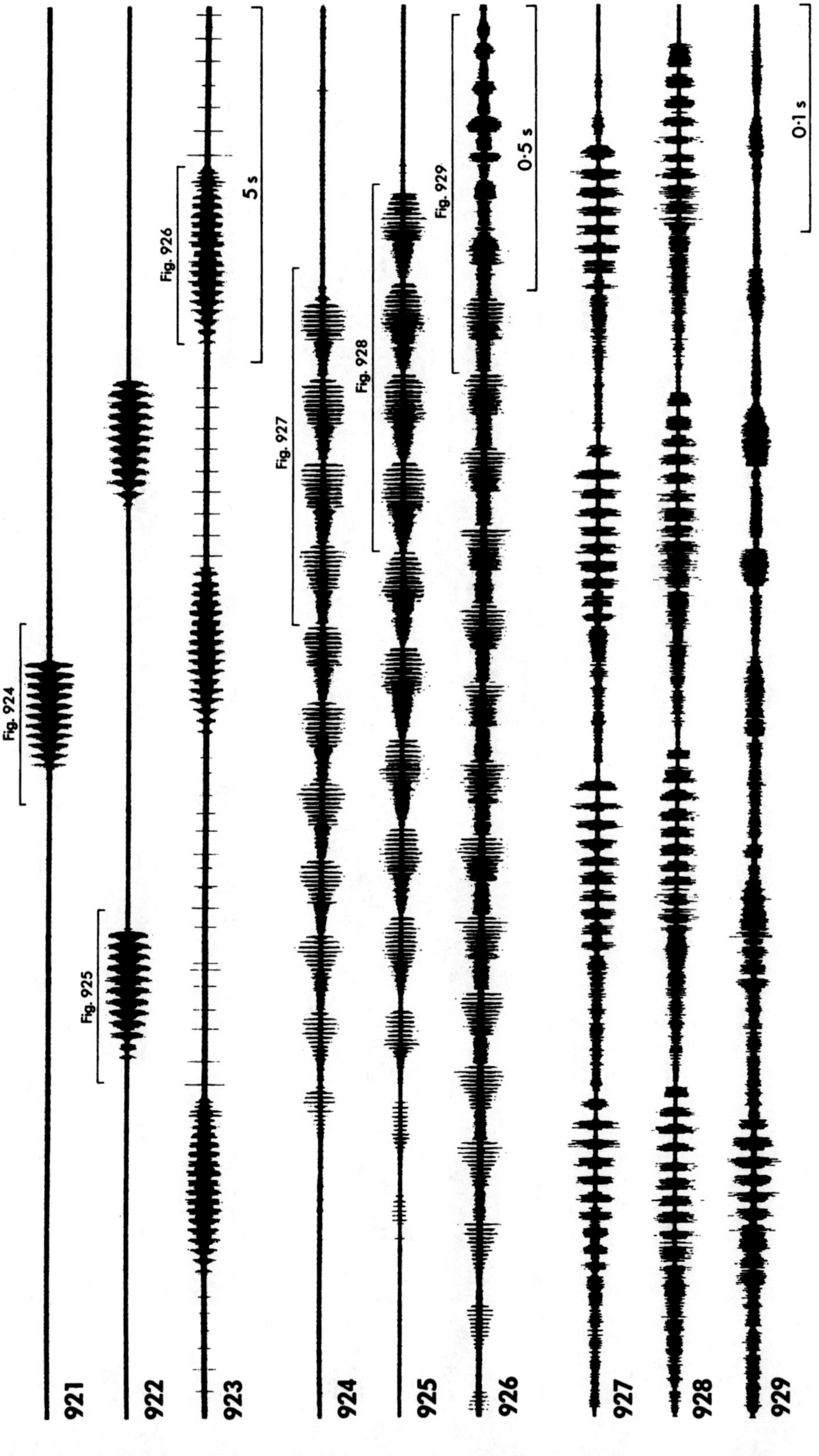

Figures 921–929 Oscillograms at three different speeds of the songs of three males of *Omocestus uhagonii*. **921, 922, 924, 925, 927, 928**. Calling songs. **923, 926, 929**. Part of the courtship song. (From Ragge, 1986)

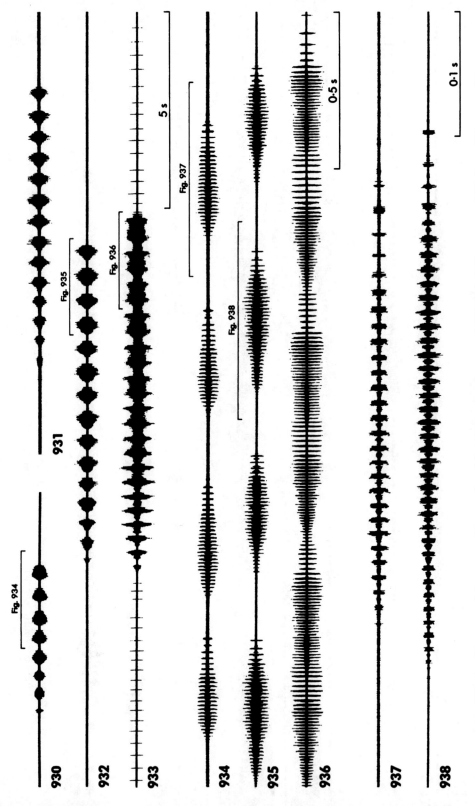

Figures 930–938 Oscillograms of the songs of four males of *Omocestus minutissimus*. **930–932**. Calling songs. **933**. Part of the courtship song. **934–938**. Faster oscillograms of the indicated parts of the songs shown in Figs 930–933. (The oscillograms of calling songs are from Ragge, 1986)

begins and ends quietly, giving a spindle-shaped oscillogram (Figs 934, 935). Each of the later echemes in the sequence lasts about 300–550 ms and contains about 20–50 syllables repeated at the rate of about 75–110/s.

In the presence of a female the male produces an echeme-sequence of a quite different kind (Figs 933, 936). It begins quietly rather like the calling song, but as it becomes louder it rapidly develops into a sequence in which the sound is continuous, though regularly fluctuating in loudness and in syllable repetition rate. The whole sequence lasts about 8–10 s, with the syllable repetition rate fluctuating from about 25/s to about 90/s. In the intervals between successive echeme-sequences of this kind the male produces 'ticks' at the rate of about 4/s. We have observed the courtship song in only one male from Torre de Tamurcia, Lérida province, and it is quite possible that it varies in different parts of Spain.

The calling song of *O. minutissimus* is similar to that of *Myrmeleotettix maculatus*, though quieter and, on average, of shorter duration and composed of fewer echemes; the courtship song, however, is quite different.

DISTRIBUTION. Widely distributed in mountains in the eastern half of Spain, including the Sistema Central, where it occurs as far west as the Sierra de Gredos.

Omocestus femoralis Bolívar

Omocestus femoralis Bolívar, 1908: 317.

Stripe-legged Grasshopper.

REFERENCES TO SONG. **Oscillogram**: Clemente, 1987. **Verbal description only**: Clemente *et al.*, 1990.

RECOGNITION. This species is brachypterous, the fore wings not reaching the hind knees. It may be distinguished from most other brachypterous species of *Omocestus* by having the hind wings, when flexed, reaching as far back as the tips of the flexed fore wings (often slightly beyond them in dried specimens). From *O. minutissimus*, in which the hind wings sometimes also reach the fore wing tips, it may be distinguished by its larger size and the much less strongly sigmoid ventral profile of the lower valves of the ovipositor; in the field the male calling song is also quite different from that of *O. minutissimus*.

SONG (Figs 939, 940, 942, 943, 945, 946. CD 2, track 24). The male calling song is an echeme lasting about 1·5–3·0 s and composed of about 17–40 syllables repeated at the rate of about 12–15/s. The echeme begins quietly, soon reaching maximum intensity. Oscillographic analysis shows that there are about 2–4 gaps in each downstroke hemisyllable and that they are maintained throughout the echeme or almost so. The echemes are repeated at irregular intervals, varying from a few seconds to more than a minute.

DISTRIBUTION. Known only from mountains in south-east Spain, at altitudes above 1000 m.

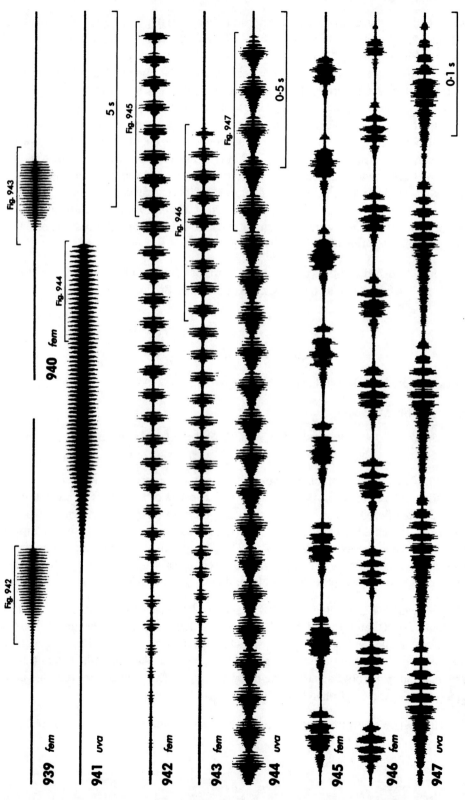

Figures 939–947 Oscillograms at three different speeds of the male calling songs of (939, 940, 942, 943, 945, 946) *Omocestus femoralis* and (941, 944, 947) *O. uvarovi*.

Omocestus uvarovi Zanon

Omocestus uvarovi Zanon, 1926: 181.

Uvarov's Grasshopper.

REFERENCES TO SONG. **Oscillogram**: Schmidt, 1996; Waeber, 1989. **Leg-movement**: Waeber, 1989.

RECOGNITION. This species is unique in *Omocestus* in having an enlarged medial area in the fore wings. This is particularly conspicuous in the male, in which the cross-veins of the medial area are regularly ladder-like and Cu_1 runs very close to Cu_2, the two veins sometimes coalescing in places. The medial area of the female fore wings is also enlarged, but less conspicuously so; Cu_1 and Cu_2 are not quite so close to each other and the medial cross-veins are not arranged entirely regularly. The species is otherwise rather nondescript and brown-coloured.

SONG (Figs 941, 944, 947. CD 2, track 25). The rather quiet male calling song is an echeme lasting about 6–8 s and composed of about 60–70 syllables repeated at the rate of about 9–10/s. The echeme begins quietly, reaching maximum intensity during the first quarter of its duration. Oscillographic analysis shows that there are about 4–5 gaps during each downstroke hemisyllable and that they are maintained throughout the echeme. The echemes are repeated at irregular intervals.

The courtship song is described by Waeber (1989).

DISTRIBUTION. Known only from Italy, where it occurs locally from Tuscany southwards.

Stenobothrus lineatus (Panzer)

Gryllus lineatus Panzer, 1796: 9.

Stripe-winged Grasshopper; (F) Criquet de la Palène; (D) Heidegrashüpfer; (NL) Zoemertje; (SF) Juovaheinäsirkka.

REFERENCES TO SONG. **Oscillogram**: Bukhvalova & Zhantiev, 1993; Clemente, 1987; Dumortier, 1963c; Elsner, 1974a, 1975; Elsner & Popov, 1978; Grein, 1984; Haskell, 1957, 1958, 1961; Kleukers *et al.*, 1997; Ragge, 1987a; Vedenina & Zhantiev, 1990; Waeber, 1989. **Diagram**: Bellmann, 1985a, 1988, 1993a; Bellmann & Luquet, 1995; Duijm & Kruseman, 1983; Haskell, 1957; Jacobs, 1950b, 1953a; Luquet, 1978; Ragge, 1965. **Leg-movement**: Elsner, 1974a, 1975; Elsner & Popov, 1978; Waeber, 1989. **Frequency information**: Dumortier, 1963c; Haskell, 1957, 1958; Meyer & Elsner, 1996. **Musical notation**: Yersin, 1854b. **Verbal description only**: Baier, 1930; Broughton, 1972b; Chopard, 1922; Faber, 1928, 1953a; Harz, 1957; Haskell, 1955a; Weber, 1984. **Disc recording**: Andrieu & Dumortier, 1963 (LP), 1994 (CD); Bellmann, 1993c (CD); Bonnet, 1995 (CD); Grein, 1984 (LP); Odé, 1997 (CD); Ragge *et al.*, 1965 (LP). **Cassette recording**: Bellmann, 1985b, 1993b; Burton & Ragge, 1987.

RECOGNITION (Plate 3: 6). The dozen or so western European species of *Stenobothrus* can be recognized reliably as members of this genus only by the ovipositor, which has an extra lateral tooth on all four valves. Males can sometimes be confused with *Omocestus*, but in most cases can be recognized as belonging to *Stenobothrus* by the widened medial area of the fore wings. Both sexes of most species have somewhat broader pale bands along the pronotal lateral carinae than are shown by *Omocestus* and other related genera. See Ragge (1987*a*) for further discussion and a morphological key to the principal western European species of *Stenobothrus*.

Males of *S. lineatus* may be distinguished from the rather similar species *S. nigromaculatus* and *S. fischeri* by the cerci, which are simply conical (laterally compressed towards the tip in the other two species). In both sexes the veins Cu_1 and Cu_2 are more completely fused together in the fore wings than in *S. nigromaculatus* and *S. fischeri*, and the hind wings are more strongly darkened. In the field males may be recognized at once by their highly characteristic calling song.

SONG (Figs 948–960. CD 2, track 26). The male calling song of this species is unique among the western European Gomphocerinae in both the slowness of the leg-movements and the quality of the sound produced. It consists of an echeme lasting about 10–25 s and composed of about 15–35 syllables repeated at the rate of about 1·2–1·5/s. The echeme usually begins quietly but reaches maximum intensity within a few syllables. The sound is usually continuous, with a highly characteristic, rather wheezy quality, and showing slight fluctuations in intensity corresponding with the leg-movements – though partly masked by the fact that one hind leg always moves slightly in advance of the other.

In the presence of a female the male usually produces longer echemes (with a more gradual crescendo at the beginning), sometimes lasting up to a minute or more but otherwise similar to the calling song. Between these echemes the male produces a quite different 'ticking' sound (Figs 951, 952, 956, 957) by moving the hind legs through a much smaller angle than in the calling song and at a much higher rate (usually 5–7 syllables/s). This ticking sound lasts typically for about half a minute before another echeme of the calling song type is produced, but it may be greatly prolonged, sometimes lasting for even longer than an hour. The ticking sound and the 'calling song' echemes usually alternate several times before the male attempts to copulate with the female. The courtship song of this species and the leg-movements that produce it have been described in great detail by Elsner (1974*a*, 1975).

DISTRIBUTION. Widespread in Europe south of latitude 55°N, but confined to mountains in the southern peninsulas; local in southern England. Eastwards, the range extends through Kazakhstan to southern Siberia and Mongolia.

Stenobothrus nigromaculatus (Herrich-Schäffer)

Acridium nigromaculatum Herrich-Schäffer, 1840: 10.

Black-spotted Grasshopper; (F) Sténobothre bourdonneur; (D) Schwarzfleckiger Grashüpfer.

REFERENCES TO SONG. **Oscillogram**: Bukhvalova & Zhantiev, 1993; Clemente, 1987; Grein, 1984; Ragge, 1987*a*; Schmidt & Schach, 1978; Waeber, 1989. **Diagram**: Bellmann, 1985*a*, 1988, 1993*a*; Bellmann & Luquet, 1995; Duijm & Kruseman, 1983; Jacobs, 1950*b*, 1953*a*.

Figures 948–952 Oscillograms of the songs of five males of *Stenobothrus lineatus*. **948–950**. Calling songs of males with (948, 949) two hind legs and (950) one hind leg. **951, 952**. Parts of two courtship songs. (From Ragge, 1987*a*)

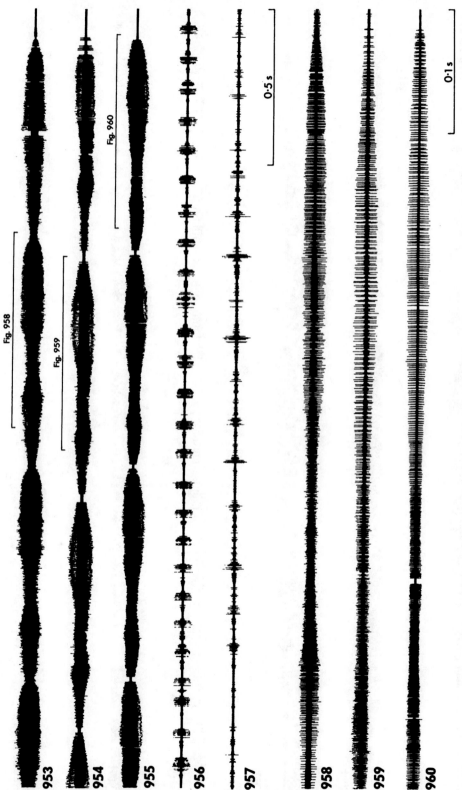

Figures 953–960 Faster oscillograms of the indicated parts of the songs of *Stenobothrus lineatus* shown in Figs 948–952. Figs 955 and 960 are from a male with only one hind leg. (From Ragge, 1987a)

Leg-movement: Waeber, 1989. **Frequency information**: Meyer & Elsner, 1996. **Verbal description only**: Faber, 1928, 1953a; Harz, 1957; Luquet, 1991. **Disc recording**: Bellmann, 1993c (CD); Bonnet, 1995 (CD); Grein, 1984 (LP). **Cassette recording**: Bellmann, 1985b, 1993b.

RECOGNITION. Males of this species may be distinguished from *S. lineatus* by the cerci, which are laterally compressed towards the tip (simply conical in *S. lineatus*). In the fore wings of both sexes the veins Cu_1 and Cu_2 are less completely fused together than in *S. lineatus*, and the stigma is nearer the wing-tip than in *S. fischeri* (see Ragge, 1987a, Figs 5, 6).

In the field males may be recognized quite easily by their calling song, in which the syllable repetition rate (about 70–110/s) is much higher than in *S. lineatus* (about 1·2–1·5/s) and *S. fischeri* (about 10–15/s).

SONG (Figs 961–974. CD 2, track 27). The male calling song consists of a series of 2–6 echemes (each beginning quietly but showing a rapid crescendo), separated by intervals of 1–3 s. Each echeme lasts about 0·8–2·0 s (rarely up to 2·5 s) and consists of about 90–200 syllables (rarely up to 250) repeated at the rate of about 70–110/s.

In the presence of a female the male produces a quite complicated courtship song, consisting of two alternating phases (Figs 965, 966, 971–974). The first phase consists of a series of fairly quiet echemes separated by intervals of about 0·6–0·9 s; during these intervals a few even quieter sounds are often produced. Each echeme lasts for about 0·3–0·9 s and is composed of about 40–80 syllables repeated at the rate of about 80–100/s. This phase lasts for an indefinite period, usually more than 30 s and often several minutes. The second phase, which follows abruptly, consists of a series of about 3–7 louder and longer echemes, each of which is preceded by a much shorter echeme. The shorter echemes last for less than half a second, and in fact after the first longer echeme they are often reduced to little more than a single syllable. The longer echemes usually last for about 0·8–2·2 s and are subdivided into two parts, one with slower syllables (about 40–60/s) and one with faster syllables (about 70–120/s). At the end of this second phase the male either attempts to copulate with the female or, after a short pause, begins the first phase again.

Note that the oscillograms shown in Figs 966, 973 and 974 were taken from a song recorded in hazy sunshine, and this may account, at least partly, for the difference in temporal pattern between these oscillograms and those shown in Figs 965, 971 and 972, which were taken from a song recorded in full sunshine.

DISTRIBUTION. Quite widespread, but local, in western Europe, not reaching as far north as northern France, the Low Countries or northern Germany; southwards, it occurs in northern Spain and as far south as Gran Sasso in Italy. Absent from the British Isles. Eastwards, the range extends through central and eastern Europe (including all but the southernmost parts of the Balkan Peninsula) to southern Siberia.

Stenobothrus fischeri (Eversmann)

Oedipoda fischeri Eversmann, 1848: 11.

Fischer's Grasshopper; (F) Sténobothre cigalin; (D) Südlicher Grashüpfer.

REFERENCES TO SONG. **Oscillogram**: Ragge, 1987a; Waeber, 1989. **Diagram**: Luquet, 1978.

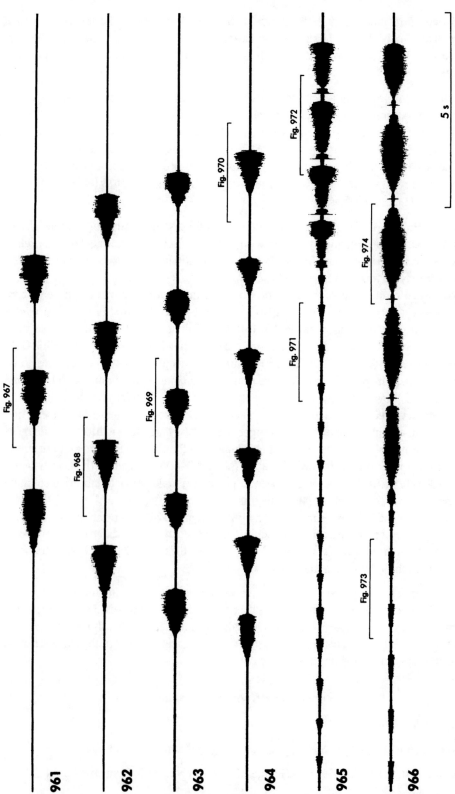

Figures 961–966 Oscillograms of the songs of six males of *Stenobothrus nigromaculatus.* **961–964.** Calling songs. **965, 966.** Parts of two courtship songs (see the remarks on p. 337). (From Ragge, 1987*a*)

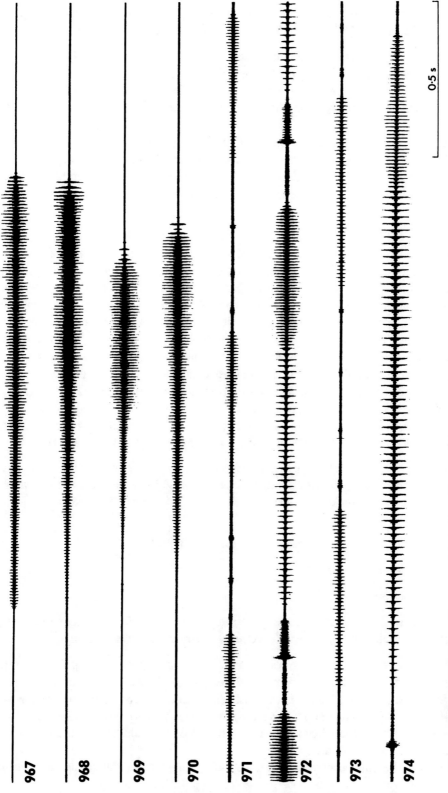

0·5 s

Figures 967–974 Faster oscillograms of the indicated parts of the songs of *Stenobothrus nigromaculatus* shown in Figs 961–966. (From Ragge, 1987a)

967 968 969 970 971 972 973 974

Leg-movement: Waeber, 1989. **Verbal description only**: Bellmann, 1993a; Bellmann & Luquet, 1995; Faber, 1953a; Harz, 1957. **Disc recording**: Bonnet, 1995 (CD).

RECOGNITION. Males of this species may be distinguished from *S. lineatus* by the cerci, which are laterally compressed towards the tip (simply conical in *S. lineatus*). In the fore wings of both sexes the veins Cu_1 and Cu_2 are less completely fused together than in *S. lineatus* and the stigma is further from the wing-tip than in *S. nigromaculatus* (see Ragge, 1987a, Figs 5, 6).

In the field the single echeme of the male calling song, which lasts for 3–5 s and is composed of more than 40 syllables, enables this species to be distinguished from *S. lineatus*, *S. nigromaculatus*, *S. grammicus* and *S. bolivarii*. The calling songs of *S. festivus* and *S. stigmaticus* are more similar, but in both these species the echeme is shorter, with fewer than 40 syllables.

SONG (Figs 975–983. CD 2, track 28). The male calling song is a single echeme lasting about 3–5 s and consisting of about 40–55 syllables repeated at the rate of about 12–13/s. The echeme begins quietly, usually reaching maximum intensity about halfway through its duration. Oscillographic analysis shows that there are gaps in each downstroke hemisyllable. Previously published accounts of the calling song of this species have been based mostly on Faber (1953a), who gave lower figures for both the duration of the echeme ('2 oder etwas mehr Sekunden') and the syllable repetition rate ('8–9½'/s), although suggesting that these figures applied to males singing in full sunshine. Faber's observations were made in the laboratory on males collected for him from Oberweiden, east of Vienna in Lower Austria, whereas ours were made in the field in southern France; in France we should expect such a low syllable repetition rate only in dull weather or at unusually low air temperatures.

In the presence of a female the male produces a long and complicated courtship song (Figs 977, 980, 983), which has been fully described by Faber (1953a: Beilage 4).

DISTRIBUTION. In western Europe this species seems to be confined to southern France and central and northern Spain. Farther east it occurs from Austria eastwards across eastern Europe (including the Balkan Peninsula) to Kazakhstan, Siberia and Mongolia.

Stenobothrus festivus Bolívar

Stenobothrus festivus Bolívar, 1887: 94.

Festive Grasshopper; (F) Sténobothre occitan.

REFERENCES TO SONG. **Oscillogram**: Clemente, 1987; Ragge, 1987a; Waeber, 1989. **Leg-movement**: Waeber, 1989. **Verbal description only**: Clemente *et al.*, 1989b.

RECOGNITION. This species is perhaps most likely to be confused with *S. grammicus* or *S. bolivarii*, which share with it the strongly inflexed pronotal lateral carinae. However, males of *S. festivus* lack the modified or characteristically coloured last palpal segments shown by the other two species, and in both sexes the pronotal lateral carinae are more sharply angled in the prozona, where they are clearly interrupted by an additional transverse sulcus on each side of the pronotum.

In the field the single echeme of the male calling song enables this species to be distin-

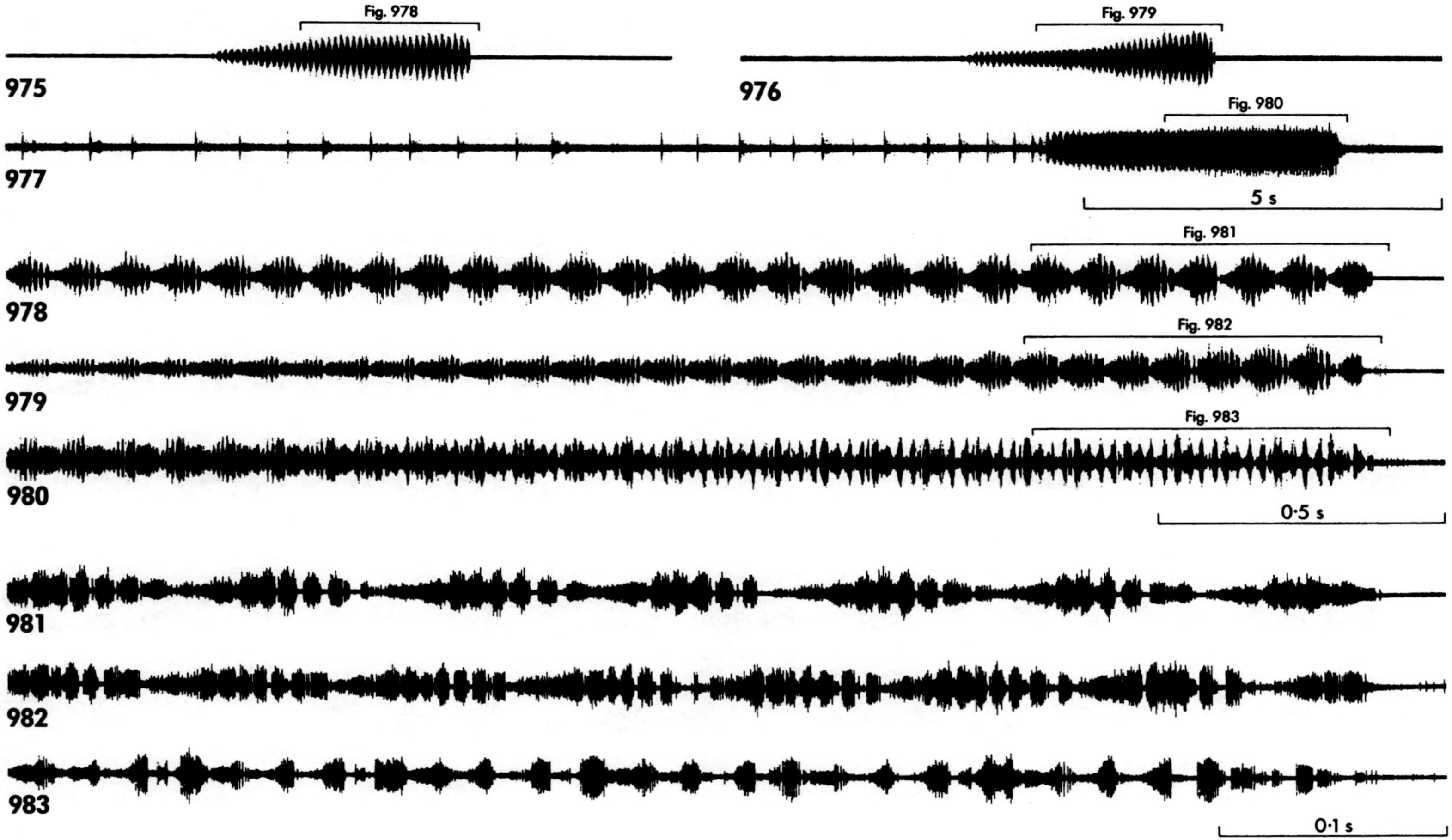

Figures 975–983 Oscillograms at three different speeds of the songs of three males of *Stenobothrus fischeri*. 975, 976, 978, 979, 981, 982. Calling songs. 977, 980, 983. Courtship song. (From Ragge, 1987a)

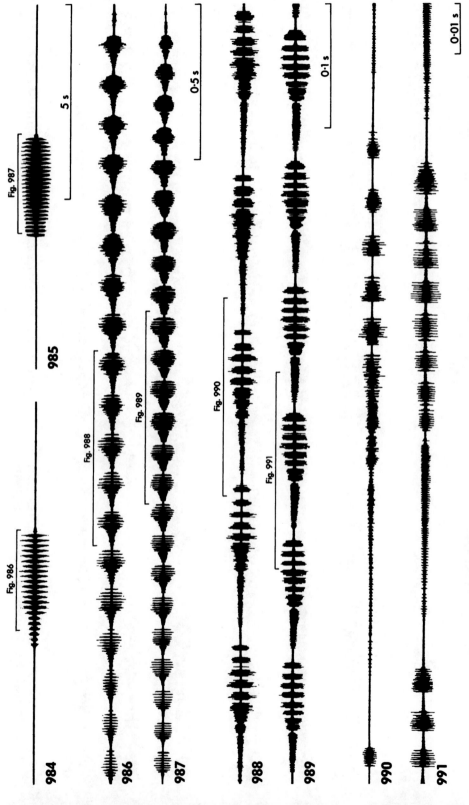

Figures 984–991 Oscillograms at four different speeds of the calling songs of two males of *Stenobothrus festivus*, one (984, 986, 988, 990) with two hind legs and the other (985, 987, 989, 991) with one hind leg. (From Ragge, 1987a)

guished easily from *S. grammicus* and *S. bolivarii*, which both produce a rapid sequence of echemes. The calling songs of *S. fischeri* and *S. stigmaticus* are more similar, but *S. fischeri* has a longer echeme, with more than 40 syllables (usually fewer than 30 in *S. festivus*) and in *S. stigmaticus* the syllable repetition rate is usually more than 12/s (less than 12/s in *S. festivus*).

SONG (Figs 984–991. CD 2, track 29). The calling song is a single echeme lasting about 2–3 s and consisting of about 20–30 syllables repeated at the rate of about 7–11/s. The echeme usually begins quietly, reaching maximum intensity about halfway through its duration or a little earlier, but sometimes it begins more abruptly and sometimes there are one or more louder syllables before the quiet beginning. Oscillographic analysis shows that there is a characteristic pattern of gaps in the downstroke hemisyllables: at the beginning of the echeme these hemisyllables are completely broken up by gaps, but as the echeme progresses the gaps gradually disappear so that there are usually no gaps in the last few syllables. In a song produced by only one hind leg the gaps often persist until rather later in the echeme (Fig. 987), since there is no tendency for the gaps in the syllables produced by one hind leg to be obscured by sounds produced by the other.

DISTRIBUTION. Widespread in the Iberian Peninsula and occurring locally in southern France.

Stenobothrus grammicus Cazurro

Stenobothrus grammicus Cazurro, 1888: 457.

Dark-palped Grasshopper; (F) Gomphocère fauve-queue; (D) Gezeichneter Grashüpfer.

REFERENCES TO SONG. **Oscillogram**: Ragge, 1987a; Waeber, 1989. **Diagram**: Luquet, 1978. **Leg-movement**: Waeber, 1989. **Verbal description only**: Bellmann, 1993a; Bellmann & Luquet, 1995.

RECOGNITION. Both sexes of this species may be recognized easily by the palps, which are black or dark brown at the tip. In addition the male antennae are curved outwards and noticeably thickened and darkened towards the tip.

In the field the distinctive calling song, consisting of about 10–30 very brief and rapidly repeated echemes, enables males to be easily distinguished from most other species of *Stenobothrus*; however, oscillographic analysis is necessary to show clearly the difference from the calling song of *bolivarii*, which has longer echemes with a rather different syllable structure (see below).

SONG (Figs 992–994, 998–1000, 1004–1006, 1010, 1011. CD 2, track 30). The calling song consists of a sequence of about 10–30 echemes lasting about 5–15 s; the echemes are thus repeated at the rate of about 2/s. Each echeme lasts about 40–80 ms and is composed of a series of 4–8 syllables. The only other species of *Stenobothrus* known at present to have a rather similar calling song is *S. bolivarii*, but oscillographic analysis shows that species to have longer echemes, lasting about 80–110 ms, with a rather different syllable structure: in *S. grammicus* the loudest syllables are towards the end of each echeme (Figs 1004, 1005), whereas in *S. bolivarii* the loudest syllable is at the beginning (Figs 1007–1009).

Luquet (1978: 427) gives the rather lower number of 6–8 echemes in each sequence of the calling song, based on observations in the Mont Ventoux area of Vaucluse in S.E. France. Our own studies of four males at two localities, also in the region of Mont Ventoux, suggest that this number is unusually low.

In the presence of a female the male produces a series of quieter and much longer echemes (Figs 994, 1000, 1006). Each of these echemes lasts for about 1–2 s and consists of a soft, continuous and rather amorphous sound. Between the echemes there are intervals of varying duration (usually between 3 and 15 s), during which quieter sounds are often produced.

DISTRIBUTION. Known only from upland areas in southern France and the Iberian Peninsula.

Stenobothrus bolivarii (Brunner)

Gomphocerus (Stenobothrus) Bolivarii Brunner, *in* Bolívar, 1876: 327.

Pink-palped Grasshopper.

REFERENCES TO SONG. **Oscillogram**: Clemente, 1987; Ragge, 1987a, Waeber, 1989. **Leg-movement**: Waeber, 1989. **Verbal description only**: Clemente *et al.*, 1989b; Defaut, 1987, 1988b.

RECOGNITION. Males of this species may be recognized by the maxillary palps, in which the last segment is bulbous and usually coloured orange-pink or reddish brown. Females are more difficult to recognize, as the last segment of the maxillary palps is more normal in shape and, although sometimes distinctively coloured like that of the male, is not always so; however, in the upland areas of the Iberian Peninsula where *bolivarii* occurs the only other species of *Stenobothrus* with which females are likely to be confused is *S. grammicus*, in which both pairs of palps have black or dark brown tips in both sexes.

In the field the short, rapidly repeated echemes of the calling song enable males to be easily distinguished from most other species of *Stenobothrus*; the calling song of *S. grammicus* is, however, rather similar and oscillographic analysis is necessary to show clearly that *S. bolivarii* has longer echemes with a different syllable structure.

For a note on the spelling of the specific name of this species see p. 10.

SONG (Figs 995–997, 1001–1003, 1007–1009, 1012–1014. CD 2, track 31). The male calling song is quite similar to that of *S. grammicus*, consisting of a sequence of about 10–20 echemes lasting about 4–8 s; the echemes are repeated at the rate of about 2–3/s. Each echeme lasts about 80–150 ms and is composed of a series of about 5–10 syllables; the echemes are thus rather longer than those of *S. grammicus* and show a different syllable structure, with the loudest sound at the beginning of each echeme (Figs 1007–1009).

DISTRIBUTION. Known only from upland areas in the Iberian Peninsula.

Stenobothrus stigmaticus (Rambur)

Gryllus stigmaticus Rambur, 1838: 93.

Lesser Mottled Grasshopper; (F) Sténobothre nain; (D) Kleiner Heidegrashüpfer; (NL) Schavertje.

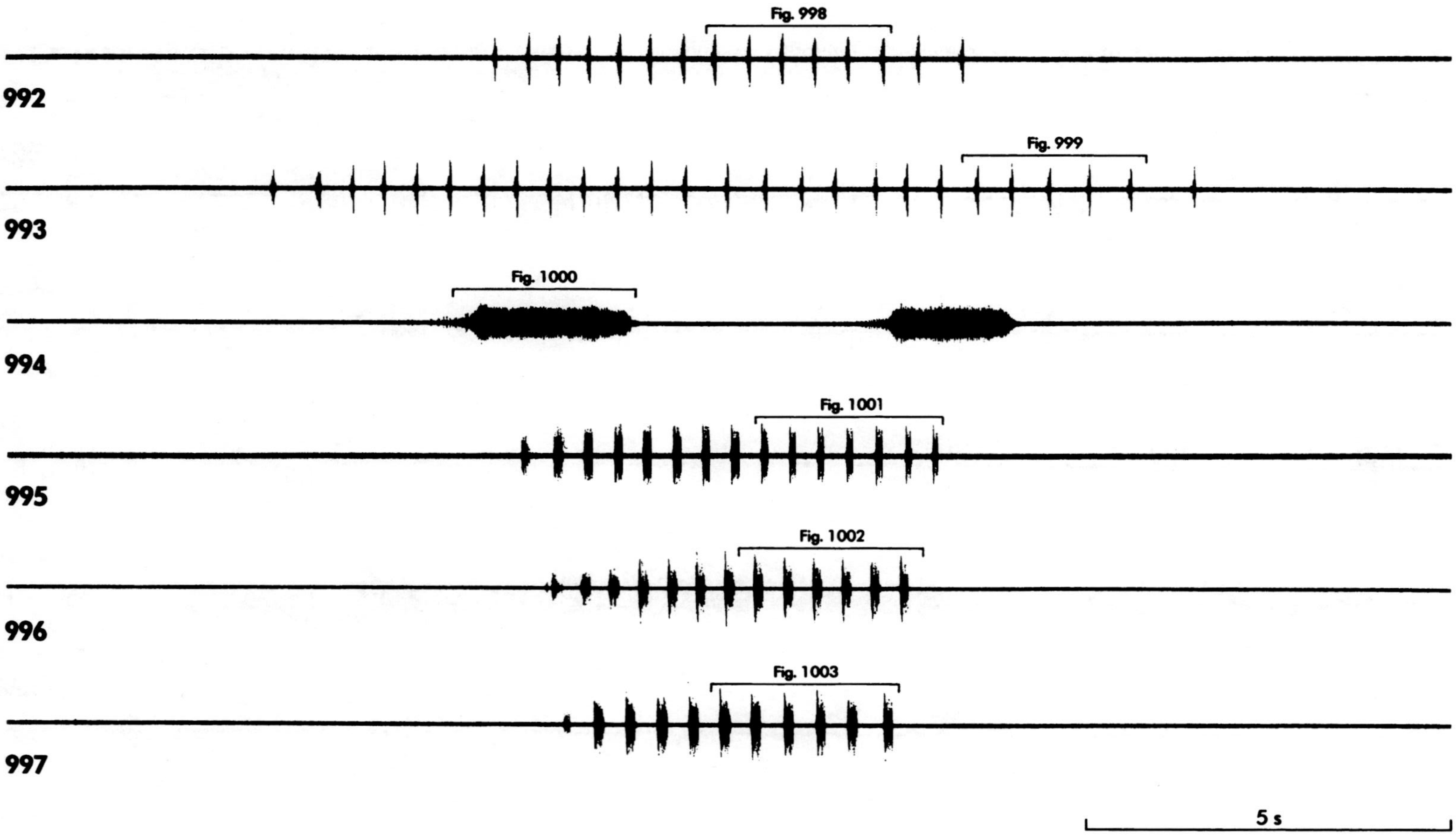

Figures 992–997 Oscillograms of the songs of (992–994) two males of *Stenobothrus grammicus* and (995–997) three males of *S. bolivarii*. All are of calling songs except for Fig. 994, which shows part of a courtship song produced by the same male as the calling song shown in Fig. 992. (Figs 992–995 are from Ragge, 1987*a*)

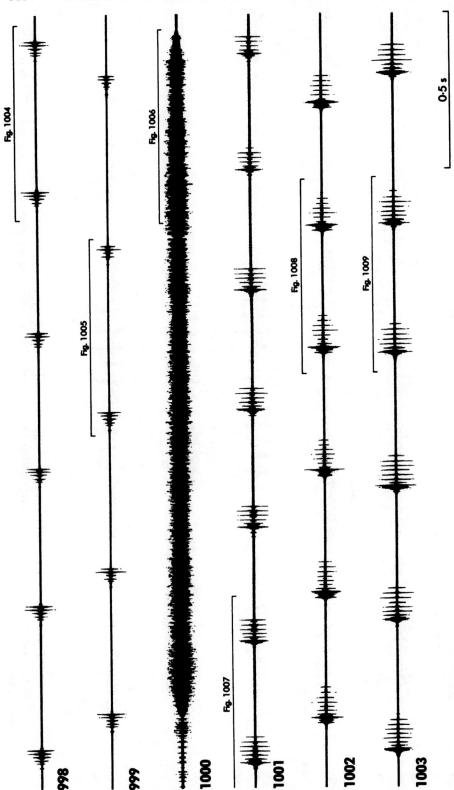

Figures 998–1003 Faster oscillograms of the indicated parts of the songs of *Stenobothrus grammicus* and *S. bolivarii* shown in Figs 992–997. (Figs 998–1001 are from Ragge, 1987a)

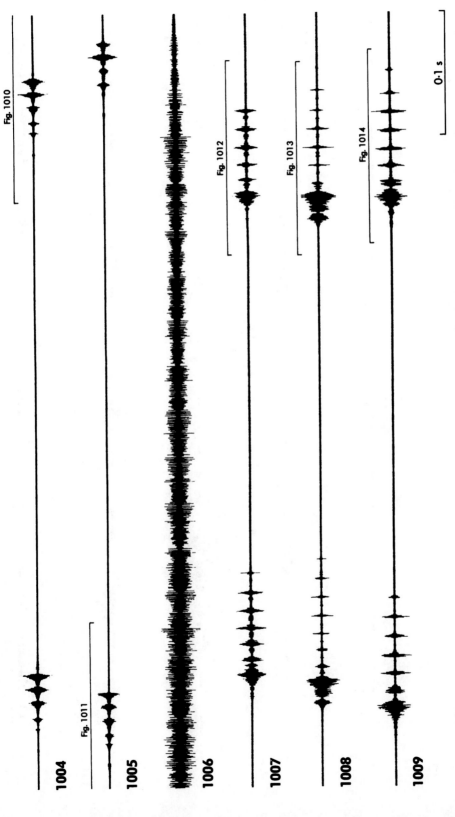

Figures 1004–1009 Faster oscillograms of the indicated parts of the songs of *Stenobothrus grammicus* and *S. bolivarii* shown in Figs 998–1003. (Figs 1004–1007 are from Ragge, 1987a)

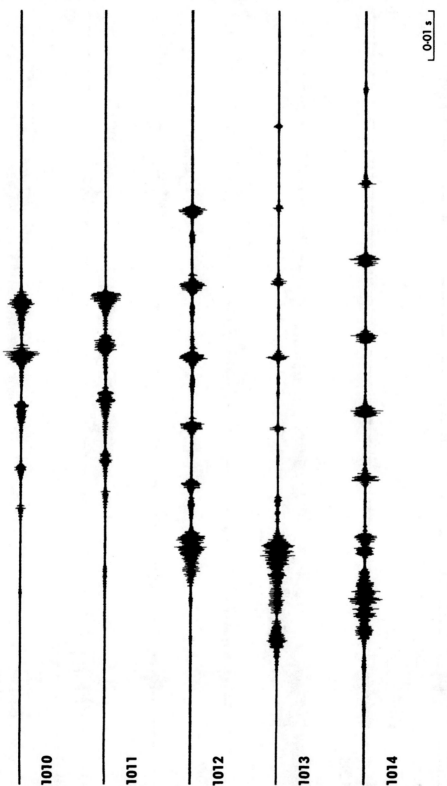

0·01 s

Figures 1010–1014 Faster oscillograms of the indicated parts of the songs of *Stenobothrus grammicus* and *S. bolivarii* shown in Figs 1004, 1005, 1007–1009. (Figs 1010–1012 are from Ragge, 1987*a*)

REFERENCES TO SONG. **Oscillogram**: Clemente, 1987; Grein, 1984; Kleukers *et al.*, 1997; Ragge, 1987*a*; Waeber, 1989. **Diagram**: Bellmann, 1985*a*, 1988, 1993*a*; Bellmann & Luquet, 1995; Duijm & Kruseman, 1983; Jacobs, 1950*b*, 1953*a*; Ragge, 1965. **Leg-movement**: Waeber, 1989. **Frequency information**: Meyer & Elsner, 1996. **Verbal description only**: Broughton, 1972*b*; Defaut, 1987, 1988*b*; Faber, 1928, 1953*a*; Harz, 1957; Weber, 1984. **Disc recording**: Bellmann, 1993*c* (CD); Bonnet, 1995 (CD); Grein, 1984 (LP); Odé, 1997 (CD); Ragge *et al.*, 1965 (LP). **Cassette recording**: Bellmann, 1985*b*, 1993*b*; Burton & Ragge, 1987.

RECOGNITION. *S. stigmaticus* is conspicuously smaller than any other fully winged species of *Stenobothrus* occurring in western Europe. Were it not for the size difference it could be confused with *S. nigromaculatus*, but the veins Cu$_1$ and Cu$_2$ are much more widely separated in the fore wings than in that species.

In the field the male calling song, usually consisting of a single echeme lasting 1–3 s, enables this species to be distinguished easily from all other species of *Stenobothrus* with overlapping distributions except *S. fischeri* and *S. festivus*; these species produce single echemes of a similar kind to *S. stigmaticus*, but in *S. fischeri* the echeme is longer, with more than 40 syllables (fewer than 40 in *S. stigmaticus*), and in *S. festivus* the syllable repetition rate is less than 12/s (usually more than 12/s in *S. stigmaticus*). Both the calling and courtship songs are extremely similar to those of *S. apenninus*, but as that species is confined to the Apennines, where *S. stigmaticus* does not occur, confusion between the songs of these two species is unlikely. (See also the remarks on p. 73.)

SONG (Figs 1015–1031. CD 2, track 32). The male calling song is a rather quiet echeme lasting about 1–3 s and consisting of about 25–40 syllables repeated at the rate of about 10–20/s (usually 14–18/s). The echeme usually begins quietly, reaching maximum intensity within about one second. Oscillographic analysis shows that there are usually gaps in the downstroke hemisyllables, especially near the beginning of the echeme, but they are not always very clear. The echemes are usually produced singly and repeated at irregular intervals, but occasionally two echemes are produced within a few seconds of each other (Fig. 1017).

In the presence of a female the male produces a series of echemes separated by intervals of about 2–7 s (Figs 1019, 1024, 1029). Each echeme is similar to that of the calling song, but is quieter, longer (usually lasting 3–5 s and composed of about 50–70 syllables) and with one hind leg moving more vigorously than the other, at least at the beginning and end of the echeme; the last few syllables are rather modified and then the hind leg that has been moving more vigorously produces a short, loud hemisyllable by a very quick downstroke. The more vigorously moving hind leg alternates from one side to the other in successive echemes. Sometimes the loud syllable at the end of each echeme is omitted.

DISTRIBUTION. Quite widespread in Europe south of latitude 55°N, but apparently absent from the southern half of Spain and from peninsular Italy, where it is replaced by *S. apenninus*. In the British Isles known only from the Isle of Man. Farther east it occurs in the more northerly parts of the Balkan Peninsula, Asia Minor, the Ukraine and the southern parts of European Russia. Also recorded from the Rif mountains in Morocco.

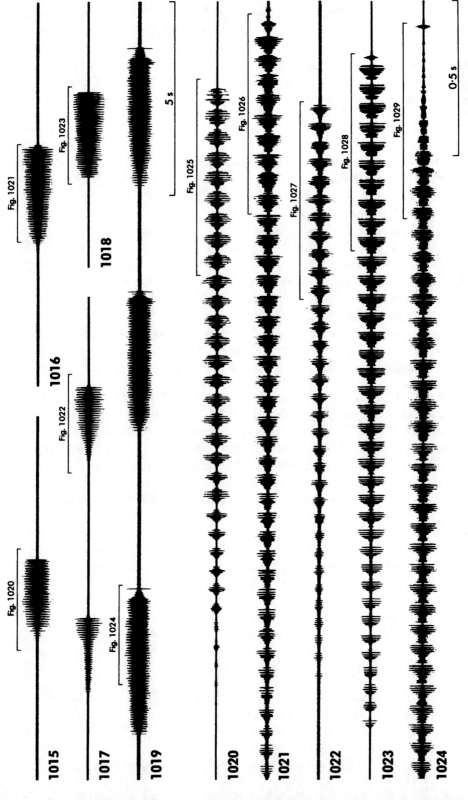

Figures 1015–1024 Oscillograms of the songs of five males of *Stenobothrus stigmaticus*. **1015–1018.** Calling songs of males with (1015–1017) two hind legs and (1018) one hind leg. **1019.** Part of the courtship song. **1020–1024.** Faster oscillograms of the indicated parts of the songs shown in Figs 1015–1019. (From Ragge, 1987a)

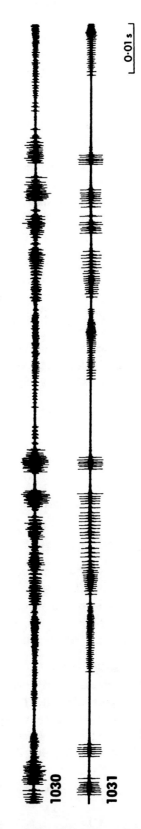

Figures 1025–1031 Faster oscillograms of the indicated parts of the songs of *Stenobothrus stigmaticus* shown in Figs 1020–1024. Figs 1028 and 1031 are from a male with only one hind leg. (From Ragge, 1987a)

Stenobothrus apenninus Ebner

Stenobothrus apenninus Ebner, 1915: 551.

Apennine Grasshopper.

REFERENCES TO SONG. **Oscillogram**: Ragge, 1987*a*; Waeber, 1989 (as *stigmaticus*). **Leg-movement**: Waeber, 1989 (as *s.*).

RECOGNITION. This montane species is endemic to the Apennine range in the Italian Peninsula, where it may be easily distinguished from other species of *Stenobothrus* by the vestigial hind wings of both sexes and the much reduced fore wings of the female. In the field the calling song of the males also enables this species to be distinguished easily from other Italian species of the genus occurring in the Apennines. (See also the remarks on p. 73.)

SONG (Figs 1032–1040. CD 2, track 33). The male calling song is a single echeme lasting about 1–2 s and consisting of about 20–30 syllables repeated at the rate of about 15–20/s. The echeme usually begins quietly but reaches maximum intensity within 0·5 s. Oscillographic analysis shows that there are clear gaps in the downstroke hemisyllables throughout the echeme.

In the presence of a female the male produces a quieter courtship song (Figs 1034, 1037, 1040) very similar to that of *S. stigmaticus* (see p. 349) but with shorter echemes (often lasting about 2 s and composed of about 30 syllables).

DISTRIBUTION. Known only from the Apennine range in Italy.

Stenobothrus ursulae Nadig

Stenobothrus ursulae Nadig, 1986: 214.

Ursula's Grasshopper; (D) Aostatal-Grashüpfer.

REFERENCES TO SONG. **Oscillogram** and **leg-movement**: Waeber, 1989. **Verbal description only**: Bellmann, 1993*a*.

RECOGNITION. This montane species is known only from the Italian Alps, and both sexes can be easily distinguished from other species of *Stenobothrus* occurring in this region by the much reduced fore wings and vestigial hind wings. The antennae are distinctly thickened towards the tip, especially in the male.

For remarks on the taxonomy of this species see p. 74.

SONG (Figs 1041–1050. CD 2, track 34). The male calling song is an echeme lasting about 2–3 s and consisting of about 18–30 syllables repeated at the rate of about 8–10/s. The echeme begins quietly, reaching maximum intensity within about 1·0–1·5 s. Oscillographic analysis shows that there are gaps in the downstroke hemisyllables throughout the echeme. The intervals between successive echemes vary from a few seconds to half a minute or more. (Waeber (1989) gives the longer echeme duration of 5–7 s for the calling song of this species.)

In the presence of a female the male produces quieter echemes in which there is no

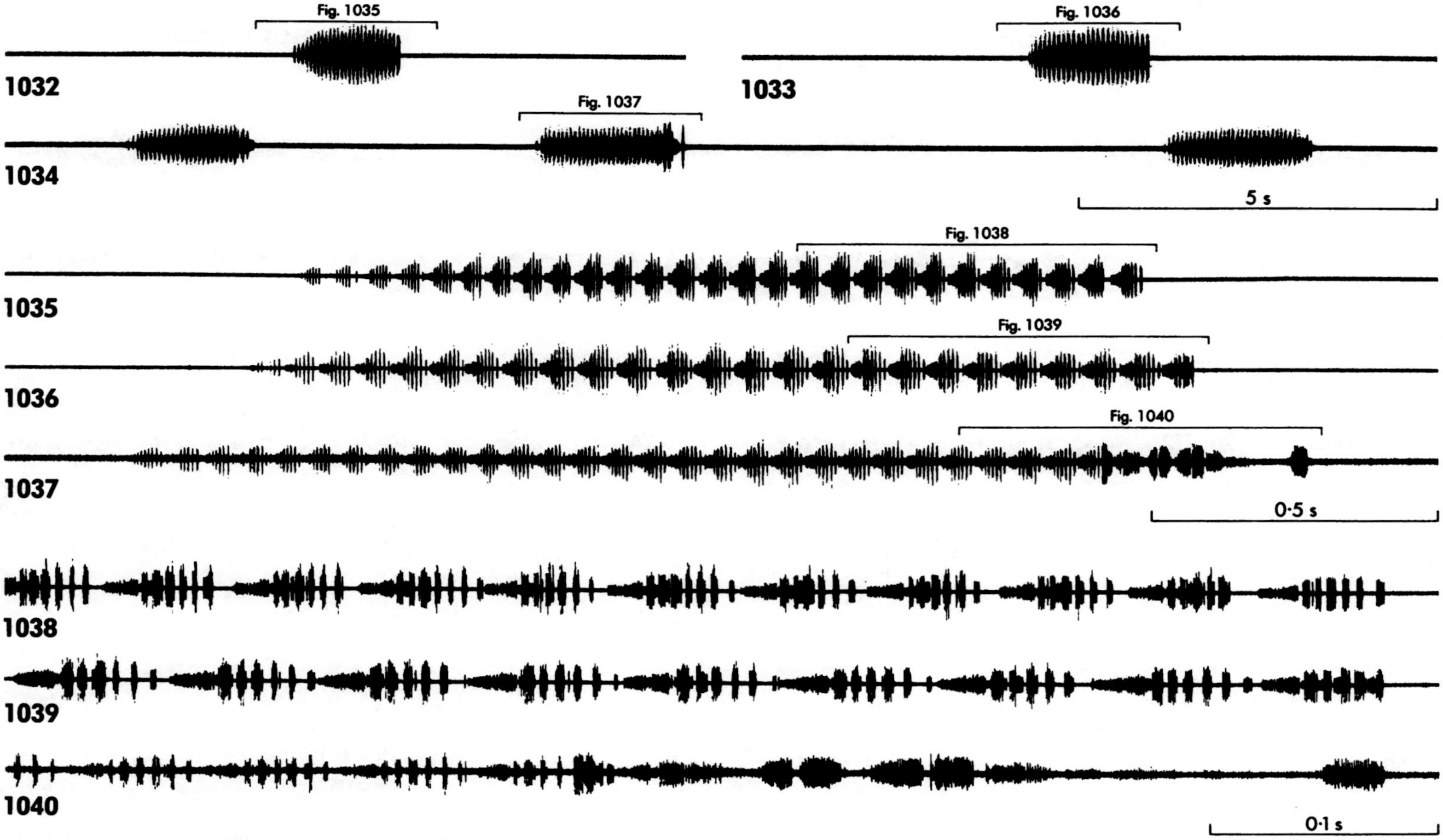

Figures 1032–1040 Oscillograms at three different speeds of the songs of three males of *Stenobothrus apenninus*. **1032, 1033, 1035, 1036, 1038, 1039.** Calling songs. **1034, 1037, 1040.** Part of the courtship song. (From Ragge, 1987a)

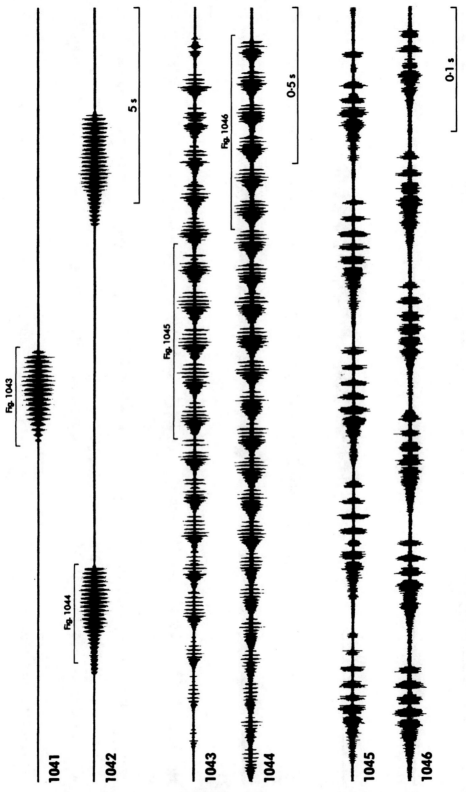

Figures 1041–1046 Oscillograms at three different speeds of the calling songs of two males of *Stenobothrus ursulae*.

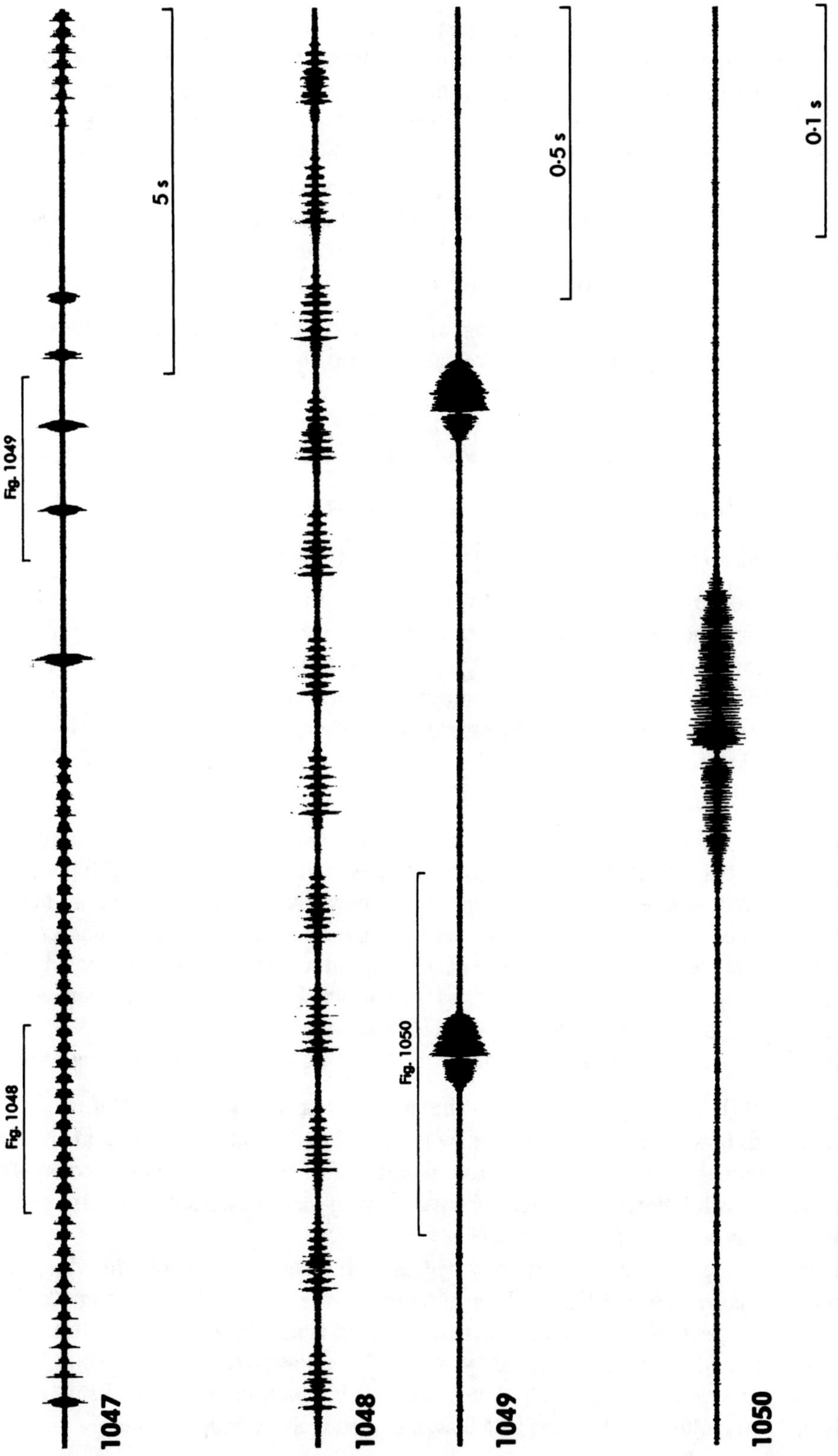

Figures 1047–1050 Oscillograms at three different speeds of part of the courtship song of a male of *Stenobothrus ursulae*.

crescendo and the syllable repetition rate is slower, about 4–7/s (Figs 1047–1050). The duration of the echemes is very variable, but usually within the range 3–8 s. Oscillographic analysis shows that the downstroke hemisyllables have even more gaps than in the calling song, often 5–7 per syllable. The echemes are repeated at intervals varying from 2 to 8 s, and during these intervals there are often isolated diplosyllables of a quite different kind: they have no gaps in either hemisyllable, are produced by only one leg and the sound has a different, rather 'hissy' quality (Figs 1049, 1050). Sometimes these syllables are in small groups of 2 or more in quick succession, and sometimes such a group occurs at the end of an echeme, in a manner reminiscent of the courtship song of *S. apenninus*.

DISTRIBUTION. Known at present only from a few localities in the Italian Alps, in the region between Turin and Aosta, all at altitudes above 1000 m.

Stenobothrus rubicundulus Kruseman & Jeekel

Stenobothrus rubicundulus Kruseman & Jeekel, 1967b: 79.

Wing-buzzing Grasshopper; (F) Sténobothre alpin; (D) Bunter Alpengrashüpfer.

REFERENCES TO SONG. **Oscillogram**: Elsner, 1974b (as *rubicundus*); Elsner & Popov, 1978 (as *rubicundus*); Elsner & Wasser, 1995a, 1995b, 1995c (as *rubicundus*); Ragge, 1987a; Waeber, 1989. **Leg-movement**: Elsner & Popov, 1978 (as *rubicundus*); Elsner & Wasser, 1995a, 1995b, 1995c (as *rubicundus*); Waeber, 1989. **Wing-movement**: Elsner & Wasser, 1995a, 1995b, 1995c (as *rubicundus*). **Frequency information**: Meyer & Elsner, 1996 (as *rubicundus*). **Verbal description only**: Bellmann, 1993a; Bellmann & Luquet, 1995; Chopard, 1952 (as *rubicundus*); Faber, 1953a (as *rubicundus*); Harz, 1957 (as *rubicundus*).

RECOGNITION. The very broad fore wings, with unbranched M, regular ladder-like cross-veins in the medial area, and Cu_1 and Cu_2 quite clearly separated, enable males of this species to be recognized quite easily. In the female these characters of the fore wing are not so clearly developed, but the hind wings of both sexes have an unbranched M and greatly enlarged medial area, and the more distal part of the anterior margin is heavily sclerotized and conspicuously dark in colour. *S. cotticus* is superficially similar, but in the fore wings Cu_1 and Cu_2 are fused together except towards the base and M is bifurcate.

In the field males may be easily recognized by their highly characteristic calling song.

SONG (Figs 1051–1064. CD 2, track 35). The male calling song is composed of two quite different kinds of sound: one is produced by the usual femoro-alary stridulation and the other by wing-vibration, either on the ground or in flight. The song sometimes consists of femoro-alary stridulation immediately followed by wing-vibration, but each kind of sound production may occur without the other.

The femoro-alary stridulation usually begins with a series of syllables in which the upstroke and downstroke of the hind legs produce clearly separated, simple sounds (Figs 1053, 1055); the syllable repetition rate is variable, but is usually 8–13/s. After a few seconds the pattern is often complicated by some of the downstroke hemisyllables being prolonged and broken up by gaps (Figs 1054, 1056). This femoro-alary echeme usually lasts for about 6–10 s, after which it may be followed immediately by wing-vibration.

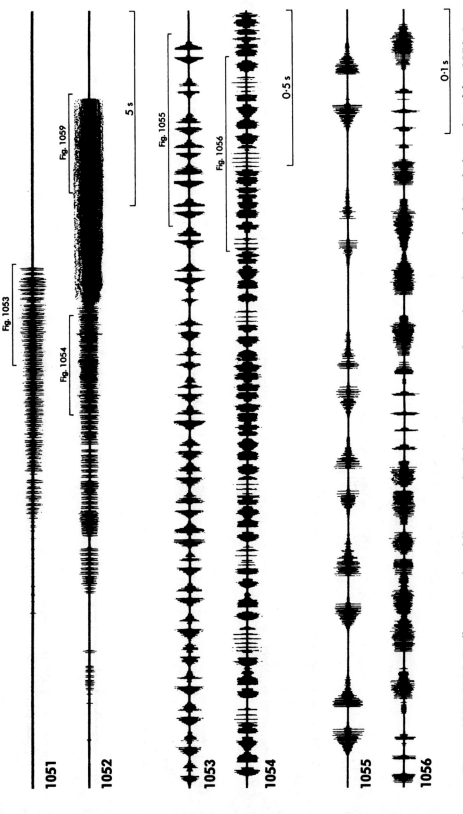

Figures 1051–1056 Oscillograms at three different speeds of the calling songs and wing-buzz of two males of *Stenobothrus rubicundulus*. **1051**. An echeme of purely femoro-alary stridulation. **1052**. A femoro-alary echeme followed immediately by a wing-buzz while the male remained on the ground. **1053–1056**. Faster oscillograms of the indicated parts of the femoro-alary stridulation shown in Figs 1051 and 1052. (From Ragge, 1987a)

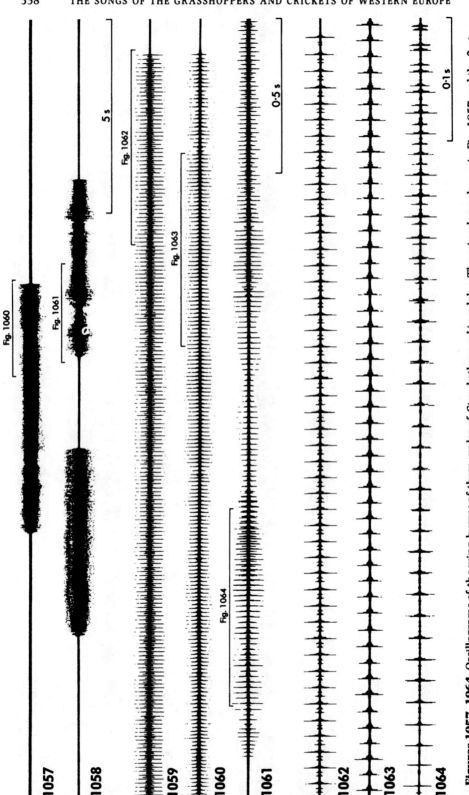

Figures 1057–1064 Oscillograms of the wing-buzzes of three males of *Stenobothrus rubicundulus*. The wing-buzz shown in Fig. 1057 and the first wing-buzz shown in Fig 1058 were produced while the male was on the ground, but part of the second wing-buzz shown in Fig. 1058 was produced during hovering flight. (From Ragge, 1987a)

The wing-vibration produces a loud buzzing sound (Figs 1057–1064) usually lasting about 4–10 s. According to Elsner (1974*b*) the sound is produced by the strongly sclerotized anterior margins of the two hind wings being beaten against each other. Oscillographic analysis shows the buzzing to consist of a series of sharply distinct short-lived sounds repeated at the rate of about 55–85/s (Figs 1059–1064). The whole of the wing-buzz may be produced while the male is sitting on the ground, but often the male takes off and continues to produce the buzzing sound in flight, often hovering above the ground (Fig. 1058, second wing-buzz).

In the presence of a female the male begins with femoro-alary syllables similar in kind to those of the calling song, but the echeme is often greatly prolonged, sometimes lasting for more than 30 minutes. This is followed by wing-vibration and there is then often an alternation of shorter echemes of femoro-alary stridulation and bursts of wing-vibration. A sequence of this kind may end in the male taking off during a burst of wing-vibration and finally an attempt to copulate with the female.

Elsner & Wasser (1995*a*, 1995*b*, 1995*c*) have shown that the acoustic behaviour of this species differs significantly in Greece from that in all other regions where the song has been studied. The prolonged, 'gapped' downstroke hemisyllables that normally occur in the second part of a femoro-alary echeme are replaced by short bursts of wing-vibration. In spite of this behavioural difference, the sound-pattern produced remains very similar to the usual one. These observations were made on courtship songs, but it seems likely that the calling song shows the same difference.

DISTRIBUTION. This montane species is known only from the Alps, higher Apennines and mountains of the Balkan Peninsula.

Stenobothrus cotticus Kruseman & Jeekel

Stenobothrus (Stenobothrodes) cotticus Kruseman & Jeekel, 1967*a*: 1.

Cottian Grasshopper; (F) Sténobothre cottien; (D) Cottischer Grashüpfer.

REFERENCES TO SONG. **Oscillogram** and **leg-movement**: Waeber, 1989. **Verbal description only**: Bellmann, 1993*a*; Bellmann & Luquet, 1995; Kruseman & Jeekel, 1967*a*; Walther *in* Harz, 1975*b*.

RECOGNITION. See under *S. rubicundulus*. In the field males can be easily recognized by their characteristic calling song.

SONG (Figs 1065–1072. CD 2, track 36). The male calling song consists of a series of echemes repeated regularly at the rate of about 1–2/s, the whole series continuing for up to a minute or more. Each echeme lasts about 150–300 ms and shows a marked crescendo from a quiet beginning. Oscillographic analysis shows that a 'two-legged' echeme consists of about 45–70 syllables, but Waeber (1989) has shown that the hind legs are moved alternately, so that each hind leg produces only half this number of syllables.

Both sexes can crepitate in flight (see Figs 1067, 1070). Recordings taken from two males showed the crepitation rate to be 50–65/s at air temperatures of 13–16°C. When courting a female the male makes short crepitating flights between periods of producing echeme-sequences of the calling song type while sitting near the female.

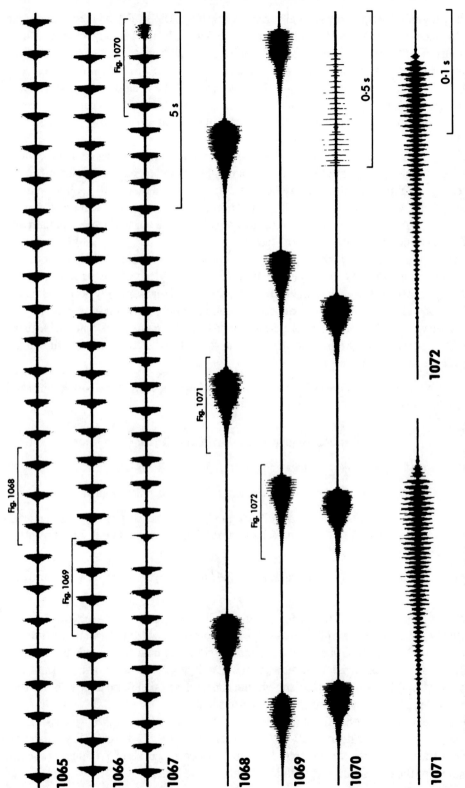

Figures 1065–1072 Oscillograms at three different speeds of the calling songs of three males of *Stenobothrus cotticus*. There is a short burst of crepitation during flight at the end of the oscillograms shown in Figs 1067 and 1070.

DISTRIBUTION. Known at present only from the French Alps, particularly the vicinity of the Cottian Alps, at altitudes of 1700–3000 m.

Myrmeleotettix maculatus (Thunberg)

Gomphocerus maculatus Thunberg, 1815: 221.

Mottled Grasshopper; (F) Gomphocère tacheté; (D) Gefleckte Keulenschrecke; (NL) Knopsprietje; (DK) Køllegræshoppe; (S) Liten Klubbgräshoppa; (SF) Nuijaheinäsirkka.

REFERENCES TO SONG. **Oscillogram**: Clemente, 1987; Elsner, 1983*b*; Elsner & Popov, 1978; Grein, 1984; Holst, 1970, 1986; Kleukers *et al.*, 1997. **Diagram**: Bellmann, 1985*a*, 1988, 1993*a*; Bellmann & Luquet, 1995; Duijm & Kruseman, 1983; Holst, 1970; Jacobs, 1950*b*, 1953*a*; Luquet, 1978; Ragge, 1965; Wallin, 1979. **Leg-movement**: Elsner, 1983*b*; Elsner & Popov, 1978. **Frequency information**: Meyer & Elsner, 1996. **Musical notation**: Yersin, 1954*b* (as *biguttatus*). **Verbal description only**: Broughton, 1972*b*; Bull, 1979; Clemente *et al.*, 1989*b*; Faber, 1928, 1953*a*; Harz, 1957. **Disc recording**: Andrieu & Dumortier, 1963 (LP), 1994 (CD); Bellmann, 1993*c* (CD); Bonnet, 1995 (CD); Grein, 1984 (LP); Odé, 1997 (CD); Ragge *et al.*, 1965 (LP). **Cassette recording**: Bellmann, 1985*b*, 1993*b*; Burton & Ragge, 1987; Wallin, 1979.

RECOGNITION. Males of this small species may be easily recognized in western Europe by their strongly clubbed antennae; the broadened antennal tip is clearly defined, angled slightly to the side and uniformly dark-coloured. In the female the antennae are broadened towards the tip but not strongly clubbed. In both sexes the pronotal lateral carinae are strongly inflexed, forming an angle in the prozona, and there is no bulge near the base of the anterior (ventral) margin of the fore wings. *Gomphocerus sibiricus* and *Gomphocerippus rufus* also have clubbed antennae, but these species are larger and have a bulge near the base of the anterior margin of the fore wings; in addition, males of *G. sibiricus* have conspicuously swollen fore tibiae and in *G. rufus* the antennae are pale-tipped.

SONG (Figs 1073–1096. CD 2, track 37). The male calling song consists of a sequence of about 10–25 (rarely up to 30) echemes, typically lasting about 8–15 s. The sequence begins quietly, successive echemes becoming louder until maximum intensity is reached by about half to two-thirds of the duration of the sequence; occasionally the crescendo continues until the last echeme. The echeme repetition rate is typically about 1·5–2·5/s at the beginning of the sequence, slowing down to about 1·2–2·0/s by the end of the sequence. Oscillographic analysis shows that each echeme typically lasts about 200–350 ms in the early part of the sequence, lengthening to about 400–500 ms at the end of the sequence. The later echemes are composed of about 20–30 diplosyllables repeated at the rate of about 50–65/s (rarely up to 75/s), becoming a little slower towards the end of the echeme; all the echemes begin and end quietly, giving a spindle-shaped oscillogram. Elsner (1983*b*) has shown that the two hind legs are moved up and down alternately, so that the upstroke hemisyllable produced by one leg coincides with the downstroke hemisyllable produced by the other. The hemisyllables are so well-defined and the leg-alternation so regular that the syllable structure is clear in fast oscillograms of most 'two-legged' songs; in oscillograms of 'one-legged' songs it can be seen that the downstroke hemisyllables are louder than the upstroke ones (Fig. 1088).

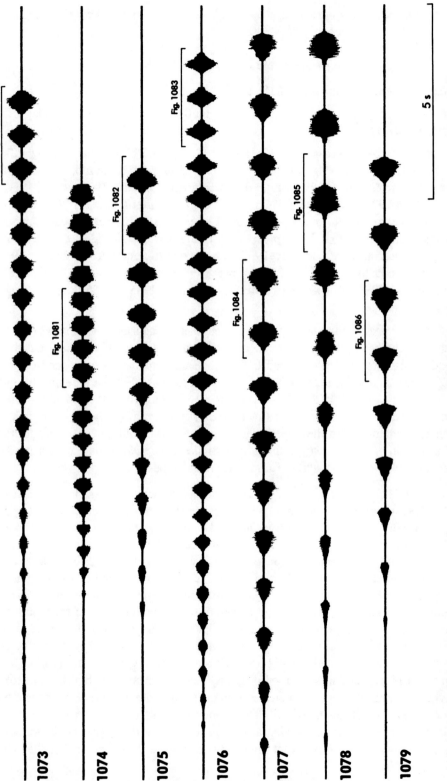

Figures 1073–1079 Oscillograms of the calling songs of seven males of *Myrmeleotettix maculatus*. Figs 1073, 1074 and 1076 are of typical songs. Fig. 1075 is from a central Italian male and Figs 1077–1079 are from Spanish males, showing the markedly slower tempo. Fig. 1074 is from a male with only one hind leg.

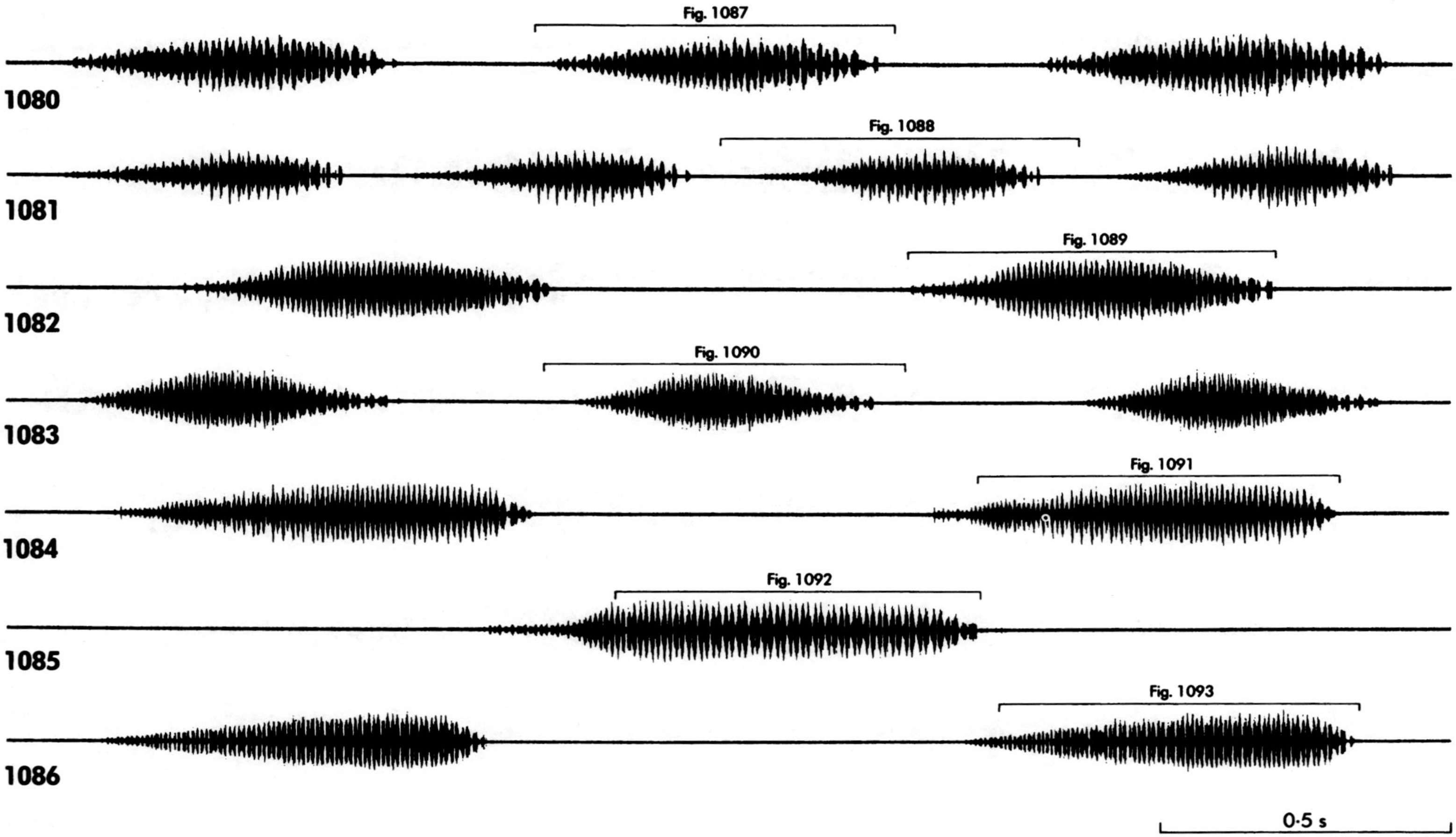

Figures 1080–1086 Faster oscillograms of the indicated parts of the songs of *Myrmeleotettix maculatus* shown in Figs 1073–1079. Fig. 1081 is from a male with only one hind leg.

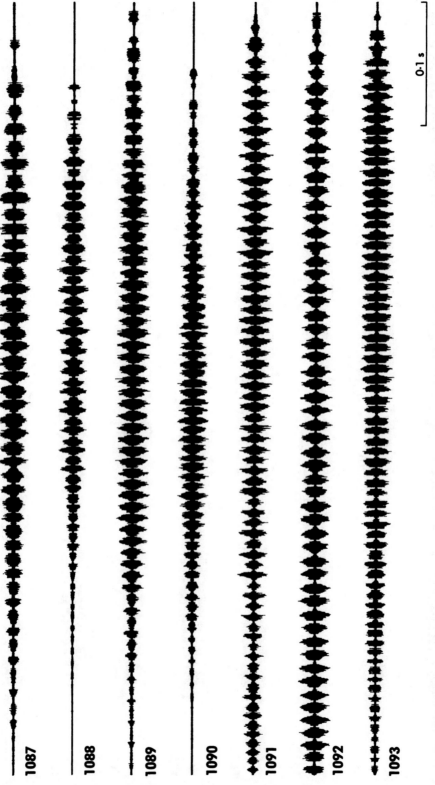

0·1 s

Figures 1087–1093 Faster oscillograms of the indicated parts of the songs of *Myrmeleotettix maculatus* shown in Figs 1080–1086. Fig. 1088 is from a male with only one hind leg.

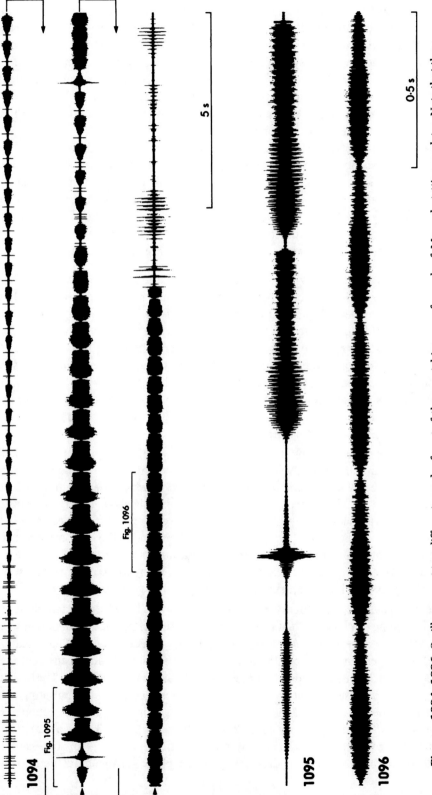

Figures 1094–1096 Oscillograms at two different speeds of part of the courtship song of a male of *Myrmeleotettix maculatus*. Note that the oscillogram shown in Fig. 1094 is split between three lines; the high amplitude 'spikes' near the beginning and end of the second line are produced by the down-jerks of the hind legs that mark the beginning of phases 2 and 3, respectively (see text); the third line ends in the loud syllables that precede an attempt to copulate.

In the Iberian Peninsula the echemes of the calling song tend to be longer and to be repeated more slowly than further north (Figs 1077–1079): about 0·7–1·0/s at the beginning of the sequence, where they last about 300–500 ms, and about 0·5–0·8/s towards the end of the sequence, where they last about 500–800 ms. The whole sequence tends to last longer, sometimes up to 18 s or more. The number of syllables in an echeme is also greater, about 30–40 (rarely up to 50) towards the end of the sequence, but the syllable repetition rate remains about the same. This tendency towards a slower tempo is also shown, though to a lesser extent, in the Italian Peninsula (Fig. 1075).

In the presence of a female the male produces an elaborate courtship song (Figs 1094–1096), one of the most complex ones known at present from any western European grasshopper. It is less rigidly stereotyped than other well-known courtship songs, such as those of *Gomphocerippus rufus*, *Chorthippus albomarginatus* and *Stenobothrus stigmaticus*, and does not usually consist of numerous repetitions of an identical cycle of behaviour. The male is also very easily disturbed, especially by movements of the female.

A fully developed courtship song consists of the following three phases:

1. Courtship usually begins with a series of songs similar to the calling song, with a similar echeme repetition rate but usually quieter and often more prolonged, with a larger number of echemes (often 30–50, rarely more than 100). The male soon begins to sway from side to side in the intervals between the songs and then during the songs themselves; Bull (1979) refers to this body-swaying stage of phase 1 as 'phase 1b'. The male often makes ticking sounds in the intervals between the echeme-sequences, and sometimes makes a clearly audible tick between successive echemes in a sequence (in a manner reminiscent of *Chorthippus mollis*). The swaying becomes more and more pronounced, often including vertical as well as lateral movement, and the male shows every sign of becoming increasingly excited.

2. Eventually, at the end of a phase 1 echeme-sequence, the hind legs are suddenly jerked downwards, making a relatively loud sound, and the antennae are flung backwards. This is followed by a new kind of echeme-sequence, in which the echemes are louder than in phase 1 and repeated rather more slowly (usually about 1/s or a little faster). The hind legs are vibrated in a higher position in the earlier part of each echeme and in a lower position in the later part; while in the higher position the hind legs are opened out somewhat, so that the tips of the tibiae move away from the femora. There are most often 10–20 echemes in phase 2, but sometimes as many as 30 or more. The way in which a phase 2 echeme-sequence ends varies greatly, as discussed below.

3. The third phase begins with exactly the same jerk of the hind legs and antennae as at the beginning of phase 2. Then follows yet a third kind of echeme-sequence, in which the echemes are repeated more rapidly than in both phases 1 and 2 (often about 2/s). During each echeme the vibrating hind legs are raised slightly and opened out (as in phase 2) and the head is swung to the left and right in alternate echemes. The hind legs vibrate continuously, so that the sound is continuous though usually pulsating to the rhythm of the movements of the head and the changes in the angle of vibration of the hind femora. There are usually about 20–30 echemes in phase 3, which is almost always followed by the male attempting to mate with the female; if unsuccessful, the male usually begins another courtship sequence at phase 1. Attempts at mating some-

times occur in earlier phases of the courtship song. Exceptionally phase 3 reverts to phase 2 without pause and occasionally it is followed by a hind leg and antennal jerk and a repetition of phase 3.

In our experience it is unusual for phase 3 to follow immediately after the first appearance of phase 2 in a courtship sequence. Phase 2 can end in a number of different ways. Sometimes it simply stops after a short series of echemes (occasionally with a prolongation of the second part of the last echeme); in this case the male usually begins a phase 1 echeme-sequence after a short pause. Quite often the phase 2 echemes gradually revert to phase 1 echemes, which in turn lead to another hind leg and antennal jerk followed by a repeat of phase 2. Frequently there is a repeat of phase 2 without any reversion to phase 1 echemes or with only a slight reversion. Phase 1 and phase 2 quite often alternate several times before the appearance of phase 3. Rarely, phase 3 follows immediately after a phase 1 echeme-sequence. Fig. 1094 shows a courtship song in which phase 3 follows phase 2 after a brief reversion to phase 1.

The leg and body movements of this species during courtship are described in greater detail by Bull (1979).

The calling song of this species is similar to that of *O. minutissimus*, but the sequences are usually longer and composed of more echemes, and the syllable repetition rate is lower. The courtship song is completely different.

DISTRIBUTION. Widespread in Europe, from near the Arctic Circle in the north to the southern peninsulas, where it is largely confined to mountains. It occurs widely in the British Isles and the range extends eastwards to Siberia.

Gomphocerus sibiricus (Linnaeus)

Gryllus (Locusta) sibiricus Linnaeus, 1767: 701.

Club-legged Grasshopper; (F) Gomphocère des alpages; (D) Sibirische Keulenschrecke.

REFERENCES TO SONG. **Oscillogram**: Elsner, 1974a, 1975; Grein, 1984. **Diagram**: Bellmann, 1985a, 1988, 1993a; Bellmann & Luquet, 1995; Jacobs, 1950b, 1953a. **Leg-movement**: Elsner, 1974a, 1975. **Frequency information**: Meyer & Elsner, 1996. **Musical notation**: Yersin, 1854b. **Verbal description only**: Chopard, 1952; Faber, 1953a; Harz, 1957; Defaut, 1987, 1988b. **Disc recording**: Andrieu & Dumortier, 1994 (CD); Bellmann, 1993c (CD); Bonnet, 1995 (CD); Grein, 1984 (LP). **Cassette recording**: Bellmann, 1985b, 1993b.

RECOGNITION (Plate 3: 7). Males of this species may be recognized at once by their greatly swollen fore tibiae. Both sexes have clubbed, but not pale-tipped, antennae, and there is a bulge near the base of the anterior (ventral) margin of the fore wings. The prozona of the pronotum is usually strongly swollen in the male, forming a hump, but only slightly so in the female.

For a note on the generic assignment of this species see p. 11.

SONG (Figs 1097–1108. CD 2, track 38). The male calling song begins with a loud echeme lasting about 10–25 s (occasionally up to 40 s or more) and consisting of about 55–120

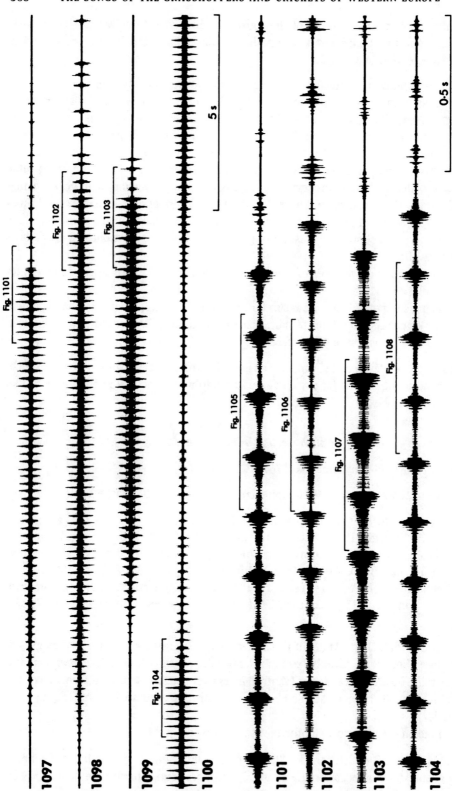

Figures 1097–1104 Oscillograms of the songs of four males of *Gomphocerus sibiricus*. **1097–1099.** Calling songs. **1100.** Part of the courtship song. **1101–1104.** Faster oscillograms of the indicated parts of the songs shown in Figs. 1097–1100.

Figures 1105–1108 Faster oscillograms of the indicated parts of the songs of *Gomphocerus sibiricus* shown in Figs 1101–1104.

0·1 s

1105

1106

1107

1108

syllables repeated at the rate of about 4·5–5·5/s. The echeme begins quietly, reaching maximum intensity by about a third to half of its duration. Most of the sound consists of downstroke hemisyllables, and oscillographic analysis shows that during each of these there are 3–5 very brief gaps (often partly obscured in 'two-legged' songs). This main echeme is usually followed immediately by a series of quieter, very short echemes, each consisting of a small number of syllables produced by a low-amplitude vibration of the hind legs. These echemes are repeated at the rate of about 3·5–4·5/s, usually a little more slowly than the syllables of the main echeme. The number of echemes in this 'aftersong' is very variable, but is usually within the range 10–50.

In the presence of a female the male often produces much longer echemes (of the louder kind, described above), sometimes lasting as long as 3 minutes or more, especially when courtship begins. This frequently develops into a succession of such echemes, now lasting 20–30 s and tending to become progressively shorter, linked by periods of about 10 s during which a series of very short, quieter echemes is produced (as in the 'aftersong' described above) (Fig. 1100). This may in turn develop into quite a complex display involving movements of the antennae and palps; this behaviour has been described in detail by Faber (1953a), Jacobs (1953a) and Elsner (1974a).

DISTRIBUTION. In western Europe this species is found only in mountains, usually at altitudes above 1000 m. It occurs in the Cantabrian Mountains, Pyrenees, Vosges, Alps and higher parts of the central Apennines. Eastwards, the range extends through eastern Europe, including the Balkan Peninsula, to Asia Minor, Kazakhstan and much of central and eastern Asia, including Mongolia and palaearctic China.

Gomphocerippus rufus (Linnaeus)

Gryllus (Locusta) rufus Linnaeus, 1758: 433.

Rufous Grasshopper; (F) Gomphocère roux; (D) Rote Keulenschrecke; (NL) Rosse Sprinkhaan; (S) Stor Klubbgräshoppa; (SF) Osi Nuijaheinäsirkka.

REFERENCES TO SONG. **Oscillogram**: Elsner, 1968, 1973, 1974a, 1975, 1983b; Elsner & Huber, 1969; Grein, 1984; Holst, 1986; Kleukers et al., 1997; Loher & Huber, 1964, 1966; Riede, 1983, 1986; Schmidt, 1989; Schmidt & Baumgarten, 1977; Vedenina & Zhantiev, 1990. **Diagram**: Bellmann, 1985a, 1988, 1993a; Bellmann & Luquet, 1995; Duijm & Kruseman, 1983; Jacobs, 1950b, 1953a; Loher & Huber, 1964; Ragge, 1965; Wallin, 1979. **Leg-movement**: Elsner, 1973, 1974a, 1975, 1983b. **Frequency information**: Meyer & Elsner, 1996. **Musical notation**: Yersin, 1984b. **Verbal description only**: Baier, 1930; Broughton, 1972b; Defaut, 1987, 1988b; Faber, 1928, 1953a; Harz, 1957; Wadepuhl, 1983. **Disc recording**: Bellmann, 1993c (CD); Bonnet, 1995 (CD); Grein, 1984 (LP); Odé, 1997 (CD); Ragge et al., 1965 (LP). **Cassette recording**: Bellmann, 1985b, 1993b; Burton & Ragge, 1987.

RECOGNITION (Plate 3: 8). The conspicuously pale-tipped, clubbed antennae provide an easy means of recognizing both sexes of this species.

For a note on the generic assignment of this species see p. 11.

SONG (Figs 1109–1130. CD 2, track 39). The male calling song is a single, rather quiet,

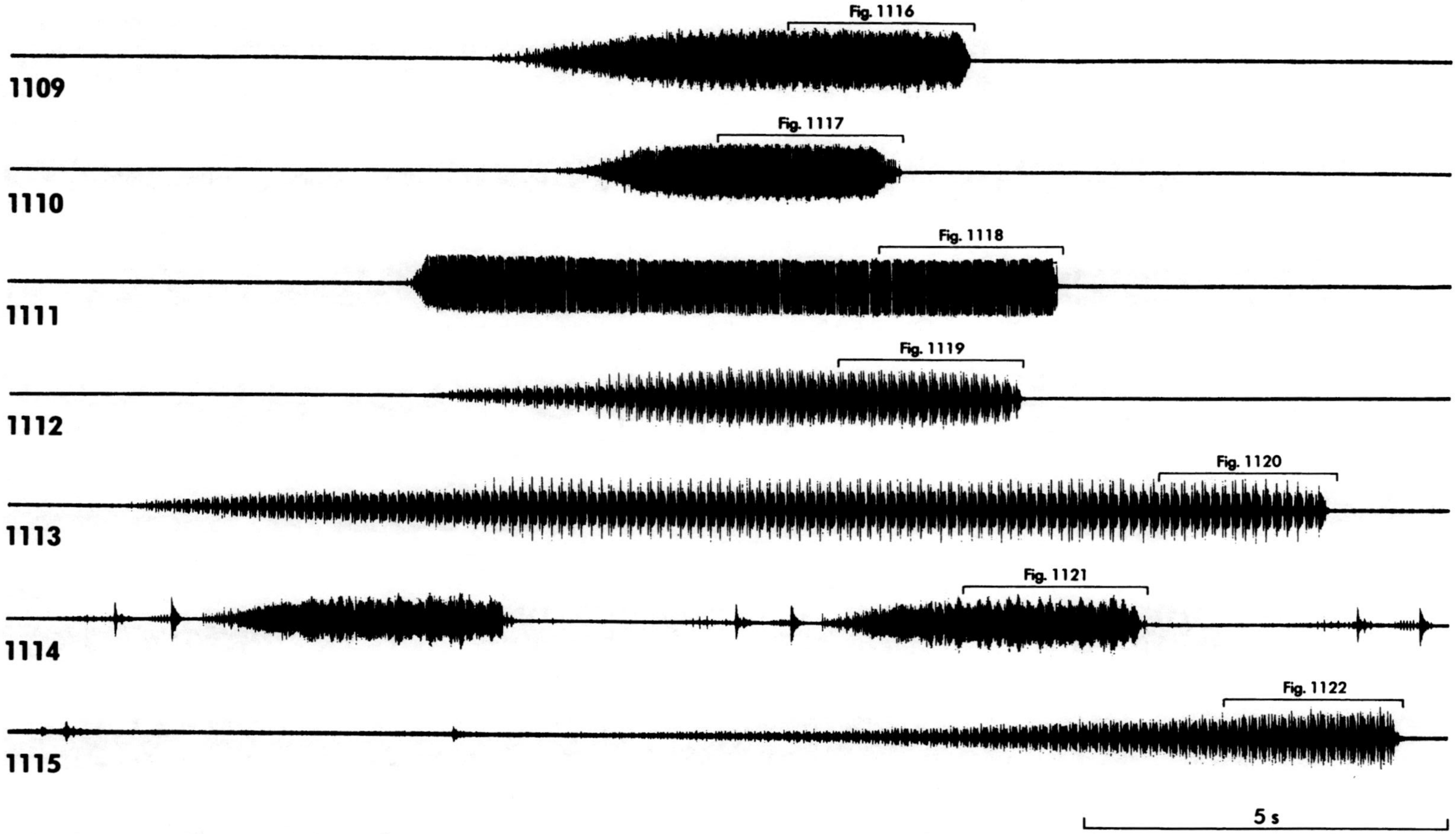

Figures 1109–1115 Oscillograms of the songs of seven males of *Gomphocerippus rufus*. 1109–1111. Typical calling songs from males with (1109, 1110) two hind legs and (1111) one hind leg. 1112, 1113. Calling songs of two males from the Italian Apennines. 1114. Part of a typical courtship song. 1115. Part of the courtship song of a male from near Florence, Italy.

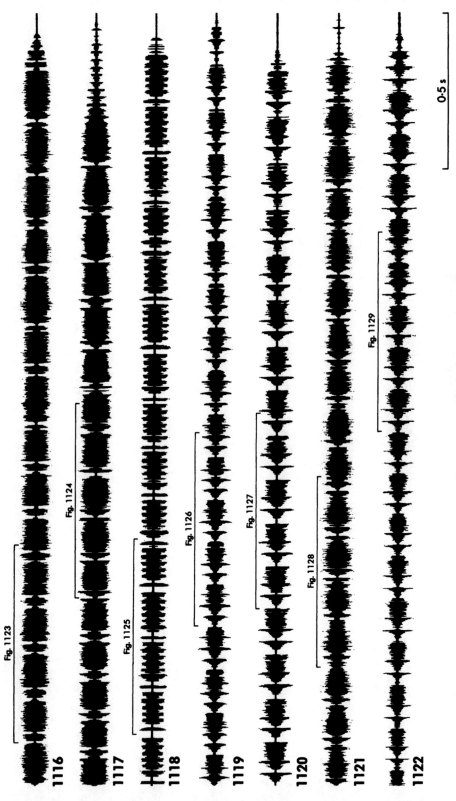

Figures 1116–1122 Faster oscillograms of the indicated parts of the songs of *Gomphocerippus rufus* shown in Figs 1109–1115. Fig. 1118 is from a male with only one hind leg.

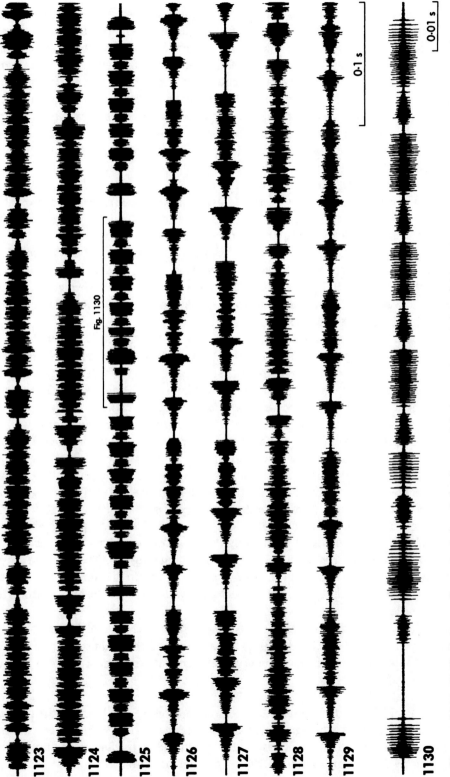

Figures 1123–1130 Faster oscillograms of the indicated parts of the songs of *Gomphocerippus rufus* shown in Figs 1116–1122. Figs 1125 and 1130 are from a male with only one hind leg.

dense echeme-sequence, usually lasting about 3–10 s (but see below) and composed of about 20–80 echemes. The sequence usually begins very quietly, reaching maximum intensity about halfway through its duration; however, the crescendo sometimes lasts throughout the sequence and occasionally the intensity is uniform from start to finish (Fig. 1111). In warm, sunny weather the echeme repetition rate is about 7–9/s (heard as a slight unevenness in the sound), usually decreasing somewhat during the course of the sequence, but can be much lower in cooler conditions. Oscillographic analysis shows that each echeme is typically composed of a single diplosyllable or downstroke hemisyllable followed after a very brief pause by about 4–7 evenly spaced diplosyllables; there is then a slightly longer pause before the next echeme follows. These pauses are usually no more than about 10 ms, so that the sequence sounds continuous to the human ear. As in *Chorthippus biguttulus*, the syllable structure is rather obscure in oscillograms of 'two-legged' songs, but can be seen clearly in a fast oscillogram of a 'one-legged' song (Figs 1118, 1125, 1130). In warm conditions each echeme lasts about 90–150 ms, usually becoming a little longer as the sequence progresses.

In the Apennines the echeme structure tends to be slightly different: the first two syllables of each echeme are often a little more widely spaced (and usually louder) than the remaining ones (Figs 1119, 1120, 1126, 1127). In one locality, near Montemignaio, east of Florence, the sequences of the calling song were exceptionally long, ranging from 13 s to 20 s in one of the males studied (Fig. 1113).

In the presence of a female the male produces an elaborate courtship display, with visual as well as acoustic components (Figs 1114, 1121, 1128). It consists of regularly repeated echeme-sequences of the calling song type, in the intervals between which is a highly stereotyped sequence of behaviour. First the head is rocked from side to side (about 1–2 complete left and right movements per second) and the palps are waggled to the same rhythm. Each time the head is moved in either direction, a quiet and very brief echeme (usually di- or trisyllabic) is usually produced by the hind legs. After about 2–7 s the frequency of the head, palp and hind leg movements is suddenly increased for about 0·5–2·0 s and then the antennae are flung backwards while a louder sound is produced, initiating a short period (of the order of 0·5 s) of rapid hind leg vibration. After a similar short period, during which further quiet sounds are usually produced by slower hind leg movements, there is a repetition of the antennal jerk accompanied by a louder sound and further rapid leg vibration. (Occasionally this repetition is omitted; rarely there are two repetitions.) There is then a short pause followed by an echeme-sequence of the calling song type. The complete cycle, from the end of one echeme-sequence of the calling song type to the end of the next one, usually lasts 8–20 s, and the cycle is often repeated many times before the male attempts to mate with the female.

In one courtship display observed in sunny conditions in the Monti Reatini, central Apennines, (Figs 1115, 1122, 1129) the period between the end of a main echeme-sequence and the first antennal jerk was 10 s, the interval between the two antennal jerks was 5 s and the complete cycle lasted 27 s.

DISTRIBUTION. Widespread in western Europe, from northern Scandinavia to the central Apennines. Local in southern England, but absent from the Iberian Peninsula. Eastwards, the range extends through much of eastern Europe, including the Balkan Peninsula (except for most of Greece), to Kazakhstan, Siberia and northern China.

Stauroderus scalaris (Fischer de Waldheim)

Oedipoda scalaris Fischer de Waldheim, 1846: 317.

Large Mountain Grasshopper; (F) Criquet jacasseur; (D) Gebirgsgrashüpfer; (DK) Skærende Græshoppe; (S) Skärrande Gräshoppa.

REFERENCES TO SONG. **Oscillogram**: Bukhvalova & Zhantiev, 1993; Elsner & Popov, 1978; Grein, 1984. **Diagram**: Bellmann, 1985a, 1988, 1993a; Bellmann & Luquet, 1995; Jacobs, 1950b, 1953a; Luquet, 1978; Wallin, 1979; Xi et al., 1992. **Leg-movement**: Elsner & Popov, 1978. **Musical notation**: Yersin, 1854b (as *melanopterus*). **Verbal description only**: Chopard, 1952; Defaut, 1987, 1988b; Faber, 1928 (as *morio*), 1932 (as *mor.*), 1953a; Fischer, 1853 (as *mel.*); Harz, 1957; Karny, 1908 (as *mor.*). **Disc recording**: Bellmann, 1993c (CD); Bonnet, 1995 (CD); Deroussen, 1993 (CD); Grein, 1984 (LP). **Cassette recording**: Bellmann, 1985b, 1993b; Wallin, 1979.

RECOGNITION (Plate 3: 9). This typically montane species, particularly common in the Alps, may be easily recognized by its large size and smoky hind wings. In the male fore wings the costal and medial areas are both conspicuously expanded, with a ladder-like arrangement of cross-veins; in the female fore wings the medial area is also expanded, but has an irregular network of veinlets. In the fore wings of both sexes veins Cu_1 and Cu_2 are fused together except near the base. In the field the loud calling song and rattling flight are also very distinctive.

For remarks on the taxonomic status of this species see p. 9.

SONG (Figs 1131–1145. CD 2, track 40). The loud and highly characteristic male calling song consists of an echeme-sequence lasting about 10–30 s and composed of about 20–50 echemes repeated at the rate of about 1–3/s (usually 1·5–2·0/s). Each echeme lasts about 300–700 ms and is composed of a series of rapidly repeated syllables followed by a much slower upstroke hemisyllable, during which the hind legs move jerkily to a higher position. There are about 6–12 syllables in the series that begins each echeme; they are usually diplosyllables, each lasting about 25–50 ms, and are repeated at the rate of about 20–30/s. The first echeme in a sequence usually begins with a larger number (up to 20 or more) of these syllables, and there is often a tendency for the number of syllables per echeme (and thus the duration of the echeme) to decrease slightly as the sequence progresses. The upstroke hemisyllable that concludes each echeme lasts about 100–200 ms and consists of about 5–10 short-lived ticks, each usually lasting about 3–5 ms; the ticks are repeated at the rate of about 30–60/s, and Halfmann & Elsner (in Elsner & Popov, 1978) have shown that they are produced by short upward movements of the jerkily rising hind legs. Each echeme thus consists of an alternation of sibilant buzzing and harder rattling (the rattling usually sounding louder), and is a common and unmistakable sound in the uplands of central and southern Europe. The two hind legs are well synchronized and there is little difference between 'one-legged' and 'two-legged' oscillograms (cf. Figs 1142, 1143).

Males typically produce a calling song immediately on landing after a flight. However, if a male sits still for any length of time before doing so, it usually produces as a prelude a series of short echemes similar to the buzzing component of the calling song, but quieter and usually shorter (see Figs 1131–1133). These echemes are usually composed of 5–8

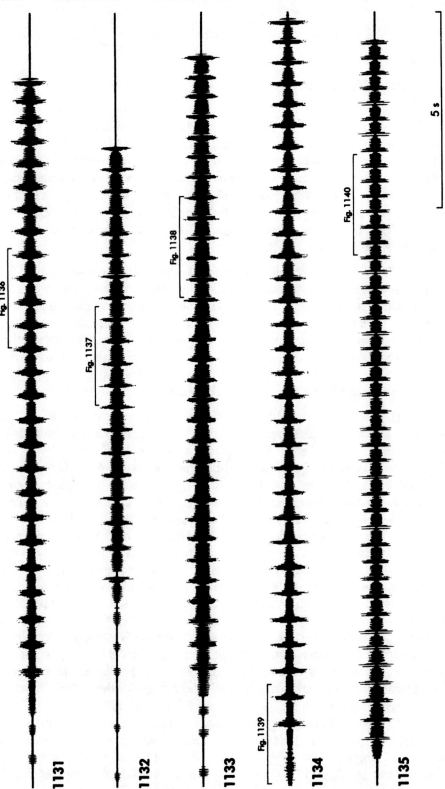

Figures 1131–1135 Oscillograms of the calling songs of five males of *Stauroderus scalaris*. Fig. 1133 is from a male with only one hind leg. There is a short burst of crepitation during flight at the beginning of the oscillogram shown in Fig. 1134.

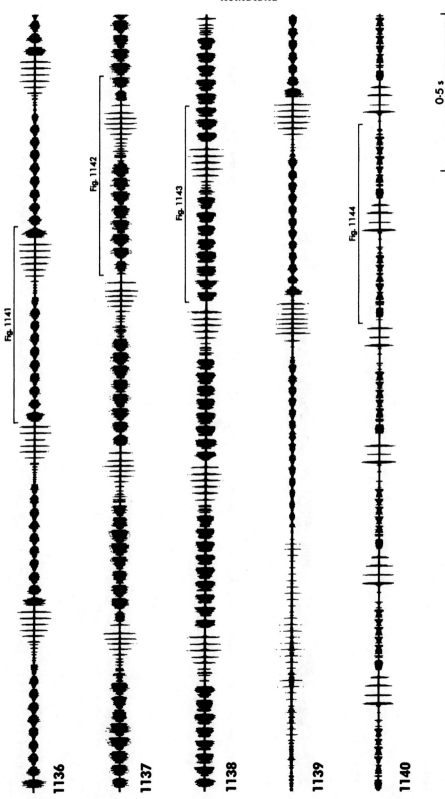

Figures 1136–1140 Faster oscillograms of the indicated parts of the songs of *Stauroderus scalaris* shown in Figs 1131–1135. Fig. 1138 is from a male with only one hind leg.

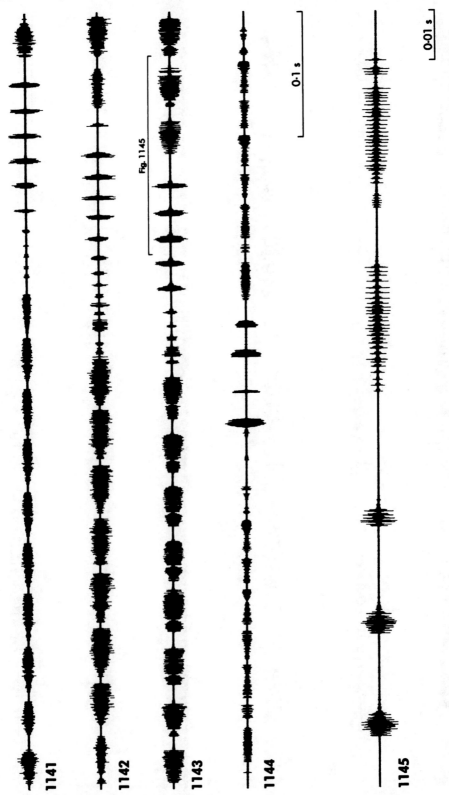

Figures 1141–1145 Faster oscillograms of the indicated parts of the songs of *Stauroderus scalaris* shown in Figs 1136–1140. Figs 1143 and 1145 are from a male with only one hind leg.

syllables, repeated at the same rate as in the fully developed calling song, and are separated by intervals of about 0·3–1·0 s.

Both sexes crepitate in flight, the crepitation rate being about 50/s (see Figs 1134, 1139).

DISTRIBUTION. Widespread in the mountains of central and southern Europe, including the southern peninsulas. Also recorded from one or two lowland localities in Belgium and even as far north as the island of Öland, Sweden, but absent from the British Isles. Eastwards, the range extends to Siberia, Mongolia and China.

Chorthippus apricarius (Linnaeus)

Gryllus (Locusta) apricarius Linnaeus, 1758: 433.

Upland Field Grasshopper; (F) Criquet des adrets; (D) Feld-Grashüpfer; (NL) Locomotiefje; (S) Solgräshoppa.

REFERENCES TO SONG. **Oscillogram**: Bukhvalova & Zhantiev, 1993; Grein, 1984; Holst, 1970, 1986; Kleukers *et al.*, 1997; Vedenina & Zhantiev, 1990; Xi *et al.*, 1992. **Diagram**: Bellmann, 1985a, 1988, 1993a; Bellmann & Luquet, 1995; Duijm & Kruseman, 1983; Holst, 1970; Jacobs, 1950b, 1953a; Wallin, 1979. **Frequency information**: Meyer & Elsner, 1996; Vedenina & Zhantiev, 1990. **Musical notation**: Yersin, 1854b. **Verbal description only**: Defaut, 1987, 1988b; Faber, 1928, 1953a; Fischer, 1850; Harz, 1957. **Disc recording**: Bellmann, 1993c (CD); Bonnet, 1995 (CD); Grein, 1984 (LP); Odé, 1997 (CD). **Cassette recording**: Bellmann, 1985b, 1993b; Wallin, 1979.

RECOGNITION. *Chorthippus* is the largest genus in the Gomphocerinae, including a total of over 150 species, of which about two dozen occur in western Europe. It may be distinguished from *Omocestus* and *Stenobothrus* by the presence of a bulge towards the base of the anterior (ventral) margin of the fore wings, and from *Gomphocerus*, *Gomphocerippus* and *Myrmeleotettix* by the filiform, rather than clubbed, antennae. From *Euchorthippus* it may be distinguished by the lack of any longitudinal carinulae on the vertex.

C. *apricarius* may be easily recognized by the enlarged medial area of the fore wings, veins Cu_1 and Cu_2 being fused together except near the base. In the male the costal area is also enlarged, and both the costal and medial areas have a ladder-like arrangement of cross-veins. In the field the male calling song is also highly distinctive.

SONG (Figs 1146–1158. CD 2, track 41). The male calling song consists of an echeme-sequence lasting about 10–35 (usually 15–25) s. The sequence begins very quietly and shows a gradual crescendo during the first third to half of its duration; it then remains at a constant intensity until its abrupt end. The sequence is composed of about 50–150 echemes of a highly characteristic kind: each consists of a 'tick' lasting about 10–20 ms followed by two syllables lasting, together, about 100–300 (usually 150–250) ms. Each whole echeme lasts about 150–400 (usually 200–300) ms and the echeme repetition rate is about 2·5–5·5/s. The gradual crescendo, the presence of ticks and the duration of the sequence are all reminiscent of the song of C. *mollis*, but the structure of each echeme, and thus the sound-pattern heard by the human ear, is completely different (see p. 414).

In the presence of a female the male produces a series of echeme-sequences of the call-

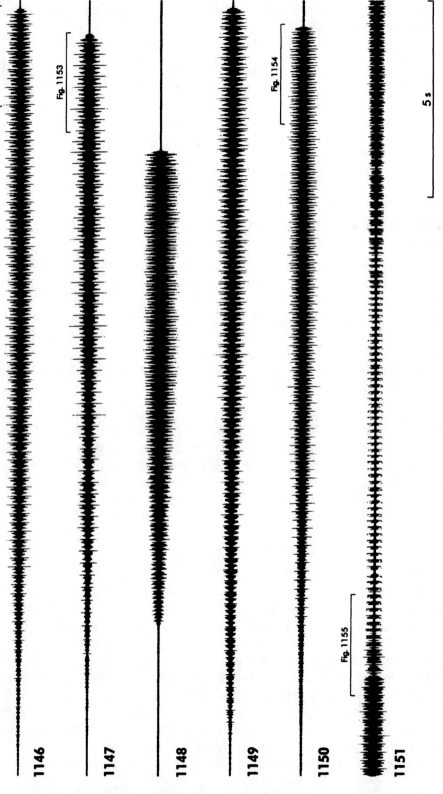

Figures 1146–1151 Oscillograms of the songs of six males of *Chorthippus apricarius*. 1146–1150. Calling songs. 1151. Part of the courtship song.

Figures 1152–1158 Faster oscillograms of the indicated parts of the songs of *Chorthippus apricarius* shown in Figs 1146–1151.

ing song type (but sometimes longer, up to 50 s or more) separated by shorter periods (usually about 7–10 s) during which a simple series of single syllables is produced at a repetition rate of about 4–7 syllables/s (Figs 1151, 1155).

DISTRIBUTION. Local in western Europe, from southern Sweden and Denmark in the north to the Pyrenees (including the Spanish side), Alps and central Apennines in the south. Eastwards, the range extends through central and eastern Europe (including the Balkan Peninsula) to Kazakhstan, Siberia, Mongolia and China. Absent from the British Isles and very local in the northern half of its range in western Europe; in the southern half mainly confined to uplands.

Chorthippus corsicus (Chopard)

Omocestus corsicus Chopard, 1923: 268.

Corsican Grasshopper; (F) Sténobothre corse.

REFERENCES TO SONG. **Oscillogram** and **leg-movement**: Waeber, 1989 (as *pascuorum*). **Verbal description only**: Pfau, 1984 (comparative remarks only).

RECOGNITION. The small size of this species (total length less than 13 mm in the male, 15 mm in the female), and the fact that it occurs only on mountains in Corsica, make identification quite easy. It is sometimes rather brachypterous and the bulge on the anterior margin of the fore wings is not always clearly developed.
    For a discussion of the taxonomic problems presented by this species, see p. 74.

SONG (Figs 1159–1167. CD 2, track 42). The male calling song is an echeme lasting about 4–10 s and consisting of about 30–80 syllables repeated at the rate of about 7–9/s. In cooler conditions the echeme sometimes lasts as long as 15 s, with a syllable repetition rate as low as 4/s. The echeme begins quietly, gradually becoming louder until maximum intensity is reached about halfway through its duration. Oscillographic analysis shows that there are usually gaps in the downstroke hemisyllables, especially in the earlier part of the echeme.

DISTRIBUTION. Known only from Corsica, where it is widespread, especially at altitudes above 1000 m.

Chorthippus cazurroi (Bolívar)

Stenobothrus (Stauroderus) Cazurroi Bolívar, 1898a: 8.

Cazurro's Grasshopper.

REFERENCES TO SONG. **Oscillogram**: Reynolds, 1987.

RECOGNITION. This species is known only from high altitudes in the Picos de Europa in northern Spain, where it may be easily distinguished from most other gomphocerine grasshoppers by being brachypterous, the fore wings not reaching the hind knees. From *C. parallelus* it may be distinguished by the angled pronotal lateral carinae and the relatively well developed hind wings, which reach the tips of the flexed fore wings. The males usually lack the bulge on the anterior (ventral) margin of the fore wings that is normally characteristic of

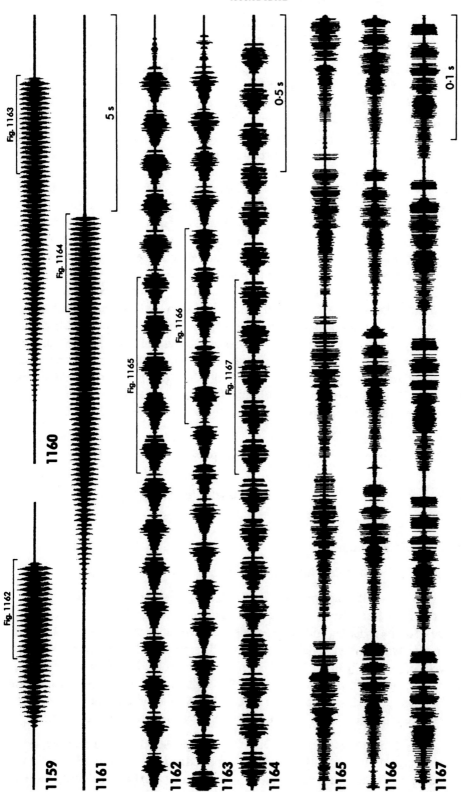

Figures 1159–1167 Oscillograms at three different speeds of the calling songs of three males of *Chorthippus corsicus*. Figs 1160, 1163 and 1166 are from a male with only one hind leg.

Chorthippus, but the bulge is almost always present in the females. In the field males may be easily recognized by the remarkably prolonged calling song.

SONG (Figs 1168–1183. CD 2, track 43). The male calling song is an echeme-sequence lasting about 20–90 s and composed of about 150–450 echemes repeated at the rate of about 4–7/s. The echeme repetition rate is highest at the beginning of the sequence, about 6–7/s, gradually lessening to about 4–6/s by the end. The sequence begins very quietly, gradually increasing in loudness for the first third to two-thirds of its duration, and then remaining at constant intensity until the abrupt end; occasionally the last few echemes are a little quieter. In some songs the echemes are alternately quieter and louder for part or all of the sequence (Figs 1168, 1172, 1176, 1180). Oscillographic analysis shows that in the early part of the sequence there are usually only two or three quick syllables in each echeme, but in the later part each echeme usually comprises three or four quick, loud syllables followed by a slower, quieter one; each of these later echemes usually lasts about 140–280 ms.

In the presence of a female the male produces echeme-sequences similar to those of the calling song but even longer (up to 3 minutes) and modified towards the end (Figs 1171, 1175, 1179, 1183). When the maximum intensity has been reached (after about 1·0–2·5 minutes) an extra-slow syllable is produced in place of the slow syllable that normally concludes an echeme and lasting about twice as long as a normal slow syllable. Thereafter, such extra-slow syllables occur every 2–5 echemes until the end of the sequence. Note that the rather slow tempo of the courtship song shown in the oscillograms (in comparison with the calling song oscillograms) could have resulted, at least partly, from the fact that they were taken from a song recorded in dull conditions, at an air temperature of 20°C; the calling songs were recorded in sunny conditions at air temperatures of 25–30°C.

DISTRIBUTION. Known at present only from the western and central massifs of the Picos de Europa in the Cantabrian Mountains of northern Spain.

Chorthippus nevadensis Pascual

Chorthippus nevadensis Pascual, 1978: 167.

Sierra Nevadan Grasshopper.

REFERENCES TO SONG. **Oscillogram**: Clemente *et al.*, 1994.

RECOGNITION. This small, brachypterous, dark-coloured grasshopper is known only from high altitudes in the Sierra Nevada in southern Spain. In the male the fore wings fall well short of the tip of the abdomen and in the female they reach at most to the third abdominal tergite; the hind wings are vestigial in both sexes. The pronotal lateral carinae are very poorly developed, especially in the posterior part of the prozona. This pronotal character, together with the bulge on the anterior (ventral) margin of the fore wings, enables both sexes to be distinguished from the otherwise rather similar brachypterous species of *Omocestus* that occur in the Sierra Nevada: *O. bolivari*, *O. minutissimus* and *O. uhagonii*.

SONG (Figs 1184–1189. CD 2, track 44). The male calling song is an echeme lasting about 1·5–2·5 s and consisting of about 10–15 syllables repeated at the rate of about 5–8/s (slower in cool conditions). The echeme begins quietly, soon reaching maximum intensity. Oscillo-

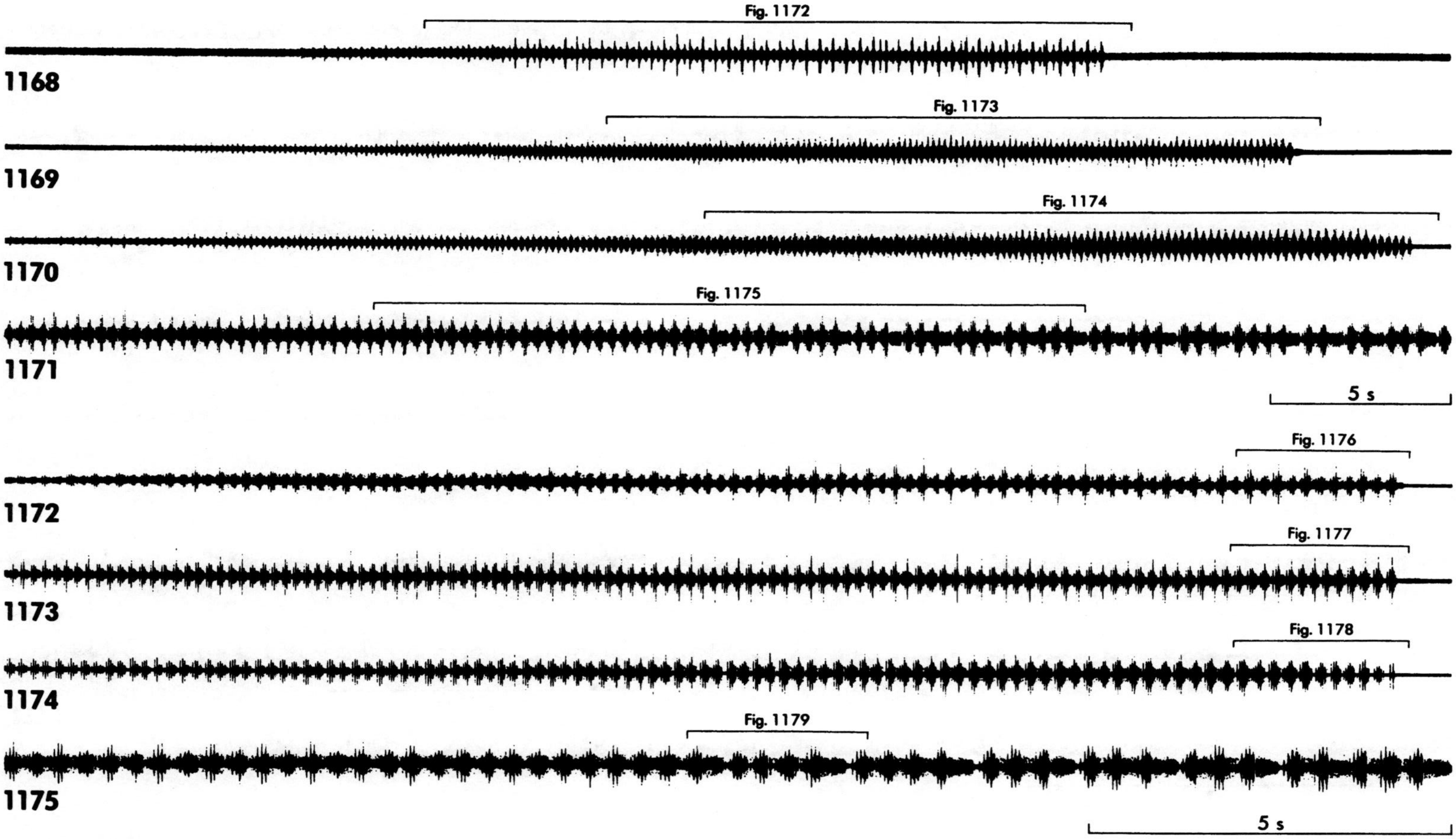

Figures 1168–1175 Oscillograms of the songs of four males of *Chorthippus cazurroi*. **1168–1170.** Calling songs. **1171.** Part of the courtship song. **1172–1175.** Faster oscillograms of the indicated parts of the songs shown in Figs 1168–1171. (From Reynolds, 1987)

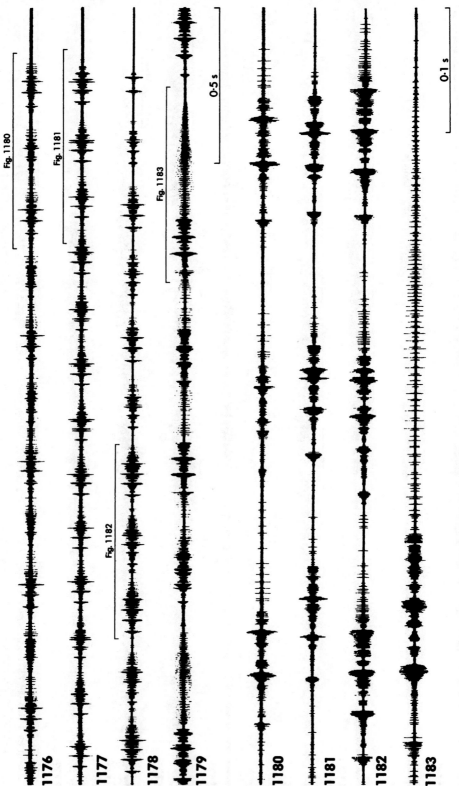

Figures 1176–1183 Faster oscillograms of the indicated parts of the songs of *Chorthippus cazurroi* shown in Figs 1172–1175. (From Reynolds, 1987)

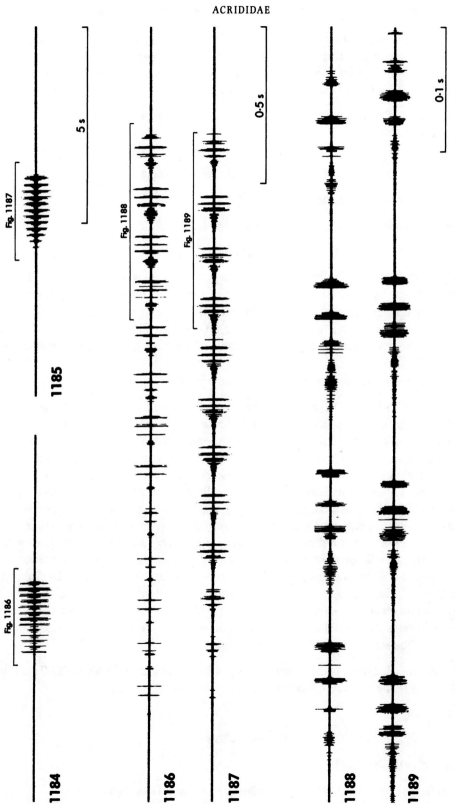

Figures 1184–1189 Oscillograms at three different speeds of the calling songs of two males of *Chorthippus nevadensis*.

graphic analysis shows that there are typically two large gaps in each downstroke, which thus usually consists of three well-separated sounds; this 'gappy' nature of the syllables, which persists throughout the echeme, gives the song a characteristic rasping, scratchy quality. The echemes are repeated at irregular intervals.

DISTRIBUTION. Known only from the Sierra Nevada in southern Spain, mostly at altitudes above 2000 m.

Chorthippus cialancensis Nadig

Chorthippus cialancensis Nadig, 1986: 218.

Piedmont Grasshopper; (D) Cialancia-Grashüpfer.

REFERENCES TO SONG. No published work known to us.

RECOGNITION. This small brachypterous grasshopper occurs only at altitudes above 2000 m in the Piedmont Alps, and is the only brachypterous species of *Chorthippus* to occur in these mountains at such high altitudes. The fore wings fall well short of the hind knees in the male and are even shorter in the female, reaching only to the fourth or fifth abdominal tergite; the hind wings are vestigial in both sexes. The hind femora are red or yellow ventrally.

For remarks on the taxonomy of this species see p. 74.

SONG (Figs 1190–1198. CD 2, track 45). The male calling song is an echeme-sequence lasting about 5–12 s and composed of about 10–30 echemes repeated at the rate of about 2–3/s, the rate lessening slightly during the course of the sequence. The sequence begins quietly, reaching maximum intensity after about a third to two-thirds of the duration; occasionally there is a decline in intensity during the last few echemes. The stridulatory leg-movements of this species have not yet been studied, but oscillographic analysis suggests that, by the time the middle of the sequence is reached, each echeme consists of at least 7–10 rapidly repeated syllables, of which only the first 5–7 are at all clear, the remaining ones taking the form of a series of short-lived sounds of roughly equal amplitude; these rapid syllables are then followed by a single slower and quieter syllable, varying in duration from about 30–100 ms at the Colle di Sampeyre (Figs 1194, 1195) to about 100–200 ms at the Cialancia site (Fig. 1193). Each of these later echemes lasts a total of about 300–400 ms at the Colle di Sampeyre and about 400–500 ms at Cialancia. In the structure of the echemes the song shows a remarkable resemblance to that of *C. cazurroi*, but the echeme-sequence is much shorter and the echeme repetition rate much lower.

DISTRIBUTION. Known at present only from a few localities at altitudes above 2000 m in the Cottian Alps in the Piedmont region of north-western Italy.

Chorthippus pullus (Philippi)

Gryllus pullus Philippi, 1830: 38.

Gravel Grasshopper; (F) Criquet des iscles; (D) Kiesbank-Grashüpfer.

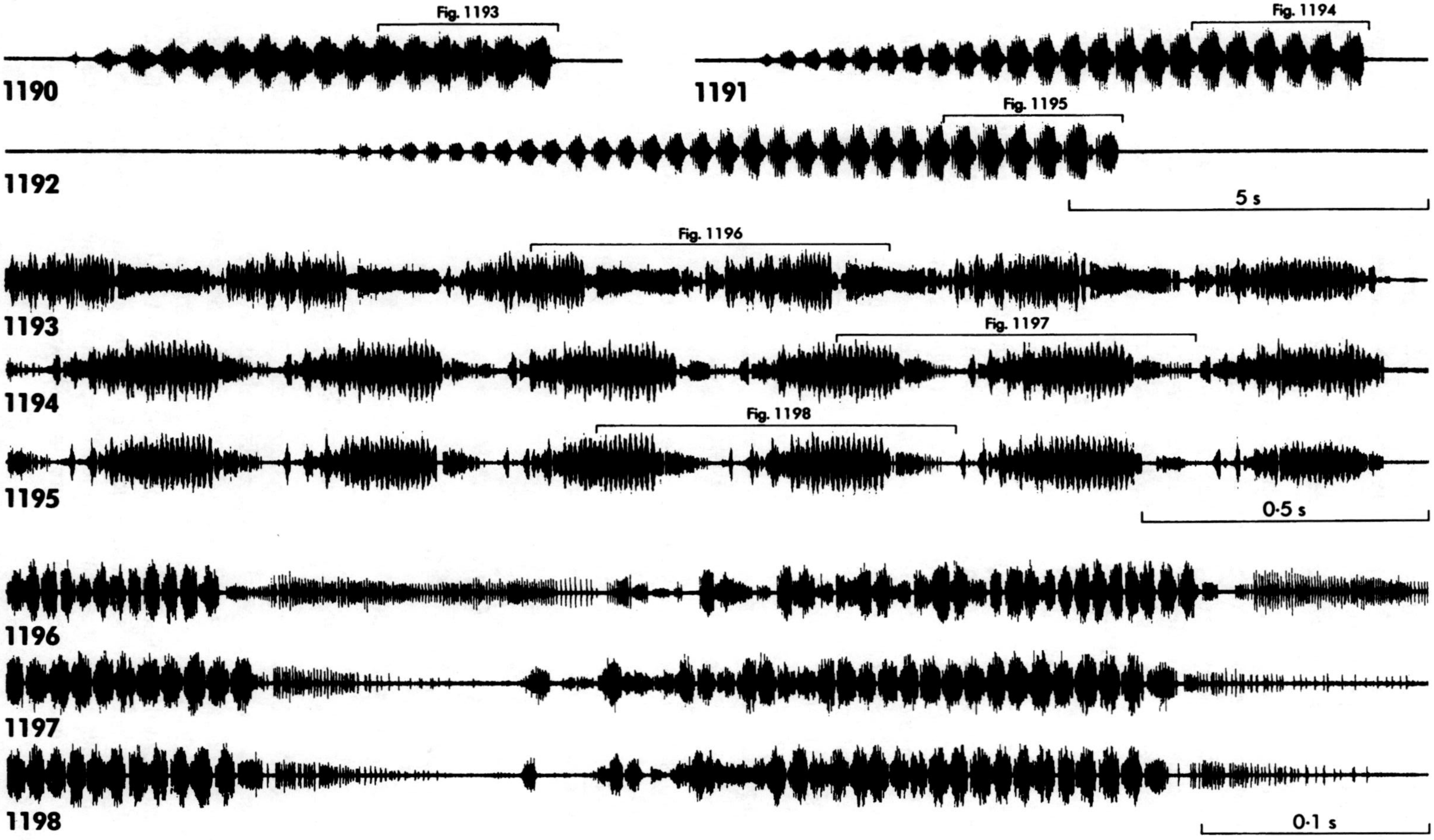

Figures 1190–1198 Oscillograms at three different speeds of the calling songs of three males of *Chorthippus cialancensis*, one (1190, 1193, 1196) from near the Passo della Cialancia and the other two (1191, 1192, 1194, 1195, 1197, 1198) from the Colle di Sampeyre.

REFERENCES TO SONG. **Oscillogram**: Grein, 1984. **Diagram**: Bellmann, 1985*a*, 1988, 1993*a*; Bellmann & Luquet, 1995; Jacobs, 1950*b*, 1953*a*. **Verbal description only**: Chopard, 1952; Faber, 1953*a*; Harz, 1957; Princis, 1935. **Disc recording**: Bellmann, 1993*c* (CD); Grein, 1984 (LP). **Cassette recording**: Bellmann, 1985*b*, 1993*b*.

RECOGNITION. In western Europe this is a rare species, occurring very locally on the gravel-banks of alpine rivers and streams, and on sandy heathland. It is rather brachypterous (apart from a rare macropterous form), the fore wings not reaching the hind knees in the male and reaching only to the third to fifth abdominal tergite in the female; the venation of the male fore wings is characteristic, with the distal part of the costal area markedly expanded. The tympanal apertures are open, oval, and the hind tibiae are red. Some of the small montane forms of *C. binotatus* are superficially similar, but in that species the tympanal apertures are slit-like and the costal area of the male fore wings less expanded.

SONG (Figs 1199–1204. CD 2, track 46). In its most complete form the male calling song is an echeme lasting about 2–4 s and composed of two clearly different parts. The first part, lasting about 1·5–2·5 s, is a whizzing sound, beginning quietly but rapidly becoming louder, produced by a rapid, small amplitude vibration of the hind legs. The second part, which follows immediately, lasts about 1–2 s, is a more broken sound, produced by a slower, larger amplitude vibration of the hind legs, and usually becomes quieter towards the end. Sometimes the second part is omitted.

There has been no published study of the precise relationship between the leg-movements and the sounds they produce in this species. The oscillograms suggest that the leg-movements are complex and perhaps asynchronous.

DISTRIBUTION. In western Europe this species occurs only very locally and is now probably restricted to the vicinity of the Alps, in France, Switzerland, Italy, southern Bavaria and Austria. Farther east it occurs widely, but locally, in eastern Europe, reaching as far north as Russian Finland and as far south as Romania and the northern Caucasus.

Chorthippus alticola Ramme

Chorthippus (Stenobothrus) alticola Ramme, 1921*b*: 246.

Eastern Alpine Grasshopper; (D) Höhengrashüpfer.

REFERENCES TO SONG. **Verbal description only**: Bellmann, 1993*a*.

RECOGNITION. This brachypterous species is known only from the southern side of the Alps, from about longitude 10°E eastwards, especially at altitudes above 1500 m. The fore wings fall well short of the hind knees in the male and are even shorter in the female; the hind wings are vestigial in both sexes. The tympanal apertures are open, oval. The hind knees are conspicuously dark-coloured, especially in the male, and the hind tibiae are yellowish to reddish. The species gives the impression of being a brachypterous derivative from *C. vagans*, which has an extremely similar song. There is some resemblance to *C. pullus*, but the male fore wings lack the expanded costal area shown by that species and the song is completely different.

Figures 1199–1204 Oscillograms at three different speeds of the calling songs of two males of *Chorthippus pullus*, one (1199, 1201, 1203) with two hind legs and the other (1200, 1202, 1204) with one hind leg.

SONG (Figs 1205–1210. CD 2, track 47). The male calling song is an echeme lasting about 10–15 s or more and consisting of about 50–80 or more syllables repeated at the rate of about 4–5/s. The echeme begins very quietly and gradually increases in loudness until it reaches maximum intensity about halfway through its duration. Oscillographic analysis shows that, by the time maximum intensity is reached, there are about 6–7 gaps in each downstroke hemisyllable.

In the presence of a female the male produces longer echemes, lasting up to 25 s or longer.

DISTRIBUTION. Known only from the southern side of the Alps, from the region between Lakes Iseo and Garda, through the eastern Italian Alps to the Karawanken range (both the Austrian and Slovenian sides) and the northern Julian Alps. It is seldom found at altitudes below 1200 m.

Chorthippus modestus (Ebner)

Stauroderus modestus Ebner, 1915: 555.

Reatine Grasshopper.

REFERENCES TO SONG. No published work known to us.

RECOGNITION. This species, known only from the Monti Reatini in the central Apennines, shows some resemblance to the *biguttulus*-group, but differs from it in having shorter wings, not usually reaching the hind knees, and less well-developed, rather shallow foveolae on the head. It may be distinguished from the brachypterous species described above in having well-developed hind wings, reaching the tips of the flexed fore wings, and from C. *vagans* in having half-closed, slit-like tympanal apertures. At the point where the pronotal lateral carinae are angled in the prozona they are clearly interrupted by an additional transverse sulcus on each side of the pronotum, as in *Stenobothrus festivus*. The hind tibiae are reddish in the male and the hind knees are dark-coloured in both sexes.

The male calling song is quite distinctive, showing an interesting combination of the overall pattern of the song of C. *rubratibialis* with the even syllable-sequences that are typical of C. *yersini*.

SONG (Figs 1211–1219. CD 2, track 48). The male calling song is a short sequence of dense echemes following immediately after one another, the whole sequence lasting about 3·0–4·4 s and usually including 1–4 echemes. The first echeme in the sequence lasts about 1·0–3·0 s and is usually longer than the remaining ones; it begins quietly, shows a gradual crescendo during about the first second of its duration and has an abrupt ending. The remaining echemes last about 0·7–1·0 s and begin with a more rapid crescendo; the last echeme often has a less abrupt ending than the earlier ones. There has been no published study of the rapid stridulatory leg-movements of this species, but oscillographic analysis suggests that the echemes are composed of diplosyllables repeated at the rate of about 70–80/s (see especially Fig. 1218, taken from a 'one-legged' song).

In the presence of a female the male produces somewhat modified echeme-sequences (Figs 1213, 1216, 1219), in which the echemes become progressively louder and slightly longer during the course of each sequence. In the single courtship song studied, there were

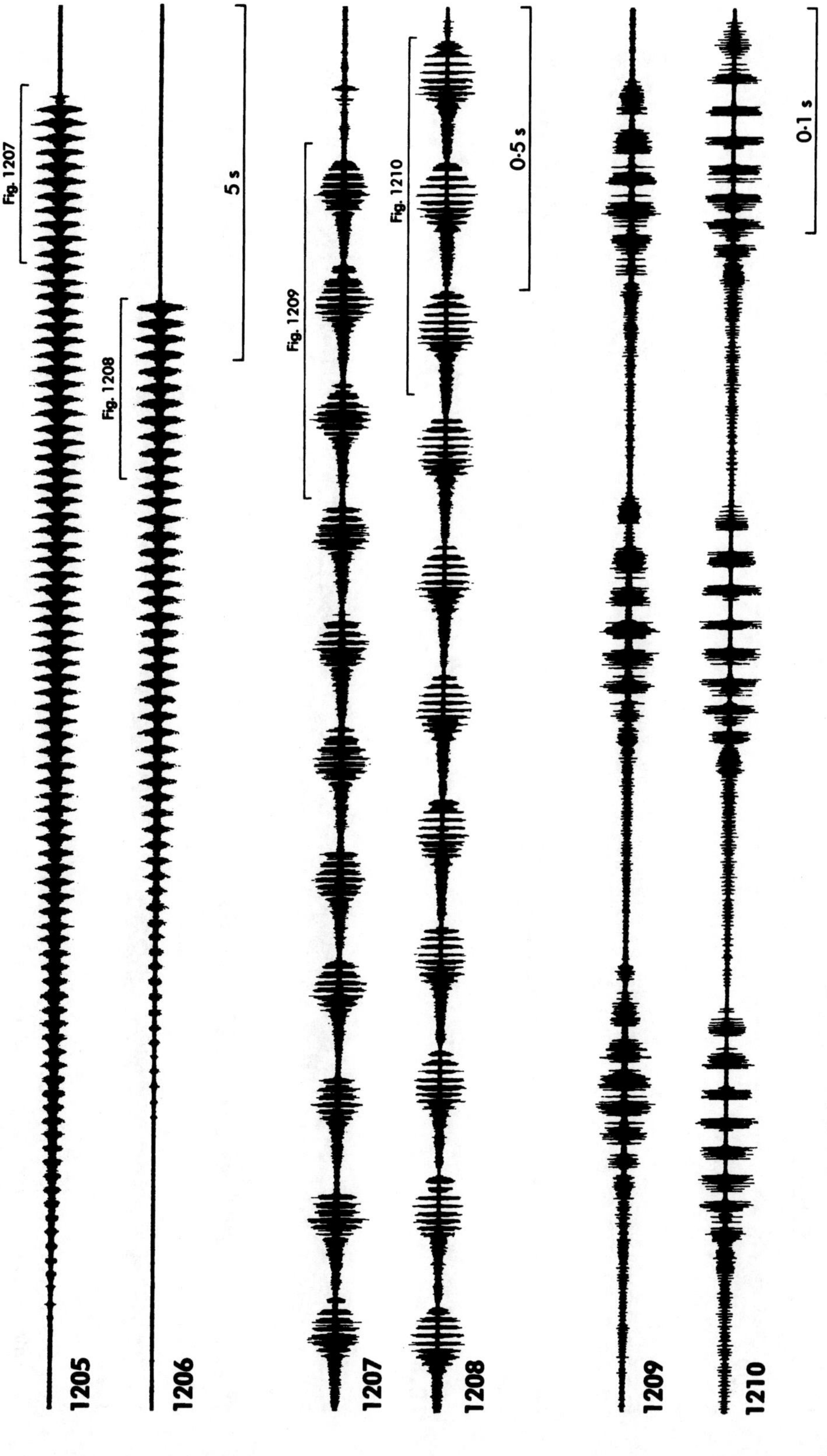

Figures 1205–1210 Oscillograms at three different speeds of the calling songs of two males of *Chorthippus alticola*.

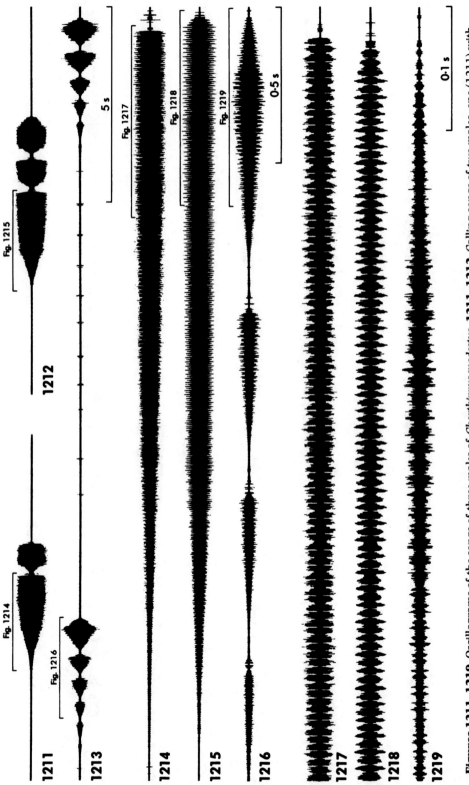

Figures 1211–1219 Oscillograms of the songs of three males of *Chorthippus modestus*. **1211, 1212.** Calling songs of two males, one (1211) with two hind legs and the other (1212) with one hind leg. **1213.** Part of the courtship song. **1214–1219.** Faster oscillograms of the indicated parts of the songs shown in Figs 1211–1213.

5–7 echemes in each sequence and the sequences were separated by intervals of 11–18 s, during which the male jerked its body from side to side while making quieter sounds.

DISTRIBUTION. Known at present only from the Monti Reatini in the central Italian Apennines, at altitudes above 1500 m.

Chorthippus vagans (Eversmann)

Oedipoda vagans Eversmann, 1848: 12.

Heath Grasshopper; (F) Criquet des Pins; (D) Steppengrashüpfer; (NL) Doorkijksprinkhaan.

REFERENCES TO SONG. **Oscillogram**: Bukhvalova & Zhantiev, 1993; Elsner & Popov, 1978; Grein, 1984; Holst, 1986; Kleukers *et al.*, 1997; Loher & Broughton, 1955; Schmidt, 1990, 1996. **Diagram**: Bellmann, 1985a, 1988, 1993a; Bellmann & Luquet, 1995; Duijm & Kruseman, 1983; Holst, 1970; Jacobs, 1950b, 1953a; Luquet, 1978; Ragge, 1965. **Sonagram**: Leroy, 1978. **Leg-movement**: Elsner & Popov, 1978. **Frequency information**: R.-G. Busnel, 1955; Loher & Broughton, 1955. **Musical notation**: Yersin, 1854b. **Verbal description only**: Broughton, 1972a, 1972b; Chopard, 1922; Defaut, 1987, 1988b; Faber, 1928, 1932, 1953a; Harz, 1957; Zeuner, 1931. **Disc recording**: Bellmann, 1993c (CD); Bonnet, 1995 (CD); Grein, 1984 (LP); Odé, 1997 (CD); Ragge *et al.*, 1965 (LP). **Cassette recording**: Bellmann, 1965b, 1993b; Burton & Ragge, 1987.

RECOGNITION. This species, predominantly brown or grey in colour, is superficially very similar to the *biguttulus*-group, but may be easily distinguished from that group by its wide open, oval tympanal apertures. The same character distinguishes it from *C. binotatus* and *C. pullus*, to which it also bears some resemblance. The tympanal apertures of *C. alticola* are almost as widely open, but that species has shorter fore wings and vestigial hind wings. For the distinction between *C. vagans* and the very similar *C. reissingeri*, see under the latter species.

The male calling song is quite different from those of any members of the *biguttulus*-group, but can easily be confused with the calling songs of *C. binotatus*, *C. alticola* and *C. reissingeri*.

SONG (Figs 1220–1241. CD 2, track 49). The calling song of a completely isolated male is typically a simple echeme of syllables repeated at the rate of about 3–8/s (usually 5–7/s). The duration of the echeme is very variable: in England it seldom lasts longer than 10 s (usually 4–8 s), but in other parts of Europe it is often longer, especially in Spain, where calling songs lasting 20–30 s (occasionally 35 s or longer) are quite common. The echeme usually begins quietly, but in England the maximum intensity is usually reached very quickly, within 1–3 s; elsewhere in Europe there is often a more gradual crescendo, commonly lasting for a third to a half of the echeme, occasionally for longer and rarely for virtually the entire echeme. In the early part of the echeme the syllables are sometimes alternately louder and softer. Oscillographic analysis shows that there are usually 3–8 gaps in each downstroke hemisyllable, and that the gaps tend to become larger as the hemisyllable progresses. Although frequently continuous from start to finish, the echeme is quite often interrupted by very brief pauses of about 0·3–1·0 s or sometimes a little longer; in some songs there are

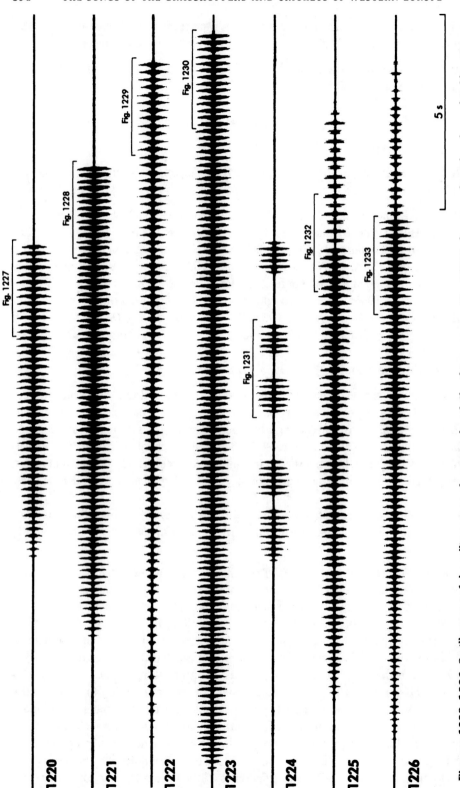

Figures 1220–1226 Oscillograms of the calling songs of seven males of *Chorthippus vagans*. Fig. 1220 is from a male with only one hind leg. Fig. 1224 shows the short echemes sometimes produced by an isolated male (but much more frequently by two or more males reacting with one another). Figs 1225 and 1226 show the calling songs of two males from the Sierra Nevada in southern Spain.

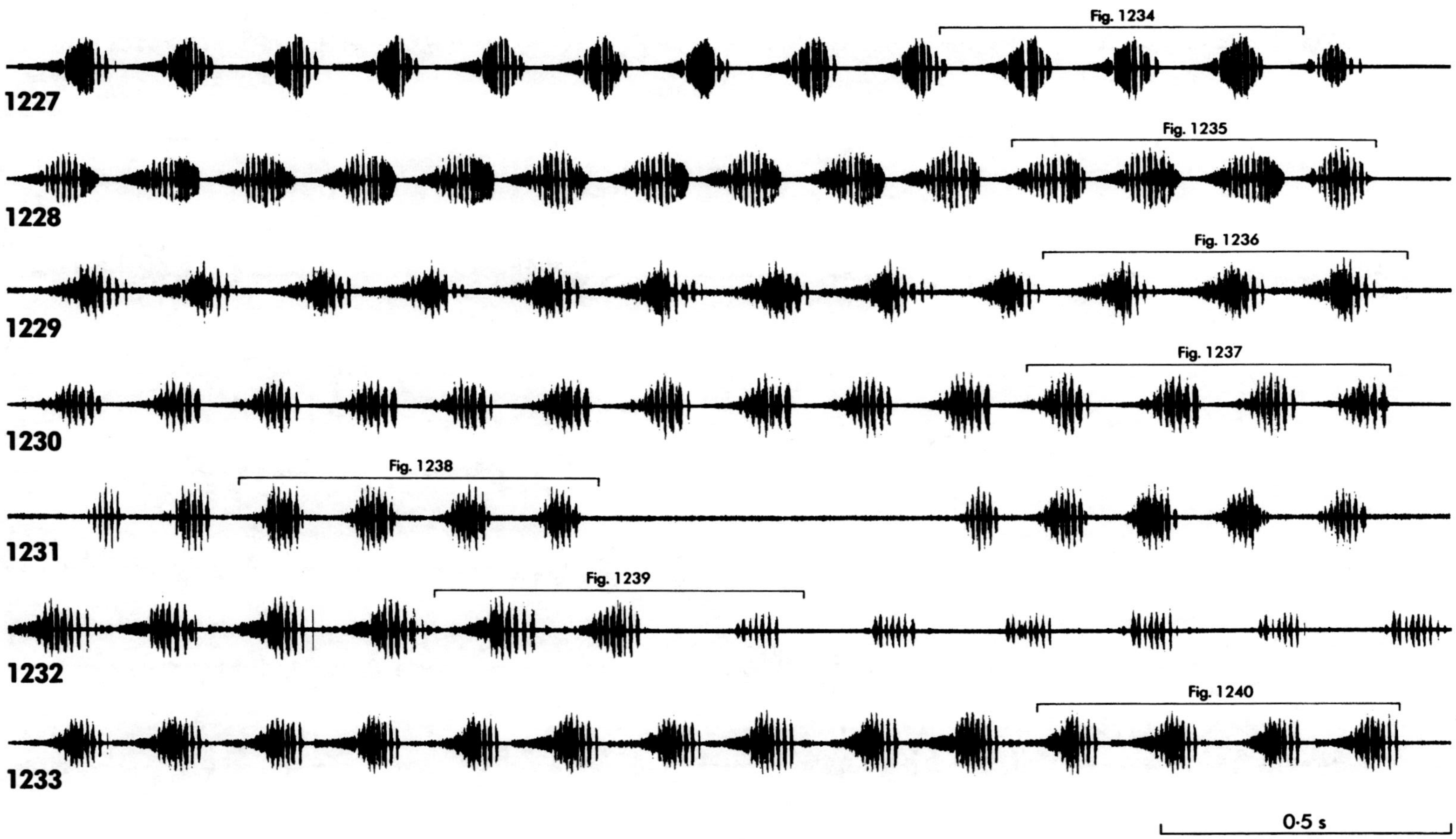

Figures 1227–1233 Faster oscillograms of the indicated parts of the songs of *Chorthippus vagans* shown in Figs 1220–1226. Fig 1227 is from a male with only one hind leg.

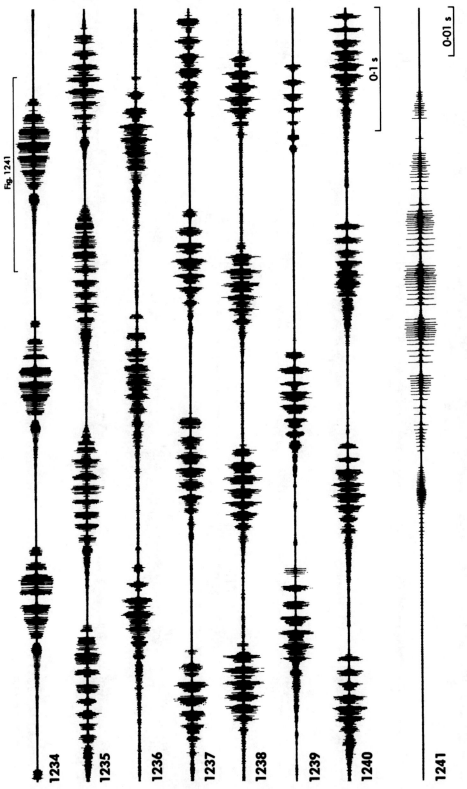

Figures 1234–1241 Faster oscillograms of the indicated parts of the songs of *Chorthippus vagans* shown in Figs 1227–1233. Figs 1234 and 1241 are from a male with only one hind leg.

only one or two of these pauses, but in others the song is broken up into short segments of about 0·5–1·5 s by frequent hesitations (Fig. 1224). When two or more males are together they commonly sing in this way, producing short echemes of about 1–6 syllables, all of equal loudness; two males often produce short echemes of this kind alternately.

In the Sierra Nevada in southern Spain the calling song shows an interesting modification (Figs 1225, 1226, 1232, 1233, 1239, 1240): an 'aftersong' is usually added immediately after the main echeme, lasting about 3–5 s and consisting of syllables produced by each hind leg alternately. These syllables have rather more gaps (usually 5–11) than those of the main echeme, and follow one another at a slightly slower tempo, taking both left-leg and right-leg syllables into account. One male observed producing a calling song of this kind had only one hind leg; it maintained the usual tempo of the aftersong by moving the remaining leg up and down twice as frequently as in a 'two-legged' aftersong.

There is no special courtship song, but in the presence of a female the male often produces more prolonged echemes, occasionally lasting as long as a minute. Between echemes, courting males often produce single syllables in which there are no gaps.

DISTRIBUTION. Widespread in Europe from Denmark southwards; common in the Iberian Peninsula, much less so in Italy and not reaching the southernmost part of the Balkan Peninsula. Very local in southern England. Farther east it occurs in the Ukraine, southern Russia and Kazakhstan. Also recorded from mountains in Morocco and Algeria.

Chorthippus reissingeri Harz

Chorthippus mollis reissingeri Harz, 1972: 130.

Reissinger's Grasshopper.

REFERENCES TO SONG. No published work known to us.

RECOGNITION. This species, known at present only from Alicante Province in south-east Spain, is extremely similar to *C. vagans* in both morphology and song. The most reliable character for separating males of these species is perhaps the penis (revealed by pulling the subgenital plate downwards), of which the dorsal valves extend clearly beyond the ventral ones in *C. reissingeri*, whereas the dorsal and ventral valves extend to about the same point in *C. vagans*. In the female the ventral valves of the ovipositor provide a more easily accessible character, when viewed from below: in *C. reissingeri* there is a distinct angle in the lateral profile where the broad base meets the narrow, pointed tip, whereas in *C. vagans* this profile is a sigmoid curve without an angle. There is also a difference in the pronotal lateral carinae, at the point where they approach each other most closely in the prozona: in *C. reissingeri* they are clearly interrupted at this point by an additional transverse sulcus on each side of the pronotum, whereas in *C. vagans* this sulcus is poorly developed or absent. The tympanal aperture is not quite as broad in *C. reissingeri* (usually more than twice as long as broad) as in *C. vagans* (usually less than twice as long as broad).

SONG (Figs 1242–1247. CD 2, track 50). The male calling song is an echeme lasting about 5–10 s and composed of about 35–70 syllables repeated at the rate of about 5–7/s. The echeme usually begins quietly, reaching maximum intensity within 1–3 s. There are about

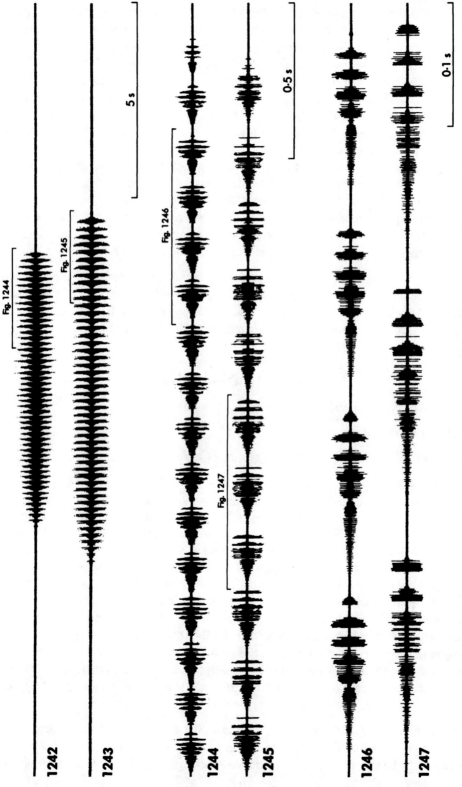

Figures 1242–1247 Oscillograms at three different speeds of the calling songs of two males of *Chorthippus reissingeri*.

3–5 gaps in each downstroke hemisyllable. We have been unable to find any significant difference between the calling songs of *C. reissingeri* and *C. vagans*.

DISTRIBUTION. Known at present only from a few localities in Alicante province, southeast Spain.

Chorthippus biguttulus (Linnaeus)

Gryllus (Locusta) biguttulus Linnaeus, 1758: 433.

Bow-winged Grasshopper; (F) Criquet mélodieux; (D) Nachtigall-Grashüpfer; (NL) Ratelaar; (S) Slåttergräshoppa; (SF) Ahoheinäsirkka.

REFERENCES TO SONG. **Oscillogram**: Bukhvalova, 1993b; Busnel & Loher, 1954a, 1954c; Elsner, 1974a, 1975, 1983a, 1983b; Elsner & Popov, 1978; Faber, 1957a; Grein, 1984; Halfmann & Elsner, 1978; Hedwig, 1990; Heinrich & Elsner, 1997; Helversen, 1972, 1979, 1989, 1993; Helversen & Elsner, 1977; Helversen & Helversen, 1975a, 1981, 1983, 1994; Holst, 1970, 1986; Ingrisch, 1995; Jacobs, 1957, 1963; Kleukers *et al.*, 1997; Loher, 1957; Loher & Broughton, 1955; Lottermoser, 1952; Meyer & Elsner, 1997; Perdeck, 1957; Ragge, 1976, 1981, 1984, 1987b; Ragge & Reynolds, 1988; Ragge *et al.*, 1990; Ronacher, 1989; Schmidt, 1978, 1987b, 1989, 1990; Schmidt & Schach, 1978; Vedenina, 1990; Vedenina & Zhantiev, 1990; Zhantiev, 1981. **Diagram**: Bellmann, 1985a, 1988, 1993a; Bellmann & Luquet, 1995; Duijm & Kruseman, 1983; Holst, 1970; Jacobs, 1950b, 1953a; Luquet, 1978; Wallin, 1979; Weih, 1951. **Leg-movement**: Elsner, 1974a, 1975, 1983a, 1983b; Elsner & Popov, 1978; Halfmann & Elsner, 1978; Hedwig, 1990; Heinrich & Elsner, 1997; Helversen, 1979, 1989; Helversen & Elsner, 1977; Helversen & Helversen, 1983, 1994; Meyer & Elsner, 1997; Ronacher, 1989; Wolf, 1985. **Frequency information**: R.-G. Busnel, 1955; Busnel & Loher, 1953, 1954c; Dumortier, 1963c; Loher, 1957; Loher & Broughton, 1955; Meyer & Elsner, 1996, 1997; Vedenina & Zhantiev, 1990. **Musical notation**: Faber, 1932, 1934; Yersin, 1854b. **Verbal description only**: Beier, 1956; Broughton, 1972a, 1972b; Busnel & Loher, 1954b; Chopard, 1952; Defaut, 1984b, 1987, 1988b; Faber, 1928, 1929b, 1932, 1953a, 1953b; Harz, 1957; Helversen & Helversen, 1990; Ramme, 1921a; Ronacher & Stumpner, 1988; Sychev, 1979; Weber, 1984; Weih, 1951; Zeuner, 1931. **Disc recording**: Andrieu & Dumortier, 1963 (LP), 1994 (CD); Bellmann, 1993c (CD); Bonnet, 1995 (CD); Burton, 1969 (LP, as French Meadow Grasshopper); Grein, 1984 (LP); Odé, 1997 (CD). **Cassette recording**: Bellmann, 1985b, 1993b; Wallin, 1979.

RECOGNITION (Plate 3: 10). This species gives its name to the *biguttulus*-group, which, in western Europe and North Africa, also includes *C. brunneus*, *C. mollis*, *C. jacobsi*, *C. yersini*, *C. rubratibialis* and *C. marocanus*. The group is characterized by its narrow, slit-like tympanal apertures, angled pronotal lateral carinae, fully developed wings and lack of any striking features of colour pattern. All the members of the group are very similar morphologically but can be distinguished in the field by the male calling songs. An identification key for all the western European species except *C. rubratibialis* is given by Ragge & Reynolds (1988). For a discussion of the taxonomy of this group see p. 75.

Males of *C. biguttulus* may be distinguished from *C. brunneus*, *C. mollis* and *C. jacobsi* by the broader costal and subcostal areas of the fore wings (maximum width of combined costal and subcostal areas usually exceeding 1·05 mm) and, additionally, from typical *C. brunneus* by the larger number of stridulatory pegs (usually more than 80). Females can usually be distinguished from *C. brunneus* and *C. jacobsi* by their shorter fore wings (usually less than 16 mm), but there seems to be no reliable character for separating females of *C. biguttulus* and *C. mollis*.

C. biguttulus is also very similar morphologically to *C. yersini, C. rubratibialis* and *C. marocanus*, but none of these species overlap with one another geographically and so there is no problem in practice in distinguishing between them.

In the field *C. biguttulus* can be recognized at once from the highly characteristic male calling song, one of the commonest summer sounds of the European countryside from Scandinavia to the Pyrenees and Alps.

Natural hybrids occasionally occur between this species and *C. brunneus* (Klingstedt, 1939: 370; Perdeck, 1957: 15; Faber, 1957a: 11; Ragge, 1976). In the field male hybrids can be recognized easily by their calling song (see p. 49 and Figs 25, 28, 31).

SONG (Figs 1248–1269. CD 2, track 51). The male calling song consists of a series of echeme-sequences, each beginning quietly and ending abruptly; the crescendo in the first echeme-sequence is typically more gradual than in the remaining ones. The number of echeme-sequences in the series is usually within the range 1–4 and is most often 3. The first echeme-sequence is nearly always longer than the others, typically lasting about 2·5–6·0 s, and each of the others usually lasts about 1·5–2·5 s. When there are three or more echeme-sequences, the intervals between them usually become longer as the song progresses; in a song of three echeme-sequences the first interval is usually about 0·5–1·5 s and the second about 1·0–3·0 s. There are usually 35–80 echemes in the first echeme-sequence, and 15–45 in the remaining ones. Each echeme usually lasts 40–80 ms and the echeme repetition rate is usually 10–20/s.

Each echeme consists of a small number of syllables, typically three but occasionally two and sometimes four or more. Elsner (1974a) and later workers have shown that they are diplosyllables, both the downstrokes and upstrokes of the hind legs producing sounds. Because the two hind legs are slightly out of phase the syllable structure is not clear in a 'two-legged' song (Fig. 1264), but in a 'one-legged' song the six hemisyllables typically forming each echeme may be seen quite clearly in a fast oscillogram (Fig. 1265). The first (downstroke) hemisyllable in each echeme is normally louder than the remaining ones and this, together with a slight irregularity in rhythm, results in the echemes being just distinguishable to the unaided ear and imparting a characteristic uneven quality (sometimes described as 'metallic') to the sound of each echeme-sequence.

Local populations are sometimes found in which the song is rather atypical. As Jacobs (1957) and Ingrisch (1995) have observed, the song often consists of a larger number of shorter echeme-sequences (even up to ten or more) in parts of the southern Alps, from Aosta in the west to the Dolomites and southern Carinthia in the east (Figs 1254, 1255); this form is sometimes regarded as being subspecifically or even specifically distinct (see p. 76). Jacobs (1963) also found that the populations in Jutland have an unusually short first echeme-sequence (averaging about 2 s in duration) and a rather different quality of sound. One of us (DRR) has found a population near the Col de Vence in the French Alpes Maritimes in which there seem always to be only one or two echeme-sequences in each song (Fig. 1248).

Figures 1248–1255 Oscillograms of the calling songs of eight males of *Chorthippus biguttulus*. Fig. 1251 is from a male with only one hind leg. Figs 1254 and 1255 are from two males from southern Carinthia in Austria and show the larger number of echeme-sequences produced by some populations in the southern Alps.

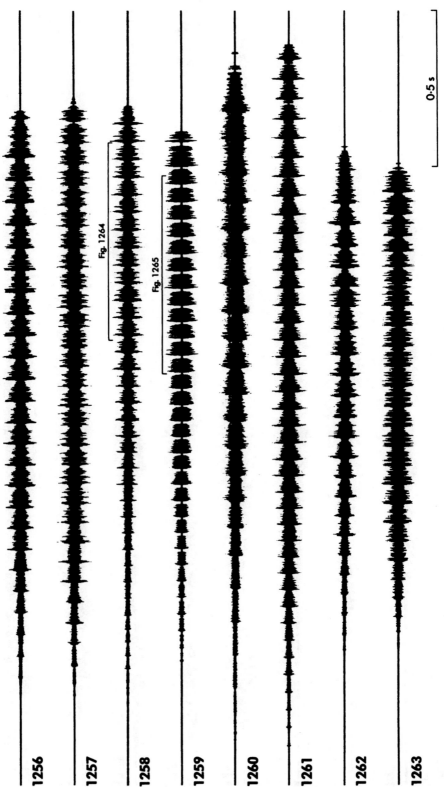

Figures 1256–1263 Faster oscillograms of the indicated parts of the songs of *Chorthippus biguttulus* shown in Figs 1248–1255. Fig. 1259 is from a male with only one hind leg.

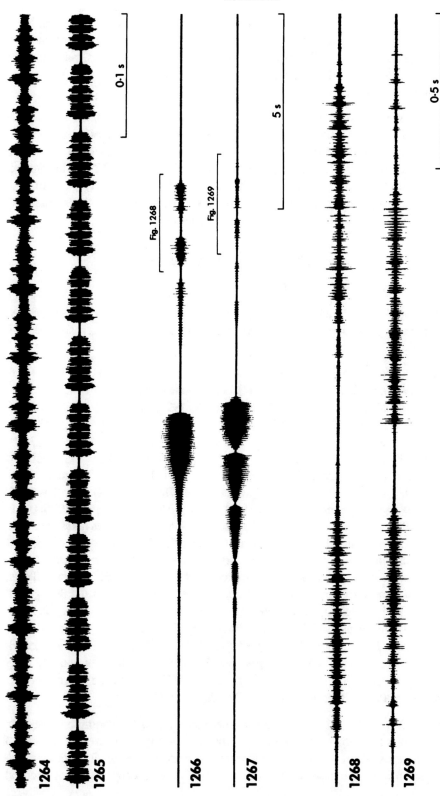

Figures 1264–1269 Oscillograms of the songs of *Chorthippus biguttulus*. **1264, 1265.** Faster oscillograms of the indicated parts of the songs shown in Figs 1258 and 1259; Fig. 1265 is from a male with only one hind leg. **1266–1269.** Oscillograms at two different speeds of the courtship songs of two males.

The courtship song (Figs 1266–1269) consists of a single echeme-sequence similar to the first echeme-sequence of the calling song (but usually quieter and with a hesitant start) followed by a series of quite different and less regular echeme-sequences (referred to by German workers as 'flatternd klingender Laute' or 'Schüttelversen') produced by the hind legs vibrating in a higher position.

DISTRIBUTION. Very common in northern and central Europe, from southern Scandinavia southwards. The southern limit runs from the Spanish side of the Pyrenees along the Mediterranean coast of France to the Italian side of the Alps, and then down the Dalmatian coast and across the Balkan Peninsula at about the level of latitude 42°N. The range is believed at present to extend eastwards across temperate Asia to China and Japan, but studies on the song are needed to confirm this. Although very common in northern France, *C. biguttulus* does not occur in the British Isles or Channel Islands.

Chorthippus brunneus (Thunberg)

Gryllus brunneus Thunberg, 1815: 256.

Field Grasshopper; (F) Criquet duettiste; (D) Brauner Grashüpfer; (NL) Tandradje; (DK) Almindelige Markgræshoppe; (S) Backgräshoppa; (N) Gråbrun Markgrashoppe; (SF) Ketoheinäsirkka.

REFERENCES TO SONG. **Oscillogram**: Broughton, 1954 (as *bicolor*), 1963a, 1976; Bukhvalova, 1993b; Busnel & Loher, 1953 (as *bic.*), 1954a (as *bic.*), 1954c (as *bic.*), 1955 (as *bic.*), 1956 (as *bic.*); Butlin *et al.*, 1985; Charalambous, 1990; Charalambous *et al.*, 1994; Dumortier, 1963c; Elsner & Popov, 1978; Faber, 1957a; Green, 1987; Grein, 1984; Halfmann & Elsner, 1978; Haskell, 1957, 1958, 1961; Helversen, 1972; Helversen & Helversen, 1981, 1994; Holst, 1970, 1986; Ingrisch, 1995; Kleukers *et al.*, 1997; Loher, 1957; Loher & Broughton, 1955 (as *bic.*); Perdeck, 1957; Ragge, 1976, 1987b; Ragge & Reynolds, 1988; Ragge *et al.*, 1990; Schmidt, 1989; Schmidt & Schach, 1978; Vedenina & Zhantiev, 1990; Young, 1971. **Diagram**: Bellmann, 1985a, 1988, 1993a; Bellmann & Luquet, 1995; Duijm & Kruseman, 1983; Haskell, 1957; Holst, 1970; Jacobs, 1950b (as *bic.*), 1953a (as *bic.*); Luquet, 1978; Ragge, 1965; Wallin, 1979. **Sonagram**: Young, 1971. **Leg-movement**: Elsner & Popov, 1978; Halfmann & Elsner, 1978; Helversen & Helversen, 1994. **Frequency information**: Busnel & Loher, 1953 (as *bic.*); Froehlich, 1989; Haskell, 1957; Meyer & Elsner, 1996. **Musical notation**: Yersin, 1854b (as *bic.*). **Verbal description only**: Broughton, 1972a, 1972b; Butlin & Hewitt, 1986; Chopard, 1952 (as *bic.*); Defaut, 1987, 1988b; Faber, 1928 (as *bic.*), 1929b (as *bic.*), 1953a, 1953b; Harz, 1957; Haskell, 1955a; Ramme, 1921a; Weber, 1984; Weih, 1951 (as *bic.*); Yersin, 1857 (as *bic.*); Zeuner, 1931 (as *bic.*). **Disc recording**: Andrieu & Dumortier, 1963 (LP), 1994 (CD); Bellmann, 1993c (CD); Bonnet, 1995 (CD); Grein, 1984 (LP); Odé, 1997 (CD); Ragge *et al.*, 1965 (LP). **Cassette recording**: Bellmann, 1985b, 1993b; Burton & Ragge, 1987; Wallin, 1979.

RECOGNITION. The most reliable morphological character for distinguishing both sexes of typical *C. brunneus* from the other members of the *biguttulus*-group is the low number of stridulatory pegs: usually fewer than 80 in *C. brunneus*, more than 80 in all the other members of the group. Males can also be distinguished from *C. biguttulus* by the narrower costal and subcostal areas of the fore wings (maximum width of combined costal and subcostal areas usually less than 1·05 mm), and both sexes from *C. mollis* by the longer fore wings (usually more than 12 mm in the male, 16 mm in the female). In northern Spain, where *C. brunneus* meets *C. jacobsi*, and in the Italian Peninsula, where it overlaps with *C. rubratibialis*, it is necessary to use the stridulatory pegs for a reliable identification from morphology.

In at least four relatively small areas of western Europe there are populations of *C. brunneus* in which the number of stridulatory pegs is larger and thus of little or no diagnostic value. In Corsica, Sardinia and associated smaller islands, there are often more than 80 pegs and occasionally as many as 100; however, *C. brunneus* is the only member of the *biguttulus*-group on these islands and so there is no difficulty in identifying it there. In Sicily there are about 105–135 pegs in the male, closely matching the number in *C. yersini*, the only other member of the group to occur on the island. Sicilian *C. brunneus* may be distinguished morphologically from Sicilian *C. yersini* by the length of the fore wings, which are longer than 13 mm in the male, 18 mm in the female (shorter than these values in *C. yersini*). Ingrisch (1995) has shown that some populations of *C. brunneus* in the southern Alps have a higher number of pegs, often more than 100, and are also closer to *C. biguttulus* and *C. mollis* in wing-venation; there seems to be no reliable way of distinguishing morphologically between these three species in these localities.

In Finland *C. brunneus* has about 90–140 stridulatory pegs in the male and usually more than 80 in the female; this form, which also shows small differences in the venation of the fore wings, is sometimes regarded as a distinct subspecies, *C. b. brevis* Klingstedt. Unlike the forms mentioned above, Finnish *C. brunneus* has a longer stridulatory file than typical *C. brunneus*, so that the density of the pegs remains normal for the species. The longer file enables males of this Finnish population to be distinguished morphologically from *C. biguttulus*, the only other member of the group to occur in Finland: the file is usually longer than 4 mm in Finnish *C. brunneus*, shorter than 4 mm in *C. biguttulus*. The costal and subcostal areas of the fore wings are a little broader in Finnish *C. brunneus* than is typical of the species, but their combined maximum width is still usually less than 1·05 mm.

Throughout its range *C. brunneus* may be easily distinguished in the field from all other members of the *biguttulus*-group by the very characteristic male calling and rivalry songs.

SONG (Figs 1270–1311. CD 2, track 52). The male calling song is a sequence of short, abrupt echemes, each typically lasting about 120–250 ms. There are usually 6–10 echemes in each sequence, occasionally fewer and rarely up to 14. The intervals between successive echemes are very variable: usually they are within the range 1·0–2·5 s, but can be as short as 0·3 s or as long as 4·5 s. Sometimes the intervals are quite uniform within one sequence, but more often they vary; there is a common tendency for them to become longer as the sequence progresses, but sometimes they vary haphazardly. The first one or two echemes in a sequence are often quieter than the remaining ones, and occasionally there is a gradual increase in loudness affecting the first three or more echemes.

Oscillographic analysis shows that each echeme consists of a rather irregular series of fairly discrete sounds, the louder ones usually numbering between 10 and 20. Halfmann & Elsner (1978) have shown that the number of syllables (i.e. complete down and up leg-movements) in each echeme is only about half this number, that the two hind legs are moved

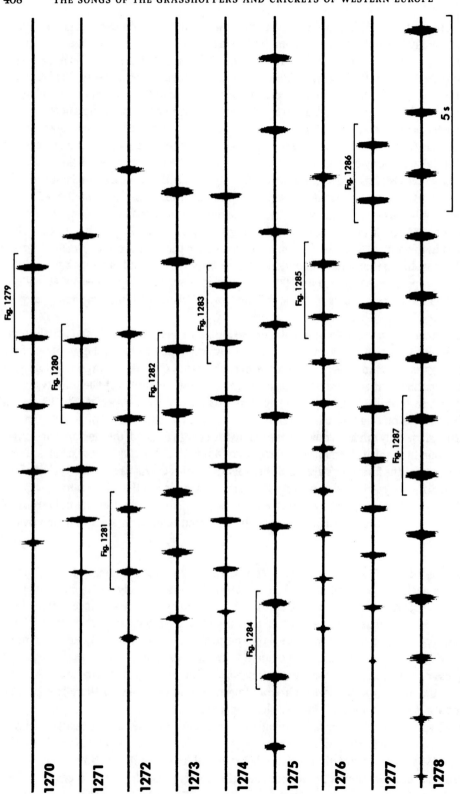

Figures 1270–1278 Oscillograms of the calling songs of nine males of *Chorthippus brunneus*. Fig. 1276 is from a male with only one hind leg.

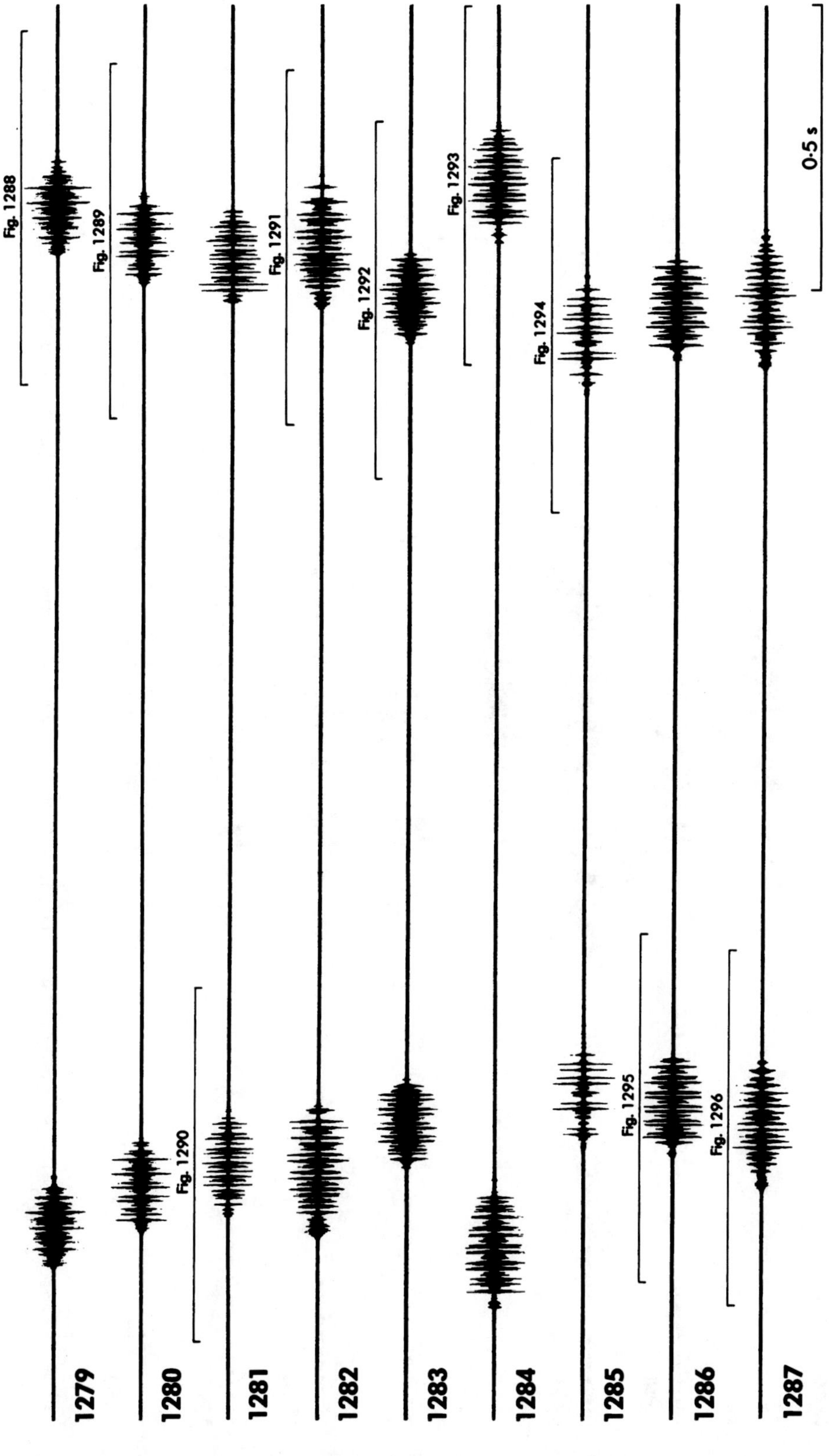

Figures 1279–1287 Faster oscillograms of the indicated parts of the songs of *Chorthippus brunneus* shown in Figs 1270–1278. Fig. 1285 is from a male with only one hind leg.

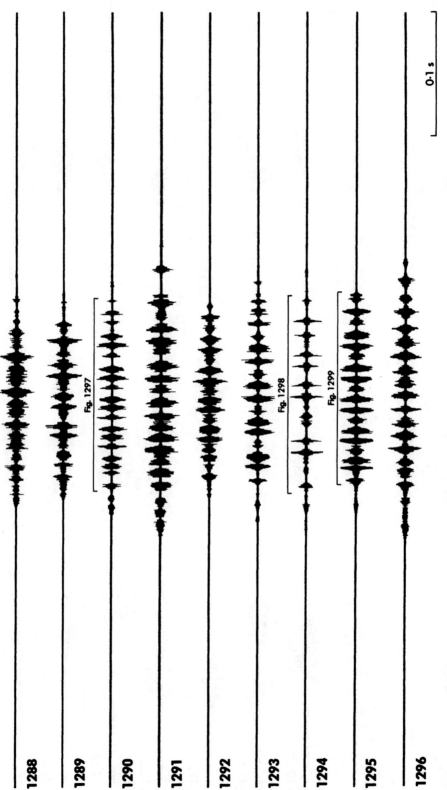

Figures 1288–1296 Faster oscillograms of the indicated parts of the songs of *Chorthippus brunneus* shown in Figs 1279–1287. Fig. 1294 is from a male with only one hind leg.

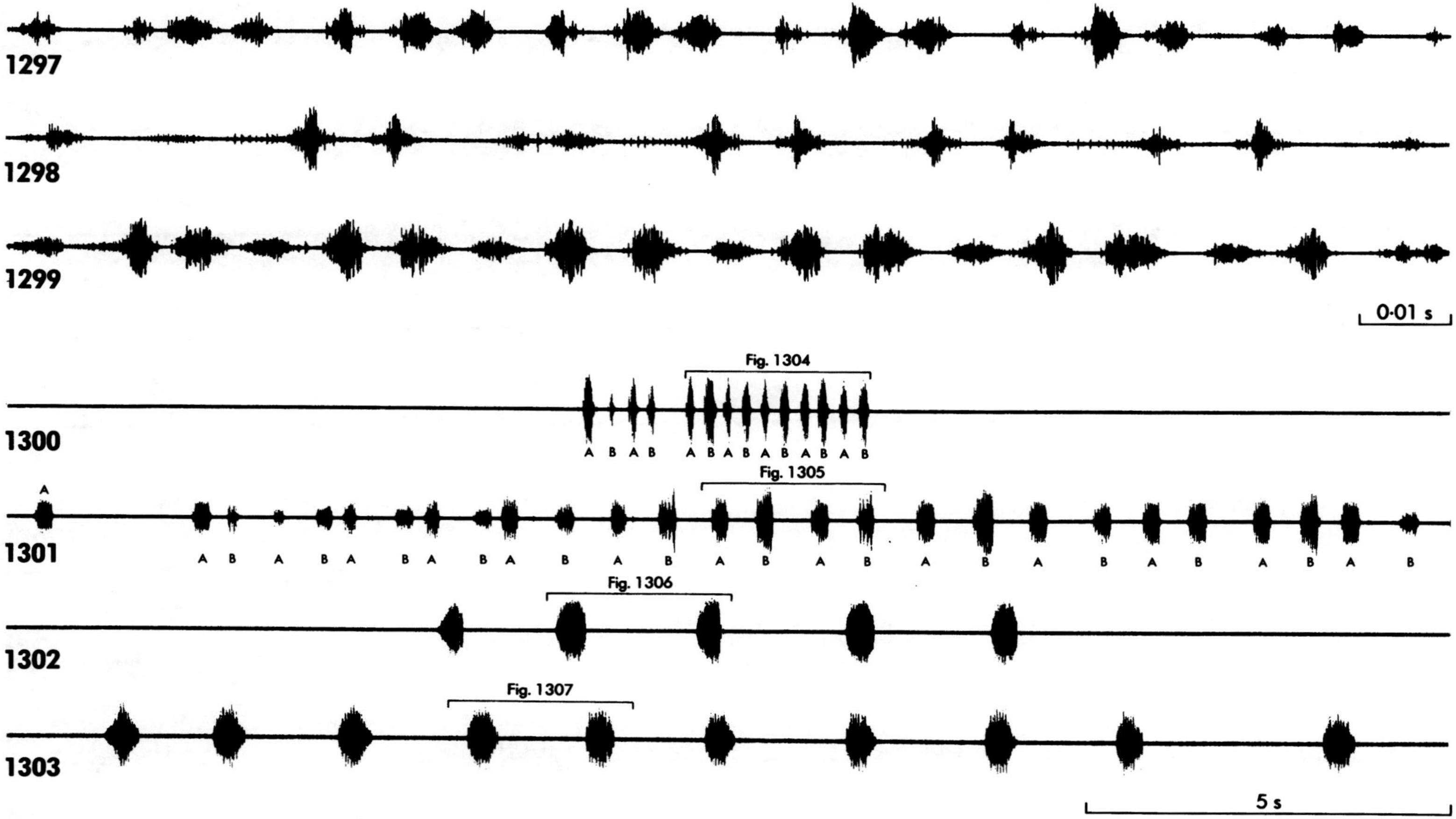

Figures 1297–1303 Oscillograms of the songs of *Chorthippus brunneus*. **1297–1299.** Faster oscillograms of the indicated parts of the songs shown in Figs 1290, 1294 and 1295; Fig. 1298 is from a male with only one hind leg. **1300, 1301.** Rivalry songs produced by two males (A and B) about 10 cm apart (1300) and about 30 cm apart (1301), interacting with each other by alternating echemes. **1302, 1303.** The calling songs of two males from Sicily, showing the unusually long echemes.

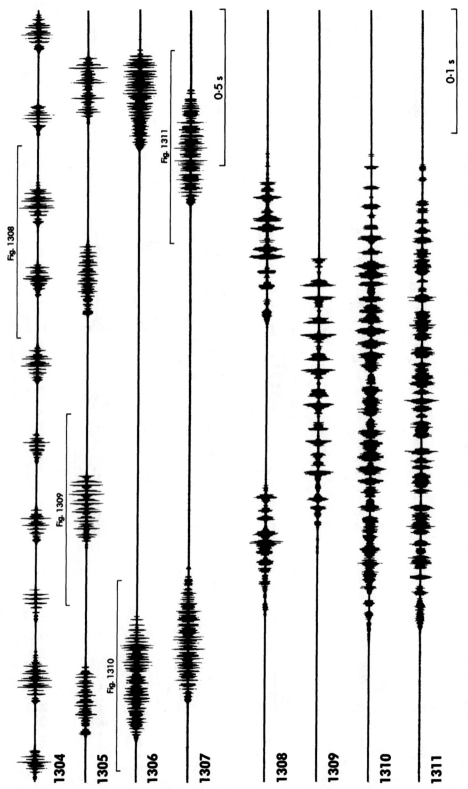

Figures 1304–1311 Faster oscillograms of the indicated parts of the songs of *Chorthippus brunneus* shown in Figs 1300–1303.

up and down alternately rather than synchronously, and that sound is produced during both downstrokes and upstrokes.

It is characteristic of this species for two males to interact with each other by alternating echemes. In such 'rivalry' songs the echemes are typically shorter than usual, about 100–150 ms, and are repeated more rapidly; sometimes more than two males will interact with one another in this way. As Weih (1951) and Loher (1957) have observed, there is a tendency for the alternation to be more rapid when the males are closer together. The oscillogram shown in Fig. 1300 is from a pair about 10 cm apart, with an echeme repetition rate for each male of about 1·9/s, and that shown in Fig. 1301 is from a pair about 30 cm apart, with an echeme repetition rate for each male of about 0·8/s.

In Sicily the echemes are longer than in typical *C. brunneus*, lasting about 300–700 ms in the calling song and about 200–300 ms in the rivalry song; oscillographic analysis suggests that there are more syllables per echeme, in proportion to the longer duration (Figs 1302, 1303, 1306, 1307, 1310, 1311). According to Perdeck (1957) the same is true of Finnish *C. brunneus*, and Ingrisch (1995) has observed songs of this kind in some populations of *C. brunneus* in the southern Alps. These songs are quite similar to the songs of hybrids between typical *C. brunneus* and *C. biguttulus* (Figs 25, 28, 31), but each echeme sounds less uniform and oscillograms show the more complex internal structure typical of *C. brunneus*.

There is no well-developed courtship song in this species, but in the presence of a female the male often produces rather softer and longer echemes with somewhat slower leg-movements.

DISTRIBUTION. Widespread in northern and central Europe, reaching northwards to all but the northernmost parts of Scandinavia. Southwards, the range includes the extreme north of the Iberian Peninsula, from Galicia to the Spanish Pyrenees, all of France, Corsica, Sardinia, the whole of Italy, including Sicily, and the northern part of the Balkan Peninsula down to about the level of latitude 42°N. Widespread in the British Isles. The range is believed at present to extend eastwards across temperate Asia to China, but this needs to be confirmed by studies on the song.

Chorthippus mollis (Charpentier)

Gryllus mollis Charpentier, 1825: 164.

Lesser Field Grasshopper; (F) Criquet des jachères; (D) Verkannter Grashüpfer; (NL) Snortikker.

REFERENCES TO SONG. **Oscillogram**: Bukhvalova, 1993*b*; Bukhvalova & Zhantiev, 1993; Elsner, 1970, 1974*a*, 1975; 1983*a*, 1983*b*; Elsner & Popov, 1978; Faber, 1957*a*; Grein, 1984; Halfmann & Elsner, 1978; Hedwig, 1990; Heinrich & Elsner, 1997; Helversen & Elsner, 1977; Helversen & Helversen, 1975*a*, 1981, 1994; Holst, 1970, 1986; Ingrisch, 1995; Kleukers *et al.*, 1997; Loher & Broughton, 1955; Meyer & Elsner, 1997; Ragge, 1981, 1984, 1987*b*; Ragge & Reynolds, 1988; Ragge *et al.*, 1990; Schmidt, 1990; Schmidt & Schach, 1978; Vedenina & Zhantiev, 1990; Zhantiev, 1981. **Diagram**: Bellmann, 1985*a*, 1988, 1993*a*; Bellmann & Luquet, 1995; Duijm & Kruseman, 1983; Holst, 1970; Jacobs, 1950*b*, 1953*a*; Luquet, 1978. **Leg-movement**: Elsner, 1970, 1974*a*, 1975, 1983*a*, 1983*b*; Elsner & Popov, 1978; Halfmann

& Elsner, 1978; Hedwig, 1990; Heinrich & Elsner, 1997; Helversen & Elsner, 1977; Helversen & Helversen, 1994; Meyer & Elsner, 1997. **Frequency information**: M.-C. Busnel, 1953; R.-G. Busnel, 1955; Loher & Broughton, 1955; Meyer & Elsner, 1996, 1997. **Musical notation**: Yersin, 1854*b*. **Verbal description only**: Broughton, 1972*a*, 1972*b*; Chopard, 1952; Dumortier, 1963*d*; Faber, 1928, 1929*b*, 1953*a*, 1953*b*; Harz, 1957, 1975*b*; Ramme, 1921*a*, 1923; Sychev, 1979; Weber, 1984; Zeuner, 1931. **Disc recording**: Bellmann, 1993*c* (CD); Bonnet, 1995 (CD); Grein, 1984 (LP); Odé, 1997 (CD). **Cassette recording**: Bellmann, 1985*b*, 1993*b*.

RECOGNITION. Both sexes of typical *C. mollis* may be distinguished from *C. brunneus* and mainland Iberian *C. jacobsi* by their shorter fore wings: usually less than 12 mm in the male, 16 mm in the female. Males may be distinguished from *C. biguttulus* by the narrower costal and subcostal areas of the fore wings (maximum width of combined costal and subcostal areas usually less than 1 mm), but there seems to be no reliable way of distinguishing morphologically between females of these two species.

In much of the Alps there is a larger form of *C. mollis*, in which the fore wings are longer and the costal and subcostal areas of the male fore wings broader, even in relation to the length of the fore wings. There seems to be no reliable morphological way of distinguishing this form, usually regarded as a distinct subspecies, *C. m. ignifer* Ramme, from *C. biguttulus* and the southern Alpine form of *C. brunneus* with a higher number of stridulatory pegs.

In the field *C. mollis* may be recognized at once from its highly characteristic male calling song.

SONG (Figs 1312–1353. CD 2, tracks 54, 55).   The male calling song of typical *C. mollis* (Figs 1312–1327) is a long echeme-sequence, beginning very quietly and gradually increasing in loudness until near the end, when the echemes normally become more spaced out, less regular in structure and progressively quieter. All but these last echemes begin with a characteristic 'tick', produced by the downstroke of one hind leg only. The number of audible echemes in the sequence is quite variable but is usually within the range 40–80; the sequence usually lasts for 15–30 s. The duration of the individual echemes (measured between successive ticks) increases somewhat during the course of the sequence, the early ones lasting about 250–350 ms and the later ones about 350–450 ms; there is, however, often a slight decrease in echeme duration during the first few echemes, before the steady increase begins, and we have recorded some songs in which the echeme duration remained almost constant throughout the sequence. The concluding echemes that lack ticks vary greatly in number: there are seldom more than 10, sometimes only 1 or 2, and rarely they are completely absent.

Oscillographic analysis shows that, after the initial tick, each echeme begins and ends quietly, reaching maximum loudness in the middle, and consists of syllables repeated at the rate of about 60–80/s; the syllable repetition rate often increases a little during the course of the echeme-sequence. The number of syllables per echeme also increases during the echeme-sequence, beginning at about 8–12 and reaching about 20–25 or more at the loudest part of the sequence. The left and right hind legs are usually slightly out of phase (sometimes even in opposite phase, or almost so), so that the syllable structure is not completely clear in a 'two-legged' song (this is particularly true of the concluding, tickless echemes). However, in a fast oscillogram of a 'one-legged' song (Figs 1346, 1352) it can be

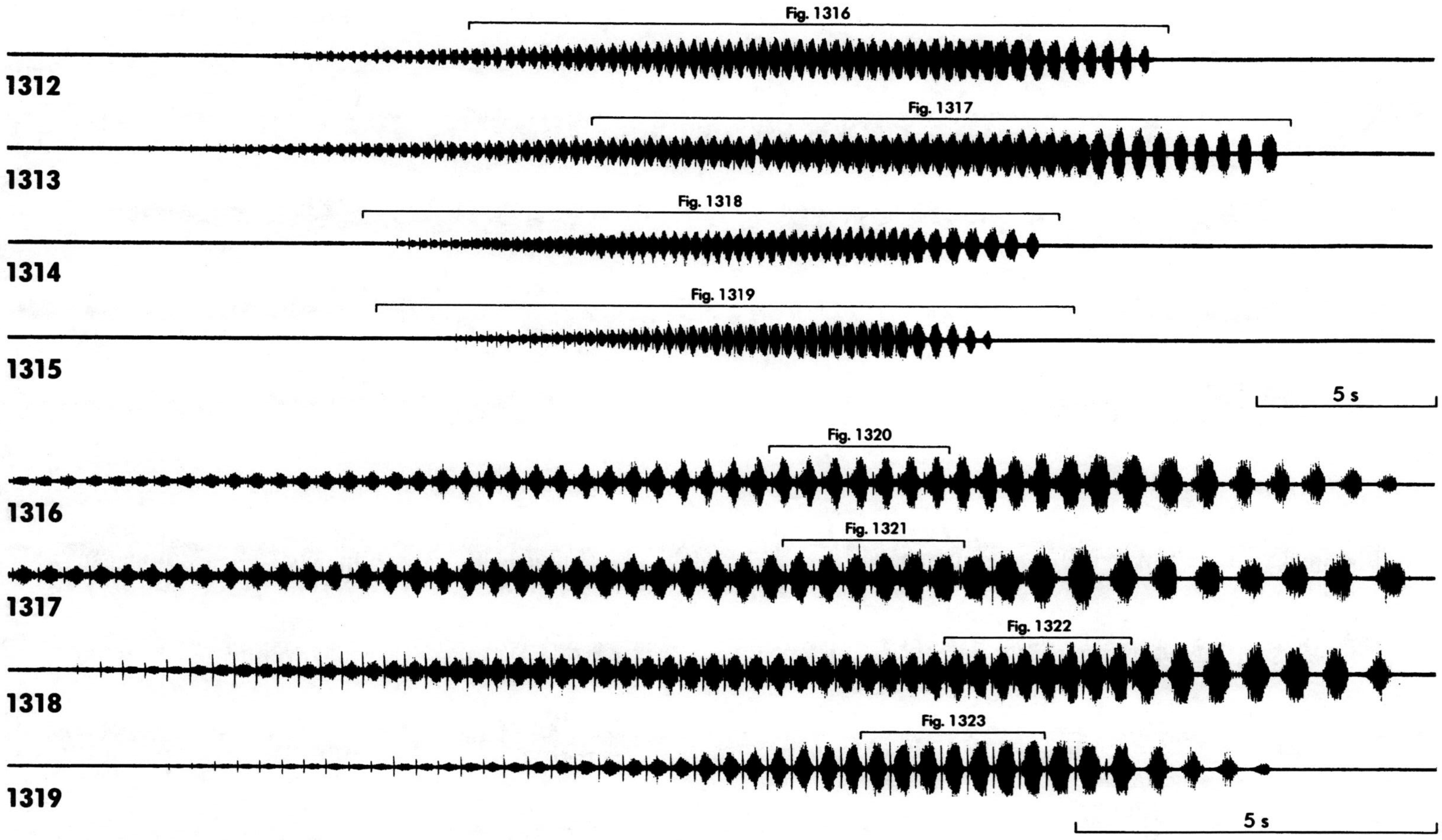

Figures 1312–1319 Oscillograms at two different speeds of the calling songs of four males of typical *Chorthippus mollis*.

Figures 1320–1327 Faster oscillograms of the indicated parts of the songs of *Chorthippus mollis* shown in Figs 1316–1319.

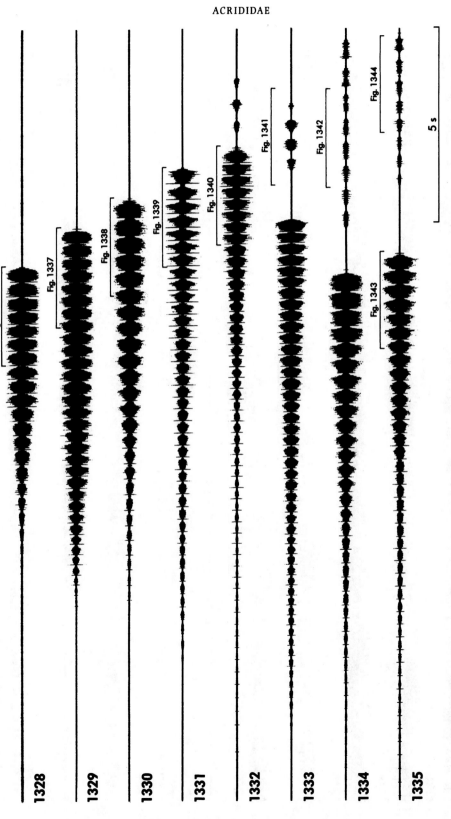

Figures 1328–1335 Oscillograms of the songs of eight males of *Chorthippus mollis ignifer*. **1328–1332.** Calling songs of males with (1328, 1330–1332) two hind legs and (1329) one hind leg. **1333–1335.** Courtship songs of males with (1333, 1334) two hind legs and (1335) one hind leg.

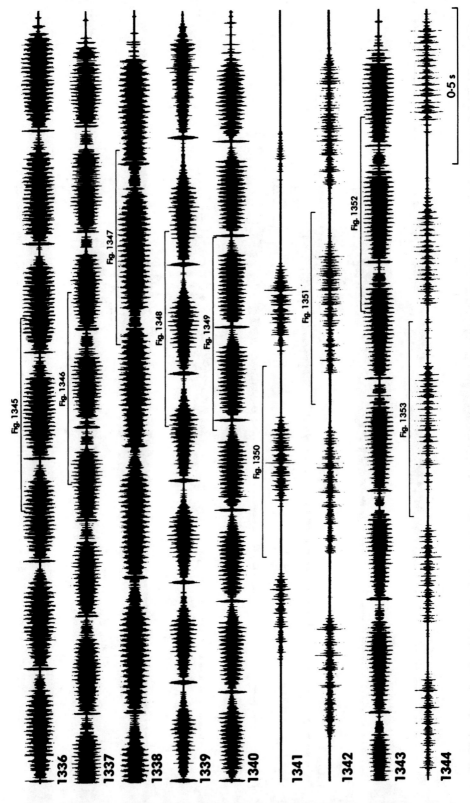

Figures 1336–1344 Faster oscillograms of the indicated parts of the songs of *Chorthippus mollis ignifer* shown in Figs 1328–1335. Figs 1337, 1343 and 1344 are from males with only one hind leg.

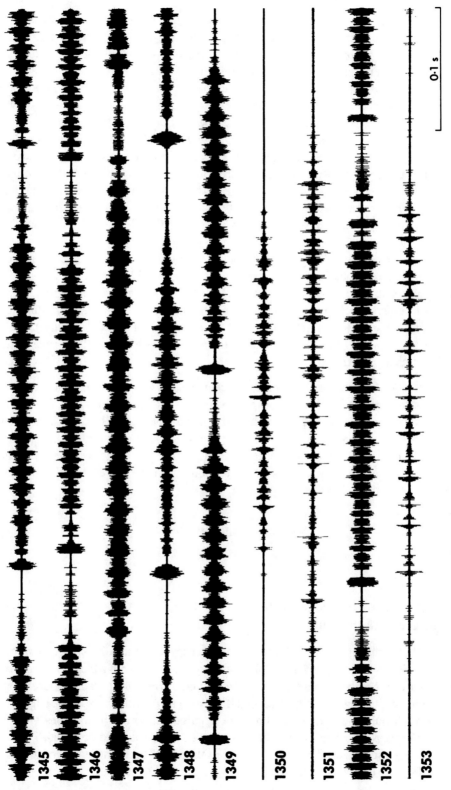

0·1 s

Figures 1345–1353 Faster oscillograms of the indicated parts of the songs of *Chorthippus mollis ignifer* shown in Figs 1336–1344. Figs 1346, 1352 and 1353 are from males with only one hind leg.

seen that there are alternate louder and quieter sounds, and Elsner (1970) has shown that these are hemisyllables produced by the downstroke and upstroke, respectively, of the hind leg; in an oscillogram of a 'two-legged' song it is usually only the louder, downstroke hemisyllables that are clearly distinguishable (Figs 1324–1327).

There is no well-developed courtship song in typical *C. mollis*, but in the presence of a female the male usually produces longer echeme-sequences (lasting for about 25–35 s) composed of a larger number of echemes (about 60–100).

In the Alpine form, *C. m. ignifer*, the male calling song (Figs 1328–1331) is usually shorter (about 7–15 s), with fewer echemes (about 15–40), and often has less well-marked ticks at the beginning of the echemes and a more uniform loudness within each echeme; it is especially characterized by having an abrupt ending, without the more spaced out, quietening terminal echemes of the typical calling song (Figs 1316–1335). Sometimes the main sequence is followed by a series of quite different 'aftersongs' (Fig. 1332), as in the courtship song (see below).

In the presence of a female the male of this form produces longer echeme-sequences, usually composed of about 35–60 echemes and lasting about 12–20 s (Figs 1333–1335). The main sequence is then followed, either immediately or after a pause of up to 2 s, by a series of quite different echemes or echeme-sequences, produced by the hind legs vibrating in a more upright position (very reminiscent of the courtship song of *C. biguttulus*). There are usually about 3–9 of these sounds, each lasting about 0·3–0·5 s, and they are repeated at the rate of about 1·5–2·2/s.

DISTRIBUTION. Widespread, but rather local, in central Europe, from France eastwards and Denmark southwards; absent from the Scandinavian Peninsula and British Isles. The southern limit is very similar to that of *C. biguttulus* (see p. 406). Farther eastwards the range is believed to extend into temperate Asia, but this cannot be confirmed without song studies. The Alpine form, *C. m. ignifer*, occurs widely in the Alps, and is particularly common on the southern side; in the northern Alps it is more local and tends to be nearer, in both morphology and song, to typical *C. mollis*.

Chorthippus jacobsi Harz

Chorthippus (Glyptobothrus) jacobsi Harz, 1975b: 890.

Jacobs's Grasshopper.

REFERENCES TO SONG. **Oscillogram**: Ragge, 1987b; Ragge & Reynolds, 1988; Ragge *et al.*, 1990. **Verbal description only**: Defaut, 1984b (as *yersini*); Harz, 1975b.

RECOGNITION. For distinguishing this species from *C. biguttulus* and *C. brunneus*, see under those species. Separating *C. jacobsi* morphologically from *C. yersini*, with which it is sympatric over a large part of the Iberian Peninsula, is often difficult. The apical part of the fore wing, beyond the stigma, is longer than in *C. yersini*: the distance from the centre of the stigma to the wing-tip is usually more than 0·38 times the length of the fore wing in males, 0·40 in females (usually less than these values in *C. yersini*). In males the maximum width of the combined costal and subcostal areas is usually less than 0·07 times the length of the fore wing; usually more than this in *C. yersini*.

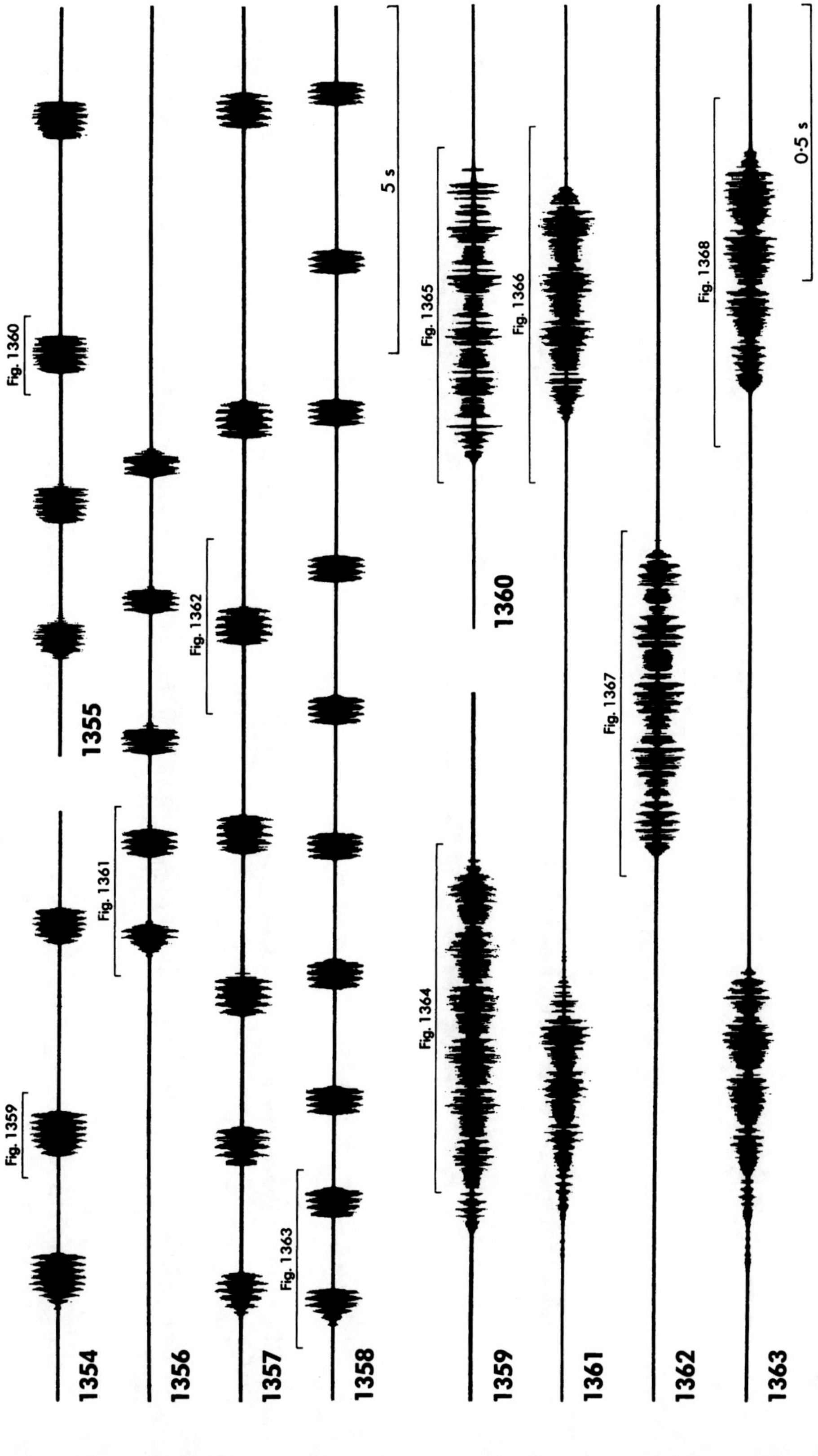

Figures 1354–1363 Oscillograms at two different speeds of the calling songs of five males of *Chorthippus jacobsi*. Figs 1355 and 1360 are from a male with only one hind leg. (From Ragge & Reynolds, 1988)

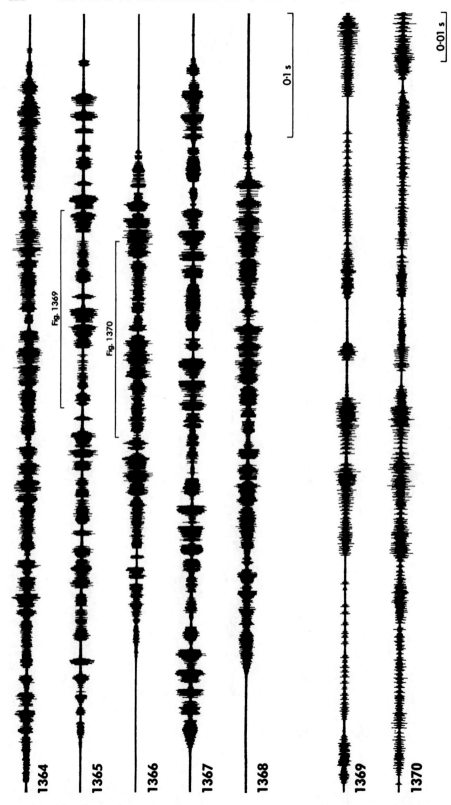

Figures 1364–1370 Faster oscillograms of the indicated parts of the songs of *Chorthippus jacobsi* shown in Figs 1354–1363. Figs 1365 and 1369 are from a male with only one hind leg. (From Ragge & Reynolds, 1988)

In the field *C. jacobsi* may be recognized easily by the male calling song, in which the syllable repetition rate is lower than in the other western European members of the group, so that the syllables can be clearly distinguished by the human ear.

This species has two generations per year and thus has an exceptionally long season, adults occurring from May to November. Adults of the first generation are most numerous in June and those of the second generation in September.

SONG (Figs 1354–1370. CD 2, track 56). The male calling song is a sequence of up to about 13 (usually 2–6) short echemes, each lasting about 0·3–1·2 s and composed of about 4–13 syllables. The echemes are separated by intervals of about 0·5–3·5 s, the later echemes in the sequence being typically more widely spaced than the earlier ones. The first echeme in the sequence almost always begins more quietly than it ends and usually has one or two more syllables than the remaining echemes, which lack the quiet beginning. The syllable repetition rate within each echeme is about 8–14/s and remains constant during each echeme-sequence; the syllables are thus repeated much more slowly than in the other western European members of the *biguttulus*-group and can be clearly distinguished by the human ear.

In one locality in the western Pyrenees (Navarra province), one of us (DRR) has recorded songs from two males in which the syllables were much less clearly distinct than usual, the echemes were rather short for *C. jacobsi* (about 0·3–0·4 s) and the number of stridulatory pegs rather low (about 100); all these characteristics suggest the possibility of hybridization with *C. brunneus*, which also occurs in this region. However, at another locality in Navarra both species were found together, each with typical morphology and song, and with no sign of hybridization.

No well-developed courtship song has yet been observed.

DISTRIBUTION. Known only from the Iberian Peninsula, where it is widespread, and Majorca. The northern limit of the range corresponds closely with the southern limit of *C. brunneus*, with a slight overlap in one or two places.

Chorthippus yersini Harz

Chorthippus (Glyptobothrus) biguttulus yersini Harz, 1975b: 895.

Yersin's Grasshopper.

REFERENCES TO SONG. **Oscillogram**: Bukhvalova, 1993b; Bukhvalova & Zhantiev, 1993; Ragge, 1987b (as 'N. Iberian *biguttulus*'); Ragge & Reynolds, 1988; Ragge *et al.*, 1990.

RECOGNITION. See under *C. biguttulus*, *C. brunneus* and *C. jacobsi*.

In the field typical *C. yersini* may be recognized quite easily by the male calling song; even in the north-eastern part of its range in Spain, where the song becomes more similar to that of *C. biguttulus*, it is usually still quite distinctive (see below).

SONG (Figs 1371–1411. CD 2, track 57). The male calling song of typical *C. yersini* is a sequence of echemes, each beginning quietly and ending abruptly; in southern Spain the beginning of each echeme tends to be more abrupt, sometimes almost as abrupt as the end. The number of echemes in the sequence is almost always within the range 1–6 and is

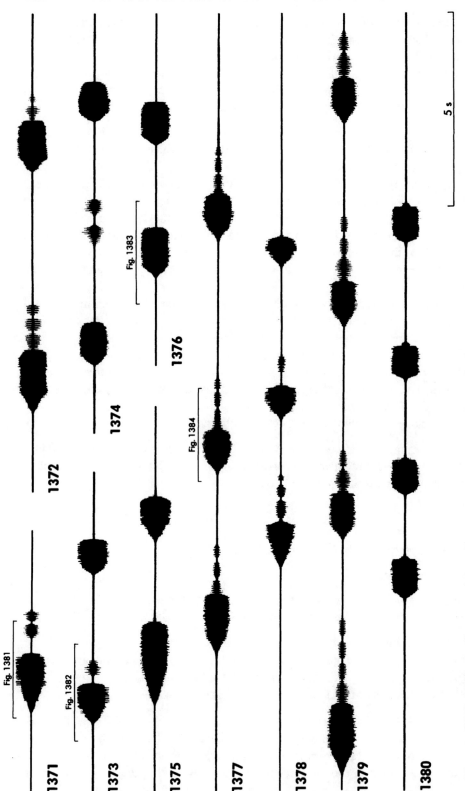

Figures 1371–1380 Oscillograms of the calling songs of ten males of *Chorthippus yersini*. Figs 1371, 1372 and 1380 are from males with only one hind leg. The male used for Fig. 1371 was from central Spain, those used for Figs 1372, 1375, 1377, 1378 and 1379 were from north-western Spain (including three – Figs 1372, 1377 and 1379 – from the type locality) and those used for Figs 1373, 1374, 1376 and 1380 were from southern Spain. (From Ragge & Reynolds, 1988)

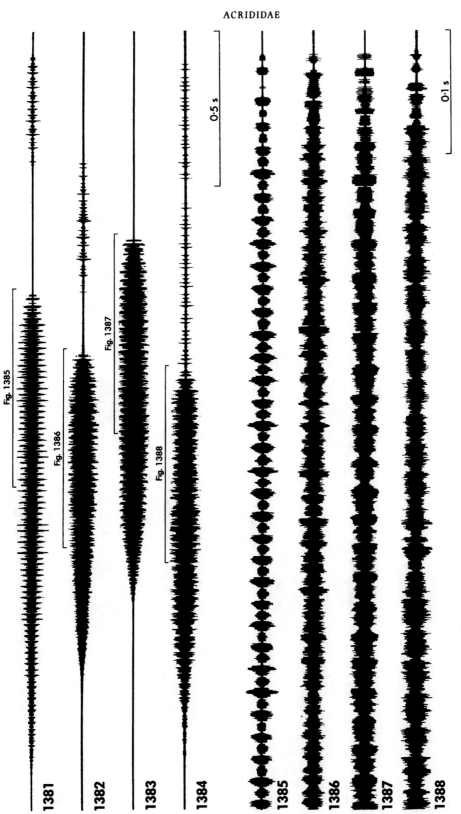

Figures 1381–1388 Faster oscillograms of the indicated parts of the songs of *Chorthippus yersini* shown in Figs 1371–1377. Figs 1381 and 1385 are from a male with only one hind leg. (From Ragge & Reynolds, 1988)

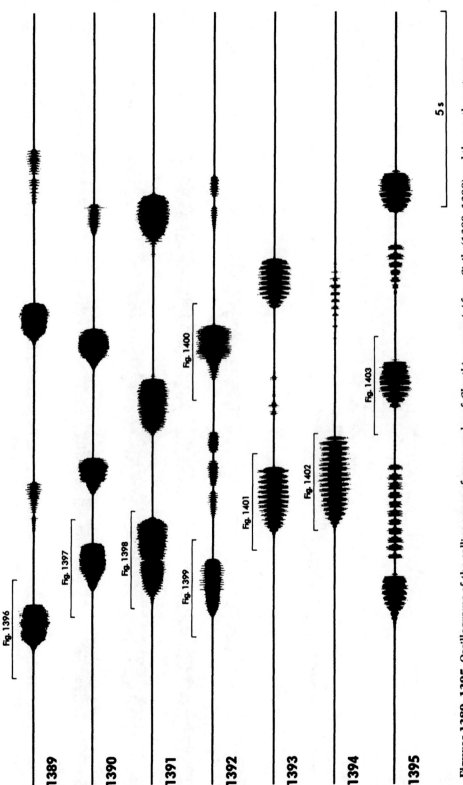

Figures 1389–1395 Oscillograms of the calling songs of seven males of *Chorthippus yersini* from Sicily (1389, 1390) and the north-eastern mountains of Spain (1391–1395). The males used for Figs 1391 and 1392 were from the Sierra de las Hormazas (north-west of Soria), and those used for Figs 1393–1395 were from the mountains of Teruel and Cuenca provinces.

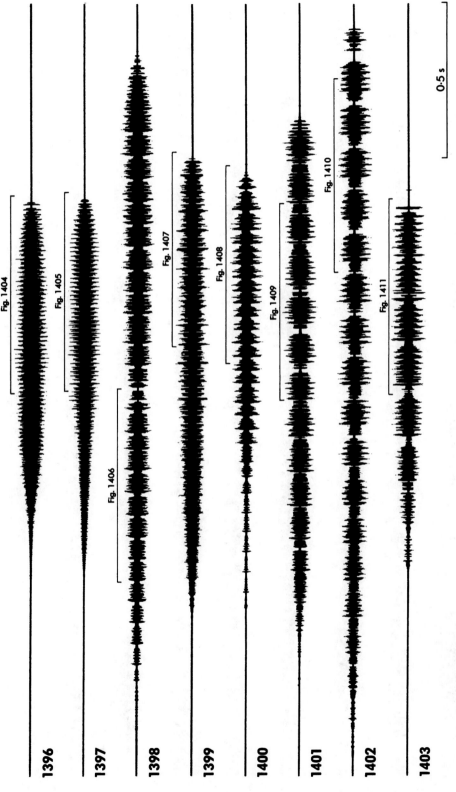

Figures 1396–1403 Faster oscillograms of the indicated parts of the songs of *Chorthippus yersini* shown in Figs 1389–1395.

Figures 1404–1411 Faster oscillograms of the indicated parts of the songs of *Chorthippus yersini* shown in Figs 1396–1403.

usually fewer than 5. When there are two or more echemes, the first one is almost always longer than the others, usually lasting 1·0–2·0 s; each of the remaining echemes usually lasts about 0·7–1·5 s (this difference tends to be less pronounced in southern Spain). Oscillographic analysis shows that each echeme consists of a series of syllables repeated at the rate of about 35–75/s; there are about 40–100 syllables in the first echeme, fewer in the later echemes in proportion to their shorter duration. They are diplosyllables, the up- and downstrokes of the hind legs producing clearly distinct sounds. In 'two-legged' songs the two hind legs are seldom exactly in phase and so the syllable structure is often somewhat obscured (e.g. Figs 1382–1384, 1386–1388), but in a 'one-legged' song the hemisyllables may be seen very clearly in faster oscillograms (Figs 1381, 1385). The hemisyllables are usually alternately louder and softer, the louder ones probably being produced by the downstrokes of the hind legs. The syllables typically form long, even, unbroken series, thus contrasting with the short groups of syllables in the calling song of C. biguttulus.

Over most of the range of this species each main echeme of the song is typically followed by a series of quieter echemes of a different kind (e.g Figs 1371, 1372), very reminiscent of the courtship song of C. biguttulus. There are usually 2–4 of these quieter echemes, but sometimes only one and occasionally none at all; typically they follow each of the main echemes, but sometimes only the early ones. Each of the quieter echemes consists of about 15–30 diplosyllables, usually grouped into twos or threes (in which case the groups are strictly echemes, in turn grouped into echeme-sequences). Sometimes the first quieter echeme follows immediately after a main echeme, but more often there is a short interval of less than 1 s. The interval between the last quiet echeme and the next main echeme (or between successive main echemes when there are no quiet echemes) varies from about 0·5 s to about 4·5 s.

In southern Spain and Sicily these quieter echemes are often absent (e.g. Figs 1376, 1380); when present they often occur after only one of the main echemes (Figs 1373, 1374), and are often separated from it by a longer interval (up to 3 s).

In the north-eastern part of the range of C. yersini in Spain, from the Sierra de la Demanda to the Montes Universales, the male calling song is more similar to that of C. biguttulus. At the western end of this region we have studied the song at two localities in the Sierra de las Hormazas, near the border between Logroño and Soria provinces. At both of these, the main echemes typical of C. yersini showed a clear tendency to be subdivided into groups of 2–4 syllables (Figs 1406–1408), like the echeme-sequences of C. biguttulus; the quieter echeme-sequences that are typically found after the main echemes of C. yersini were sometimes present and sometimes absent at both localities. At one of these localities the first echeme-sequence usually included one or two slight hesitations (Figs 1391, 1398), and at the other each song of the single male studied usually consisted of two main echeme-sequences, of which the second was, curiously, always louder than the first (Fig. 1392).

In the more easterly part of this region, in the mountains of Teruel and Cuenca provinces, the song consists of about 1–5 echeme-sequences, each composed of about 5–12 (rarely up to 22) clearly distinct echemes (Figs 1393–1395). Each echeme contains about 5–8 syllables, the last echeme in a sequence often containing more, up to about 20. The echeme repetition rate is about 5–9/s and the syllable repetition rate within an echeme about 40–70/s. As in the Sierra de las Hormazas, quieter 'aftersong' echeme-sequences are often present but sometimes absent. When the song comprises two or more main echeme-sequences, the intervals between them are sometimes longer than the intervals between the main echemes in typical *C. yersini*, occasionally as long as 10 s.

Although these songs show a strong trend towards the song of *C. biguttulus*, they also show a clear affinity to the typical song of *C. yersini*, particularly in the frequent presence of the quieter echeme-sequences. In Teruel and Cuenca provinces the number of echemes in each echeme-sequence is generally much lower than in *C. biguttulus*, and there are more syllables per echeme. As we have suggested previously (Ragge, 1987b; Ragge & Reynolds, 1988), we consider *C. yersini* to be a fairly recent derivative of *C. biguttulus*, and are therefore not surprised that, in the part of the range of *C. yersini* nearest to the southern limit of *C. biguttulus*, there should be a song showing characteristics intermediate between the typical songs of these two species. We prefer to regard these populations as local forms of *C. yersini* rather than to treat them formally as distinct taxa, even at the subspecific level.

We have not observed any special courtship song in this species.

DISTRIBUTION. In western Europe *C. yersini* is at present known only from the Iberian Peninsula and Sicily. In both these regions it is an upland species, usually occurring at altitudes above 1000 m. It occurs widely in the Iberian Peninsula but, unlike *C. jacobsi*, does not apparently occur in the Balearic Islands or north of the Ebro valley, which separates it from the southern limit of *C. biguttulus*. Bukhvalova (1993b) has recently recorded *C. yersini* from widely scattered localities in south-east Europe and Asia; she has studied the calling songs of males from North Ossetia (Caucasus) and Tuva, and has found them to be similar to those of the western European males studied by us.

Chorthippus rubratibialis Schmidt

Chorthippus (Glyptobothrus) biguttulus rubratibialis Schmidt, 1978: 257.

Red-shinned Grasshopper.

REFERENCES TO SONG. **Oscillogram**: Ingrisch, 1995; Ragge, 1987b (as 'Apennine *biguttulus*'); Ragge *et al.*, 1990; Schmidt, 1978, 1989.

RECOGNITION. *C. rubratibialis* occurs only in the Italian Apennines, where the only other member of the *biguttulus*-group is *C. brunneus*. These two species are very similar in appearance and the only way of separating them morphologically that is at all reliable is by counting the stridulatory pegs: there are usually more than 90 in *C. rubratibialis*, usually fewer than 90 in mainland Italian *C. brunneus*. The hind tibiae of *C. rubratibialis* are usually more reddish in colour than those of *C. brunneus* and there are small differences in the shape and venation of the male fore wings, but these characters are less reliable than the stridulatory pegs.

In the field these species can be separated easily by the contrasting male calling songs.

SONG (Figs 1412–1426. CD 2, track 58). The male calling song is typically an echeme-sequence beginning quietly and ending abruptly. The sequence is almost always broken up by about 1–8 (mostly 1–3) very brief interruptions, usually of less than 100 ms, giving it a characteristic hesitant quality. Sometimes the last hesitation is prolonged into an interval of about 200–500 ms, in which case the song can be regarded as comprising two echeme-sequences, of which the second is short (usually about 1 s) and uninterrupted. Rarely there

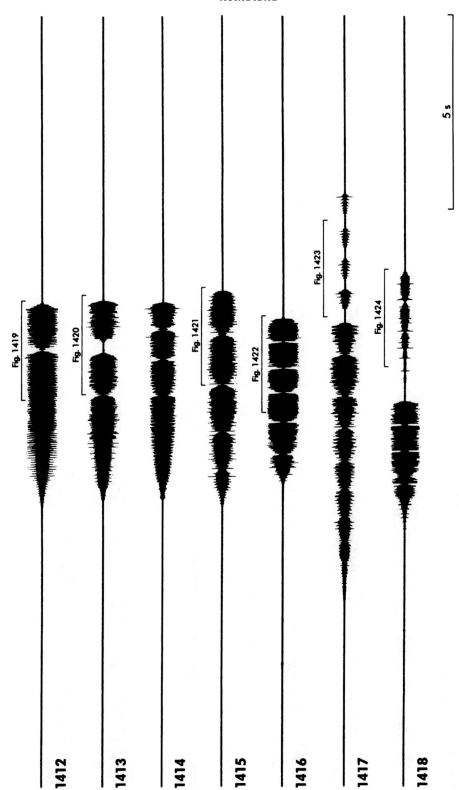

Figures 1412–1418 Oscillograms of the calling songs of seven males of *Chorthippus rubratibialis*.

5 s

1412

1413

1414

1415

1416

1417

1418

Fig. 1419

Fig. 1420

Fig. 1421

Fig. 1422

Fig. 1423

Fig. 1424

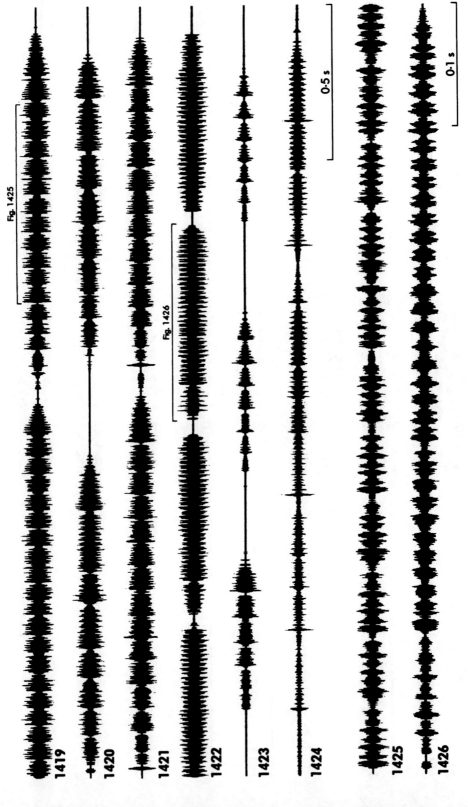

Figures 1419–1426 Faster oscillograms of the indicated parts of the songs of *Chorthippus rubratibialis* shown in Figs 1412–1418.

are two hesitant sequences separated by a longer interval of up to 4 s or more. The duration of the first (or only) echeme-sequence is almost always within the range 2–5 s.

Each sequence is composed of echemes repeated at the rate of about 10–18/s. The number of syllables in each echeme often varies during the course of the sequence; it is usually within the range 4–8, but towards the end of the sequence the echemes are often prolonged and may contain as many as 20 or more syllables. Occasionally a large part of the sequence consists of ungrouped syllables but, even in these cases, the quiet opening of the sequence usually shows the syllables grouped into small echemes.

Rarely the main sequence is followed by a quieter 'aftersong' (Figs 1417, 1418). This may normally be a courtship song, although no female was present when these aftersongs were observed.

In comparison with *C. biguttulus*, the calling song of *C. rubratibialis* differs in usually consisting of only one echeme-sequence, in having one or more brief interruptions in the sequence, and in having more syllables per echeme. The resemblance in song, however, as well as the close morphological similarity, suggests that *C. rubratibialis* is a fairly recent derivative from *C. biguttulus*.

DISTRIBUTION. Found throughout the Italian Apennines, from the Ligurian Apennines in the north to at least as far south as the Sila mountains in Calabria; absent from Sicily.

Chorthippus marocanus Nadig

Chorthippus (Glyptobothrus) biguttulus marocanus Nadig, 1976: 649.

Moroccan Grasshopper.

REFERENCES TO SONG. **Oscillogram**: Ragge, 1987b (as 'Moroccan *biguttulus*'); Ragge & Reynolds, 1988; Ragge et al., 1990.

RECOGNITION. This species is the only member of the *biguttulus*-group known from North Africa, to which it is confined, and so in practice there is no problem in recognizing it. Both sexes may in any case be distinguished from all the western European members of the group (except for the Finnish form of *C. brunneus*) by the longer stridulatory file: more than 4·8 mm in the male, usually more than 5·9 mm in the female.

In the field the male calling song is quite distinctive and the 'hissy' component of the rivalry song, produced when two or more males are together, is unique among the western European members of the *biguttulus*-group.

SONG (Figs 1427–1435. CD 2, track 59). The male calling song is a series of echeme-sequences, each beginning fairly quietly and ending abruptly. There are usually 2 or 3 echeme-sequences in each series, occasionally more and sometimes only one. Each echeme-sequence usually lasts about 1·5–2·5 s and consists of about 20–30 echemes repeated at the rate of about 10–15/s. The interval between successive echeme-sequences is very variable but is usually between 2 and 6 s. Oscillographic analysis suggests that there are 2–4 syllables in each echeme, but they do not form a regular pattern and it is likely that the two hind legs are not synchronized.

When two or more males are near to one another they produce a characteristic rivalry

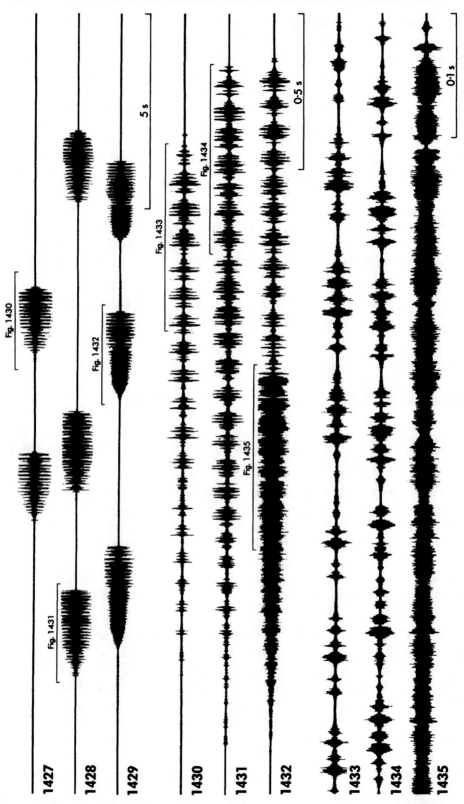

Figures 1427–1435 Oscillograms at three different speeds of the songs of two males of *Chorthippus marocanus*. **1427, 1428, 1430, 1431, 1433, 1434.** Calling songs. **1429, 1432, 1435.** Rivalry song of the same male as Figs 1427, 1430 and 1433. (From Ragge & Reynolds, 1988)

song (Figs 1429, 1432, 1435). Typically this consists of 1–4 echeme-sequences, each beginning with a simple echeme of about 15–30 syllables lasting about 0·5–2·0 s and with a distinctive 'hissy' quality, followed immediately by a sequence of about 10–20 echemes of the calling song type lasting for about the same time. Sometimes isolated 'hissy' echemes are produced without any further echemes of the calling song type. The 'hissy' echemes begin quietly, rapidly increasing in intensity, and are produced with the hind legs vibrating at a steeper angle than when producing echemes of the calling song type.

Courtship behaviour has not yet been observed.

DISTRIBUTION. Known only from the mountains of Morocco and Algeria. The presence of this species in Algeria has not yet been confirmed by studies on the song.

Chorthippus binotatus (Charpentier)

Gryllus binotatus Charpentier, 1825: 158.

Two-marked Grasshopper; (F) Criquet des Ajoncs; (D) Zweifleckiger Grashüpfer; (NL) Gaspeldoornsprinkhaan.

REFERENCES TO SONG. **Oscillogram**: García *et al.*, 1995; Kleukers *et al.*, 1997. **Verbal description only**: Bellmann, 1993*a*; Bellmann & Luquet, 1995; Defaut, 1987, 1988*b*. **Disc recording**: Bonnet, 1995 (CD); Odé, 1997 (CD).

RECOGNITION. This species is very variable, some of the extreme forms looking superficially as if they belong to different species. Some of these variants are currently regarded as distinct subspecies, but their distribution suggests that they are better treated as local or ecological forms. Over much of its range *C. binotatus* is a large, fully winged grasshopper, and is most easily recognized by its habit of living on shrubs, such as broom or gorse, and its colour-pattern: the hind femora usually have two conspicuous dark transverse bands across their dorsal surface, and the hind tibiae and tarsi are usually red. The fore wings have no markedly expanded areas and are brown in colour; the general body colour is brown in the central mountains of Spain, often greenish elsewhere.

In many mountain areas (e.g. parts of the Pyrenees, Cevennes and French Alps) the wings are reduced, so that the fore wings fall short of the hind knees and the hind wings often fall short of the tips of the flexed fore wings; the general size is often also smaller and the insects tend to live nearer ground-level, often on heather or similar low vegetation. The colour-pattern of these forms is often less striking: a more even brown, with the hind tibiae and tarsi often pinkish or orange rather than red. These forms could be confused with *C. pullus* or *C. vagans*, but can be distinguished from both by the tympanal apertures, which are half-closed and slit-like, especially in the female.

In the higher parts of the Sierra Nevada in southern Spain, where this species is very common, the fore wings usually reach the hind knees, but the hind tibiae and tarsi, although sometimes red, are more often yellowish in colour.

The male calling song varies less than the morphology and is a useful aid to field identification. It is similar to the calling songs of *C. vagans* and *C. reissingeri*, but the syllable repetition rate is faster.

See also the remarks on p. 78.

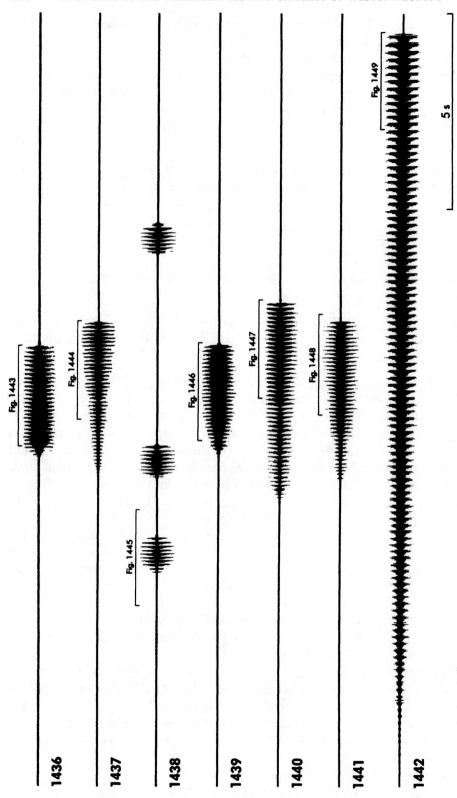

Figures 1436–1442 Oscillograms of the songs of seven males of *Chorthippus binotatus*. The song shown in Fig. 1442 was produced in the presence of a female.

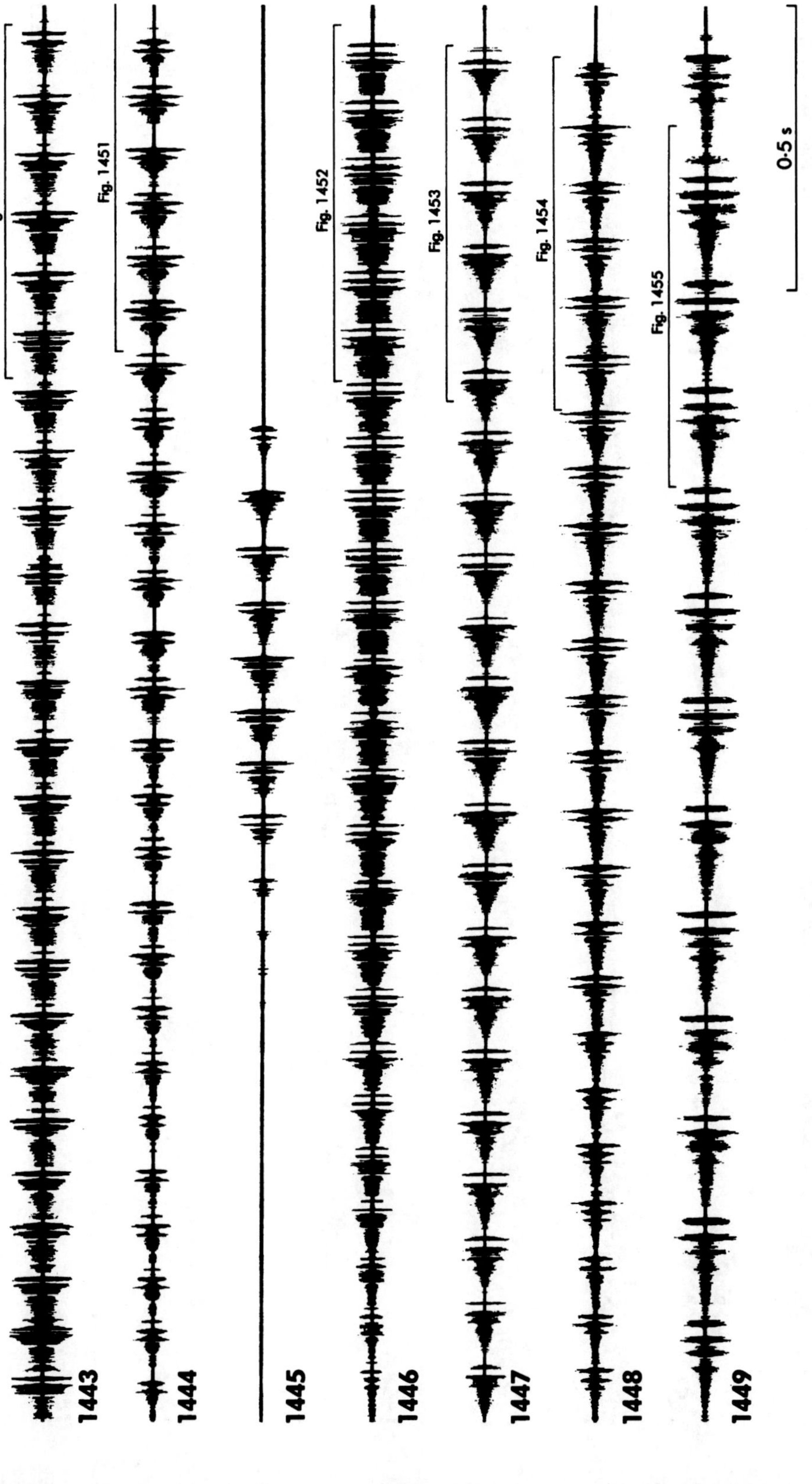

Figures 1443–1449 Faster oscillograms of the indicated parts of the songs of *Chorthippus binotatus* shown in Figs 1436–1442.

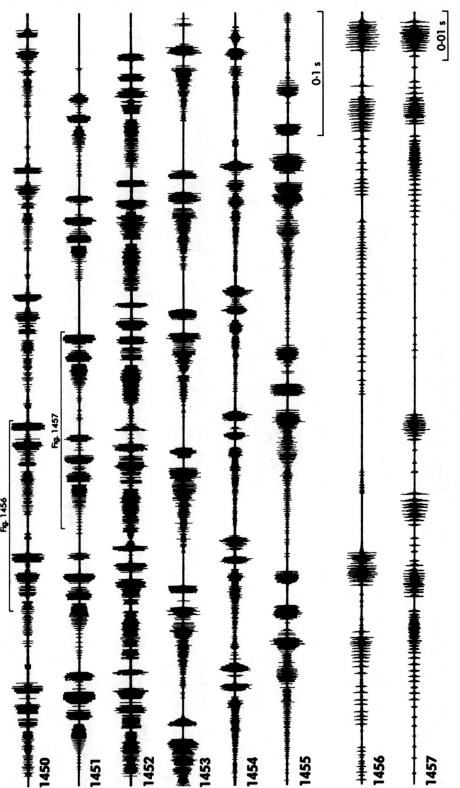

Figures 1450–1457 Faster oscillograms of the indicated parts of the songs of *Chorthippus binotatus* shown in Figs 1443–1449.

SONG (Figs 1436–1457. CD 2, track 60). The calling song of a completely isolated male is a simple echeme of syllables repeated at the rate of about 8–12/s. The duration of the echeme is usually within the range 1–5 s. There is almost always a quiet beginning; maximum intensity is often reached after the first 2–4 syllables, but sometimes the crescendo lasts for most of the duration of the echeme (Fig. 1437). The echemes are usually separated by quite long intervals, but sometimes two or three are produced in fairly quick succession (Fig. 1438). Oscillographic analysis shows that each syllable lasts for about 60–100 ms and that there are usually 1–3 gaps in each downstroke hemisyllable.

As in *C. vagans*, two or more males often react with one another by alternately producing short echemes. In the presence of a female the male usually produces longer echemes, lasting up to 30 s or longer, with a more gradual crescendo, often lasting for most of the duration of the echeme (Fig. 1442).

DISTRIBUTION. Known at present only from France (especially western and southern), the Iberian Peninsula and the Atlas Mountains in Morocco.

Chorthippus apicalis (Herrich-Schäffer)

Acridium apicale Herrich-Schäffer, 1840: 10.

June Grasshopper; (F) Criquet hyalin.

REFERENCES TO SONG. **Verbal description only**: Defaut, 1987, 1988*b*.

RECOGNITION. This common Iberian species has a very early season, adults being most numerous in June and few remaining by late July. The only other Iberian grasshopper that is adult as early as this and that could be confused with *C. apicalis* is *C. jacobsi*, in the first of its two annual generations. However, *C. apicalis* is often yellowish green in colour and has wide open tympanal apertures; *C. jacobsi* is usually brown in colour and has slit-like tympanal apertures. In the field the striking contrast between the calling songs of these two species prevents any confusion.

SONG (Figs 1458–1466. CD 2, track 61). The male calling song is a series of simple echemes, each lasting about 0·8–2·2 s and composed of about 15–25 syllables repeated at the rate of about 12–22/s. The echemes are repeated quite regularly at intervals of about 2·5–6·5 s for indefinite periods. Each echeme begins quietly, reaching maximum intensity about halfway through its duration. Oscillographic analysis shows that most of the sound consists of downstroke hemisyllables lasting about 20–35 ms; the upstroke hemisyllables last about 30–50 ms and are usually very quiet, especially the final one, which is sometimes silent or almost so.

The calling song of this species is quite similar to that of *Chrysochraon dispar*, but confusion in the field is unlikely as *C. dispar* does not occur south of the Pyrenees.

DISTRIBUTION. Common and widespread in the Iberian Peninsula; also found in mountains in Morocco, and recorded from Sardinia.

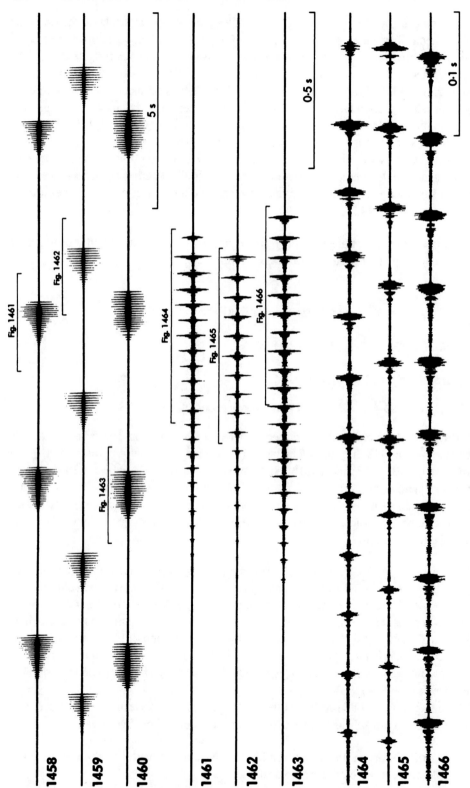

Figures 1458–1466 Oscillograms at three different speeds of the calling songs of three males of *Chorthippus apicalis.*

Chorthippus jucundus (Fischer)

Stenobothrus jucundus Fischer, 1853: 315.

Large Green Grasshopper; (F) Criquet des marais.

REFERENCES TO SONG. **Oscillogram**: Busnel & Loher, 1954c, 1955; Loher, 1957; Loher & Broughton, 1955. **Frequency information**: M.-C. Busnel, 1953; R.-G. Busnel, 1955; Busnel & Loher, 1954c; Loher, 1957; Loher & Broughton, 1955. **Verbal description only**: Busnel & Loher, 1954a; Dumortier, 1963b; Defaut, 1987, 1988b.

RECOGNITION. This species is easily recognized by its large size and bright green general colouring, contrasting with the red or orange hind tibiae and tarsi. It is usually found in damp places, with lush grass or herbs.

SONG (Figs 1467–1472. CD 2, track 62). The male calling song is a sequence of about 2–7 (usually 3 or 4, occasionally only 1) echemes separated by intervals of about 1–4 s. The quality of the sound is 'hissy' and rather amorphous, the syllables not being easily distinguishable by the unaided human ear and not always clearly distinct even in a fast oscillogram (e.g. Fig. 1471). Each echeme lasts about 0·4–1·0 s and consists of about 8–13 syllables repeated at the rate of about 15–22/s.

Two or more males will often react with one another by alternating single echemes in a manner reminiscent of *C. brunneus* but with a slower rhythm.

DISTRIBUTION. Southern France, much of the Iberian Peninsula and Morocco.

Chorthippus albomarginatus (De Geer)

Acrydium albo-marginatum De Geer, 1773: 480.

Lesser Marsh Grasshopper; (F) Criquet marginé; (G) Weißrandiger Grashüpfer; (NL) Kuntsprinkhaan; (DK) Strandengræshoppe; (S) Strandängsgräshoppa.

REFERENCES TO SONG. **Oscillogram**: Bukhvalova & Zhantiev, 1993; Elsner, 1983b; Grein, 1984; Helversen, 1986; Holst, 1970, 1986; Kleukers et al., 1997; Schmidt & Schach, 1978; Xi et al., 1992. **Diagram**: Bellmann, 1985a, 1988, 1993a; Bellmann & Luquet, 1995; Brown, 1955; Duijm & Kruseman, 1983; Holst, 1970; Jacobs, 1950b, 1953a; Ragge, 1965; Wallin, 1979. **Sonagram**: Leroy, 1978. **Leg-movement**: Elsner, 1983b; Helversen, 1986. **Frequency information**: Meyer & Elsner, 1996. **Musical notation**: Yersin, 1854b (as *elegans*). **Verbal description only**: Broughton, 1972a, 1972b; Defaut, 1987, 1988b; Faber, 1928 (as *e.*), 1953a; Harz, 1957; Perdeck, 1951; Weber, 1984. **Disc recording**: Bellmann, 1993c (CD); Grein, 1984 (LP); Odé, 1997 (CD); Ragge et al., 1965 (LP). **Cassette recording**: Bellmann, 1985b, 1993b; Burton & Ragge, 1987; Wallin, 1979.

RECOGNITION. This species is very similar in appearance to *C. dorsatus* and *C. dichrous*. Males differ from both these species in their fore wings: the bulge normally present in *Chorthippus* near the base of the anterior (ventral) margin is absent or almost so, the costal area is narrower and there is a sudden expansion of the radial area about halfway along the wing. Females may usually be distinguished from *C. dorsatus* by having a white stripe along the

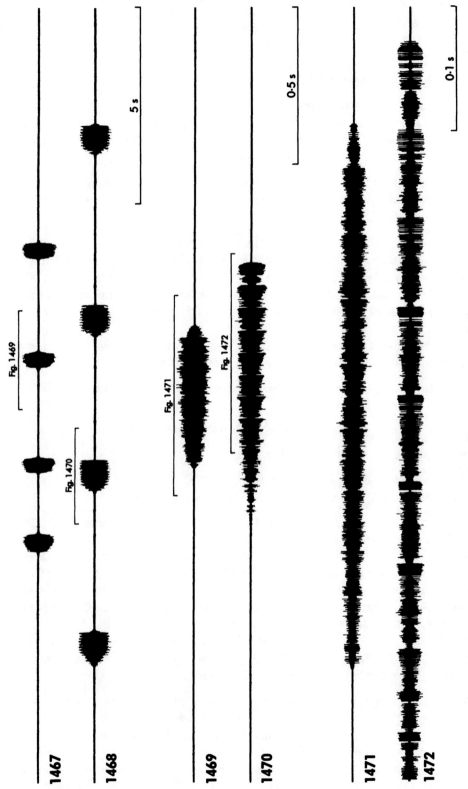

Figures 1467–1472 Oscillograms at three different speeds of the calling songs of two males of *Chorthippus jucundus*.

costal area of their fore wings, and from both *C. dorsatus* and *C. dichrous* (which sometimes has a similar stripe) by a sudden expansion of the radial area similar to that shown by the males. In both sexes of *C. albomarginatus* the pronotal lateral carinae are usually straighter and less divergent than in the other two species, but this character is variable and not easy to use.

In the field males of *C. albomarginatus* and *C. dorsatus* may be distinguished easily by their calling songs. The calling songs of *C. albomarginatus* and *C. dichrous* are quite similar, but the echemes of *C. albomarginatus* have a less hissy quality and a more distinct internal structure than those of *C. dichrous*.

SONG (Figs 1473–1489. CD 2, track 63). The male calling song is a sequence of 2–6 (most often 3 or 4) brief echemes separated by intervals of about 1·5–5·0 s. In warm conditions each echeme lasts about 300–800 ms and consists of about 17–30 syllables repeated at the rate of about 35–50/s; in cooler conditions the echemes may last twice as long, with the syllables repeated half as rapidly. The echemes often begin with a small group of rather more rapidly repeated syllables; both the beginning and end of each echeme are fairly abrupt, so that it is rather like an echeme of the calling song of *C. brunneus*, but longer and more regular in structure. Elsner (1983*b*) and Helversen (1986) have shown that the two hind legs move alternately, the downstroke of one coinciding with the upstroke of the other. As the downstroke hemisyllables are usually much louder than the upstroke ones, a 'one-legged' oscillogram of a calling song echeme shows half the number of loud sounds of a 'two-legged' one (cf. Figs 1483, 1484).

Two or more males sometimes react with one another by alternating single echemes, rather in the manner of *C. jucundus*.

In the presence of a female the male produces one of the most complex courtship songs of the western European grasshoppers, composed of elements all quite different from the calling song. Courtship begins with low amplitude vibration of the hind legs, producing very quiet sounds. This soon develops into an alternation of two different kinds of echeme, one short-lived with the hind legs held in a low position and vibrating more slowly (about 15–30/s) and the other longer with the legs in a higher position and vibrating more rapidly (about 40–80/s); in each case the two hind legs move synchronously or almost so. The intensity of the sound gradually increases, with the 'low' echemes becoming rather longer and the 'high' echemes rather shorter; the two together usually last about 1·0–1·5 s. This gradually intensifying alternation of 'low' and 'high' echemes forms a long prelude to courtship proper, sometimes lasting for several minutes.

Eventually the fully developed courtship song begins (Figs 1485–1489). It consists of a highly stereotyped repeated cycle, composed of the following parts:

1. A short 'low' echeme, composed of about 4–8 syllables, followed immediately by a 'high' echeme of about 25–40 syllables, the two together lasting about 0·7–1·4 s. The syllable repetition rates are as given above and the two hind legs are quite well synchronized.

2. The legs are moved into an extra-high position and (sometimes after a short burst of syllables similar to those of the 'high' echeme) are vibrated in a complex pattern producing a sequence of trisyllabic echemes in which the middle syllable is much louder than the other two (Figs 1487, 1488). These echemes are produced alternately by the left and right hind legs, each leg producing about 10–20 echemes at the rate of about

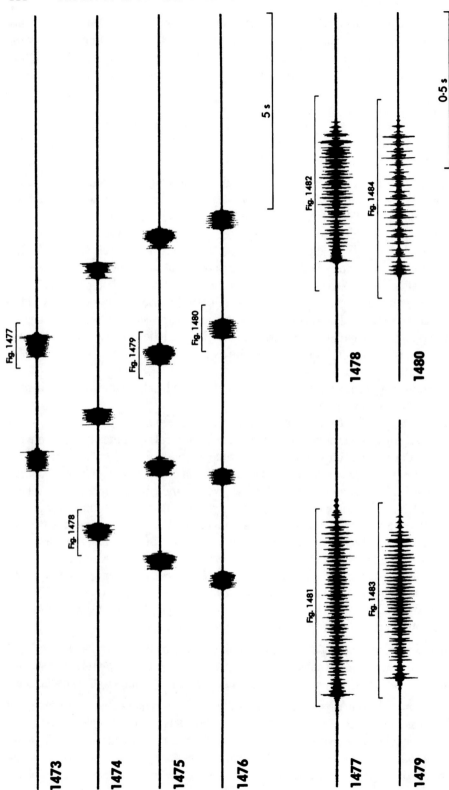

Figures 1473–1480 Oscillograms at two different speeds of the calling songs of four males of *Chorthippus albomarginatus*. Figs 1476 and 1480 are from a male with only one hind leg.

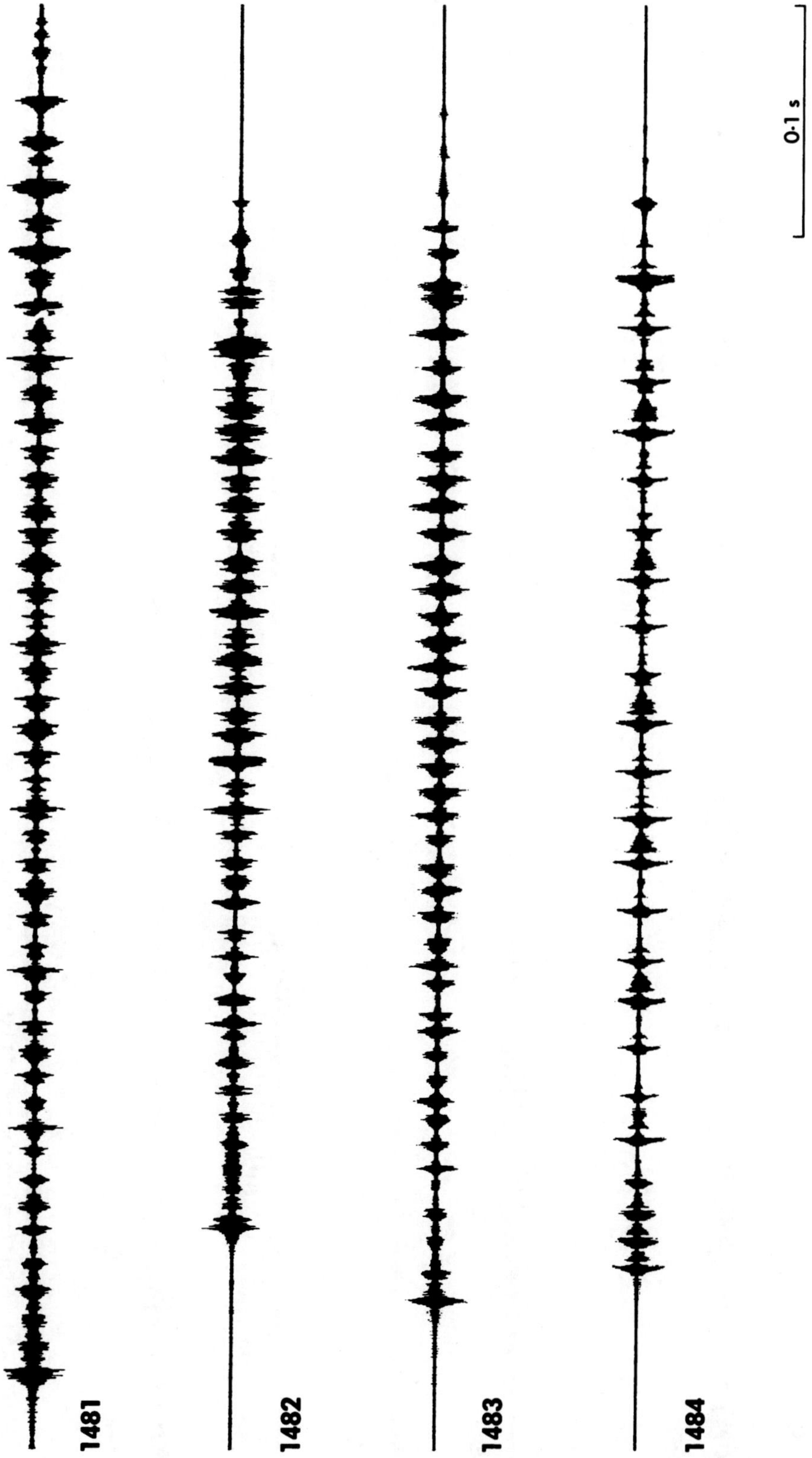

0·1 s

Figures 1481–1484 Faster oscillograms of the indicated parts of the songs of *Chorthippus albomarginatus* shown in Figs 1477–1480. Fig. 1484 is from a male with only one hind leg.

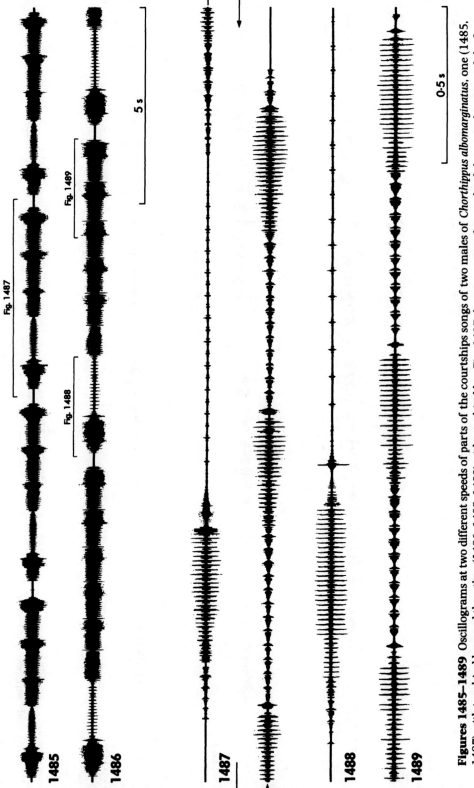

Figures 1485–1489 Oscillograms at two different speeds of parts of the courtships songs of two males of *Chorthippus albomarginatus*, one (1485, 1487) with two hind legs and the other (1486, 1488, 1489) with one hind leg. Fig. 1487 shows one complete cycle of the courtship song of the first male (as indicated in Fig. 1485). Figs 1488 and 1489 show the indicated parts of the courtship song of the second male.

10–15/s; in a 'two-legged' song there are thus about 20–40 of these echemes, repeated at about 20–30/s. This sequence usually lasts about 1·0–1·5 s.

3. There is then a series of alternating 'low' and 'high' echemes, similar to those of the courtship prelude described above. The 'low' echemes last about 400–800 ms, and the 'high' echemes are usually slightly shorter, about 300–600 ms; the syllable repetition rates are as given above and the legs are vibrated synchronously. The 'high' echemes usually become somewhat louder during the course of the series. The number of 'low/high' pairs of echemes in the series varies from one courtship song to another, and sometimes even within one courtship song; it is usually within the range 3–6 and seldom more than 8. The series sometimes ends with a 'high' echeme, but more often with a (usually short) 'low' echeme.

There is then a short pause of about 300–500 ms and the cycle begins again. The duration of the complete cycle naturally depends on the number of 'low/high' pairs of echemes in part 3 above; if there are 3 or 4 of these pairs the cycle usually lasts about 5–6 s, but with 8 of them it sometimes lasts more than twice as long as this. The cycle is usually repeated many times (occasionally more than 100) before the male attempts to mate with the female.

As the summer progresses this species tends to aggregate into small groups including both sexes, and in such groups singing males produce only the courtship song. Because of this, and the fact that the courtship song is so prolonged, it is usually this song that is heard in late summer, when the calling song may be a relatively uncommon sound.

DISTRIBUTION. According to Helversen (1986) this species has a more restricted distribution than was previously believed. Its range extends northwards almost to the Arctic Circle and southwards to the Pyrenees, including the Spanish side. It occurs locally in Ireland and southern Britain, much of France, the Low Countries, Germany, central Europe and probably parts of Russia, but the eastern and south-eastern limits have yet to be reliably established. It seems to be largely absent from the southern peninsulas, and probably also from North Africa.

Chorthippus dorsatus (Zetterstedt)

Gryllus dorsatus Zetterstedt, 1821: 82.

Steppe Grasshopper; (F) Criquet verte-échine; (D) Wiesengrashüpfer; (NL) Weide-sprinkhaan; (S) Syd-ängsgräshoppa.

REFERENCES TO SONG. **Oscillogram**: Bukhvalova & Zhantiev, 1993; Faber, 1957a; Grein, 1984; Holst, 1970, 1986; Kleukers *et al.*, 1997; Komarova & Dubrovin, 1973; Ronacher, 1990, 1991; Schmidt & Maumgarten, 1977; Schmidt & Schach, 1978; Stumpner & Helversen, 1992, 1994; Vedenina, 1990; Vedenina & Zhantiev, 1990. **Diagram**: Bellmann, 1985a, 1988, 1993a; Bellmann & Luquet, 1995; Duijm & Kruseman, 1983; Holst, 1970; Jacobs, 1950b, 1953a; Luquet, 1978; Wallin, 1979. **Leg-movement**: Ronacher, 1990, 1991; Stumpner & Helversen, 1992, 1994. **Frequency information**: Meyer & Elsner, 1996. **Musical notation**: Yersin, 1854b. **Verbal description only**: Chopard, 1952; Defaut, 1987,

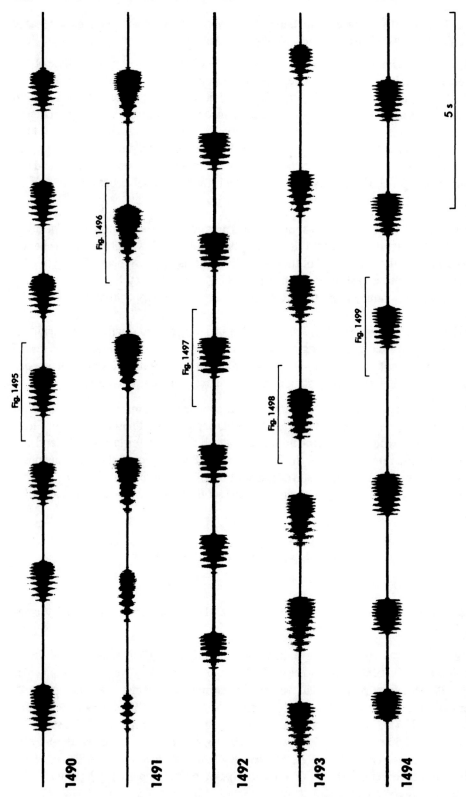

Figures 1490–1494 Oscillograms of the calling songs of five males of *Chorthippus dorsatus*. Fig. 1491 shows the beginning of a series of echemes. Fig. 1494 is from a male with only one hind leg.

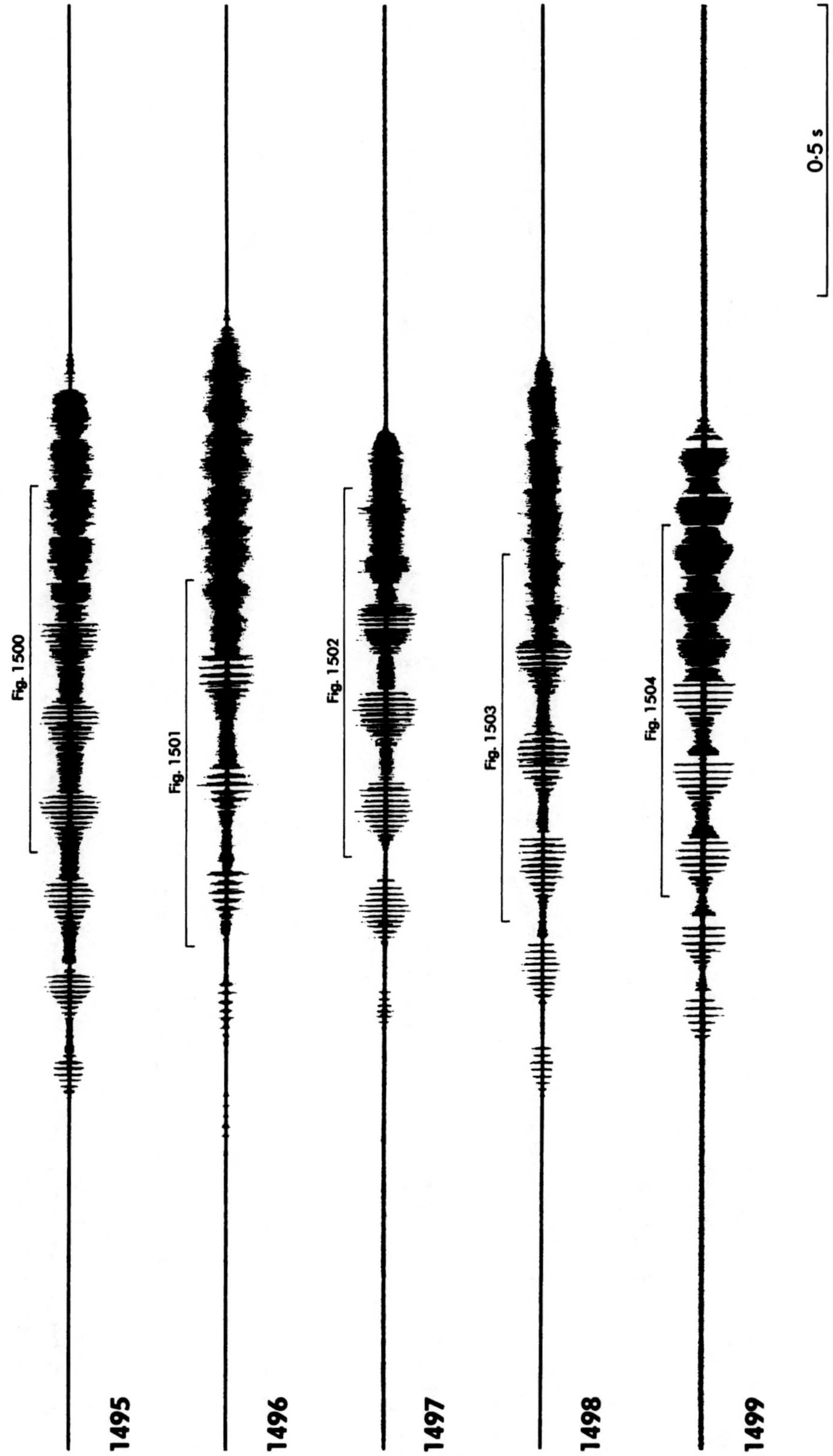

Figures 1495–1499 Faster oscillograms of the indicated parts of the songs of *Chorthippus dorsatus* shown in Figs 1490–1494. Fig. 1499 is from a male with only one hind leg.

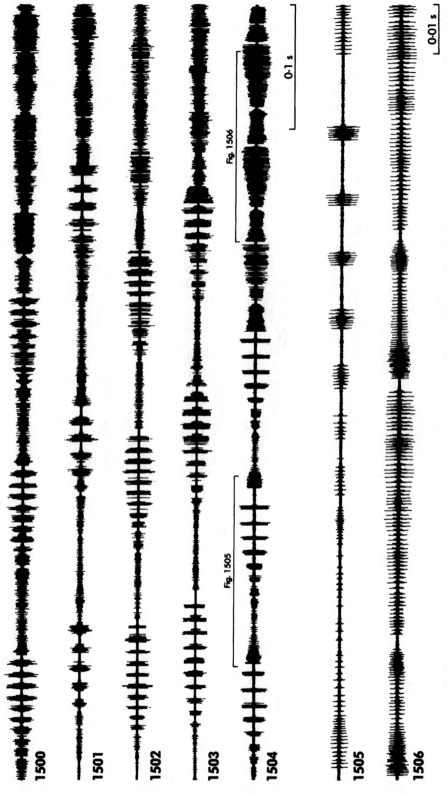

Figures 1500–1506 Faster oscillograms of the indicated parts of the songs of *Chorthippus dorsatus* shown in Figs 1495–1499. Figs 1504–1506 are from a male with only one hind leg.

1988*b*; Faber, 1928, 1929*b*, 1953*a*; Fischer, 1853; Harz, 1957; Taschenberg, 1871; Weber, 1984; Weih, 1951. **Disc recording**: Bellmann, 1993*c* (CD); Grein, 1984 (LP); Odé, 1997 (CD). **Cassette recording**: Bellmann, 1985*b*, 1993*b*; Wallin, 1979.

RECOGNITION. For the distinction between this species and *C. albomarginatus*, see under that species. Separating *C. dorsatus* from *C. dichrous* is more difficult, but within the area of our study *C. dichrous* occurs only in the extreme east of Austria and a few localities in southern Italy, and so the possibility of confusion seldom arises. The fore wings are shorter and broader in *C. dorsatus*, especially in Italy, where they usually fall well short of the hind knees; in *C. dichrous* the fore wings are narrower and almost always extend beyond the hind knees. The females of *C. dorsatus* very rarely have a white stripe along the costal area of the fore wings, but this stripe is often present in *C. dichrous*.

In the field *C. dorsatus* can be easily recognized by the very characteristic male calling song.

SONG (Figs 1490–1506. CD 2, track 64). The male calling song consists of a series of regularly spaced echemes, each lasting about 0·8–1·5 s and separated from adjacent ones by intervals of about 1–3 s or occasionally longer; the series is of indefinite duration, though commonly lasting 15–25 s. Each echeme has a highly characteristic structure, being divided into two distinct parts (it could thus be regarded as two echemes, but is more conveniently treated as one). The first part consists of about 3–7 syllables repeated at the rate of about 6–10/s, with the two hind legs moving synchronously. This is followed immediately by a second part in which the hind legs move up and down alternately, each producing about 2–6 syllables repeated at the rate of about 10–17/s; this second part often develops gradually during the first few echemes of a series, and is sometimes completely lacking from the first one or two (Fig. 1491). The syllables of the first part can be easily heard and are reminiscent of the calling song of *C. parallelus*; those of the second part are not clearly distinguishable becuase of the asynchrony, and merge into a rather amorphous hissy sound more reminiscent of the calling songs of *C. dichrous* and *C. jucundus*. There is usually a crescendo during the first part. Oscillographic analysis shows that all the syllables except the first one or two are diplosyllables; the closing hemisyllables of the first part of the echeme have clearly defined gaps, usually 4–8 per syllable, but both the opening and closing hemisyllables of the second part produce continuous sounds. The structure of the second part of the echeme can be seen clearly in oscillograms of a 'one-legged' song (Figs 1499, 1504–1506), in which the syllables are no longer obscured by asynchrony.

The change from the first part to the second part is often sudden and complete, but in some echemes the two parts are separated by a syllable that is intermediate, both in duration and in the structure of the downstroke hemisyllable, between typical first-part and second-part syllables.

DISTRIBUTION. Widespread in Europe from southern Sweden southwards; mainly confined to mountains in the southern peninsulas. Eastwards, the range extends to Siberia. Absent from the British Isles, and probably also from North Africa.

Chorthippus dichrous (Eversmann)

Oedipoda dichroa Eversmann, 1859: 132.

Two-coloured Grasshopper; (D) Ungarischer Grashüpfer.

REFERENCES TO SONG. **Oscillogram**: Bukhvalova & Zhantiev, 1993; Komarova & Dubrovin, 1973; Schmidt, 1987*b*; Schmidt & Schach, 1978; Stumpner & Helversen, 1994. **Leg-movement**: Stumpner & Helversen, 1994.

RECOGNITION. See under *C. albomarginatus* and *C. dorsatus*.

SONG (Figs 1507–1512. CD 2, track 65). The male calling song consists of a brief, simple echeme lasting about 300–600 ms, with a rather hissy quality, reminiscent of the second (asynchronous) part of an echeme from the calling song of *C. dorsatus*. The echeme is not repeated regularly, as in *C. dorsatus*, but at very variable intervals, usually between 3 and 10 s but occasionally longer. Oscillographic analysis shows that the echeme usually begins with a few very short-lived, isolated sounds (during which the hind legs tend to move synchronously) and then consists of about 5–8 diplosyllables repeated at the rate of about 15–25/s (during which the hind legs move more or less asynchronously); there are no gaps in either upstroke or downstroke hemisyllables.

DISTRIBUTION. In the area of our study this species occurs only at the extreme eastern end of Austria (especially in the vicinity of Neusiedler See) and in a few localities in southern Italy. Farther east it occurs in the Balkan Peninsula, Ukraine and across the southern part of European Russia to the Urals; the range is also believed to extend into much of the southern part of temperate Asia, but this needs to be confirmed by studies on the song.

Chorthippus parallelus (Zetterstedt)

Gryllus parallelus Zetterstedt, 1921: 85.

Meadow Grasshopper; (F) Criquet des pâtures; (D) Gemeiner Grashüpfer; (NL) Krasser; (DK) Enggræshoppe; (S) Kortningad Ängsgräshoppa; (SF) Nurmiheinäsirkka.

REFERENCES TO SONG. *C. p. parallelus*. **Oscillogram**: Broughton, 1954, 1955*a*, 1963*a*, 1976; Bukhvalova & Zhantiev, 1993; Butlin, 1989; Butlin & Hewitt, 1985; Faber, 1957*a* (as *longicornis*), 1958 (as *l.*), 1960; Grein, 1984; Haskell, 1955*b*, 1957, 1961; Hewitt, 1993; Holst, 1970 (as *l.*), 1986; Jacobs, 1953*a*; Kleukers *et al.*, 1997; Lewis *et al.*, 1971; Reynolds, 1980; Schmidt, 1989; Vedenina, 1990; Vedenina & Zhantiev, 1990; Zhantiev, 1981. **Diagram**: Bellmann, 1985*a*, 1988, 1993*a*; Bellmann & Luquet, 1995; Diujm & Kruseman, 1983; Haskell, 1957; Holst, 1970 (as *l.*); Jacobs, 1950*b*, 1953*a*; Luquet, 1978; Ragge, 1965; Wallin, 1979. **Frequency information**: R.-G. Busnel, 1955 (as *l.*); Froehlich, 1989; Haskell, 1957; Lewis *et al.*, 1971; Meyer & Elsner, 1996. **Musical notation**: Yersin, 1854*b* (as *pratorum*). **Verbal description only**: Bauer & Helversen, 1987; Broughton, 1972*a*, 1972*b*; Chopard, 1952 (as *l.*); Defaut, 1983, 1987, 1988*a*, 1988*b*; Faber, 1928, 1929*a*, 1929*b*; Harz, 1957; Haskell, 1955*a*, 1958; Helversen, 1979 (as *l.*); Helversen & Helversen, 1981 (as *l.*); Lux, 1961; Tschuch & Köhler, 1990; Weber, 1984; Weih, 1951. **Disc recording**: Andrieu &

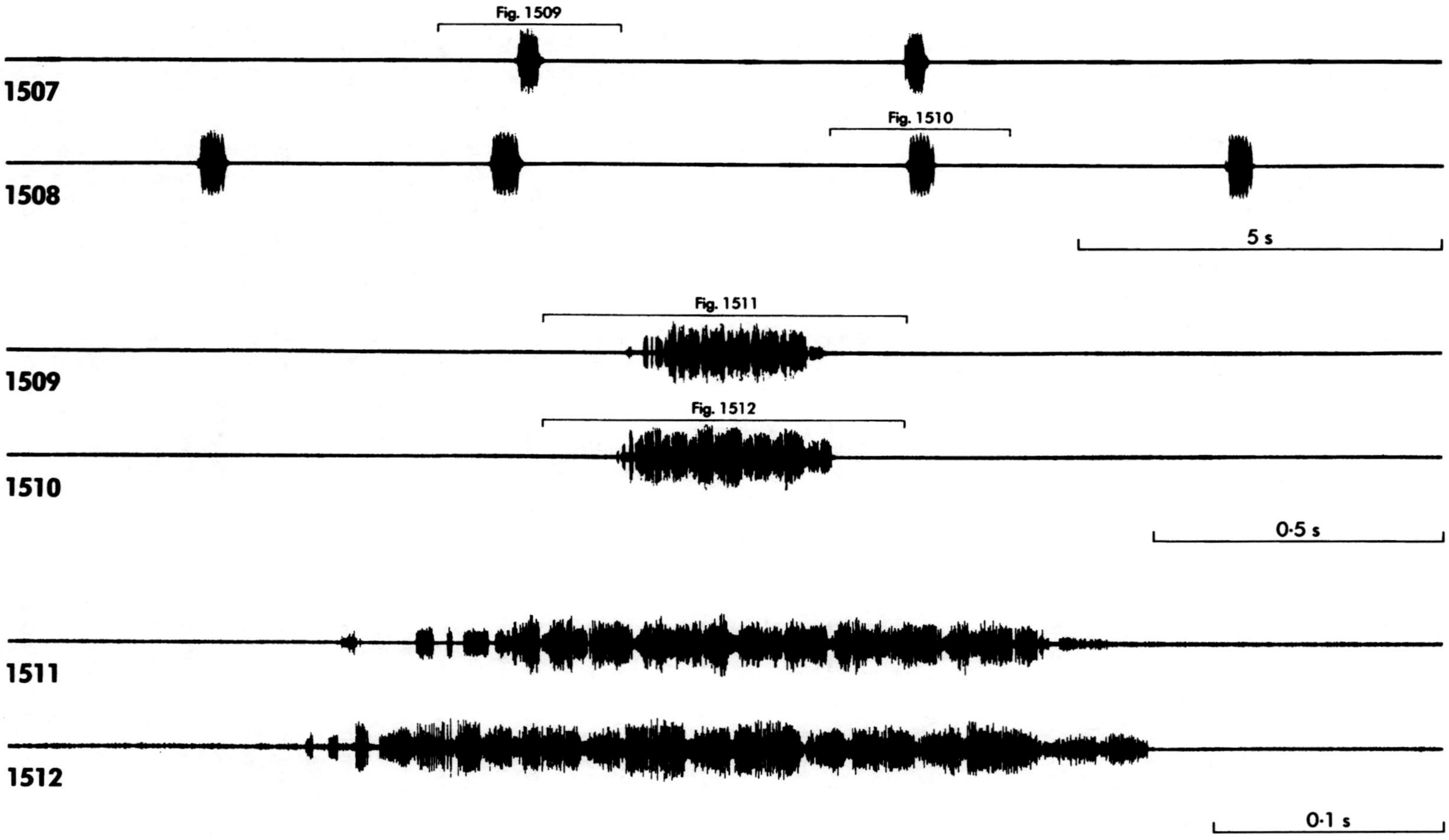

Figures 1507–1512 Oscillograms at three different speeds of the calling songs of two males of *Chorthippus dichrous.*

Dumortier, 1994 (CD); Bellmann, 1993c (CD); Bonnet, 1995 (CD); Grein, 1984 (LP); Odé, 1997 (CD); Ragge *et al.*, 1965 (LP). **Cassette recording**: Bellmann, 1985b, 1993b; Burton & Ragge, 1987; Wallin, 1979.

C. p. erythropus. **Oscillogram**: Butlin, 1989; Butlin & Hewitt, 1985; Faber, 1958, 1960; Reynolds, 1980. **Verbal description only**: Butlin & Hewitt, 1988; Defaut, 1987, 1988b.

RECOGNITION (Plate 3: 11). This very common species may be distinguished from most of its relatives by being brachypterous. The hind wings are vestigial in both sexes; the male fore wings fall well short of the hind knees, often not reaching the tip of the abdomen, and the female fore wings usually reach only to the third or fourth abdominal tergite. The closely related, much rarer species, *C. montanus*, is also brachypterous, but slightly less so: the male fore wings usually reach the tip of the abdomen, sometimes almost to the hind knees, and the female fore wings usually reach to the fifth or sixth abdominal tergite. In *C. parallelus* the male cercus is usually shorter than 0·65 mm and the visible part of the ventral valves of the ovipositor usually shorter than 1·3 mm; in *C. montanus* these measurements usually exceed these values. Fully winged forms of both these species sometimes occur, but are usually accompanied by the normal brachypterous form and are unlikely to cause confusion.

Typical *C. parallelus* is replaced in the Iberian Peninsula and parts of the Pyrenees by a form currently regarded as a distinct subspecies, *C. p. erythropus* Faber. This form is most clearly characterized by its acoustic behaviour (see below), but it has red hind tibiae (usually yellow or brown in typical *C. parallelus*) and a higher number of stridulatory pegs (usually more than 110 in *C. p. erythropus*, usually fewer than 110 in typical *C. parallelus*).

For remarks on the taxonomy of *C. parallelus* and *C. montanus* see p. 78.

SONG (Figs 1513–1554. CD 2, tracks 66, 67). The male calling song consists of a series of fairly regularly spaced echemes, each typically lasting about 1–2 s and separated from adjacent ones by intervals of about 3–5 s or sometimes longer; the series is of indefinite duration. Each echeme is composed of about 10–20 syllables repeated at the rate of about 7–12/s. These values are all affected by temperature; in cool conditions, dull weather or at night the echemes may last up to 5 s, with the syllables repeated as slowly as 2/s, and are usually separated by longer intervals of up to 10 s or more. The echemes always begin quietly, reaching maximum intensity by about a third to half of their duration. Most of the sound is produced by the downstrokes of the hind legs, and oscillographic analysis shows that there are about 2–3 gaps in each downstroke hemisyllable (often more in a 'one-legged' song); each diplosyllable typically lasts about 100–150 ms (up to 300 ms or longer in cool conditions).

Two or more males commonly react with one another by producing a rivalry song consisting of shorter echemes of fewer (about 6–12), more rapidly repeated syllables (up to about 20/s). Sometimes the hind legs are moved synchronously, as in the calling song, but in typical *C. parallelus* they are more often moved up and down alternately, at least in the later part of the echeme, so that the syllable structure becomes obscured (Fig. 1519, 1526, 1533).

Typical *C. parallelus* lacks a special courtship song, but *C. p. erythropus* has a well-developed one (Figs 1538–1540), consisting of echemes of about 2–15 syllables of a quite different kind from those of the calling song. The syllables are repeated at the rate of about 15–25/s;

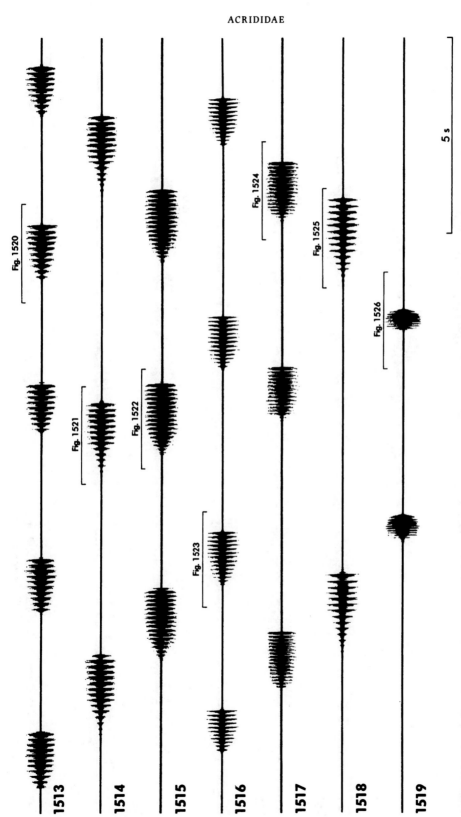

Figures 1513–1519 Oscillograms of the songs of seven males of typical *Chorthippus parallelus*. **1513–1518.** Calling songs. **1519.** Rivalry song. Fig. 1517 is from a male with only one hind leg, and Fig. 1518 is from a macropterous male.

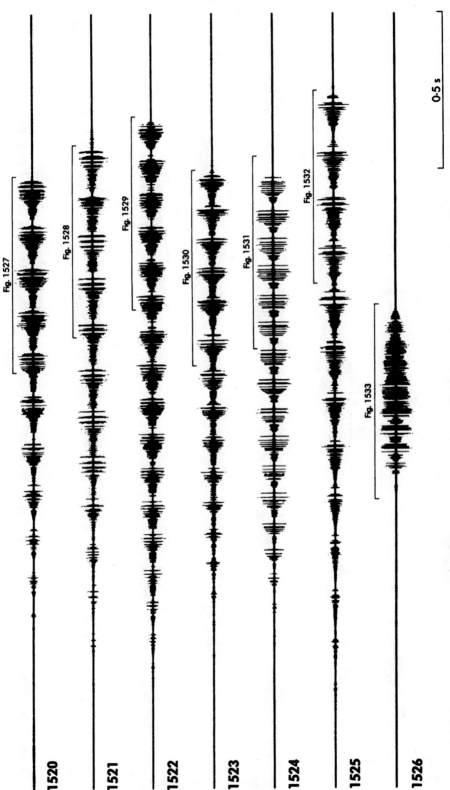

Figures 1520–1526 Faster oscillograms of the indicated parts of the songs of *Chorthippus parallelus* shown in Figs 1513–1519. Fig. 1524 is from a male with only one hind leg, and Fig. 1525 is from a macropterous male.

0·1 s

1527

1528

1529

1530

1531

1532

1533

Figures 1527–1533 Faster oscillograms of the indicated parts of the songs of *Chorthippus parallelus* shown in Figs 1520–1526. Fig. 1531 is from a male with only one hind leg, and Fig. 1532 is from a macropterous male.

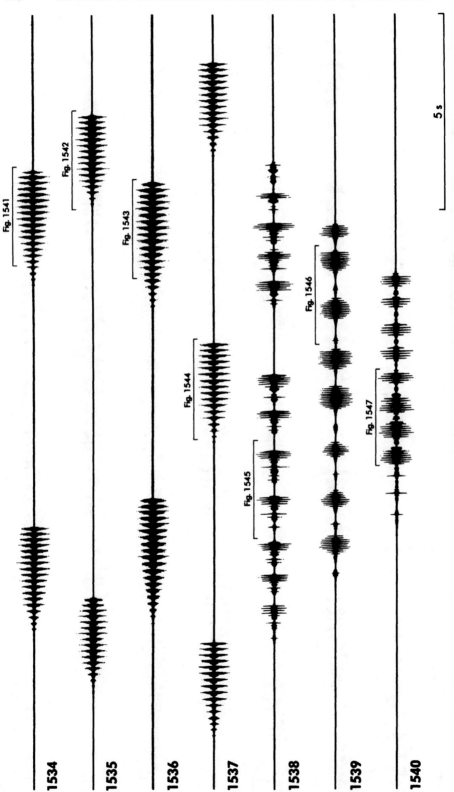

Figures 1534–1540 Oscillograms of the songs of seven males of *Chorthippus parallelus erythropus*. 1534–1537. Calling songs. 1538–1540. Courtship songs.

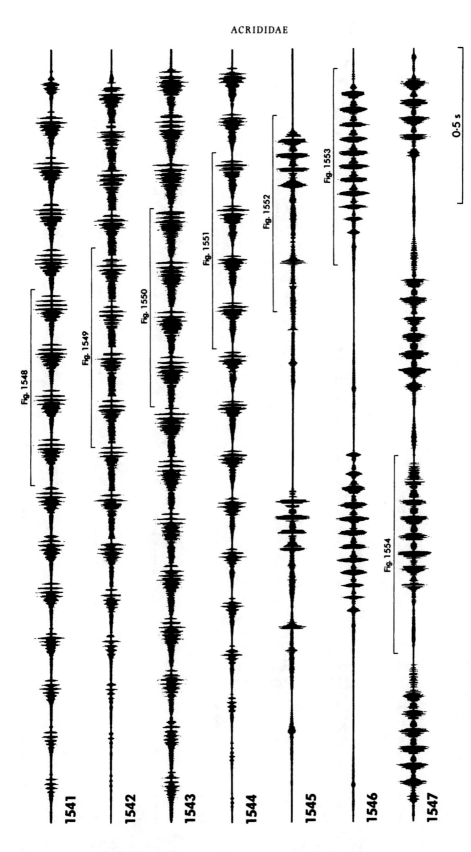

Figures 1541–1547 Faster oscillograms of the indicated parts of the songs of *Chorthippus parallelus erythropus* shown in Figs 1534–1540.

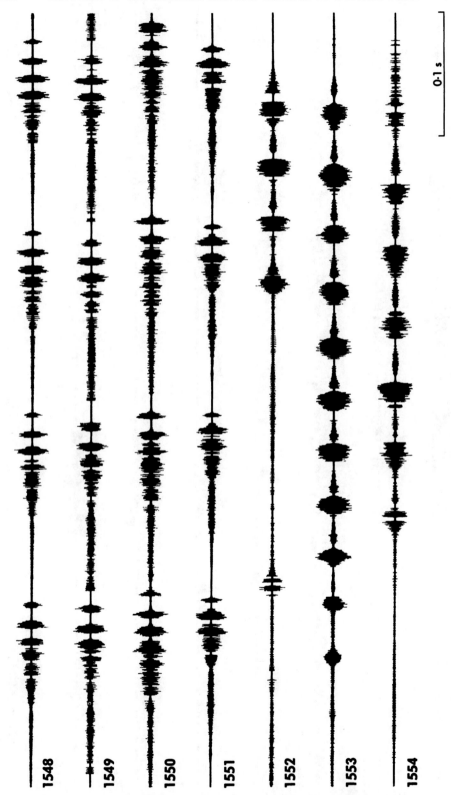

0·1 s

1548

1549

1550

1551

1552

1553

1554

Figures 1548–1554 Faster oscillograms of the indicated parts of the songs of *Chorthippus parallelus erythropus* shown in Figs 1541–1547.

each lasts about 30–50 ms and usually consists of well-defined hemisyllables, the downstroke one being much louder than the upstroke one. The sound often appears to be made mainly with only one of the hind legs. The echemes are grouped into sequences of about 3–10, in which they are repeated at the rate of about 0·5–2·0/s and often interspersed with quieter sounds produced by much slower movements of the hind leg.

The calling song of *C. p. erythropus* (Figs 1534–1537) tends to be intermediate between that of typical *C. parallelus* and that of *C. montanus*: the echemes (and syllables) tend to last a little longer than in typical *C. parallelus*, to be separated by rather longer intervals and to be grouped into shorter sequences. The rivalry song is similar to that of typical *C. parallelus*, but the hind legs have been moved synchronously in all the rivalry songs so far observed.

Unlike most other European grasshoppers, *C. parallelus* quite often sings during the night when it is not too cold. The echemes are always of the kind described above for cool conditions.

DISTRIBUTION. Common and widespread in Europe from southern Scandinavia and Finland southwards, including Britain but not Ireland. Eastwards, the range extends through temperate Asia to Mongolia. *C. p. erythropus* replaces typical *C. parallelus* in the Iberian Peninsula and parts of the Pyrenees; a form resembling *C. p. erythropus* in appearance also occurs in the Landes region of south-west France, but the song has not yet been studied in this area.

Chorthippus montanus (Charpentier)

Gryllus montanus Charpentier, 1825: 173.

Water-meadow Grasshopper; (F) Criquet palustre; (D) Sumpfgrashüpfer; (NL) Zompsprinkhaan; (S) Myrgräshoppa; (SF) Nevaheinäsirkka.

REFERENCES TO SONG. **Oscillogram**: Bukhvalova & Zhantiev, 1993; Faber, 1957a; Grein, 1984; Helversen & Helversen, 1994; Holst, 1986; Jacobs, 1953a; Kleukers et al., 1997; Reynolds, 1980; Schmidt & Baumgarten, 1977; Schmidt & Schach, 1978. **Diagram**: Bellmann, 1985a, 1988, 1993a; Bellmann & Luquet, 1995; Duijm & Kruseman, 1983; Jacobs, 1950b, 1953a; Wallin, 1979; Weih, 1951. **Leg-movement**: Helversen & Helversen, 1994. **Frequency information**: Meyer & Elsner, 1996. **Verbal description only**: Bauer & Helversen, 1987; Broughton, 1972a; Chopard, 1952; Faber, 1928, 1929a, 1929b, 1953a, 1958, 1960; Helversen, 1979; Helversen & Helversen, 1981; Harz, 1957; Korn-Kremer, 1963; Lux, 1961; Tschuch & Köhler, 1990; Weber, 1984. **Disc recording**: Andrieu & Dumortier, 1963 (LP), 1994 (CD); Bellmann, 1993c (CD); Grein, 1984 (LP); Odé, 1997 (CD). **Cassette recording**: Bellmann, 1985b, 1993b; Wallin, 1979.

RECOGNITION. See under *C. parallelus*.

SONG (Figs 1555–1563. CD 2, track 68). The male calling song is similar to that of *C. parallelus*, consisting of a series of simple echemes, but at the same temperature the tempo is slower in every respect. The echemes usually last about 2·0–4·5 s and the intervals between them are about 4–7 s or sometimes longer. They are composed of about 12–22 syllables repeated at the rate of about 4–6/s. Oscillographic analysis shows that the syllables are

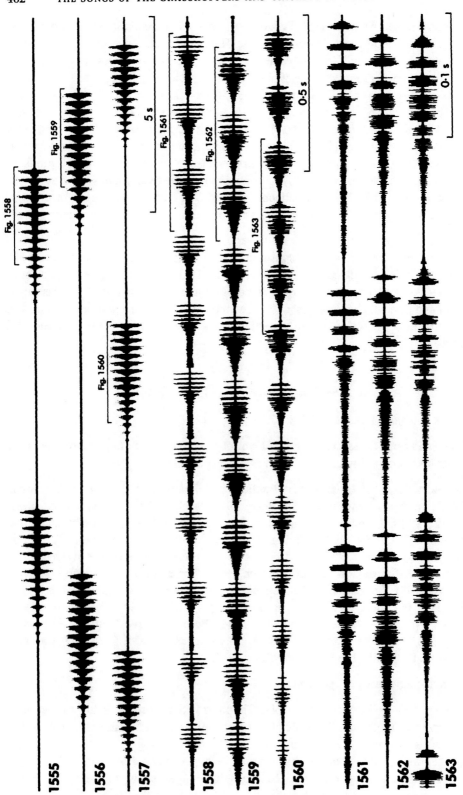

Figures 1555–1563 Oscillograms at three different speeds of the calling songs of three males of *Chorthippus montanus*.

similar in structure to those of the calling song of *C. parallelus*, but there are more gaps (usually 3–5) during the downstroke hemisyllables; each syllable lasts about 150–250 ms. As in *C. parallelus*, the echemes begin quietly, reaching maximum intensity about halfway through their duration.

The rivalry song is also similar to that of *C. parallelus*, but the tempo is again slower and the hind legs are always moved synchronously. There is no well-developed courtship song.

DISTRIBUTION. In northern Europe the distribution is similar to that of *C. parallelus*, but *C. montanus* is much more local and largely confined to marshy habitats. Its range extends a little farther north in Sweden and Finland, but it does not occur in the British Isles and is largely absent from the southern peninsulas. Eastwards, the range extends through central Europe to Siberia, Mongolia and Manchuria.

Euchorthippus declivus (Brisout)

Acridium declivum Brisout, 1849: 420.

Sharp-tailed Grasshopper; (F) Criquet des mouillères; (D) Dickkopf-Grashüpfer.

REFERENCES TO SONG. **Oscillogram**: Descamps, 1968; Ragge & Reynolds, 1984; Schmidt & Schach, 1978; Schmidt, 1989. **Diagram**: Luquet, 1978. **Musical notation**: Yersin, 1854*b*. **Verbal description only**: Bellmann, 1993*a*; Bellmann & Luquet, 1995; Defaut, 1987, 1988*b*; Yersin, 1857. **Disc recording**: Bonnet, 1995 (CD).

RECOGNITION (Plate 3: 12). *Euchorthippus* is generally similar in appearance to *Chorthippus*, but the foveolae on the head are weakly developed, and from each of them a feeble carinula extends backwards along the vertex. The lateral pronotal carinae are usually straight or only slightly incurved, and in western Europe the general colouring is always brown, straw-coloured or grey. Fuller information and a morphological key to the western European and North African species are given by Ragge & Reynolds (1984). See also the notes on p. 79.

The male calling song is remarkably uniform in the European species of *Euchorthippus* and, as a field character, is more useful in recognizing the genus than in identifying the species. It consists of sequences of brief echemes repeated at a fairly constant rate of about 0·5–1·5/s, usually for indefinite periods of up to a minute or more; in western Europe each echeme is composed of 6–8 (occasionally 4, 5 or 9) syllables with a crescendo during the first 3–5. The calling song of *Chorthippus brunneus* shows some similarity, but the echemes begin more abruptly and (being produced by the two hind legs moving alternately) do not show a clear syllable structure. Confusion is more likely with the calling songs of some species of Tettigoniidae, especially such decticines as *Metrioptera brachyptera*, *Platycleis sabulosa* and *P. albopunctata*, but the echeme repetition rate is more constant and usually higher than 2/s in these bush-crickets, and the sequences are usually maintained for longer periods, often many minutes.

E. declivus may be distinguished from the other European species by the relatively short hind wings, which, when flexed, fall short of the flexed fore wings by about 1–2 mm in the male and 0·5–1·5 mm in the female; in all the other species the tips of the flexed fore and hind wings are coincident or almost so. In addition, the male subgenital plate is long and acutely pointed, in marked contrast to all the other European species except *E. pulvinatus*.

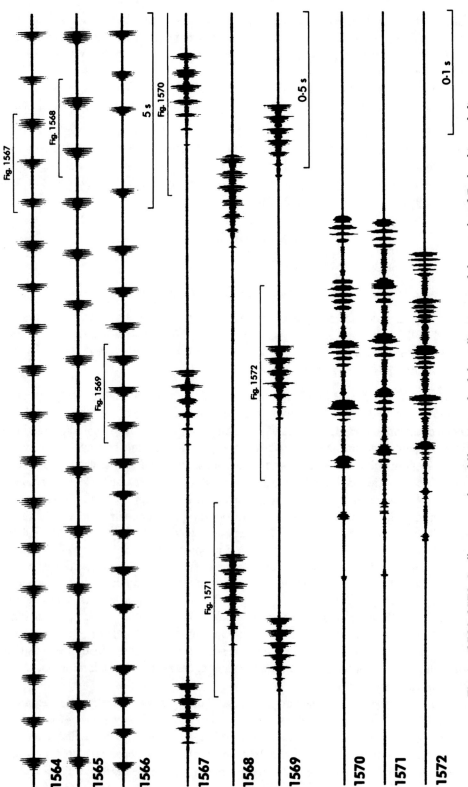

Figures 1564–1572 Oscillograms at three different speeds of the calling songs of three males of *Euchorthippus declivus*.

SONG (Figs 1564–1572. CD 2, track 69). The calling song consists of a sequence of brief echemes repeated at the rate of about 0·7–1·3/s in warm, sunny weather. The duration of the sequence is very variable, but an undisturbed male may produce a sequence lasting up to a minute or more. Oscillographic analysis shows that each echeme begins quietly, usually reaching maximum intensity about halfway through its duration, and is composed of 6–8 (occasionally 4, 5 or 9) syllables. In warm, sunny weather each echeme lasts about 200–300 ms and each syllable about 35–45 ms, but in cooler conditions they can last up to twice as long. The syllable repetition rate within an echeme is usually between 20 and 30/s, but can be as low as 10/s in cool, dull weather. In the louder part of the echeme there are conspicuous gaps in the downstroke hemisyllables, usually three (rarely four) in each of the last few syllables; the total number of such gaps in each echeme is usually 10–15, more than in the other western European species of *Euchorthippus*.

DISTRIBUTION. Most of France except for parts of the north and north-west, northern Spain, southern and eastern Alps, Italy, southern Slovakia, the whole of the Balkan Peninsula except for the extreme south, western Ukraine and Asia Minor.

Euchorthippus pulvinatus (Fischer de Waldheim)

Oedipoda pulvinata Fischer de Waldheim, 1846: 305.

Straw-coloured Grasshopper; (F) Criquet glauque; (D) Gelber Grashüpfer.

REFERENCES TO SONG. **Oscillogram**: Broughton, 1954; Bukhvalova & Zhantiev, 1993; Descamps, 1968; Ragge & Reynolds, 1984; Schmidt & Schach, 1978. **Diagram**: Luquet, 1978. **Frequency information**: Busnel, 1953. **Verbal description only**: Bellmann & Luquet, 1995; Broughton, 1972a; Chopard, 1922; Faber, 1953a; Harz, 1957; Yersin, 1857. **Disc recording**: Andrieu & Dumortier, 1994 (CD); Bonnet, 1995 (CD). **Cassette recording**: Burton & Ragge, 1987.

RECOGNITION. In western Europe *E. pulvinatus* overlaps in range with *E. declivus* and *E. chopardi*. Both sexes may be distinguished from *E. declivus* by the longer hind wings, which, when flexed, reach the tips of the flexed fore wings. The male subgenital plate, although somewhat elongate and conical in shape, is not produced into an acute point as in *E. declivus*. Distinguishing *E. pulvinatus* morphologically from *E. chopardi* in southern France and the Iberian Peninsula is much more difficult and can be achieved reliably only by plotting various measurements on graphs (see Ragge & Reynolds, 1984). In the field males of these two species can usually be distinguished by their calling songs: in warm, sunny weather the echeme repetition rate is usually more than 0·8/s in *E. pulvinatus*, usually slower than this in *E. chopardi*.

SONG (Figs 1573–1593. CD 2, track 70). In western Europe the male calling song sounds very similar, to the unaided ear, to that of *E. declivus*, consisting of sequences of brief echemes repeated at the rate of about 0·8–1·5/s in warm, sunny weather. However, oscillographic analysis shows that both the echemes and syllables are of shorter duration than in *E. declivus*, the echemes usually lasting about 120–200 ms and the syllables about 20–30 ms. The syllable repetition rate within an echeme is usually between 35 and 50/s, much

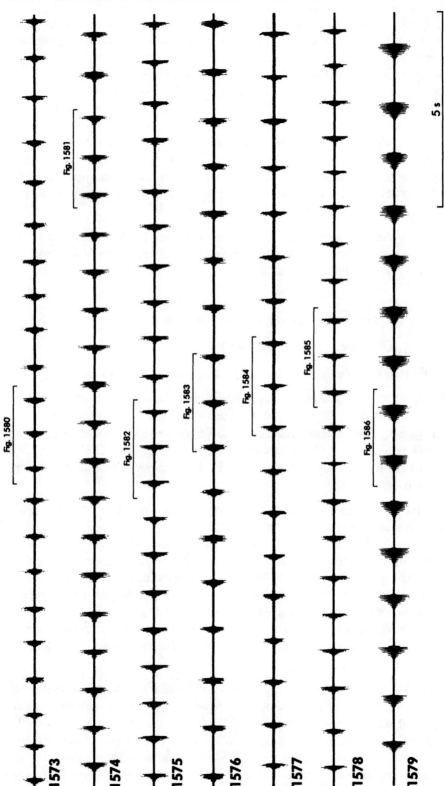

Figures 1573–1579 Oscillograms of the calling songs of seven males of *Euchorthippus pulvinatus*. 1573–1575. *E. p. pulvinatus*. 1576–1578. *E. p. gallicus*. 1576–1578. *E. p. elegantulus* from (1576) Jersey, Channel Islands and (1577, 1578) Brittany, France. 1579. *E. p. pulvinatus*.

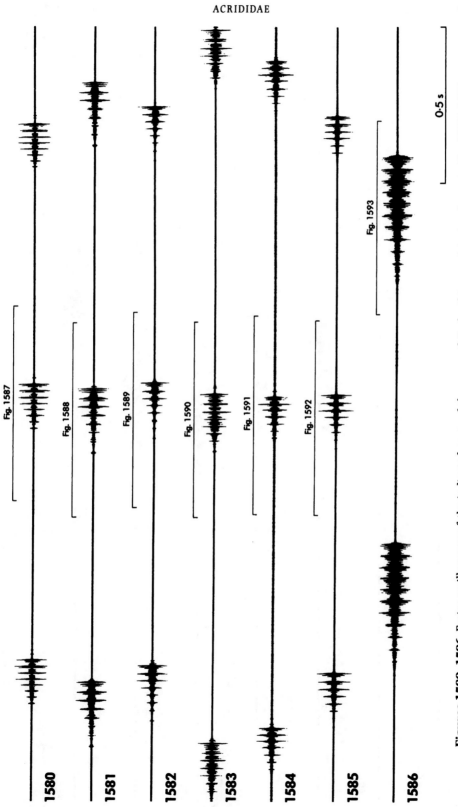

Figures 1580–1586 Faster oscillograms of the indicated parts of the songs of *Euchorthippus pulvinatus* shown in Figs 1573–1579.

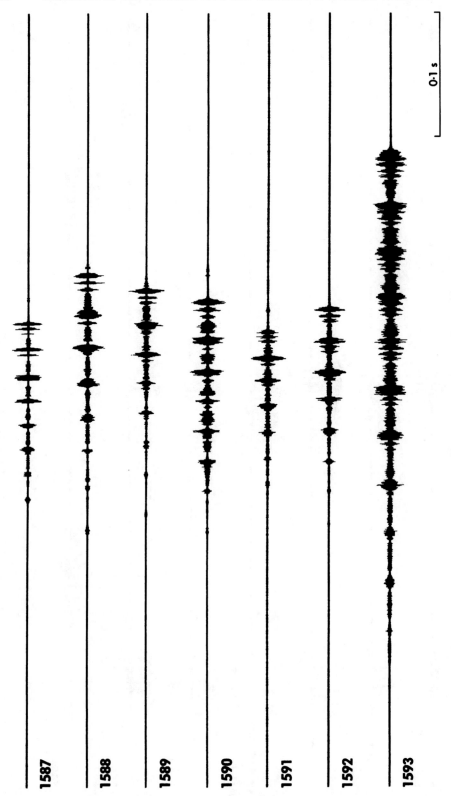

Figures 1587–1593 Faster oscillograms of the indicated parts of the songs of *Euchorthippus pulvinatus* shown in Figs 1580–1586.

faster than in *E. declivus*. There are also fewer gaps in the downstroke hemisyllables, usually only one or two (rarely three) in the last two syllables, so that the total number of gaps in each echeme is usually only 3–5, less than half the number in *E. declivus*. As in *E. declivus*, all these figures except the number of gaps become altered in cooler conditions, when the hind legs are moved more slowly.

This description applies to the two western subspecies of *E. pulvinatus*, *E. p. gallicus* Mařan and *E. p. elegantulus* Zeuner. Little information is at present available on the song of the nominate subspecies, *E. p. pulvinatus*, which occurs only in eastern Europe and Asia. Faber's (1953*a*) verbal description, which was based on males from Mödling, near Vienna, and which should therefore apply to this subspecies, suggests no significant difference from the calling songs of French and Spanish males. Schmidt & Schach (1978) gave brief notes (with oscillograms) on the courtship song only, based on males from near Neusiedler See, south-east of Vienna. In the calling song of a male captured at Lake Mladost, south-east of Skopje in Macedonia (recorded in warm conditions by our colleague Dr F. Willemse, who kindly gave us a copy of the recording – see Figs 1579, 1586, 1593), the echemes are markedly different from those produced in western Europe, each lasting 350–400 ms and composed of 10–11 syllables repeated at the rate of about 25/s; this agrees with the information recently given by Bukhvalova & Zhantiev (1993) for the calling song of Russian males of this subspecies.

DISTRIBUTION. Represented in western Europe mainly by the subspecies *E. p. gallicus*, which occurs in central and southern France and the Iberian Peninsula. A small form, showing no difference in calling song, occurs in southern Brittany and Jersey, where it is currently treated as the subspecies *E. p. elegantulus*. *E. pulvinatus* is apparently absent from the Alps and Italy, but occurs as the nominate subspecies from the Czech Republic, the extreme east of Austria, Hungary and the Balkan Peninsula eastwards to Central Asia and perhaps even China.

Euchorthippus chopardi Descamps

Euchorthippus chopardi Descamps, 1968: 8.

Chopard's Grasshopper; (F) Criquet du Bragalou.

REFERENCES TO SONG. **Oscillogram**: Descamps, 1968; Ragge & Reynolds, 1984. **Diagram**: Luquet, 1978. **Verbal description only**: Bellmann & Luquet, 1995; Defaut, 1987, 1988*b*. **Disc recording**: Bonnet, 1995 (CD).

RECOGNITION. For distinguishing this species from *E. declivus* and *E. pulvinatus*, with both of which it overlaps in range, see under those species.

SONG (Figs 1594–1602. CD 2, track 71). The male calling song sounds quite similar to those of *E. declivus* and *E. pulvinatus*, but the echeme repetition rate is slower, usually 0·5–0·7/s in warm, sunny weather. Oscillographic analysis shows that each echeme lasts about 230–350 ms and each syllable about 40–55 ms, both much longer than in *E. pulvinatus*. The syllable repetition rate within an echeme is about 15–25/s. There are usually 2–3 gaps in each of the last 2–4 downstroke hemisyllables, making the total number of gaps in each echeme about 5–10, more than in *E. pulvinatus* but fewer than in *E. declivus*.

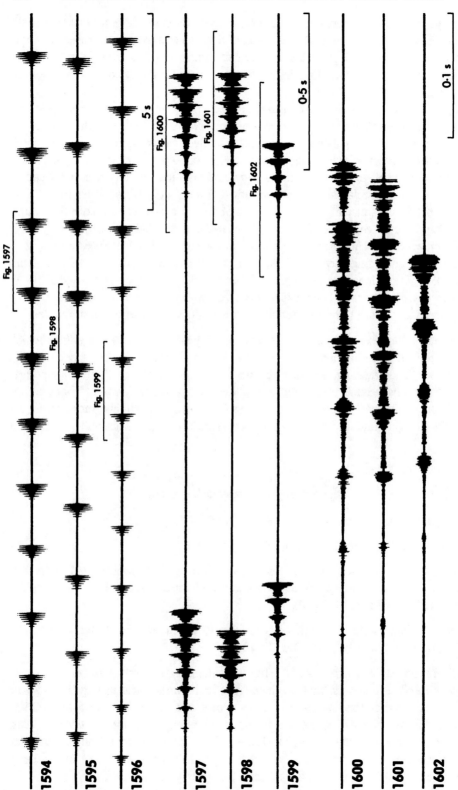

Figures 1594–1602 Oscillograms at three different speeds of the calling songs of three males of *Euchorthippus chopardi*. Figs 1596, 1599 and 1602 are from a male with only one hind leg.

DISTRIBUTION. Most of the Iberian Peninsula, and southern France from the eastern Pyrenees to Vaucluse and Var.

Euchorthippus albolineatus (Lucas)

Oedipoda albo lineata Lucas, 1849: 38.

White-lined Grasshopper.

REFERENCES TO SONG. No published work known to us on the Sicilian subspecies. Defaut (1987, 1988*b*) makes a brief reference to the song of the nominate subspecies at Rabat, Morocco.

RECOGNITION. *E. albolineatus* occurs in the area of our study only on the island of Sicily, where it is currently treated as the subspecies *E. a. siculus* Ramme. The only other species of *Euchorthippus* recorded from Sicily is *E. declivus*, which can be easily distinguished from *E. a. siculus* by the characters given on p. 463.

SONG (Figs 1603–1608. CD 2, track 72). To judge from the three males studied, the calling song in Sicily tends to be intermediate in most respects between those of *E. declivus* and *E. pulvinatus*. It consists of echemes repeated for indefinite periods at the rate of about 0·6–0·8/s. Each echeme lasts about 200–250 ms and each syllable about 30–35 ms. The syllable repetition rate within an echeme is about 30–35/s. There are 1–3 gaps in each of the last 2–3 downstroke hemisyllables, so that the total number of gaps in each echeme is about 4–7.

DISTRIBUTION. The nominate subspecies, *E. a. albolineatus*, occurs in North Africa, from western Morocco to Tripolitania. *E. a. siculus* is known only from Sicily.

Euchorthippus angustulus Ramme

Euchorthippus angustulus Ramme, 1931: 191.

Balearic Grasshopper.

REFERENCES TO SONG. **Oscillogram**: Ragge & Reynolds, 1984.

RECOGNITION. This species is found only in the Balearic Islands, where it is the only species of *Euchorthippus* known to occur.

SONG (Figs 1609–1617. CD 2, track 73). The male calling song follows the usual pattern for the genus, but in all the five males we have studied the echeme-sequences were, on average, of shorter duration than in the other western European species of *Euchorthippus*, often lasting less than 10 s and composed of fewer than 10 echemes. The longest sequence produced by these males (shown in Fig. 1609) lasted 13 s and was composed of 12 echemes. The song is thus superficially similar to that of *Chorthippus brunneus*, but that species is not known from the Balearic Islands and so confusion is unlikely. The echeme repetition rate is usually between 0·5 and 1·0/s.

Oscillographic analysis shows that each echeme lasts about 130–220 ms and each syllable about 28–35 ms. The syllable repetition rate within an echeme is about 30–36/s. There are usually 2–3 gaps in each of the last few downstroke hemisyllables, and the total number

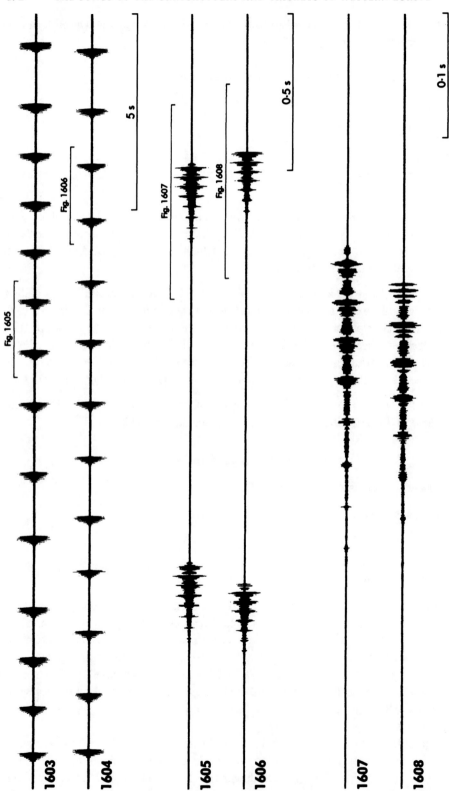

Figures 1603–1608 Oscillograms at three different speeds of the calling songs of two males of *Euchorthippus albolineatus siculus*.

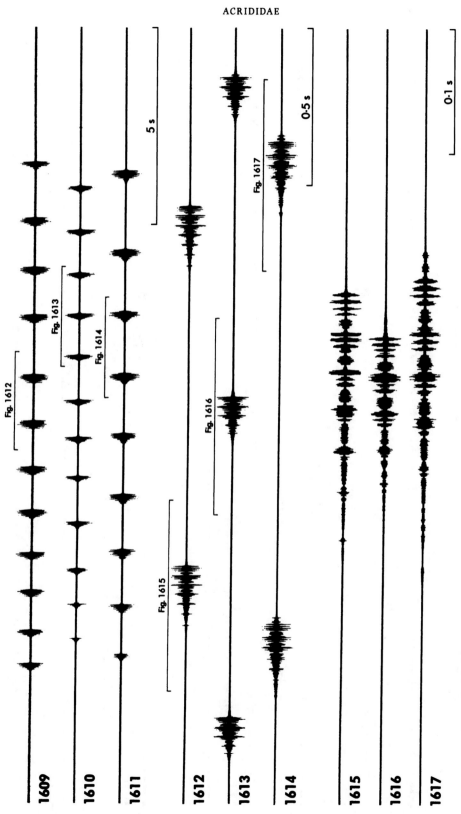

Figures 1609–1617 Oscillograms at three different speeds of the calling songs of three males of *Euchorthippus angustulus*.

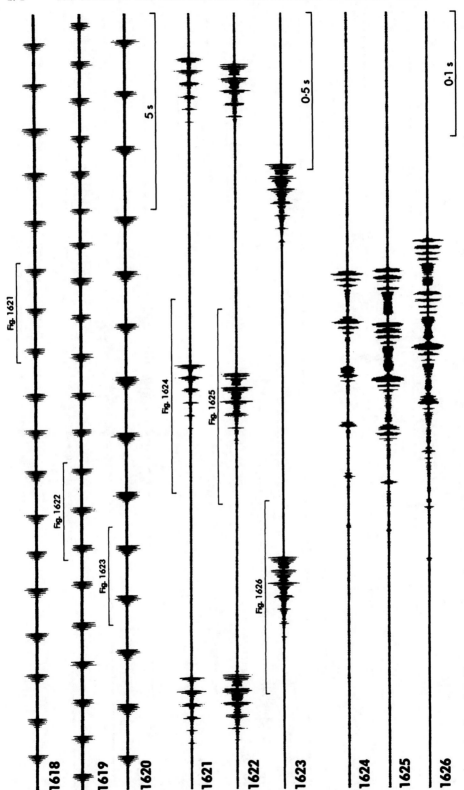

Figures 1618–1626 Oscillograms at three different speeds of the calling songs of three males of *Euchorthippus sardous*.

of gaps in each echeme is about 5–8. In the two males studied from Formentera, the echemes were shorter (130–150 ms) and composed of fewer syllables (4–5) than in those from Majorca and Ibiza (in which the echemes lasted 180–220 ms and were composed of 6–7 syllables).

DISTRIBUTION. Known only from the Balearic Islands, where it occurs on Majorca, Ibiza and Formentera.

Euchorthippus sardous Nadig

Euchorthippus sardous Nadig, *in* Nadig & Nadig, 1934: 18.

Sardinian Grasshopper.

REFERENCES TO SONG. No published work known to us.

RECOGNITION. This species is known only from Sardinia at altitudes above 1000 m, where we have seen no other species of *Euchorthippus*. It is in any case smaller than any of the other European species of *Euchorthippus*, the pronotum being less than 2·1 mm long in the male, less than 3·0 mm in the female (longer than these measurements in all the other European species).

SONG (Figs 1618–1626. CD 2, track 74). The male calling song is similar in most respects to that of *E. angustulus*, except that the echeme-sequences usually last for longer periods, as is usual in *Euchorthippus*. The echeme repetition rate is usually between 0·4 and 1·0/s. Oscillographic analysis shows that each echeme lasts about 150–250 ms and each syllable about 30–45 ms. The syllable repetition rate within an echeme is usually between 23 and 31/s. There are usually 2–3 gaps in at least the last two downstroke hemisyllables, and the total number of gaps in each echeme is about 5–9.

DISTRIBUTION. Known only from Sardinia, where it occurs in the Monti del Gennargentu and Monti Limbara, always at altitudes above 1000 m.

Chapter 10

Other animal sounds that could be confused with orthopteran songs

Orthoptera produce such a wide variety of songs, some continuous, some in short bursts, some harsh and others musical, that it would be surprising if some of them were not similar to the sounds produced by other animals. Among the insects, the only other group that produces conspicuously loud sounds is the Cicadoidea. Cicada sounds are not likely to be confused with orthopteran songs, if only because they are usually much louder (to human ears) than any bush-cricket song they might otherwise resemble, but we nevertheless thought it would be useful to give a brief account of the songs of three species of cicada commonly heard in southern Europe. To these we have added a fourth, mainly because it occurs, albeit very rarely, in Britain. The songs of all four are included at the end of the second companion CD.

The only other animals that produce sounds resembling orthopteran songs are amphibians and birds. We have listened carefully to the songs of a number of European species of both these groups and have selected those of seven (two amphibians and five birds) that we consider to be sufficiently similar to the songs of Orthoptera to be quite easily confused with them. We have included recordings of these at the end of the second companion CD and give oscillograms of them in Figs 1631–1657. Analyses of bird song usually take the form of audiospectrograms (or sonagrams, see p. 22) and ornithologists may find these oscillograms of particular interest.

Burton & Johnson (1984) have given a most useful account, written particularly from an ornithologist's viewpoint, of a number of resemblances between the songs of insects, amphibians and birds. They suggest that the tettigoniid-like songs of some of the *Locustella* warblers may confer an advantage on these birds, a kind of acoustic camouflage that could deceive predators. They even raise the possibility that mimicry of bush-cricket songs may play a part in the development of the songs of these warblers.

Cicadas

Cicadas are daytime singers, usually from shrubs or trees. Only the males sing, producing sound in a quite different way from Orthoptera: by 'popping' in and out the membrane of a tymbal organ on each side of the base of the abdomen. They can sing in flight or while walking about as well as while stationary.

Cicada orni Linnaeus (Fig. 1627. CD 2, track 75)

The loud, harsh and monotonous song of this common southern European species

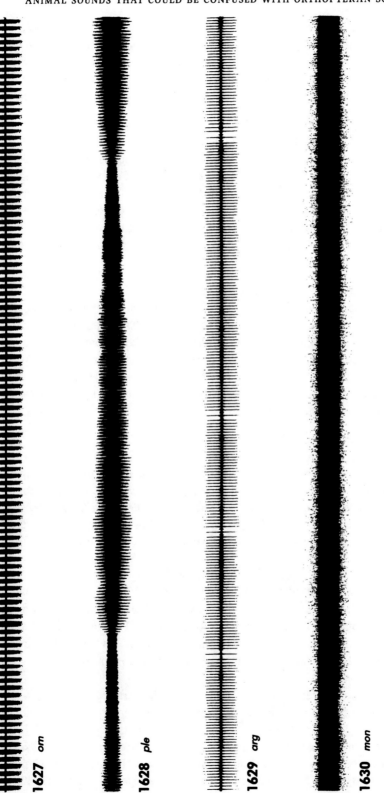

Figures 1627–1630 Oscillograms of the male calling songs of four cicadas. **1627.** *Cicada orni.* **1628.** *Lyristes plebejus.* **1629.** *Tettigetta argentata.* **1630.** *Cicadetta montana.*

consists of short bursts of sound repeated rapidly for long periods. The rate of repetition of these bursts is usually fairly constant, about 5–6/s in mid-song, but is slower and less regular when a male first begins to sing. The song is rhythmically similar to that of the cricket *Modicogryllus bordigalensis*, but the sound quality is harsher and confusion is unlikely in practice as the cicada sings from bushes and trees during the day, whereas the cricket sings from ground level in the evening and at night.

Lyristes plebejus (Scopoli) (Fig. 1628. CD 2, track 76)
The song of this equally common central and southern European species is also loud and continued for long periods, but consists of two kinds of sound alternating with each other. One component consists of rapidly repeated bursts of sound of a rather more sibilant quality than those of *Cicada orni* and with a higher repetition rate, usually about 8–12/s; this alternates with a much denser buzzing sound. The relative durations of these two components and the frequency of their alternation are often quite constant, but are sometimes less regular.

Tettigetta argentata (Olivier) (Fig. 1629. CD 2, track 77)
The song of this species, which is smaller than *C. orni* and *L. plebejus*, is a rapid ticking sound, at the rate of about 10–15/s, punctuated, more or less frequently, by slight hesitations. The song of the bush-cricket *Decticus verrucivorus* shows some resemblance to the sound made by this cicada, but the echeme repetition rate is lower, usually 8–10/s, there are no hesitations and the song is always delivered from the ground vegetation rather than shrubs or trees. *T. argentata* is at present known only from southern France, the Iberian and Italian peninsulas, and Sicily.

Cicadetta montana (Scopoli) (Fig. 1630. CD 2, track 78)
This species, similar in size to *T. argentata*, is the only cicada to occur in Britain, where it is very local in the New Forest; elsewhere it is widespread in continental Europe. Its song is a very dense, high-pitched, continuous buzz, lasting up to a minute or more, and could conceivably be confused with the songs of such bush-crickets as *Metrioptera roeselii* and *Gampsocleis glabra*, which are also daytime singers. *C. montana*, however, has an early season and is unlikely to be heard from mid-July onwards, when these bush-crickets begin to sing.

Amphibians
We consider the cricket-like calls of the Green Toad (*Bufo viridis*) and Midwife Toad (*Alytes obstetricans*) to be the only amphibian sounds in western Europe likely to be confused with the songs of Orthoptera. The calls of the Fire-bellied Toad (*Bombina bombina* (Linnaeus)) and Yellow-bellied Toad (*Bombina variegata* (Linnaeus)) are too low-pitched to be confused with cricket songs, though they can be mistaken for the hooting of owls.

Bufo viridis Laurenti, Green Toad (Figs 1631–1633. CD 2, track 79)
The call of this mainly eastern European species is certainly cricket-like, though not matching the song of any western European cricket. It is in bursts of up to about 10 s, each beginning quietly and composed of pulses repeated at the rate of about 15–20/s. The carrier frequency is about 1·0–1·5 kHz. The overall pattern of the call is quite similar to an echeme of the song of the Marsh Cricket (*Pteronemobius heydenii*), but the repetition rate

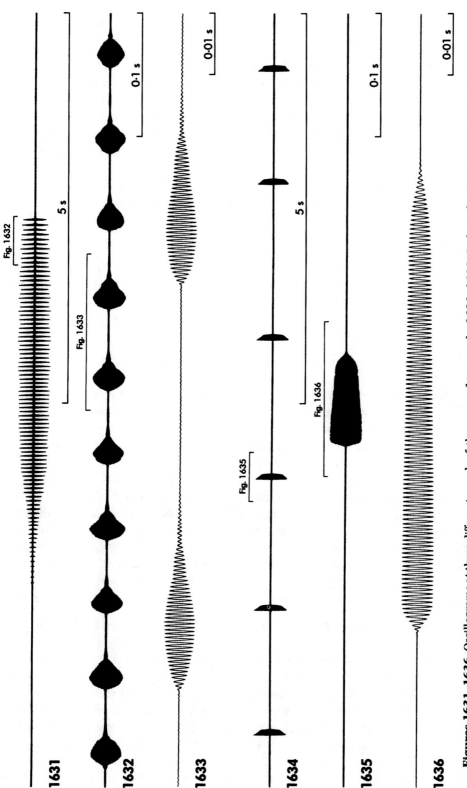

Figures 1631–1636 Oscillograms at three different speeds of the songs of two toads. **1631–1633.** *Bufo viridis*. **1634–1636.** *Alytes obstetricans.*

and carrier frequency are both much lower. The Green Toad is found in Europe mainly from Germany and Italy eastwards.

Alytes obstetricans (Laurenti), Midwife Toad (Figs 1634–1636. CD 2, track 80)
The short nocturnal calls of this toad, each consisting of a pure musical note of about $1\cdot0$–$1\cdot5$ kHz and lasting about 70–150 ms, have been aptly likened to the radio time signals broadcast by the BBC. They are remarkably cricket-like in quality and could be confused with the single syllables often produced by *Eugryllodes pipiens*. The cricket syllables are, however, usually much shorter than the calls of the toad, and the carrier frequency is higher (about $3\cdot0$–$3\cdot5$ kHz).

A. *obstetricans* occurs in a variety of habitats in the Low Countries, Germany, France and the Iberian Peninsula, but not in Britain except for rare introductions.

Birds
Of the four species of *Locustella* discussed below, *L. fluviatilis* occurs only in the extreme east of our area of study and *L. lanceolata* only as a vagrant, but we nevertheless felt they were of sufficient interest to be worth including. *Otus scops* (Linnaeus) (Scops Owl) has a call quite similar to that of *Alytes obstetricans* (Midwife Toad), but each call of the owl begins with a short higher-pitched note, thus making it distinctly bird-like and not eligible for inclusion here.

Locustella naevia (Boddaert), Grasshopper Warbler (Figs 1637, 1639, 1641, 1643. CD 2, track 81)
This species occurs widely in Europe, from southern Sweden to northern Spain and northern Italy (and including Britain), in a variety of open habitats. Its continuous, reeling song, although quite tettigoniid-like, does not closely resemble the song of any western European bush-cricket; the repetition rate of its repeated double elements is about 30/s, lying midway between the buzzing songs of *Ruspolia nitidula*, *Metrioptera roeselii* and *Gampsocleis glabra*, and the much slower, ticking songs of such species as *Tettigonia viridissima* and *Decticus verrucivorus*. It could, however, be confused with the songs of the crickets *Nemobius sylvestris*, which has a similar syllable repetition rate, and *Pteronemobius heydenii*, which is quite similar in quality although having a higher syllable repetition rate. The songs of both these crickets, however, are broken up into short bursts, even those of *P. heydenii* seldom lasting as long as 4 s, whereas *L. naevia* usually sings in bursts of about half a minute or more. When these crickets are singing in chorus, so that the intervals between individual bursts of song are difficult to detect, they could nevertheless be confused with this warbler.

Locustella luscinioides (Savi), Savi's Warbler (Figs 1638, 1640, 1642, 1644. CD 2, track 82)
This species occurs in marshy places in southern Europe, from northern Germany southwards and including, very locally, southern Britain. Its song is quite similar to that of *L. naevia*, but the repetition rate of its double elements is higher, about 45–55/s. It could also be confused with *Pteronemobius heydenii*, but is perhaps closest to the song of *Gryllotalpa vineae*, which has a similar syllable repetition rate. The long-lasting bursts of the song of *L. luscinioides*, up to a minute or more, again distinguish it from *P. heydenii*, but not from *G. vineae*, which also sings in long bursts. This mole-cricket does not occur in marshy places, however, and its song always comes from the ground, just inside its burrow.

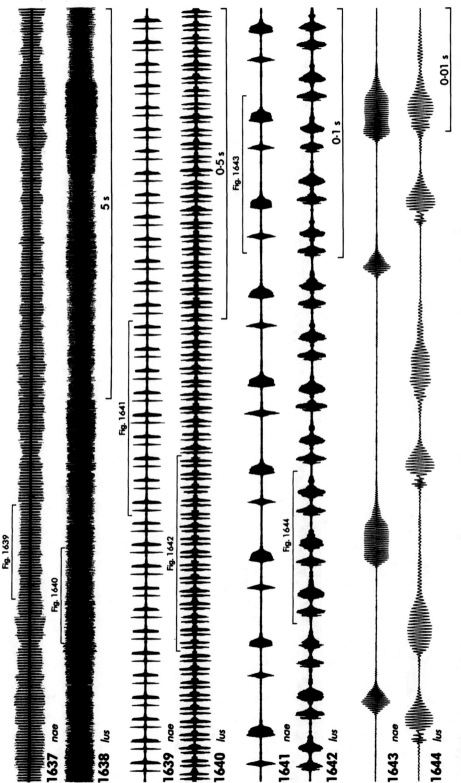

Figures 1637–1644 Oscillograms at four different speeds of the songs of two warblers. 1637, 1639, 1641, 1643. *Locustella naevia.* 1638, 1640, 1642, 1644. *L. luscinioides.*

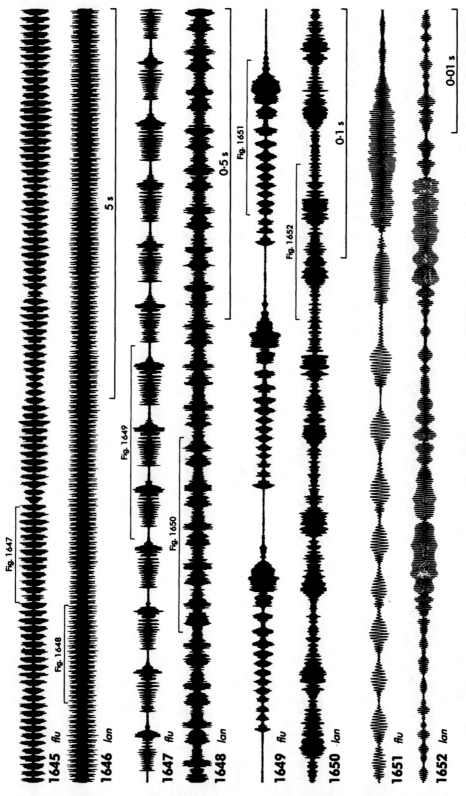

Figures 1645–1652 Oscillograms at four different speeds of the songs of two more warblers. **1645, 1647, 1649, 1651.** *Locustella fluviatilis.* **1646, 1648, 1650, 1652.** *L. lanceolata.*

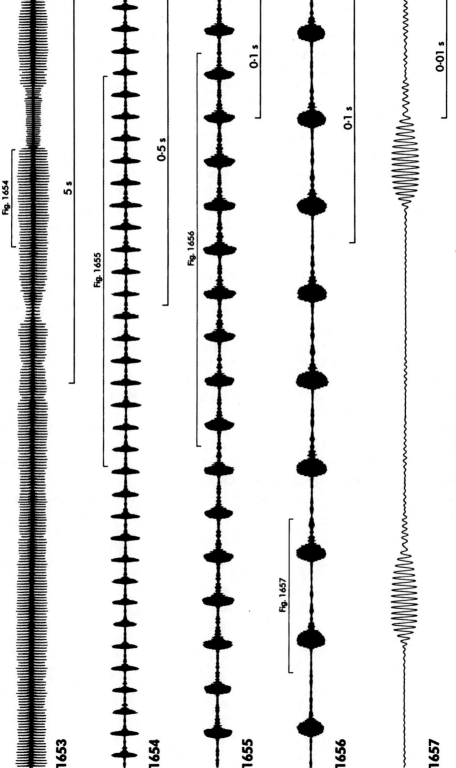

Figures 1653–1657 Oscillograms at five different speeds of the song of the Nightjar, *Caprimulgus europaeus*.

Locustella fluviatilis (Wolf), River Warbler (Figs 1645, 1647, 1649, 1651. CD 2, track
  83)

In the area of our study this species, typically a bird of moist habitats, occurs only in eastern Germany. Its prolonged song consists of quite complex elements repeated at the rate of
about 10–15/s and sometimes sounding rather like a cicada, especially *Lyristes plebejus*.
Among Orthoptera it is perhaps most likely to be confused with *Tettigonia viridissima*, but
lacks the very brief pauses that typically punctuate the song of this bush-cricket.

Locustella lanceolata (Temminck), Lanceolated Warbler (Figs 1646, 1648, 1650, 1652.
  CD 2, track 84)

This bird of marshland and water-meadows occurs only as a vagrant in western Europe. Its
song sounds quite similar to that of *L. fluviatilis*, but with its double elements repeated at
the rather higher rate of about 15–20/s. It could again be confused with *T. viridissima*, but
has a more musical quality and usually continues without interruption for long periods of
up to a minute or more.

Caprimulgus europaeus Linnaeus, Nightjar (Figs 1653–1657. CD 2, track 85)

The song of this species consists of a prolonged, continuous churring, composed of pulses
of sound repeated at the rate of about 25–30/s and with a carrier frequency of about 2 kHz.
It is quite similar to the song of *Gryllotalpa gryllotalpa*, although the pulse repetition rate is
slightly lower and the carrier frequency slightly higher. Oscillographic analysis shows that
the resemblance is not superficial but stems from a remarkable similarity in the wave-form
of the sound (cf. Figs 701–704 and 1655–1657). As an aid to distinguishing between these
two songs, the listener should bear in mind that the mole-cricket always sings from just
inside its burrow and is seldom heard after the end of June.

A chorus of *Nemobius sylvestris* could also be confused with a distant Nightjar, but the
resemblance is less close and careful listening to the cricket song, even in chorus, will always reveal the frequent interruptions in the song of each male; the churring of *C. europaeus*
continues for long periods without any interruptions of this kind.

C. europaeus occurs throughout western Europe from southern Scandinavia southwards,
including Britain.

The Colour Plates

Note. The insects illustrated on Plates 1-3
are not shown to the same scale.

Plate 1. Tettigoniidae (Bush-crickets)

1 *Phaneroptera falcata* (Sickle-bearing Bush-cricket)
Photo: K. G. & R. A. Preston-Mafham

2 *Tylopsis lilifolia* (Slender Bush-cricket)
Photo: M. Chinery

3 *Leptophyes punctatissima* (Speckled Bush-cricket)
Photo: C. G. Butler

4 *Polysarcus denticauda* (Large Saw-tailed Bush-cricket)
Photo: M. J. Skelton

5 *Poecilimon ornatus* (Ornate Bush-cricket)
Photo: The Natural History Museum, London

6 *Meconema thalassinum* (Oak Bush-cricket)
Photo: P. A. Bowman

7 *Conocephalus dorsalis* (Short-winged Cone-head)
Photo: P. T. Chadd

8 *Ruspolia nitidula* (Large Cone-head)
Photo: The Natural History Museum, London

9 *Tettigonia viridissima* (Great Green Bush-cricket)
Photo: Ron & Christine Foord

10 *Decticus verrucivorus* (Wart-biter)
Photo: Ron & Christine Foord

11 *Decticus albifrons* (White-faced Bush-cricket)
Photo: P. T. Chadd

12 *Platycleis sepium* (Sepia Bush-cricket)
Photo: P. T. Chadd

All adult males

Plate 2. Tettigoniidae, Gryllidae (Bush-crickets, True Crickets)

1 *Platycleis tessellata* (Brown-spotted Bush-cricket)
Photo: The Natural History Museum, London

2 *Metrioptera brachyptera* (Bog Bush-cricket)
Photo: P. A. Bowman

3 *Metrioptera roeselii* (Roesel's Bush-cricket)
Photo: Ron & Christine Foord

4 *Pholidoptera griseoaptera* (Dark Bush-cricket)
Photo: A. Beaumont

5 *Eupholidoptera chabrieri* (Chabrier's Bush-cricket)
Photo: M. Chinery

6 *Anonconotus alpinus* (Small Alpine Bush-cricket)
Photo: P. T. Chadd

7 *Antaxius pedestris* (Pyrenean Bush-cricket)
Photo: M. Chinery

8 *Ephippiger ephippiger* (Saddle-backed Bush-cricket)
Photo: M. Chinery

9 *Uromenus stalii* (Stål's Bush-cricket)
Photo: The Natural History Museum, London

10 *Gryllus campestris* (Field-cricket)
Photo: C. G. Butler

11 *Nemobius sylvestris* (Wood-cricket)
Photo: C. G. Butler

12 *Oecanthus pellucens* (Tree-cricket)
Photo: The Natural History Museum, London

All adult males

Plate 3. Gryllotalpidae, Acrididae (Mole-crickets, Grasshoppers)

1 *Gryllotalpa gryllotalpa* (Mole-cricket)
Photo: B. C. Pickard

2 *Stethophyma grossum* (Large Marsh Grasshopper)
Photo: M. J. Skelton

3 *Arcyptera fusca* (Large Banded Grasshopper)
Photo: P. T. Chadd

4 *Euthystira brachyptera* (Small Gold Grasshopper)
Photo: P. T. Chadd

5 *Omocestus viridulus* (Common Green Grasshopper)
Photo: P. T. Chadd

6 *Stenobothrus lineatus* (Stripe-winged Grasshopper)
Photo: The Natural History Museum, London

7 *Gomphocerus sibiricus* (Club-legged Grasshopper)
Photo: E. C. M. Haes

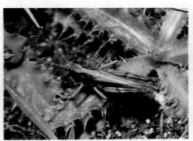

8 *Gomphocerippus rufus* (Rufous Grasshopper)
Photo: K. G. & R. A. Preston-Mafham

9 *Stauroderus scalaris* (Large Mountain Grasshopper)
Photo: The Natural History Museum, London

10 *Chorthippus biguttulus* (Bow-winged Grasshopper)
Photo: P. H. Ward

11 *Chorthippus parallelus* (Meadow Grasshopper)
Photo: D. E. Kimmins

12 *Euchorthippus declivus* (Sharp-tailed Grasshopper)
Photo: P. T. Chadd

All adult males except fig.2, which shows both sexes

Appendix 1

Check-list of the species included

For a brief discussion of the classification we have adopted, see p. 9.

Order **ORTHOPTERA**
Suborder **Ensifera**
Family **Tettigoniidae**
Subfamily Phaneropterinae
   Phaneroptera falcata (Poda) Sickle-bearing Bush-cricket[†]
   Phaneroptera nana Fieber Southern Sickle-bearing Bush-cricket
   Tylopsis lilifolia (Fabricius) Slender Bush-cricket[†]
   Acrometopa servillea (Brullé) Long-legged Bush-cricket
   Isophya pyrenaea (Serville) Large Speckled Bush-cricket
   Isophya kraussii Brunner Krauss's Bush-cricket
   Barbitistes serricauda (Fabricius) Saw-tailed Bush-cricket
   Barbitistes obtusus Targioni-Tozzetti Blunt Bush-cricket
   Barbitistes fischeri (Yersin) Southern Saw-tailed Bush-cricket
   Leptophyes punctatissima (Bosc) Speckled Bush-cricket[†]
   Leptophyes albovittata (Kollar) Striped Bush-cricket
   Leptophyes boscii Fieber Eastern Speckled Bush-cricket
   Leptophyes laticauda (Frivaldsky) Long-tailed Bush-cricket
   Polysarcus denticauda (Charpentier) Large Saw-tailed Bush-cricket[†]
   Polysarcus scutatus (Brunner) Shielded Bush-cricket
   Poecilimon ornatus (Schmidt) Ornate Bush-cricket[†]
   Poecilimon jonicus (Fieber) Ionian Bush-cricket
Subfamily Meconematinae
   Meconema thalassinum (De Geer) Oak Bush-cricket[†]
Subfamily Conocephalinae
   Conocephalus conocephalus (Linnaeus) Southern Cone-head
   Conocephalus discolor Thunberg Long-winged Cone-head
   Conocephalus dorsalis (Latreille) Short-winged Cone-head[†]
   Ruspolia nitidula (Scopoli) Large Cone-head[†]
Subfamily Tettigoniinae
   Tettigonia viridissima (Linnaeus) Great Green Bush-cricket[†]
   Tettigonia hispanica (Bolívar) Spanish Green Bush-cricket

[†] *Illustrated in colour on Plates 1–3*

Tettigonia cantans (Fuessly) Upland Green Bush-cricket
Tettigonia caudata (Charpentier) Eastern Green Bush-cricket
Subfamily Decticinae
Decticus verrucivorus verrucivorus (Linnaeus) Wart-biter[†]
Decticus verrucivorus assiduus Ingrisch, Willemse & Heller Wart-biter, Spanish form
Decticus albifrons (Fabricius) White-faced Bush-cricket[†]
Platycleis albopunctata (Goeze) Grey Bush-cricket
Platycleis sabulosa Azam Sand Bush-cricket
Platycleis affinis Fieber Tuberous Bush-cricket
Platycleis romana Ramme Roman Bush-cricket
Platycleis iberica Zeuner Iberian Bush-cricket
Platycleis falx (Fabricius) Falcate Bush-cricket
Platycleis intermedia (Serville) Intermediate Bush-cricket
Platycleis sepium (Yersin) Sepia Bush-cricket[†]
Platycleis tessellata (Charpentier) Brown-spotted Bush-cricket[†]
Platycleis veyseli Koçak Veysel's Bush-cricket
Platycleis nigrosignata (Costa) Black-marked Bush-cricket
Platycleis montana (Kollar) Steppe Bush-cricket
Platycleis stricta (Zeller) Italian Bush-cricket
Metrioptera brachyptera (Linnaeus) Bog Bush-cricket[†]
Metrioptera saussuriana (Frey-Gessner) Saussure's Bush-cricket
Metrioptera buyssoni (Saulcy) Buysson's Bush-cricket
Metrioptera caprai Baccetti Capra's Bush-cricket
Metrioptera abbreviata (Serville) Basque Bush-cricket
Metrioptera burriana Uvarov Burr's Bush-cricket
Metrioptera bicolor (Philippi) Two-coloured Bush-cricket
Metrioptera roeselii (Hagenbach) Roesel's Bush-cricket[†]
Pholidoptera griseoaptera (De Geer) Dark Bush-cricket[†]
Pholidoptera aptera (Fabricius) Alpine Dark Bush-cricket
Pholidoptera fallax (Fischer) Fischer's Bush-cricket
Pholidoptera femorata (Fieber) Large Dark Bush-cricket
Pholidoptera littoralis (Fieber) Littoral Bush-cricket
Eupholidoptera chabrieri (Charpentier) Chabrier's Bush-cricket[†]
Anonconotus alpinus (Yersin) Small Alpine Bush-cricket[†]
Yersinella raymondii (Yersin) Raymond's Bush-cricket
Yersinella beybienkoi La Greca Bei-Bienko's Bush-cricket
Pachytrachis striolatus (Fieber) Striated Bush-cricket
Ctenodecticus siculus (Ramme) Sicilian Bush-cricket
Rhacocleis germanica (Herrich-Schäffer) Mediterranean Bush-cricket
Rhacocleis neglecta (Costa) Adriatic Bush-cricket
Antaxius pedestris (Fabricius) Pyrenean Bush-cricket[†]
Antaxius hispanicus Bolívar Mottled Bush-cricket
Antaxius spinibrachius (Fischer) Spiny-legged Bush-cricket
Thyreonotus bidens Bolívar Two-toothed Bush-cricket
Gampsocleis glabra (Herbst) Heath Bush-cricket

Subfamily Ephippigerinae
  Ephippiger ephippiger (Fiebig) Saddle-backed Bush-cricket[†]
  Ephippiger terrestris (Yersin) Alpine Saddle-backed Bush-cricket
  Ephippiger perforatus (Rossi) North Apennine Bush-cricket
  Ephippiger ruffoi Galvagni Ruffo's Bush-cricket
  Ephippigerida areolaria (Bolívar) Spanish Mountain Bush-cricket
  Ephippigerida taeniata (Saussure) Large Striped Bush-cricket
  Uromenus rugosicollis (Serville) Rough-backed Bush-cricket
  Uromenus elegans (Fischer) Elegant Bush-cricket
  Uromenus brevicollis (Fischer) Short-backed Bush-cricket
  Uromenus catalaunicus (Bolívar) Catalan Bush-cricket
  Uromenus asturiensis (Bolívar) Asturian Bush-cricket
  Uromenus stalii (Bolívar) Stål's Bush-cricket[†]
  Uromenus perezii (Bolívar) Pérez's Bush-cricket
  Uromenus martorellii (Bolívar) Martorell's Bush-cricket
  Uromenus andalusius (Rambur) Andalusian Bush-cricket
  Baetica ustulata (Rambur) Sierra Nevadan Bush-cricket
  Callicrania selligera (Charpentier) Lusitanian Bush-cricket
  Platystolus martinezii (Bolívar) Martínez's Bush-cricket
  Platystolus faberi Harz Faber's Bush-cricket
Subfamily Pycnogastrinae
  Pycnogaster jugicola Graells Fat-bellied Bush-cricket
  Pycnogaster sanchezgomezi Bolívar Sánchez Gómez's Bush-cricket
  Pycnogaster inermis (Rambur) Unarmed Bush-cricket
Family **Gryllidae**
Subfamily Gryllinae
  Gryllus campestris Linnaeus Field-cricket[†]
  Gryllus bimaculatus De Geer Two-spotted Cricket
  Acheta domesticus (Linnaeus) House-cricket
  Modicogryllus bordigalensis (Latreille) Bordeaux Cricket
  Eugryllodes pipiens (Dufour) Mountain-cricket
Subfamily Nemobiinae
  Nemobius sylvestris (Bosc) Wood-cricket[†]
  Pteronemobius heydenii (Fischer) Marsh-cricket
Subfamily Oecanthinae
  Oecanthus pellucens (Scopoli) Tree-cricket[†]
Family **Gryllotalpidae**
  Gryllotalpa gryllotalpa (Linnaeus) Mole-cricket[†]
  Gryllotalpa vineae Bennet-Clark Vineyard Mole-cricket
Suborder **Caelifera**
Family **Acrididae**
Subfamily Locustinae
  Psophus stridulus (Linnaeus) Rattle Grasshopper
  Bryodema tuberculata (Fabricius) Speckled Grasshopper
  Stethophyma grossum (Linnaeus) Large Marsh Grasshopper[†]

Subfamily Gomphocerinae

Brachycrotaphus tryxalicerus (Fischer) Savanna Grasshopper
Arcyptera fusca (Pallas) Large Banded Grasshopper[†]
Arcyptera tornosi Bolívar Iberian Banded Grasshopper
Arcyptera microptera Fischer de Waldheim Small Banded Grasshopper
Ramburiella hispanica (Rambur) Striped Grasshopper
Chrysochraon dispar (Germar) Large Gold Grasshopper
Euthystira brachyptera (Ocskay) Small Gold Grasshopper[†]
Dociostaurus maroccanus (Thunberg) Moroccan Locust
Dociostaurus jagoi Soltani Jago's Grasshopper
Dociostaurus hispanicus (Bolívar) Iberian Cross-backed Grasshopper
Omocestus viridulus (Linnaeus) Common Green Grasshopper[†]
Omocestus rufipes (Zetterstedt) Woodland Grasshopper
Omocestus haemorrhoidalis (Charpentier) Orange-tipped Grasshopper
Omocestus petraeus (Brisout) Rock Grasshopper
Omocestus raymondi (Yersin) Raymond's Grasshopper
Omocestus panteli (Bolívar) Pantel's Grasshopper
Omocestus antigai (Bolívar) Pyrenean Grasshopper
Omocestus bolivari Chopard Bolívar's Grasshopper
Omocestus uhagonii (Bolívar) Uhagon's Grasshopper
Omocestus minutissimus (Bolívar) Small Mountain Grasshopper
Omocestus femoralis Bolívar Stripe-legged Grasshopper
Omocestus uvarovi Zanon Uvarov's Grasshopper
Stenobothrus lineatus (Panzer) Stripe-winged Grasshopper[†]
Stenobothrus nigromaculatus (Herrich-Schäffer) Black-spotted Grasshopper
Stenobothrus fischeri (Eversmann) Fischer's Grasshopper
Stenobothrus festivus Bolívar Festive Grasshopper
Stenobothrus grammicus Cazurro Dark-palped Grasshopper
Stenobothrus bolivarii (Brunner) Pink-palped Grasshopper
Stenobothrus stigmaticus (Rambur) Lesser Mottled Grasshopper
Stenobothrus apenninus Ebner Apennine Grasshopper
Stenobothrus ursulae Nadig Ursula's Grasshopper
Stenobothrus rubicundulus Kruseman & Jeekel Wing-buzzing Grasshopper
Stenobothrus cotticus Kruseman & Jeekel Cottian Grasshopper
Myrmeleotettix maculatus (Thunberg) Mottled Grasshopper
Gomphocerus sibiricus (Linnaeus) Club-legged Grasshopper[†]
Gomphocerippus rufus (Linnaeus) Rufous Grasshopper[†]
Stauroderus scalaris (Fischer de Waldheim) Large Mountain Grasshopper[†]
Chorthippus apricarius (Linnaeus) Upland Field Grasshopper
Chorthippus corsicus (Chopard) Corsican Grasshopper
Chorthippus cazurroi (Bolívar) Cazurro's Grasshopper
Chorthippus nevadensis Pascual Sierra Nevadan Grasshopper
Chorthippus cialancensis Nadig Piedmont Grasshopper
Chorthippus pullus (Philippi) Gravel Grasshopper
Chorthippus alticola Ramme Eastern Alpine Grasshopper

Chorthippus modestus (Ebner) Reatine Grasshopper
Chorthippus vagans (Eversmann) Heath Grasshopper
Chorthippus reissingeri Harz Reissinger's Grasshopper
Chorthippus biguttulus (Linnaeus) Bow-winged Grasshopper[†]
Chorthippus brunneus (Thunberg) Field Grasshopper
Chorthippus mollis mollis (Charpentier) Lesser Field Grasshopper
Chorthippus mollis ignifer Ramme Lesser Field Grasshopper, Alpine form
Chorthippus jacobsi Harz Jacobs's Grasshopper
Chorthippus yersini Harz Yersin's Grasshopper
Chorthippus rubratibialis Schmidt Red-shinned Grasshopper
Chorthippus marocanus Nadig Moroccan Grasshopper
Chorthippus binotatus (Charpentier) Two-marked Grasshopper
Chorthippus apicalis (Herrich-Schäffer) June Grasshopper
Chorthippus jucundus (Fischer) Large Green Grasshopper
Chorthippus albomarginatus (De Geer) Lesser Marsh Grasshopper
Chorthippus dorsatus (Zetterstedt) Steppe Grasshopper
Chorthippus dichrous (Eversmann) Two-coloured Grasshopper
Chorthippus parallelus parallelus (Zetterstedt) Meadow Grasshopper[†]
Chorthippus parallelus erythropus Faber Meadow Grasshopper, Iberian form
Chorthippus montanus (Charpentier) Water-meadow Grasshopper
Euchorthippus declivus (Brisout) Sharp-tailed Grasshopper[†]
Euchorthippus pulvinatus pulvinatus (Fischer de Waldheim) Straw-coloured Grasshopper
Euchorthippus pulvinatus gallicus Maŕan Straw-coloured Grasshopper, western European form
Euchorthippus chopardi Descamps Chopard's Grasshopper
Euchorthippus albolineatus siculus Ramme White-lined Grasshopper, Sicilian form
Euchorthippus angustulus Ramme Balearic Grasshopper
Euchorthippus sardous Nadig Sardinian Grasshopper

Included non-Orthoptera whose songs could be confused with orthopteran songs

Cicadas
Cicada orni Linnaeus
Lyristes plebejus (Scopoli)
Tettigetta argentata (Olivier)
Cicadetta montana (Scopoli)

Toads
Bufo viridis Laurenti Green Toad
Alytes obstetricans (Laurenti) Midwife Toad

Birds
Locustella naevia (Boddaert) Grasshopper Warbler
Locustella luscinioides (Savi) Savi's Warbler
Locustella fluviatilis (Wolf) River Warbler
Locustella lanceolata (Temminck) Lanceolated Warbler
Caprimulgus europaeus Linnaeus Nightjar

Appendix 2

Summary of nomenclatural changes

The formal nomenclatural changes resulting from the taxonomic conclusions discussed in Chapter 5 are listed below. Some of the new synonymies listed under *Chorthippus corsicus* have been suggested previously (see p. 74), but are set out formally for the first time here.

Arcyptera microptera (Fischer de Waldheim) (see p. 72)
Oedipoda microptera Fischer de Waldheim, 1833: 384.
Arcyptera Kheili Azam, 1900: 82. **Syn. n.**
Arcyptera Carpentieri Azam, 1907: 262. **Syn. n.**

Omocestus viridulus kaestneri (Harz) **Stat. n.** (see p. 72)
Stenobothrus kaestneri Harz, 1972: 129.
Omocestus kaestneri (Harz) Harz, 1975b: 734.

Omocestus antigai (Bolívar) (see p. 72)
Stenobothrus (Omocestus) Antigai Bolívar, 1897b: 232.
Omocestus Navasi Bolívar, 1908: 319. **Syn. n.**

Omocestus minutissimus (Bolívar) (see p. 73)
Gomphocerus (Omocestus) minutissimus Bolívar, 1878: 424.
Omocestus burri Uvarov, 1936: 378. **Syn. n.**

Chorthippus corsicus (Chopard) (see p. 74)
Omocestus corsicus Chopard, 1923: 268.
Omocestus corsicus montanus Chopard, 1923: 270. [Secondary homonym (in *Chorthippus*) of *Chorthippus montanus* (Charpentier, 1825: 173).]
Omocestus pascuorum Chopard, 1923: 271. **Syn. n**
Stauroderus incertus Chopard, 1923: 272. **Syn. n**
Chorthippus chopardi Harz, 1971: 332. [Replacement name for *Omocestus corsicus montanus* Chopard, 1923.] **Syn. n.**

Chorthippus cialancensis Nadig (see p. 74)
Chorthippus cialancensis Nadig, 1986: 218.
Chorthippus sampeyrensis Nadig, 1986: 224. **Syn. n.**

Appendix 3

Data for the song recordings

Explanatory note

The Table on the following pages shows the data for the song recordings used for the oscillograms reproduced in this book and for the excerpts included on the companion compact discs. The recordings are listed in chronological order under each species. Recordists' names are abbreviated as follows (listed in alphabetical order of surnames): HCBC = H. C. Bennet-Clark, JFB = J. F. Burton, MFC = M. F. Claridge, MD = M. Duijm, KHG = K.-H. Garberding, JAG = J. A. Grant, JCH = J. C. Hartley, KGH = K.-G. Heller, PADH = P. A. D. Hollom, MI = M. Iannantuoni, SI = S. Ingrisch, NDJ = N. D. Jago, RM = R. Margoschis, DRR = D. R. Ragge, WJR = W. J. Reynolds, DJR = D. J. Robinson, IR = I. Robinson, JCR = J.-C. Roché, PR = P. Rudkin, MJS = M. J. Samways, RS = R. Savage, LS = L. Svensson, PS = P. Szöke, GFW = G. F. Wade, FW = F. Willemse. Rec. No. = recording number in the Natural History Museum Library of Recorded Insect Sounds, London; F = field recording; S = recording made in the Acoustic Laboratory of the Natural History Museum, London; R = recording made in any other indoor situation; D = recorded in very dim light or complete darkness. The Figure numbers given are those of the first (slowest) oscillograms taken from the recordings concerned. The numbers in parentheses refer to separate excerpts within one track of recorded songs on a CD. Further information on recordings made by DRR and WJR are given in Chapter 2; the specimens from which these recordings were made are almost all in the collection of the Natural History Museum in London, cross-referenced to the recorded songs.

Species	Locality	Date recorded	Recordist	Air temp.	Rec. No.	Fig. No.	CD No.	Track No.
Phaneroptera falcata	FRANCE: Maine et Loire, near Baugé	8.viii.1989	WJR	25°C	739/2SD	40	1	1 (3)
	AUSTRIA: Carinthia, W of Pressegger See	22.ix.1992	DRR	25°C	841SD	38,39	1	1 (1,2)
nana	FRANCE: Pyrénées-Orientales, near Vernet-les-Bains, Py	24.ix.1976	WJR	20°C	184/7D	47		
	ITALY: Emilia-Romagna, Po Estuary, Isola d'Ariano	2.ix.1985	DRR	24°C	604/6RD	48	1	2
Tylopsis lilifolia	FRANCE: Dordogne, near Le Bugue, Campagne	24.viii.1974	DRR	28°C	134/1F			
	Same locality	24.viii.1974	DRR	29°C	134/2F	53	1	3 (3,4)
	ITALY: Umbria, near Orvieto	5.x.1988	WJR	26°C	708SD	54	1	3 (1,2)
	CORSICA: Corse-du-Sud, 15 km N of Ajaccio, 3 km S of Sagone	2.viii.1990	WJR	28°C	791/1SD	55		
Acrometopa servillea	YUGOSLAVIA: Montenegro, 5 km NW of Cetinje	—	DJR	—	480/3RD	62	1	4 (1)
	GREECE: Peloponnese, Arcadia, Sitaina, 800 m	24.viii.1988	FW	26°C	638/11RD		1	4 (2)
	GREECE: Peloponnese, Akhaia, 3 km W of Kalavrita, 700 m	1.viii.1988	FW	27°C	715/1RD	63	1	4 (3)

Appendix 3. (cont.)

Species	Locality	Date recorded	Recordist	Air temp.	Rec. No.	Fig. No.	CD No.	Track No.
Isophya *pyrenaea*	FRANCE: Hérault, 3 km W of Ganges, Vis-Canon	13.vi.1992	SI	22·5°C	836/1R	68		
	Same locality	13.vi.1992	SI	22°C	836/2R	69	1	5
kraussii	GERMANY: Bavaria, near Würzburg	v.1981	KGH	24°C	565/1R	70		
	GERMANY: Bavaria, near Bad Brückenau, Rhön	14.vi.1990	SI	22·5°C	836/7R		1	6
Barbitistes *serricauda*	GERMANY: Baden-Würtemberg, Black Forest, Albtal	15.vii.1992	SI	25°C	836/6RD	74	1	7
obtusus	SWITZERLAND: Ticino, near Lugano, Miglieglia–M. Lema	9.ix.1980	MD	20°C	488/7R	76	1	8
	ITALY: Lombardy, near Sondrio, San Giacomo	10.vii.1990	SI	24°C	836/8RD	75		
fischeri	FRANCE: Lozère, near St-Germain-de-Calberte	28.vi.1992	SI	24°C	836/3RD	83	1	9
	Same locality	28.vi.1992	SI	25°C	836/5RD	84		
Leptophyes *punctatissima*	ENGLAND: Nottingham	2.xi.1981	WJR	25·5°C	431S	89	1	10
	SPAIN: Lérida, between Col du Portillon and Bosost, 1100 m	9.ix.1982	WJR	28·5°C	462S	90		
albovittata	ITALY: Trentino-Alto Adige, Bolzano, Val Venosta, near Mevano	2.xi.1981	WJR	29°C	432S	91	1	11

Taxon	Locality	Date	Rec.	Temp	No.		
boscii	AUSTRIA: Lower Austria, 46 km S of Vienna, Pernitz	1.viii.1984	WJR	28°C	540S	1	98
	Same locality	3.viii.1984	WJR	23°C	542S		99	12
laticauda	ITALY: Lombardy, near Lake Como, Sondrio	29.x.1981	WJR	29°C	430S	1	100	13
Polysarcus denticauda	FRANCE: Hautes-Pyrénées, near Bagnères-de-Luchon, Superbagnères, 1800 m	9.ix.1982	WJR	27°C	460S	1	107	14
	ITALY: Piedmont, Colle di Sampeyre, 2300 m	16.ix.1982	DRR	23°C	668/8F		112	
Polysarcus scutatus	FRANCE: Alpes-de-Haute-Provence, Foux d'Allos, 1800 m	30.vii.1974	MD	—	488/8R	1	117	15
	ANDORRA: NE, near Soldeu, 1 km SW of Pasqual, 1850 m	viii/ix.1990	JCH	—	812/1R		122	
Poecilimon ornatus	CROATIA: Istria, Opatija	1.viii.1984	WJR	27°C	539S	1	127	16
jonicus	CROATIA: Istria, Mt Ucka, 800 m	25.vii.1986	FW	24°C	638/20R		128	
	ITALY: Lazio, Monti Reatini, Valle della Meta, 1400 m	12.ix.1985	DRR	22°C	609/7R	1	129	17
Meconema thalassinum	ENGLAND: Hampshire, New Milton	19.viii.1970	IR	—	80RD		136	
	ENGLAND: Greater London. Totteridge	22.viii.1979	WJR	19°C	327/2SD	1	137	18
Conocephalus conocephalus	ITALY: Calabria, near Rosarno	4.x.1988	WJR	25°C	696S	1	140	
	Same locality	4.x.1988	WJR	28°C	697SD		141	19
discolor	GERMANY: Upper Bavaria, near Chiemsee, Rottau	14.viii.1973	DRR	24°C	99/6F		147	

Appendix 3. (cont.)

Species	Locality	Date recorded	Recordist	Air temp.	Rec. No.	Fig. No.	CD No.	Track No.
Conocephalus discolor (cont.)	ENGLAND: Dorset, Chapman's Pool	23.ix.1975	WJR	25°C	157S	152	1	20(2)
	FRANCE: Landes, Arcachon	15.ix.1976	WJR	18°C	180/7R	148		
	SPAIN: Lérida, 5 km W of Solsona	8.ix.1978	WJR	28°C	271/5F	153		
	ENGLAND: Dorset, Chapman's Pool	7.ix.1981	WJR	24°C	411S	149		
	FRANCE: Morbihan, base of Quiberon Peninsula	18.ix.1981	DRR	22°C	416S	154		
	SPAIN: Zamora, 88 km W of Benavente, W of Portilla de Padornelo, 1350 m	22.viii.1985	WJR	26°C	587/1F	150		
	CORSICA: Corse-du-Sud, 20 km N of Ajaccio, 1 km N of Sagone	31.vii.1990	WJR	31°C	788S	151	1	20(1)
dorsalis	ENGLAND: Dorset, near Wareham	22.viii.1962	DRR, JFB & GFW	—	70R	171		
	ENGLAND: Hampshire, Keyhaven Marshes	15.viii.1975	WJR	27°C	147S	172	1	21
	ENGLAND: Dorset, near Wareham, Morden Bog	viii.1976	RS	26·7°C	294/2R	173		
Ruspolia nitidula	FRANCE: Landes, Arcachon	15.ix.1976	WJR	17°C	180/6RD	180	1	22
	FRANCE: Morbihan, base of Quiberon Peninsula	24.ix.1981	WJR	18·5°C	422SD	181		
Tettigonia viridissima	ENGLAND: Kent, Sandwich Bay	28.viii.1967	DRR	26°C	15/1F	188		

	Locality	Date		Temp	Recording	No.		Fig.
	FRANCE: Dordogne, near Sarlat, Petite Rou de Puymartin	24.viii.1974	DRR	23°C	134/7F	189		
	ENGLAND: Berkshire, near Staines, Hythe End	27.vii.1976	WJR	25°C	169SD	191		
	FINLAND: Åland, Mariehamn	30.vii.1980	WJR	26°C	360/2SD	192		
	ENGLAND: Dorset, Lulworth Cove	8.ix.1981	WJR	23°C	412SD		1	23
	SARDINIA: Sassari, N of Monte Pedralunga, 300 m	16.vii.1990	WJR	28°C	794/10F	190		
hispanica	SPAIN: Madrid, Sierra de Guadarrama, near Puerto de Navafria. Majada de los Cardos, 1600 m	20.vii.1989	DRR	30°C	764/1F	193	1	24
	SPAIN: Ávila, Sierra de Gredos, N of Puerto del Pico, 1300 m	22.vii.1989	DRR	23°C	764/4F	194		
cantans	AUSTRIA: Tyrol, near Kufstein, Kaisertal	10.viii.1973	DRR	29°C	96/1F	209		
	Same locality	14.viii.1973	DRR	13°C	100/3FD	211		
	FRANCE: Puy-de-Dôme, near Besse-en-Chandesse, Compains	21.viii.1974	DRR	13°C	133/5FD	212	1	25 (2)
	GERMANY: Upper Bavaria, near Wolfratshausen	26.ix.1985	WJR	25°C	595S	210	1	25 (1)
caudata	GREECE: Macedonia, Thessaloniki, 4 km NE of Vertiskos, 600 m	3.viii.1986	FW	23°C	638/17R	213	1	26

Appendix 3. (cont.)

Species	Locality	Date recorded	Recordist	Air temp.	Rec. No.	Fig. No.	CD No.	Track No.
Decticus								
verrucivorus								
verrucivorus	FRANCE: Hautes-Pyrénées, near Vielle-Aure, Lac d'Aumar	17.ix.1976	WJR	19°C	181/6F	224		
	FRANCE: Hérault, 5 km E of Lunel	21.vii.1972	MJS	—	243/1F	227		
	FRANCE: Pyrénées-Orientales, near Saillagouse, 1 km S of Eyne	14.ix.1978	WJR	21°C	275/7F	225		
	ENGLAND: East Sussex, near Lewes, Castle Hill	27.ix.1979	WJR	24°C	343S	226	1	27
	ITALY: Abruzzi, Gran Sasso d'Italia, near S. Egidio, 1680 m	1.ix.1985	DRR	19°C	603/6F	228		
v. assiduus	SPAIN: Ávila, Sierra de Gredos, SW of Hoyos del Espino. Llano de Barbedillo, 1875–1950 m	1.viii.1989	FW	20°C	754/6R	232		
	Same locality	9.vii.1990	DRR	25°C	805/5F	230	1	28
	Same locality	9.vii.1990	DRR	26°C	806/1F	231		
	Same locality	26.vii.1990	WJR	30°C	783S	229		
albifrons	FRANCE: Tarn, near Puylaurens	26.viii.1974	DRR	24°C	136/1F	251	1	29
	FRANCE: Pyrénées-Orientales, St-Cyprien	28.viii.1974	DRR	24°C	136/6F	252		

	Locality	Date	Recordist	Temp	Recording no.	No.		Fig.
Platycleis albopunctata	ENGLAND: East Sussex, Eastbourne, Cow Gap	13.vii.1976	DRR & WJR	27°C	166S	258		
	WALES: West Glamorgan, Gower, Nicholaston Burrows	6.ix.1976	DRR	24°C	173S	259	1	30
	FRANCE: Vaucluse, Mont Ventoux, road from Bédoin, 1100 m	12.viii.1977	DRR	20°C	234/4F	261		
	CHANNEL ISLANDS: Jersey, Quennevais	23.viii.1977	WJR	30°C	214S	260		
	SPAIN: Madrid, Sierra de Guadarrama. Puerto de Morcuera, 1796 m	28.ix.1982	WJR	26°C	467S	262		
	ITALY: Abruzzi, Gran Sasso d'Italia. 2 km E of Passo delle Capannelle, 1500 m	2.ix.1985	DRR	21°C	605/1F	263		
sabulosa	FRANCE: Pyrénées-Orientales, St-Cyprien	31.viii.1974	DRR	24°C	137/2F	276		
	SPAIN: Madrid, Manzanares	1.x.1982	WJR	22°C	472S	277	1	31
	SPAIN: Toledo, near Oropesa, Las Ventas de San Julian	5.viii.1983	DRR	23°C	507/7FD	278		
affinis	FRANCE: Landes, 3 km NW of Morcenx	15.ix.1976	WJR	18°C	180/10R	285		
	FRANCE: Vaucluse, near Carpentras, Bédoin	13.viii.1977	DRR	22°C	235/1F	286	1	32
	SPAIN: Zamora, Benavente	21.viii.1986	WJR	27°C	586/4RD	287		
romana	ITALY: Lazio, near Rome, SE side of Colli Albani	8.ix.1985	DRR	26°C	606/9F	293	1	33
	ITALY: Calabria, Sila, near Bocca di Piazza	20.ix.1988	DRR	16°C	725/3F	294		

Appendix 3. (cont.)

Species	Locality	Date recorded	Recordist	Air temp.	Rec. No.	Fig. No.	CD No.	Track No.
Platycleis (cont.)								
iberica	SPAIN: Ávila, Sierra de Gredos, S of Hoyos del Espino, 1340 m	19.vii.1990	DRR	25°C	805/4RD	295	1	34
falx	FRANCE: Hérault. Agde. Pic St-Loup	3.viii.1972	MJS	19°C	243/8FD	304		
	FRANCE: Pyrénées-Orientales, St-Cyprien	31.viii.1974	DRR	24°C	137/1F	302		
	FRANCE: Hérault, near Montpellier, bank of Canal du Rhône	27.viii.1986	WJR	25°C	640/1F	303	1	35
intermedia	SPAIN: Granada, La Rábita	30.vii.1984	DRR	23°C	536SD	311	1	36
	SARDINIA: Nuoro. Monti del Gennargentu, 3 km NE of Aritzo, 1100 m	2.viii.1990	WJR	28°C	791/2SD	312		
sepium	FRANCE: Pyrénées-Orientales, St-Cyprien	31.viii.1974	DRR	22°C	137/4F	317		
	Same locality	31.viii.1974	DRR	22°C	137/5F	319 318.		
	CROATIA: 2 km SE of Rijeka	4.ix.1985	WJR	25°C	582SD	320	1	37
tessellata	FRANCE: Vaucluse, Mont Ventoux, near Les Bruns	11.viii.1977	DRR	24°C	233/5RD	326		
	SPAIN: Madrid. Manzanares	30.ix.1982	WJR	22°C	473S	325	1	38
veyseli	AUSTRIA: Burgenland. Neusiedler See, near Illmitz	19.viii.1973	DRR	26°C	102/2RD	331	1	39

Taxon	Locality	Date	Recorder	Temp				
nigrosignata	MACEDONIA: near Titov Veles, Lake Mladost, 170 m	4.viii.1987	FW	22°C	677/1RD	332	1	40
montana	AUSTRIA: Burgenland, Neusiedler See, near Illmitz	24.viii.1973	DRR	25°C	104/1R	337	1	41
	Same locality	26.viii.1973	DRR	25°C	104/2R	338		
stricta	ITALY: Abruzzi, Gran Sasso d'Italia, valley N of Monte della Scindarella, 1800 m	1.ix.1985	DRR	25°C	604/5R	343	1	42
Metrioptera brachyptera	ENGLAND: Dorset, Isle of Purbeck	23.viii.1962	JFB & GFW	—	71R	348		
	AUSTRIA: Tyrol, near Kufstein, Kaisertal	13.viii.1973	DRR	24°C	98/2F	349	1	
saussuriana	ENGLAND: Hampshire, near Brockenhurst, Crab Tree Bog	15.viii.1975	WJR	27°C	146S	350	1	43
	FRANCE: Puy-de-Dôme, near Besse-en-Chandesse, Courbanges	20.viii.1974	DRR	19°C	132/6F	357		
	FRANCE: Isère, Pont du St Charles	26.ix.1979	WJR	24°C	336S	358	1	44
buyssoni	FRANCE: Hautes-Pyrénées, near Bagnères-de-Luchon, Col de Peyresourde, 1500 m	9.ix.1982	WJR	27°C	461S	359	1	45
caprai	ITALY: Abruzzi, Gran Sasso d'Italia, valley N of Monte della Scindarella, 1800 m	1.ix.1985	DRR	21°C	604/3F	366	1	
	ITALY: Lazio, Monti Reatini, foot of chair-lift to Monte Terminilletto, 1700 m	11.ix.1985	DRR	18°C	608/3F	367	1	46

Appendix 3. (cont.)

Species	Locality	Date recorded	Recordist	Air temp.	Rec. No.	Fig. No.	CD No.	Track No.
Metrioptera (cont.) abbreviata	SPAIN: Huesca, Torla	3.ix.1982	DRR	23°C	454S	372	1	47
	FRANCE: Pyrénées-Atlantiques, near Sauveterre, Burgaronne	12.vii.1989	DRR	25°C	761/1F	373		
	SPAIN: Navarra, Montes de Bidasoa, near Elizondo	14.vii.1989	DRR	24°C	762/3F	374		
burriana	SPAIN: Santander, Santillana del Mer	26.vii.1989	DRR	26°C	766/8F	381	1	48
	Same locality	26.vii.1989	DRR	27°C	767/1F	382		
bicolor	AUSTRIA: Burgenland, Neusiedler See, near Illmitz	22.viii.1973	DRR	28°C	102/8F	387		
	FRANCE: Puy-de-Dôme, near Besse-en-Chandesse, Compains	17.viii.1974	DRR	25°C	130/5F	389		
	FRANCE: near Puy-de-Dôme, Puy de Pariou	18.viii.1975	WJR	25°C	148S	388	1	49 (1)
	FRANCE: Seine-Maritime, near Gamaches	16.viii.1983	WJR	27°C	491S	390	1	49 (2)
roeselii	AUSTRIA: Burgenland, Neusiedler See, near Illmitz, Fuchslochlacke	23.viii.1973	DRR	25°C	103/1F	401		
	ENGLAND: Kent, Dartford Marshes	23.vii.1974	DRR & WJR	24°C	116S	402	1	50
	ANDORRA: near Sant Julia de Loria, 1 km S of Fontaneda	13.ix.1978	WJR	23°C	275/5F	403		

Species	Locality	Date		Temp		No.		Fig.
	FINLAND: Helsinki, Seurasaari	30.vii.1980	WJR	26°C	360/1S	404		
	ENGLAND: Greater London, Hampstead Heath	28.viii.1984	DRR	26°C	545S	405		
	FRANCE: Hérault, S of Montpellier, 2 km E of Villeneuve-les-Maguelonne	13.vii.1989	WJR	32°C	757/1F	406		
Pholidoptera griseoaptera	FRANCE: Puy-de-Dôme, near Besse-en-Chandesse, Compains	21.viii.1974	DRR	13°C	133/6FD	419		
	ENGLAND: Hampshire, 3 km E of Andover, Longparish	21.ix.1979	WJR	23·5°C	334SD	420		
	ITALY: Trentino-Alto Adige, Bolzano, W of Brunico, San Sigismondo	22.ix.1980	WJR	19·5°C	372SD	421	1	51
aptera	AUSTRIA: Tyrol, near Kufstein, Kaisertal	10.viii.1973	DRR	20°C	96/2FD	428	1	52
	ITALY: Trentino-Alto Adige, near Riva del Garda, Tremalzo, 1200 m	27.viii.1985	DRR	14°C	601/4F	429		
fallax	ITALY: Lazio, Monti Reatini, Terminillo, 1625 m	10.ix.1985	DRR	11°C	607/1FD	434		
	YUGOSLAVIA: Serbia, Slavonia, Slavonski Brod	13.vii.1987	FW	29°C	677/16R	435	1	53
femorata	GREECE: Peloponnese, Arcadia, E slopes of Mt Parnon, between Platanos and Charadros, 700 m	19.vi.1986	FW	20°C	638/14F	440	1	54

Appendix 3. (cont.)

Species	Locality	Date recorded	Recordist	Air temp.	Rec. No.	Fig. No.	CD No.	Track No.
Pholidoptera (cont.) *littoralis*	ITALY: Piedmont, Varese, Monte Tre Croci, 1000 m	11.ix.1980	MD	21°C	488/3R	443	1	55
Eupholidoptera chabrieri	FRANCE: Vaucluse, Mont Ventoux, road from Bédoin, 1100 m	12.viii.1977	DRR	20°C	234/6F	447	1	56
	FRANCE: Alpes-Maritimes, near Grasse, Gréolières	22.viii.1977	DRR	18°C	240/2F	446		
Anonconotus alpinus	FRANCE: Vaucluse, Mont Ventoux	28.ix.1977	WJR	27°C	246S	452	1	57
Yersinella raymondii	ITALY: Tuscany, S of Montecarelli, 500 m	20.viii.1991	SI	25°C	836/9RD	458	1	58
beybienkoi	ITALY: Tuscany, P. Jaggio Londo, 900–1100 m	20.viii.1991	SI	22°C	836/10RD	460	1	59
	ITALY: Tuscany, E of Florence, Consuma	9.ix.1992	DRR	16°C	845/5FD	459		
Pachytrachis striolatus	SLOVENIA: NW of Trieste, near Monfalcone	20.viii.1984	WJR	28°C	544/1SD	464	1	60 (1)
	Same locality	24.viii.1984	DRR	24°C	544/2S	465	1	60 (2)
Ctenodecticus siculus	SICILY: Monti Madonie, near Piano Zucchi, 900 m	17.ix.1988	DRR	20°C	724/4R	455	1	61
Rhacocleis germanica	CROATIA: 2 km SE of Rijeka	4.ix.1985	WJR	27°C	583SD	470		
	GREECE: Akhaia, W of Kato Vlasia, 500 m	25.viii.1988	FW	20°C	715/7R	471	1	62
neglecta	ITALY: Sicily, Monti Iblei, Buccheri, Pineta Santa Maria	14.ix.1988	DRR	24°C	724/2RD	477		

Taxon	Locality	Date	Recorder	Temp.	Recording	No.		Figure
Antaxius pedestris	ITALY: Calabria, Sila, near Bocca di Piazza, 1300 m	6.x.1988	WJR	23°C	709SD	476	1	63
	FRANCE: Alpes-Maritimes, near St-Dalmas-de-Tende, La Minière de Vallaure, 1450 m	26.viii.1977	DRR	22°C	241/5RD	483	1	64
	FRANCE: Alpes-Maritimes, near Col de Vence, 940 m	20.viii.1982	WJR	25°C	106/2SD	482		
hispanicus	SPAIN: Huesca, 10 km E of Bielsa, San Juan de Plan	2.ix.1978	WJR	23·5°C	270/3RD	488	1	65 (1,2)
	SPAIN: Huesca, Torla	7.ix.1982	WJR	20·5°C	456SD	489		
	SPAIN: Barcelona, near San Celoni, Sierra de Montseny, 1700 m	8.viii.1989	WJR	25°C	741SD		1	65 (3,4)
spinibrachius	SPAIN: Madrid, Sierra de Guadarrama, Puerto de la Morcuera, 1800 m	29.ix.1982	WJR	23°C	471SD	494	1	66
	SPAIN: Ávila, Sierra de Gredos, road to Pico Almazor, 1600 m	29.vii.1989	FW	25°C	754/3RD	495		
Thyreonotus bidens	SPAIN: Madrid, near La Cabrera	28.ix.1982	WJR	24°C	468SD	500	1	67
Gampsocleis glabra	FRANCE: Lozère, Causse Méjean, near Crosgarnon	6.viii.1982	DRR	28°C	482/8F	503	1	68
Ephippiger ephippiger	AUSTRIA: Burgenland, Leitha Gebirge, near Winden	27.viii.1973	DRR	25°C	105/3F	507	1	

Appendix 3. (cont.)

Species	Locality	Date recorded	Recordist	Air temp.	Rec. No.	Fig. No.	CD No.	Track No.
Ephippiger ephippiger (cont.)	FRANCE: Hautes-Pyrénées, near Vielle-Aure, Neste de Couplan	20.ix.1976	WJR	15°C	183/3F	506		
	FRANCE: Pyrénées-Orientales, near Saillagouse, Val d'Eyne	22.ix.1976	WJR	11°C	183/7F	509		
	FRANCE: Hérault, 6 km SW of St-Chinian. Villespassans	25.viii.1977	WJR	32°C	216/1S	510		
	SPAIN: Gerona, 3 km S of Port Bou, Colera	26.viii.1977	WJR	31°C	217/1S	511	1	69 (3)
	FRANCE: Hérault, 16 km NW of Beziers, St-Chinian	26.viii.1977	WJR	32°C	217/2S	508		
	FRANCE: Pyrénés-Orientales, near Saillagouse, 1 km S of Eyne	14.ix.1978	WJR	21°C	275/6F		1	69 (2)
	FRANCE: Hautes-Alpes, Embrun–Savines road, by Lac Serre-Ponçon	26.ix.1979	WJR	22°C	335S		1	69 (1)
terrestris	FRANCE: Alpes-Maritimes, St-Martin-du-Var/Levens	3.ix.1975	WJR	28°C	153S	524	1	70 (1)
	FRANCE: Alpes-Maritimes, near Nice, Le Broc, 570 m	19.viii.1977	DRR	22°C	238/3F	525		
	FRANCE: Alpes-Maritimes, near St-Martin-Vésubie, Col St-Martin, 1400 m	26.viii.1977	DRR	22°C	241/7F	527		

Species	Locality	Date	Recorder	Temp	Catalogue	No.	Fig.
perforatus	Same locality	6.ix.1977	WJR	27°C	221S		526	
	Same locality	10.ix.1977	DRR	19°C	242/2R	1	528	70 (2)
ruffoi	ITALY: Tuscany, E of Florence, Montemignaio	11.ix.1992	DRR	22°C	845/7R	1	540	71
	Same locality	21.ix.1992	DRR	25°C	840S		539	
	ITALY: Abruzzi, Gran Sasso d'Italia, near S. Egidio, 1680 m	1.ix.1985	DRR	20°C	603/8F		547	
	ITALY: Abruzzi, Gran Sasso d'Italia, Valle Fredda, 1160 m	3.ix.1985	DRR	22°C	606/5F	1	548	72
Ephippigerida areolaria	SPAIN: Madrid, Sierra de Guadarrama, La Bola del Mundo, 2200 m	6.viii.1983	DRR	25°C	509/6F		553	
taeniata	SPAIN: Madrid, Sierra de Guadarrama, above Puerto de Navacerrada, 2000 m	29.vii.1989	FW	25°C	754/16F	1	554	73
	SPAIN: Cadiz, 36 km W of Algeciras, 2 km NE of Zahara de los Atunes	5.iii.1992	WJR	26°C	829S	1	555	74
Uromenus rugosicollis	FRANCE: Pyrénées-Orientales, near Vernet-les-Bains, Sahorre	24.ix.1976	WJR	20°C	184/8FD		562	
	FRANCE: Pyrénées-Orientales, Vernet-les-Bains	28.ix.1976	WJR	25°C	174SD	1	563	75
elegans	GREECE: Peloponnese, Messinia, near Pilos, S of Yiavolam	14.vii.1987	FW	28°C	677/12R	1	564	76

Appendix 3. (cont.)

Species	Locality	Date recorded	Recordist	Air temp.	Rec. No.	Fig. No.	CD No.	Track No.
Uromenus (cont.)								
brevicollis	SARDINIA: Nuoro, 4 km NW of Ottana	1.viii.1990	WJR	28°C	790S	565	1	77
catalaunicus	SPAIN: Huesca, 10 km E of Bielsa, San Juan de Plan	1.ix.1978	WJR	26°C	269/5F	570		
	ANDORRA: near Sant Julia de Loria, 1 km S of Fontaneda	13.ix.1978	WJR	25°C	275/3F	571		
	SPAIN: Huesca, N of Sierra de Guara, 4 km N of Nocito, 1200 m	16.ix.1986	WJR	29°C	648/6F	572	1	78
asturiensis	SPAIN: León, 40 km N of León, Collada de Carmenes	20.ix.1984	WJR	27°C	548/2S	579		
	SPAIN: Orense, 20 km S of A Rúa, Alto de Covelo, 1050 m	23.viii.1985	WJR	26°C	588/2F	581	1	79
	SPAIN: León, Picos de Europa, Puerto de Pendetrave	30.iv.1987	WJR	29·5°C	653S	580		
stalii	SPAIN: Madrid, Sierra de Guadarrama, Puerto de la Morcuera, 1796 m	28.ix.1982	WJR	23°C	466S	588		
	Same locality	1.x.1982	WJR	21°C	474S	589		
	SPAIN: Zamora, 88 km W of Benavente, W of Portillo de Padornelo, 1350 m	22.viii.1985	WJR	25°C	586/6F	590	1	80

Taxon	Locality	Date	Rec.	Temp.	Code	No.		No.
perezii	SPAIN: Tarragona, 1 km S of Hospitalet del Infante. Cala Jostell	11.ix.1984	WJR	26°C	548/1S	597	1	81
martorellii	SPAIN: Granada, N of Huescar, S of Puerto del Pinar, 1500 m	5.ix.1986	WJR	24°C	644/6F	600	1	82
andalusius	SPAIN: Granada, Sierra Nevada, Veleta, 2250 m	12.viii.1986	FW	23°C	756/10R	603	1	83
Baetica ustulata	SPAIN: Granada, Sierra Nevada, Veleta, 2500 m	19.viii.1986	FW	26°C	756/12R	609		
	SPAIN: Granada, Sierra Nevada, El Chorrillo, 2700 m	2.ix.1986	WJR	25°C	642/1F	608	1	84
Callicrania selligera	PORTUGAL: Minho, Parque Nacional de Penada-Gerés, near Melgaço, Lamas de Mouro	15.viii.1978	MD	—	488/9R	606	1	85
Platystolus martinezii	SPAIN: Ávila, Sierra de Gredos, near start of road to Pico Almanzor, 1500 m	31.vii.1989	FW	22°C	754/12R	614	1	86
faberi	SPAIN: León, 40 km N of León, Collada de Carmenes, 1350 m	28.viii.1984	WJR	23°C	552/3F	617	1	87
Pycnogaster jugicola	SPAIN: Madrid, Sierra de Guadarrama, Puerto de la Morcuera, 1796 m	27.ix.1982	WJR	20°C	465S	619	1	88

Appendix 3. (cont.)

Species	Locality	Date recorded	Recordist	Air temp.	Rec. No.	Fig. No.	CD No.	Track No.
Pycnogaster jugicola (cont.)	SPAIN: Madrid, Sierra de Guadarrama. Puerto de Navacerrada. 1900 m	6.viii.1983	DRR	24°C	510/1F	620		
	SPAIN: Zamora, 88 km W of Benavente, W of Portilla de Padornelo. 1350 m	22.viii.1985	WJR	27°C	587/6F	621		
sanchezgomezi	SPAIN: Jaén, 10 km E of Cazorla, Nava de San Pedro, 1300 m	19.vi.1987	WJR	30°C	658SD	622	1	89
inermis	SPAIN: Granada, Sierra Nevada, Veleta. 2500 m	13.viii.1989	FW	24°C	756/14RD	623	1	90
Gryllus campestris	ENGLAND: West Sussex, Arundel	11.vi.1975	DRR & WJR	23°C	142/3F	629	1	91 (1)
	Same locality	13.vi.1975	DRR & WJR	24°C	143/2S	638	1	91 (2)
	GREECE: Macedonia, Grevena, Mt Vourinos, near Ag. Pandelimon and Katafygion, 1400 m	26.vi.1986	FW	21°C	638/19FD	630		
	SPAIN: Jaén, 10 km E of Cazorla, Nava de San Pedro, 1300 m	28.v.1987	WJR	25°C	656/1SD	631		
bimaculatus	TUNISIA: Gafsa–Tozeur road	20.iii.1975	WJR	27°C	127S	643	1	92 (1)
	Same locality	20.iii.1975	DRR & WJR	27°C	128/1S	649	1	92 (2)
	Culture of unknown origin	2.ii.1991	WJR	18°C	809SD	644		

Species	Locality	Date	Recorder	Temp.	Recording	No.		Fig.
Acheta domesticus	Culture of unknown origin	17.vi.1980	WJR	29°C	356SD	660	1	93 (2)
	—	—	KHG	—	564/24	654		
Modicogryllus bordigalensis	ENGLAND: Leicestershire, near Stamford. Great Casterton	16.vii.1984	PR	—	659/13F	655	1	93 (1)
	FRANCE: Dordogne, near Les Eyzies, Meyrals	13.vi.1984	DRR	15°C	531/1FD	666		
	ITALY: Emilia-Romagna, near Ravenna, Russi	28.viii.1985	DRR	22°C	601/7FD	667	1	94
Eugryllodes pipiens	FRANCE: Alpes-Maritimes, near Col de Vence, 940 m	15.ix.1977	WJR	24°C	229SD	672	1	95 (1)
	Same locality	17.ix.1977	DRR	19°C	242/4RD	673	1	95 (2)
Nemobius sylvestris	FRANCE: Hautes-Pyrénées, near Vielle-Aure, Neste de Couplan	19.ix.1976	WJR	17°C	182/9F	678	1	96
	ENGLAND: Hampshire, New Forest, Aldridgehill Inclosure	4.ix.1981	WJR	21·5°C	410SD	679		
Pteronemobius heydenii	FRANCE: Dordogne, near Sarlat, Petite Rou de Puymartin	24.viii.1974	DRR	23°C	134/6F	684		
	SPAIN: Granada, La Rábita	24.vi.1984	DRR	23°C	533/4F	685	1	97
Oecanthus pellucens	FRANCE: Ardèche, near Valence, St-Péray	24.vii.1967	DRR	25°C	14/1FD	690		
	AUSTRIA: Burgenland, Neusiedler See. Illmitz, Schrändlsee	23.viii.1973	DRR	18°C	103/9FD	691		
	BALEARIC ISLANDS: Majorca, near Palma, Magalluf	23.ix.1981	WJR	20°C	421SD	692	1	98
	SPAIN: Granada, La Rábita	23.vi.1984	DRR	22°C	533/1FD			

Appendix 3. (cont.)

Species	Locality	Date recorded	Recordist	Air temp.	Rec. No.	Fig. No.	CD No.	Track No.
Gryllotalpa gryllotalpa	ENGLAND: Wiltshire, Hamptworth	7.iii.1968	DRR	16·5°C	24RD	700	1	99 (1)
	FRANCE: Dordogne, near Sarlat, Petite Rou de Puymartin	26.v.1975	DRR	19°C	141RD	699		
vineae	FRANCE: Dordogne, 1 km E of Meyrals	14.v.1969	HCBC	12°C	30FD	705	1	99 (2)
	PORTUGAL: Algarve, Albufeira	25.xi.1991	WJR & DRR	20°C	820SD	706		
Psophus stridulus	FRANCE: Alpes-Maritimes, Col de Bleine, 1440 m	22.viii.1977	DRR	19°C	239/5F	711	2	1
Bryodema tuberculata	GERMANY: Upper Bavaria, near Bad Tölz, Vorderriß	18.ix.1987	DRR	20°C	669/2F	713	2	2 (3)
	Same locality	18.ix.1987	DRR	20°C	669/3F		2	2 (1,2)
	Same locality	18.ix.1987	DRR	20°C	669/4F		2	2 (4)
Stethophyma grossum	FRANCE: Puy-de-Dôme, near Bagnols, Brimessange	18.viii.1974	DRR	21°C	132/2F	715	2	
	IRELAND: Co. Galway, Connemara, near Derryclare Lough	15.ix.1975	WJR	24°C	156/1S	717	2	3 (3)
	Same locality	17.ix.1975	WJR	22°C	156/2S		2	3 (2)
	ENGLAND: Hampshire, New Forest, Crab Tree Bog	29.viii.1981	WJR	26°C	403S	716	2	3 (1)
Brachycrotaphus tryxalicerus	SPAIN: Tarragona, 1 km S of Hospitalet del Infante, Cala Jostell	16.vii.1985	WJR	30°C	585/6F	721		
	Same locality	27.vii.1990	WJR	31°C	784S	722		
	Same locality	30.vii.1990	WJR	30°C	786S		2	4

Species	Locality	Date	Recorder	Temp.	Tape ref.	No.		Fig.
Arcyptera fusca	FRANCE: Alpes-Maritimes, near St-Dalmas-de-Tende, La Minière de Vallaure, 1500 m	24.viii.1977	DRR	22°C	240/10F	727	2	5
	SPAIN: Huesca, Valle de Pineta	23.vii.1983	DRR	22°C	501/2F	728		
tornosi	PORTUGAL: 28·5 km E of Coimbra.	27.vi.1977	WJR	26°C	201S	731		
	SPAIN: Ávila, Sierra de Gredos, near Hoyos del Espino 1340 m	23.vii.1989	DRR	22°C	765/5F	732	2	6(1)
	Same locality	23.vii.1989	DRR	22°C	765/6F	733	2	6(2)
microptera	FRANCE: Alpes-Maritimes, Col de Bleine, 1440 m	22.viii.1977	DRR	15°C	240/1F	740	2	7(1,2)
	FRANCE: Lozère, Causse Méjean, near Crosgarnon	31.vii.1982	DRR	20°C	482/6F	739	2	7(3)
	SPAIN: Teruel, Sierra de Gúdar, 6 km N of Alcala de la Selva	25.vii.1989	WJR	—	760/3F	738		
	Same locality	25.vii.1989	WJR	—	760/4F	737		
Ramburiella hispanica	FRANCE: Vaucluse, c. 3 km SE of Cavaillon	14.viii.1977	DRR	30°C	236/7F	753	2	8
	SPAIN: Zaragoza, 4·5 km E of Caspe	7.ix.1978	WJR	25°C	271/4R	754		
Chrysochraon dispar	FRANCE: Puy-de-Dôme, near summit, 1400 m	15.viii.1974	DRR	28°C	130/4F	759	2	9
	FRANCE: N of Puy-de-Dôme, Le Cratère	1.ix.1975	WJR	23°C	151S	760		
	FRANCE: Pyrénées-Atlantiques, near Sauveterre, Burgaronne	12.vii.1989	DRR	25°C	761/2F	761		

Appendix 3. (cont.)

Species	Locality	Date recorded	Recordist	Air temp.	Rec. No.	Fig. No.	CD No.	Track No.
Euthystira brachyptera	AUSTRIA: Tyrol, near Kufstein, Kaisertal	15.viii.1973	DRR	26°C	101/1F	768		
	ITALY: Trentino-Alto Adige, near Riva del Garda,							
	Val dei Concei	14.ix.1987	DRR	24°C	668/4F	769		
	Same locality	14.ix.1987	DRR	25°C	668/5F	770	2	10
Dociostaurus maroccanus	FRANCE: Vaucluse, near Carpentras, Bédoin	13.viii.1977	DRR	26°C	235/6F	777	2	11(1)
	SARDINIA: Nuoro, Monti del Gennargentu, 3 km N of Aritzo, 1100 m	14.vii.1990	WJR	23°C	794/4F	778	2	11(2,3)
jagoi	PORTUGAL: Algarve, São Romão	29.viii.1979	WJR	26°C	328S	785	2	12(1)
	Same locality	30.viii.1979	WJR	28°C	329/1S		2	12(2)
	SPAIN: Madrid, near Guadarrama, Embalse de la Jarosa	5.viii.1983	DRR	28°C	508/1F	786		
	SPAIN: near Almería, Balerma	23.vi.1984	DRR	23°C	532/6F	787	2	12(3)
hispanicus	SPAIN: Madrid, near Guadarrama, Embalse de la Jarosa	12.viii.1983	DRR	23°C	511/3R	788	2	13
Omocestus viridulus	ENGLAND: Dorset, near Wareham, Morden Bog	21.vii.1976	WJR	29°C	168S	800	2	14(1)
	FRANCE: Alpes-Maritimes, near St-Dalmas-de-Tende, Vallon de la Minière, 1830 m	26.viii.1977	NDJ	—	222/3F	804, 805		

	Locality	Date	Recorder	Temp.	Ref.	Rec. no.	No.	Fig.
	ENGLAND: Surrey, near Mytchett Lake	26.vii.1979	WJR	32°C	307/2S	801		
	SPAIN: Logroño/Soria, Sierra de las Hormazas	16.vii.1989	DRR	23°C	763/1F	802		
	Same locality	16.vii.1989	DRR	23°C	763/2F	803	2	14 (3)
	SPAIN: Navarra, near Leiza, N of Puerto de Usateguieta, 800 m	4.vii.1990	DRR	17°C	803/2F		2	14 (2)
rufipes	AUSTRIA: Tyrol, near Kufstein, Kaisertal	13.viii.1973	DRR	24°C	98/3F	816		
	ENGLAND: Kent, near Goudhurst, Bedgebury Forest	19.vii.1976	WJR	29°C	167/1S		2	15 (3)
	Same locality	20.vii.1976	WJR	26°C	167/2S	817	2	15 (1,2)
	FRANCE: Hautes-Pyrénées, near Vielle-Aure, Neste de Couplan	20.ix.1976	WJR	14°C	183/2F	819		
	ENGLAND: Hampshire, New Forest, Puttles Bridge	6.viii.1979	WJR	26°C	319S	818		
haemorrhoidalis	FRANCE: Pyrénées-Orientales, near Saillagouse, Val d'Eyne	29.ix.1976	WJR	26°C	177S	828	2	16
	FRANCE: Lozère, near Mende, Col de Montmirat	10.viii.1977	DRR	21°C	232/3F	829		
	FRANCE: Alpes-Maritimes, near St-Martin-Vésubie, Col St-Martin, 1400 m	26.viii.1977	DRR	22°C	241/11F	830		

Appendix 3. (cont.)

Species	Locality	Date recorded	Recordist	Air temp.	Rec. No.	Fig. No.	CD No.	Track No.
Omocestus (cont.)								
petraeus	FRANCE: Haute-Vienne, near St-Yrieix-la-Perche,	30.viii.1978	WJR	25°C	268/3F	837, 839	2	17 (1,5)
	FRANCE: Lozère, Causse Méjean, near Crosgarnon	31.vii.1982	DRR	20°C	482/5F	838	2	17 (2–4)
raymondi	SPAIN: Huesca, 8 km E of Ainsa	3.ix.1978	WJR	31°C	270/8F	846		
	SPAIN: Madrid, Puerto de Galapagar, 800 m	17.vi.1984	DRR	30°C	531/3F	848	2	18 (2)
	SPAIN: Granada, Sierra Nevada, near Capileira, 1500 m	22.vi.1984	DRR	24°C	532/4F	847	2	18 (1)
panteli	SPAIN: Valencia, near Requena, 700 m	30.vii.1979	WJR	33°C	310S	859	2	19 (5)
	Same locality	31.vii.1979	WJR	30°C	312S	855	2	19 (1–4)
	Same locality	1.viii.1979	WJR	29°C	315S	858		
	SPAIN: Madrid, near Navacerrada, Valle de Barranca	27.vii.1983	DRR	22°C	503/5F	856		
	SPAIN: Granada, Sierra Nevada, Puerto de la Ragua, 2000 m	1.viii.1983	DRR	25°C	506/3F	857		
antigai	FRANCE: Pyrénées-Orientales, near Saillagouse, Val d'Eyne, 2100 m	15.ix.1978	WJR	23°C	276/3F	873	2	20 (1)
	Same locality	19.ix.1978	WJR	25°C	265S			

	Locality	Date	Rec.	Temp.	No.	No.		Fig.
	Same locality	19.ix.1978	WJR	25°C	266S	872, 893		
	Same locality	14.viii.1985	WJR	27°C	585/1R	874	2	20 (2)
	SPAIN: Gerona, 22 km N of Ripoll, Nuria, 2100 m	16.viii.1985	WJR	30°C	585/3F	875		
	SPAIN: Barcelona, near San Celoni, Sierra de Montseny, 1700 m	31.viii.1986	WJR	27°C	640/3F	876		
	Same locality	31.viii.1986	WJR	27°C	641/1F	894		
	SPAIN: Lérida, 13 km NW of Tremp. Torre de Tamurcia, 1205 m	14.ix.1986	WJR	26°C	648/2F	877		
	SPAIN: Huesca, N of Sierra de Guara, 4 km N of Nocito, 1200 m	16.ix.1986	WJR	29°C	648/4F	897		
	Same locality	16.ix.1986	WJR	29°C	648/7F	878		
	SPAIN: Lérida, 13 km NW of Tremp. Torre de Tamurcia, 1205 m	23.ix.1986	WJR	25°C	632S	895		
	SPAIN: Huesca, N of Sierra de Guara, 4 km N of Nocito, 1200 m	24.ix.1986	WJR	30°C	634S	896		
bolivari	SPAIN: Granada, Sierra Nevada, Campos de Otero, 2300 m	31.vii.1983	DRR	21°C	505/4F	912		
	SPAIN: Granada, Sierra Nevada, Puerto de la Ragua, 2000 m	1.viii.1983	DRR	25°C	506/5F	913	2	21 (3,4)
	Same locality	1.viii.1983	DRR	25°C	507/2F	914		
	SPAIN: Granada, Sierra Nevada, El Chorrillo, 2700 m	2.ix.1986	WJR	24°C	641/6F		2	21 (1,2)

Appendix 3. (cont.)

Species	Locality	Date recorded	Recordist	Air temp.	Rec. No.	Fig. No.	CD No.	Track No.
Omocestus (cont.) *uhagonii*	SPAIN: Madrid, Sierra de Guadarrama, La Bola del Mundo, 2200 m	6.viii.1983	DRR	21°C	508/6F	921		
	Same locality	6.viii.1983	DRR	21°C	509/1F	922	2	22 (1.2)
	Same locality	6.viii.1983	DRR	24°C	509/5F	923	2	22 (3)
minutissimus	SPAIN: Granada, Sierra Nevada, near Dornajo, 1900 m	30.vii.1983	DRR	27°C	504/6F	932		
	Same locality	30.vii.1983	DRR	27°C	505/1F	931		
	SPAIN: Madrid, Sierra de Guadarrama, Puerto de los Leones, 1500 m	5.viii.1983	DRR	23°C	508/3F	930	2	
	SPAIN: Lérida, 13 km NW of Tremp. Torre de Tamurcia, 1100 m	14.ix.1986	WJR	25°C	647/8F	933	2	23 (4)
	SPAIN: E of Teruel, Sierra de Gúdar	14.vii.1990	DRR	22°C	808/1F		2	23 (1–3)
femoralis	SPAIN: Granada, N of Huescar, Puerto del Pinar, 1600 m	6.ix.1986	WJR	22°C	645/8F		2	24 (1)
	SPAIN: Murcia, Sierra Espuña, Espuña, 1540 m	8.ix.1986	WJR	23°C	646/7F	939	2	24 (4)
	Same locality	8.ix.1986	WJR	23°C	646/9F		2	24 (3)
	Same locality	9.ix.1986	WJR	27°C	646/11R	940	2	24 (2)
uvarovi	ITALY: Campania, Paestum	22.ix.1988	DRR	22°C	726/2F	941	2	25

Taxon	Locality	Date	Recordist	Temp.	Ref.	No.	n	Fig.
Stenobothrus lineatus	ENGLAND: Kent, near Wrotham, Trottiscliffe	11.viii.1978	WJR	25°C	260S	952		
	Same locality	13.viii.1978	WJR	29°C	321/1S	949	2	26(2)
	Same locality	13.viii.1978	WJR	28°C	322/1S	950	2	26(1)
	Same locality	13.viii.1978	WJR	28°C	322/2S			
	FRANCE: Seine-Maritime, near Gamaches	11.viii.1983	DRR	24°C	511/2F	948		
	ITALY: Lazio, Monti Reatini, foot of chair-lift to Monte Terminilletto, 1700 m	11.ix.1985	DRR	18°C	608/4F	951	2	26(3)
nigromaculatus	FRANCE: Pyrénées-Orientales, near Saillagouse, Val d'Eyne	22.ix.1976	WJR	16°C	184/1F	966		
	Same locality	29.ix.1976	WJR	27°C	176S	963	2	27(2)
	Same locality	19.ix.1978	WJR	25°C	267S	961	2	27(1)
	FRANCE: Lozère, Causse Méjean, near Crosgarnon	31.vii.1982	DRR	16°C	482/2F			
	Same locality	31.vii.1982	DRR	16°C	482/3F	965	2	27(3)
	ITALY: Abruzzi, Gran Sasso d'Italia, Fossa di Paganica, 1680 m	31.viii.1985	DRR	23°C	603/4F	964		
	ITALY: Marches, Monti Sibillini, Forca di Presta, 1540 m	14.ix.1985	DRR	18°C	609/10F	962		
fischeri	FRANCE: Vaucluse, Mont Ventoux, road from Bédoin, 1350 m	15.viii.1977	DRR	23°C	237/1F	976	2	28(3)
	Same locality	15.viii.1977	DRR	23°C	237/3F	977	2	28(2,4)
	FRANCE: Alpes-Maritimes, near Col de Vence, 930 m	25.viii.1977	DRR	25°C	241/2F	975	2	28(1)

Appendix 3. (cont.)

Species	Locality	Date recorded	Recordist	Air temp.	Rec. No.	Fig. No.	CD No.	Track No.
Stenobothrus (cont.)	SPAIN: Madrid,							
festivus	near Guadarrama. Embalse de la Jarosa	19.vi.1984	DRR	22°C	531/6F	984	2	29 (1.2)
	Same locality	19.vi.1984	DRR	22°C	532/1F	985	2	29 (3)
	Same locality	4.vii.1984	WJR	—	527S			
grammicus	FRANCE: Vaucluse, Mont Ventoux, near Pavillon de Rolland, 800 m	11.viii.1977	DRR	23°C	233/4F	992	2	30 (1.3)
	FRANCE: Vaucluse, Mont Ventoux, road from Bédoin, 1100 m	12.viii.1977	DRR	19°C	234/1F	993	2	30 (2)
bolivarii	SPAIN: Madrid, near Navacerrada, Embalse de Navalmedio, 1300 m	26.vii.1983	DRR	26°C	501/8F	995	2	31 (1)
	SPAIN: 30 km E of Teruel, Puerto de San Rafael, 1600 m	25.vii.1989	WJR	—	759/9F	996	2	31 (2)
	Same locality	25.vii.1989	WJR	—	759/10F	997	2	31 (3)
stigmaticus	ISLE OF MAN: Langness Peninsula	8.viii.1974	WJR	24°C	119S	1015	2	32 (5)
	Same locality	9.viii.1974	WJR	24°C	121S	1019		
	FRANCE: Haute-Vienne, near St-Yrieix-la-Perche	30.viii.1978	WJR	21°C	268/2F	1016		
	SPAIN: Madrid, near Navacerrada, Valle de Barranca	27.vii.1983	DRR	22°C	503/6F	1017		
	ISLE OF MAN: Langness Peninsula	6.ix.1983	WJR	23°C	497S	1018		

	Locality	Date	Rec.	Temp	Spectrogram	Cat.	n	Figure
apenninus	SPAIN: Logroño/Soria, Sierra de las Hormazas	15.vii.1989	DRR	23°C	762/7F		2	32(1-3)
	SPAIN: 86 km E of Teruel. Puerto de Linares, 1700 m	14.vii.1990	DRR	23°C	807/3F		2	32(4)
	ITALY: Abruzzi, Gran Sasso d'Italia. Fossa di Paganica, 1680 m	31.viii.1985	DRR	24°C	603/2F	1032	2	33(1-4)
	Same locality	31.viii.1985	DRR	19°C	603/3F	1034	2	33(5)
	ITALY: Lazio. Monti Reatini, foot of chair-lift to Monte Terminilletto, 1700 m	11.ix.1985	DRR	18°C	608/1F	1033		
ursulae	ITALY: Piedmont, near Cuorgne, S. Elisabetta, 1300 m	9.ix.1987	DRR	20°C	668/3R	1041		
	ITALY: Valle d'Aosta, 6 km S of Aosta, Lago di Chamolè, 2320 m	3.ix.1988	DRR	15°C	720/1F	1042	2	34(1-4)
	Same locality	3.ix.1988	DRR	20°C	720/2R	1047	2	34(5)
rubicundulus	FRANCE: Alpes-Maritimes, near St-Dalmas-de-Tende, La Minière de Vallaure, 1500 m	24.viii.1977	DRR	23°C	240/4F	1058		
	ITALY: Veneto, Belluno, Pordoi Pass, below cable line to Pordoi Joch	16.ix.1980	WJR	26°C	369S	1051	2	35(1,2)
	Same locality	18.ix.1980	WJR	25°C	371S	1057	2	35(3)
	ITALY: Trentino-Alto Adige, near Riva del Garda, Tremalzo, 1200 m	27.viii.1985	DRR	13°C	601/5F	1052		
cotticus	FRANCE: Hautes-Alpes, near Briançon, Col d'Izoard, 2870 m	4.ix.1987	DRR	16°C	666/2F	1067		
	Same locality	6.ix.1987	DRR	13°C	666/3F	1065		
	Same locality	6.ix.1987	DRR	13°C	666/5F	1066	2	36

Appendix 3. (cont.)

Species	Locality	Date recorded	Recordist	Air temp.	Rec. No.	Fig. No.	CD No.	Track No.
Myrmeleotettix maculatus	ENGLAND: Dorset, near Affpuddle, Cull-pepper's Dish	2.ix.1981	WJR	27°C	405S	1073	2	37 (1,2)
	Same locality	2.ix.1981	WJR	29°C	407S	1074		
	SPAIN: Madrid, Sierra de Guadarrama, La Bola del Mundo, 2200 m	6.viii.1983	DRR	22°C	509/2F	1077	2	37 (3)
	SPAIN: Zamora, 25 km NW of Puebla de Sanabria, Porto, 1350 m	23.viii.1985	WJR	22°C	587/7F	1078		
	ITALY: Abruzzi, Gran Sasso d'Italia, 2 km E of Passo delle Capannelle, 1500 m	2.ix.1985	DRR	23°C	605/7F	1075		
	ITALY: Piedmont, Colle di Sampeyre, 2300 m	8.ix.1987	DRR	18°C	667/3F	1076		
	SPAIN: Logroño/Soria, Sierra de las Hormazas	15.vii.1989	DRR	23°C	762/4F	1079		
	ENGLAND: Surrey, Oxshott Heath	24.vii.1991	DRR	27°C	813/1S	1094	2	37 (4)
Gomphocerus sibiricus	FRANCE: Isère, Pont-du-St-Charles	26.ix.1979	WJR	27°C	339S	1097	2	38 (1)
	Same locality	26.ix.1979	WJR	27°C	340S	1100		
	ITALY: Trentino-Alto Adige, 4 km S of Arabba, N side of Fedaia Pass, 2000–2300 m	12.ix.1980	WJR	24°C	367S		2	38 (2)

Locality	Date		Temp.				
FRANCE: Alpes-Maritimes, near Isola, Col de la Lombarde, 2350 m	26.vii.1982	WJR	26°C	439S	1098		
ITALY: Abruzzi, Gran Sasso d'Italia, valley N of Monte della Scindarella, 1800 m	1.ix.1985	DRR	21°C	604/4F	1099		
Gomphocerippus rufus							
AUSTRIA: Tyrol, near Kufstein, Kaisertal	15.viii.1973	DRR	26°C	100/5F	1109		
ENGLAND: Surrey, near Dorking, Juniper Bottom	11.x.1974	WJR	23°C	123S	1114	2	39 (4)
ENGLAND: Kent, Folkestone Warren	5.x.1979	WJR	26°C	347S	1110	2	39 (1–3)
Same locality	5.x.1979	WJR	25°C	348S			
ENGLAND: West Sussex, Arundel Park	5.xii.1979	WJR	25°C	349/1S	1111		
ITALY: Lazio, Monti Reatini, Valle della Meta, 1400 m	12.ix.1985	DRR	19°C	609/5F	1115		
ITALY: Marches, Monti Sibillini, near Gualdo, 1500 m	14.ix.1985	DRR	19°C	609/12F	1112		
ITALY: Tuscany, E of Florence, near Montemignaio	9.ix.1992	DRR	23°C	844/7F	1113		
Stauroderus scalaris							
FRANCE: Puy-de-Dôme, near Besse-en-Chandesse, Lac Estivadoux	14.viii.1974	DRR	26°C	129/4F	1134	2	40
Same locality	14.viii.1974	DRR	25°C	130/2F	1131		
FRANCE: Hautes-Pyrénées, near Vielle-Aure, Lac d'Orédon	18.ix.1976	WJR	13°C	181/11F	1132		

Appendix 3. (cont.)

Species	Locality	Date recorded	Recordist	Air temp.	Rec. No.	Fig. No.	CD No.	Track No.
Stauroderus scalaris (cont.)	FRANCE: Haute-Savoie, 4 km SE of Talloires, E side of Lac d'Annecy, 900 m	25.iii.1981	WJR	30°C	386S	1133		
	ITALY: Abruzzi, Gran Sasso d'Italia, near S. Egidio, 1680 m	1.ix.1985	DRR	22°C	604/2F	1135		
Chorthippus apricarius	AUSTRIA: Burgenland, near Kittsee	28.viii.1973	DRR	27°C	105/4F	1146		
	FRANCE: Puy-de-Dôme, near summit, 1400 m	15.viii.1974	DRR	28°C	130/3F	1147	2	41 (1,2)
	FRANCE: Alpes-Maritimes, near St-Martin-Vésubie, Col St-Martin, 1400 m	26.viii.1977	DRR	22°C	241/6F	1148		
	ANDORRA: near Sant Julia de Loria, 1 km S of Fontaneda	13.ix.1978	WJR	20°C	274/8F	1149		
	ITALY: Lazio, Monti Reatini, Valle della Meta, 1400 m	12.ix.1985	DRR	18°C	608/6F	1151	2	41 (3)
	Same locality	12.ix.1985	DRR	19°C	609/2F	1150		
corsicus	CORSICA: Corse-du-Sud, 6 km NE of Zonzo, Col de Bavella, 1225 m	19.vii.1990	WJR	22·5°C	796/2F		2	42 (2)
	CORSICA: Haute-Corse, Monte Cinto, Haut Asco, 1500 m	20.vii.1990	WJR	25°C	796/7F	1161	2	42 (3)
	Same locality	25.vii.1990	WJR	30°C	779S	1159	2	42 (1)
	CORSICA: Corse-du-Sud/Haute-Corse, near Evisa, Col de Vergio, 1500 m	30.vii.1990	WJR	29°C	785S	1160		

	Locality	Date		Temp				
cazurroi	SPAIN: Santander, Picos de Europa, Fuente Dé, 2000 m	26.viii.1984	WJR	20°C	551/4F	1171	2	43(3)
	Same locality	26.viii.1984	WJR	27°C	551/5F	1170	2	43(2)
	Same locality	26.viii.1984	WJR	25°C	551/6F	1169	2	43(1)
	Same locality	28.viii.1984	WJR	30°C	552/6F	1168		
nevadensis	SPAIN: Granada, Sierra Nevada, El Chorrillo, 2700 m	31.vii.1983	DRR	20°C	505/5F	1184		
	Same locality	2.ix.1986	WJR	24°C	641/5F	1185	2	44
cialancensis	ITALY: Piedmont, near Pinerola, Conca Cialancia, 2600 m	7.ix.1987	DRR	21°C	667/2F	1190	2	45(1)
	ITALY: Piedmont, Colle di Sampeyre, 2300 m	8.ix.1987	DRR	19°C	667/4F	1191	2	45(2)
	Same locality	8.ix.1987	DRR	19°C	667/6F	1192	2	45(3)
pullus	GERMANY: Upper Bavaria, Vorderriß	25.ix.1987	WJR	25–28°C	661S	1199	2	46
	Same locality	29.ix.1987	WJR	24–25°C	663/2S	1200		
alticola	ITALY: Trentino-Alto Adige, near Riva del Garda, Tremalzo, 1200 m	27.viii.1985	DRR	12°C	601/1F	1207	2	47(1)
	Same locality	27.viii.1985	DRR	12°C	601/2F	1206	2	47(2)
modestus	ITALY: Lazio, Monti Reatini, foot of chair-lift to Monte Terminilletto, 1700 m	11.ix.1985	DRR	13°C	607/2F	1213	2	48(4)
	Same locality	11.ix.1985	DRR	13°C	607/3F	1212	2	48(2)
	Same locality	11.ix.1985	DRR	14°C	607/4F	1211	2	48(1,3)
	Same locality	11.ix.1985	DRR	14°C	607/5F			

Appendix 3. (cont.)

Species	Locality	Date recorded	Recordist	Air temp.	Rec. No.	Fig. No.	CD No.	Track No.
Chorthippus (cont.) *vagans*	ENGLAND: Hampshire, near Burley, Cranes Moor	14.viii.1975	WJR	28°C	145S		2	49 (1)
	FRANCE: Calvados, near Bayeux, Forêt de Cerisy	8.ix.1976	WJR	30°C	179/4F	1223		
	SPAIN: Huesca, 10 km E of Bielsa, San Juan de Plan	1.ix.1978	WJR	28°C	269/8F	1224		
	ENGLAND: Dorset, near Wareham, Sherford	23.vii.1979	WJR	31°C	304S	1220		
	SPAIN: Madrid, near Navacerrada, Embalse de Navalmedio, 1300 m	26.vii.1983	DRR	26°C	502/4F	1221	2	49 (2)
	SPAIN: Granada, Sierra Nevada, near Dornajo, 1900 m	30.vii.1983	DRR	27°C	504/3F	1226		
	SPAIN: Granada, Sierra Nevada, Puerto de la Ragua, 2000 m	1.viii.1983	DRR	25°C	507/1F	1225	2	49 (4)
	SPAIN: Teruel, near Albarracin, Puerto de Bronchales, 1500 m	15.vii.1989	WJR	26°C	758/3F	1222	2	49 (3)
reissingeri	SPAIN: Alicante, Calpe	11.ix.1986	WJR	31°C	647/2F	1243		
	Same locality	11.ix.1986	WJR	31°C	647/6F	1242	2	50
biguttulus	AUSTRIA: Burgenland, Leitha Gebirge, near Winden	27.viii.1973	DRR	24°C	105/1F	1249		
	SPAIN: Lérida, 7 km SE of Seo de Urgel, near Ortedo	12.ix.1978	WJR	26°C	274/1F	24. 1250	2	51 (2)

Locality	Date	Recorder	Temp	Code	No.	Count	Fig.
ITALY: Veneto, Belluno, W of Arabba, 1600 m	12.ix.1980	WJR	25°C	366S	1252		
FRANCE: Alpes-Maritimes, near Col de Vence, 940 m	10.viii.1982	DRR	23°C	484/3F	1248		
FRANCE: Seine-Maritime, near Gamaches	17.viii.1983	WJR	29°C	492S	1251	2	51 (1)
Same locality	23.viii.1983	WJR	27°C	494S	1266	2	51 (3)
ITALY: Trentino-Alto Adige, near Bressanone, Varna	19.ix.1985	DRR	18°C	610/1F	1267		
GERMANY: Berlin, Konradshöhe	5.viii.1988	WJR	26°C	688S	1253		
AUSTRIA: Carinthia, N of Pressegger See, 1100 m	4.ix.1994	DRR	21°C	869/4F	1255		
AUSTRIA: Carinthia, SE of Hermagor, Dellacher-Alm, 1400 m	5.ix.1994	DRR	14°C	869/7F	1254		
brunneus — AUSTRIA: Burgenland, Parndorfer Heide, near Mönchhof	19.viii.1973	DRR	30°C	102/4F	1275		
ENGLAND: Surrey, near Caterham, Quarry Hangers	11.vii.1974	DRR & WJR	25°C	115S		2	52 (1)
CHANNEL ISLANDS: Guernsey, Castel	28.ix.1978	WJR	28°C	277/2S	1270		
ENGLAND: Surrey, near Mytchett Lake	23.vii.1979	WJR	30°C	305S	1276		
SICILY: near Messina, 4 km S of Colle San Rizzo, 1000 m	9.ix.1988	DRR	19°C	722/1F	1302		
Same locality	10.ix.1988	DRR	22°C	722/6F	1303		
SICILY: SE side of Mount Etna, near Zafferana Etnea, 1000 m	13.ix.1988	DRR	19°C	723/4F		2	52 (3)

Appendix 3. (cont.)

Species	Locality	Date recorded	Recordist	Air temp.	Rec. No.	Fig. No.	CD No.	Track No.
Chorthippus brunneus (cont.)	ITALY: Calabria, near Rosarno	19.ix.1988	DRR	23°C	724/5F	1273		
	ITALY: Umbria, near Orvieto	25.ix.1988	DRR	25°C	727/1F	1277	2	52 (2)
	SPAIN: Santander, Picos de Europa, S of Fuente Dé, 1000 m	29.vii.1989	DRR	25°C	767/4F	1272		
	SPAIN: Navarra, near Leiza, N of Puerto de Usateguieta, 800 m	4.vii.1990	DRR	19°C	804/1F	1278		
	SARDINIA: Sassari, Monti Limbara, 3 km N of summit, 1200 m	17.vii.1990	WJR	26°C	795/2F	26, 1274		
	CORSICA: Corse-du-Sud/Haute-Corse, near Evisa, Col de Vergio, 1500 m	26.vii.1990	WJR	29°C	781S	1271		
	ENGLAND: Hampshire, 13 km SW of Salisbury, Martin Down	19.ix.1990	WJR & DRR	22°C	799S	1301	2	52 (4)
	Same locality	20.ix.1990	WJR	28°C	801S	1300	2	52 (5)
biguttulus × brunneus	AUSTRIA: Tyrol, near Imst, Mils	30.viii.1973	DRR	23°C	418F	25	2	53
mollis mollis	AUSTRIA: Burgenland, Parndorfer Heide, near Mönchhof	19.viii.1973	DRR	30°C	102/5F	1312	2	54 (1)
	AUSTRIA: Burgenland, Neusiedler See, near Illmitz, Darscholacke	23.viii.1973	DRR	25°C	103/6F	1313		

Taxon	Locality	Date	Recorder	Temp.		No.		Fig.
m. ignifer	SPAIN: Lérida, 7 km SE of Seo de Urgel, near Ortedo	12.ix.1978	WJR	27°C	273/9F	1314		
	SWITZERLAND: Neuchâtel	24.ix.1991	DRR	27·5°C	819S	1315	2	54 (2)
	FRANCE: Alpes-Maritimes, near Nice, Ste-Marguerite	9.viii.1982	DRR	22·5°C	484/2F	1328		
	FRANCE: Alpes-Maritimes, near Nice, Le Broc, 570 m	16.viii.1982	WJR	25·5°C	443S	1330		
	Same locality	16.viii.1982	WJR	27°C	444S	1334		
	Same locality	17.viii.1982	WJR	29°C	446S	1335		
	ITALY: Trentino-Alto Adige, near Bressanone, Varna	25.ix.1985	WJR	25°C	591S	1333	2	55 (2)
	ITALY: Trentino-Alto Adige, near Riva del Garda, Val dei Concei	14.ix.1987	DRR	23°C	668/2F	1331		
	ITALY: Piedmont, near Cuorgne, S. Elisabetta, 1300 m	29.ix.1987	WJR	26°C	664S	1329	2	55 (1)
	Same locality	30.ix.1987	WJR	27°C	665S			
	SWITZERLAND: Valais, Saillon	24.ix.1991	DRR	26°C	818S	1332		
jacobsi	BALEARIC ISLANDS: Majorca, Albufera Marsh	27.v.1981	WJR	23°C	391S	1354		
	SPAIN: Huesca, near Sallent	3.ix.1982	WJR	25°C	455/1S	1355	2	56 (2)
	Same locality	8.ix.1982	WJR	28°C	455/4S			
	SPAIN: Madrid, near Navacerrada, El Ventorillo, 1480 m	26.vii.1983	DRR	21°C	503/4F	1357		
	SPAIN: Huesca, near Lanave	28.vi.1984	DRR	28°C	533/8F	1356		

Appendix 3. (cont.)

Species	Locality	Date recorded	Recordist	Air temp.	Rec. No.	Fig. No.	CD No.	Track No.
Chorthippus jacobsi (cont.)	SPAIN: Granada, Sierra Nevada, Puerto de la Ragua, 2000 m	5.vii.1984	WJR	27°C	528S		2	56 (1)
	SPAIN: Huesca, Valle de Ordesa, 2100 m	6.vii.1984	WJR	27°C	529S	1358		
yersini	SPAIN: Madrid, near Navacerrada, El Ventorillo, 1480 m	26.vii.1983	DRR	21°C	503/3F	1371		
	SPAIN: 40 km N of León, Collada de Cármenes, 1350 m	28.viii.1984	WJR	23°C	552/1F	1372		
	Same locality	28.viii.1984	WJR	23°C	552/2F	1379		
	Same locality	28.viii.1984	WJR	23°C	552/4F	1377	2	57 (1)
	SPAIN: 40 km N of León, near Villamanin	28.viii.1984	WJR	30°C	552/7F	1378		
	SPAIN: Zamora, W of Portilla de Padornelo, 1350 m	22.viii.1985	WJR	27°C	587/3F	1375		
	SPAIN: Granada, Sierra Nevada, 4 km NE of Capileira, 2050 m	2.ix.1986	WJR	27°C	642/7F	1380		
	Same locality	3.ix.1986	WJR	22°C	643/3F	1374		
	Same locality	3.ix.1986	WJR	24°C	643/8F	1376	2	57 (2)
	SPAIN: Granada, N of Huescar, Puerto del Pinar, 1600 m	5.ix.1986	WJR	23°C	645/2F	1373		
	SICILY: near Messina, 4 km S of Colle San Rizzo, 1000 m	9.ix.1988	DRR	20°C	721/5F	1389		
	Same locality	9.ix.1988	DRR	19°C	722/2F	1390	2	57 (3)

Subspecies	Locality	Date	Recordist	Temp.	Recording	No.		Fig.
	SPAIN: Teruel, near Albarracin, Puerto de Bronchales, 1500 m	15.vii.1989	WJR	23°C	757/9F	1395		
	SPAIN: Logroño/Soria, Sierra de las Hormazas	15.vii.1989	DRR	23°C	762/5F	1391		
	SPAIN: Cuenca/Valencia, 35 km SW of Teruel, Casa Mojón, 1400 m	1.viii.1989	WJR	27°C	734S	1393		
	Same locality	1.viii.1989	WJR	28·5°C	735S	1394		
	SPAIN: Soria, Sierra de las Hormazas, near Montenegro de Camera	6.vii.1990	DRR	19°C	804/6F	1392		
rubratibialis	ITALY: Abruzzi, Gran Sasso d'Italia, 2 km E of Passo delle Capannelle, 1500 m	2.ix.1985	DRR	22°C	605/4F	1415	2	58 (3)
	ITALY: Lazio, Monti Reatini, Valle della Meta, 1400 m	12.ix.1985	DRR	19°C	609/6F	1414	2	58 (1)
	ITALY: Marches, Monti Sibillini, Forca di Presta, 1540 m	14.ix.1985	DRR	18°C	609/8F	1412	2	58 (2)
	ITALY: Emilia-Romagna, Foresta di Campigna, Passo La Calla, 1300 m	6.ix.1988	DRR	23°C	720/6F	1413		
	Same locality	6.ix.1988	DRR	22°C	721/2F	1417		
	ITALY: Calabria, Sila, near Bocca di Piazza, 1300 m	20.ix.1988	DRR	15°C	725/1F	1418		
	Same locality	20.ix.1988	DRR	16°C	725/4F	1416		
marocanus	MOROCCO: Haut Atlas, Oukaimeden, 2650 m	5.viii.1986	WJR	25°C	577S	1428	2	59 (2)
	Same locality	5.viii.1986	WJR	27°C	578S	1427, 1429	2	59 (1,3)

Appendix 3. (cont.)

Species	Locality	Date recorded	Recordist	Air temp.	Rec. No.	Fig. No.	CD No.	Track No.
Chorthippus (cont.) *binotatus*	FRANCE: Alpes-Maritimes, Col de Bleine, 1440 m	22.viii.1977	DRR	18°C	239/3F	1439	2	60(4)
	FRANCE: Haute-Vienne, near St-Yrieix-la-Perche, 1 km N of La Roche l'Abeille	30.viii.1978	WJR	20°C	269/1F	1436	2	60(2)
	SPAIN: Lérida, 7 km SE of Seo de Urgel, near Ortedo	11.ix.1978	WJR	26°C	272/11F	1440	2	60(3)
	FRANCE: Finistère, N side of Baie de Trépassé, N of Pointe du Raz	10.ix.1981	WJR	27°C	415S	1437		
	SPAIN: Madrid, Sierra de Guadarrama, Puerto de la Morcuera, 1796 m	15.ix.1982	WJR	27°C	469S	1441	2	60(1)
	SPAIN: Granada, Sierra Nevada, near Dornajo, 1900 m	30.vii.1983	DRR	27°C	504/5F	1438		
	SPAIN: Ávila, Sierra de Gredos, Puerto del Pico, 1400 m	4.viii.1983	DRR	22°C	507/5F	1442		
apicalis	SPAIN: Madrid, Puerto de Galapagar, 800 m	17.vi.1984	DRR	28°C	531/2F	1458		
	Same locality	17.vi.1984	DRR	30°C	531/5F	1459	2	61
	SPAIN: Madrid, Puerto de Navafría, 1773 m	16.vii.1989	DRR	25°C	763/3F	1460		
jucundus	FRANCE: Vaucluse, near Carpentras, Bédoin	16.viii.1977	DRR	31°C	238/1F	1467		

	Locality	Date	Recordist	Temp.		Recording no.		Figure
albomarginatus	SPAIN: Madrid, near Guadarrama, Embalse de la Jarosa	7.viii.1983	DRR	25°C	510/3F	1468	2	62
	AUSTRIA: Burgenland, Neusiedler See, near Illmitz, Fuchslochlacke	23.viii.1973	DRR	25°C	103/3F	1474	2	63 (2)
	Same locality	23.viii.1973	DRR	25°C	103/5F	1473		
	ENGLAND: Kent, Dartford Heath	14.viii.1978	WJR	25°C	261S	1475	2	63 (1)
	Same locality	14.viii.1978	WJR	27°C	262S	1485	2	63 (3)
	Same locality	15.viii.1978	WJR	32°C	264S	1476		
	ENGLAND: Surrey, near Mytchett Lake	25.vii.1979	WJR	34°C	306S	1486		
dorsatus	GERMANY: Upper Bavaria, near Chiemsee, Rottau	14.viii.1973	DRR	24°C	99/1F	1490		
	FRANCE: Var, near Comps, Bargème	4.ix.1975	WJR	29°C	154S	1491	2	64 (1)
	ITALY: Lazio, near Rome, SE side of Colli Albani	8.ix.1985	DRR	26°C	606/7F	1492		
	SPAIN: Jaén, Sierra de Segura, 7 km W of Santiago de la Espada, 1600 m	6.ix.1986	WJR	27°C	645/6F	1494		
	Same locality	6.ix.1986	WJR	27°C	645/7F	1493	2	64 (2)
dichrous	AUSTRIA: Burgenland, Neusiedler See, near Illmitz	18.viii.1973	DRR	30°C	101/4F	1508	2	65
	Same locality	18.viii.1973	DRR	30°C	101/5F	1507		
parallelus parallelus	GERMANY: Upper Bavaria, near Chiemsee, Rottau	14.viii.1973	DRR	24°C	99/4F	1513		
	ENGLAND: Surrey, near Caterham, Quarry Hangers	10.vii.1974	DRR & WJR	25°C	114S	1514	2	66

Appendix 3. (cont.)

Species	Locality	Date recorded	Recordist	Air temp.	Rec. No.	Fig. No.	CD No.	Track No.
Chorthippus parallelus parallelus (cont.)	ENGLAND: Kent, Chatham, Coney Hill	21.viii.1975	WJR	23°C	150S	1518		
	FRANCE: Calvados, near Bayeux, Forêt de Cerisy	8.ix.1976	WJR	29°C	179/2F	1519		
	FRANCE: Hautes-Pyrénées, near Vielle-Aure, Lac d'Orédon	18.ix.1976	WJR	10°C	181/10F	1515		
	FRANCE: Vaucluse, near Carpentras, Bédoin	13.viii.1977	DRR	26°C	235/7F	1516		
	ENGLAND: Greater London, near Coulsdon, Farthing Down	23.ix.1991	DRR	27.5°C	816S	1517		
p. erythropus	FRANCE: Pyrénées-Orientales, near Saillagouse, Val d'Eyne	21.ix.1976	WJR	20°C	183/5F	1534	2	67 (1)
	SPAIN: Huesca, Parque Nacional d'Ordesa. 200 m W of car park	5.ix.1978	WJR	22°C	271/1F	1535		
	ANDORRA: near St Julia de Loria, 1 km S of Fontaneda	13.ix.1978	WJR	21°C	275/2F	1538		
	SPAIN: Santander, near San Vicente, 1 km E of Unquera	25.viii.1984	WJR	27°C	551/2F	1539	2	67 (2)
	SPAIN: Granada, Sierra Nevada, 4 km N of Capileira, 2050 m	3.ix.1986	WJR	23°C	643/5F	1536		

	Locality	Date		Temp.				
	SPAIN: Ávila, Sierra de Gredos, N of Puerto del Pico. 1300 m	22.vii.1989	DRR	22°C	765/1F	1540		
	SPAIN: Navarra, near Leiza, N of Puerto de Usateguieta. 800 m	4.vii.1990	DRR	17°C	803/1F	1537		
montanus	GERMANY: Upper Bavaria, near Chiemsee, Rottau	14.viii.1973	DRR	24°C	99/3F	1557		
	Same locality	14.viii.1973	DRR	24°C	100/2F	1556		
	AUSTRIA: Carinthia, W of Pressegger See	3.ix.1992	DRR	21°C	843/7F	1555	2	68
Euchorthippus declivus	FRANCE: Lozère, near Mende, Col de Montmirat	10.viii.1977	DRR	21°C	232/2F	1564	2	69
	FRANCE: Vaucluse, Mont Ventoux, road from Bédoin, 1100 m	12.viii.1977	DRR	18°C	233/6F	1565		
	ITALY: Marches, Monti Sibillini, Forca di Presta, 1540 m	14.ix.1985	DRR	18°C	609/11F	1566		
pulvinatus pulvinatus	MACEDONIA: Titov Veles, Lake Mladost	30.vii.1987	FW	24°C	676/7R	1579		
p. gallicus	FRANCE: Dordogne, near Le Bugue, Campagne	24.viii.1974	DRR	30°C	134/3F	1573		
	FRANCE: Vaucluse, near Carpentras, Bédoin	13.viii.1977	DRR	22°C	234/8F	1574	2	70
	SPAIN: Valencia, near Requena, 700 m	31.vii.1979	WJR	32°C	313/1S	1575		
p. elegantulus	FRANCE: Loire-Atlantique, near Guérande, Le Croisic	12.ix.1976	WJR	20°C	180/1R	1577		

Appendix 3. (cont.)

Species	Locality	Date recorded	Recordist	Air temp.	Rec. No.	Fig. No.	CD No.	Track No.
Euchorthippus p. elegantulus (cont.)	FRANCE: Loire-Atlantique, near Guérande, Marais Salants	12.ix.1976	WJR	20°C	180/2R	1578		
	CHANNEL ISLANDS: Jersey, Quennevais	8.viii.1977	WJR	27°C	212S	1576		
chopardi	FRANCE: Vaucluse, c. 3 km SE of Cavaillon	14.viii.1977	DRR	28°C	236/6F	1594		
	SPAIN: Valencia, near Requena, 700 m	30.vii.1979	WJR	30°C	311/1S	1595	2	71
	SPAIN: Granada, N of Huescar, Puerto del Pinar, 1600 m	5.ix.1986	WJR	23°C	645/3F	1596		
albolineatus siculus	SICILY: Monti Iblei, Buccheri, Pineta Santa Maria	14.ix.1988	DRR	26°C	724/1F	1603		
	SICILY: Monti Madonie, near Piano Zucchi, 900 m	17.ix.1988	DRR	17°C	724/3F	1604	2	72
angustulus	BALEARIC ISLANDS: Majorca, Palma Nova, near Oratorio	16–18.ix.1981	WJR	29°C	417S	1609	2	73 (2)
	BALEARIC ISLANDS: Formentera, C'an Miguel Marti	12.ix.1983	WJR	29°C	499S	1610	2	73 (1)
	BALEARIC ISLANDS: Ibiza, 1 km SE of Portinatx	15.ix.1983	WJR	26°C	512/2F	1611		
sardous	SARDINIA: Nuoro, Monti del Gennargentu, Monti d'Iscudu, 1400 m	13.vii.1990	WJR	22°C	793/5F	1618	2	74

Species	Locality	Date	Recorder	Temp	Tape	No.		Page
Cicada orni	SARDINIA: Nuoro, Monti del Gennargentu, 3 km NE of Aritzo, 1100 m	14.vii.1990	WJR	23°C	794/6F	1619		
	SARDINIA: Sassari, Monti Limbara, 5 km N of summit, 1200 m	17.vii.1990	WJR	28°C	795/4F	1620		
Lyristes plebejus	FRANCE: Dordogne, near Le Bugue, Campagne	24.viii.1974	DRR	27°C	133/7F	1627	2	75
	FRANCE: Vaucluse, near Carpentras, St-Pierre-de-Vassol	8.viii.1975	MFC	—	192/8F	1628	2	76
Tettigetta argentata	SPAIN: Huesca, between Bielsa and Ainsa	24.vii.1983	DRR	27°C	501/7F	1629	2	77
Cicadetta montana	ENGLAND: Hampshire, New Forest	26.vii.1972	JAG	—	848F	1630	2	78
Bufo viridis	GREECE: Arta	iv.1960	JCR	—	847F	1631	2	79
Alytes obstetricans	FRANCE: Charente, near Angoulême, Roullet	30.vi.1984	DRR	23°C	534FD	1634	2	80
Locustella naevia	ENGLAND: Norfolk, near Brandon, Weeting	20.vii.1979	RM	—	828/1F	1637	2	81
luscinioides	NETHERLANDS: Flevoland, c. 10 km SW of Lelystad	5.v.1982	PADH	—	828/2F	1638	2	82
fluviatilis	HUNGARY	—	PS	—	828/3F	1645	2	83
lanceolata	RUSSIA: Bratsk	vi.1988	LS	—	828/4F	1646	2	84
Caprimulgus europaeus	ENGLAND: Lincolnshire, Stapleford Wood	16.vii.1983	MI	—	824F	1653	2	85

Glossary

These definitions give the meanings of the terms as used in this book and are not always suitable for general use. Some of the meanings are more easily appreciated by referring to Figs 4–7 for acoustic terms or Figs 36 and 37 for morphological terms. Italicized terms used in the definitions have their own separate alphabetical entries. Latin group-names (e.g. Ensifera, Gomphocerinae) and their English derivatives (e.g. ensiferan, gomphocerine) are not included in this glossary; their meanings are made clear in the check-list on p. 485.

abdomen (adjective: **abdominal**) The *posterior* part of the body of an insect, behind the thorax.
alary Of the wings.
allopatric Not occurring in the same area.
amplitude The maximum deviation of an oscillation (e.g. a sound wave) from its mean value. The amplitude of audible sound waves is perceived by human ears as the loudness of the sound.
analogue recorder An instrument for making *analogue recordings* of sound.
analogue recording (on magnetic tape) A recording in which the sound is represented by a continuously variable magnetic charge (as opposed to a *digital recording*).
antennae The feelers, a pair of *appendages* arising on the head, near the eyes (Figs 36, 37).
anterior In or towards the front.
appendage Any part attached by a joint to the body or to any other main structure.
audiospectrogram A visual record of the *frequency spectrum* of a sound on a time-scale (obtained from an *audiospectrograph*).
audiospectrograph Equipment for producing an *audiospectrogram*.
auricle A flap sometimes partly concealing the *tympanum*.
binary digit One of the two digits of the binary scale, i.e. 0 and 1, represented in electronic circuits by 'off' and 'on'.
bioacoustics The science of sound-production and hearing in living organisms.
brachypterous With short wings.
calling song The song produced by an isolated male.
carina (plural: **carinae**) A keel or ridge.
carinula (plural: **carinulae**) A very fine *carina*.
carnivorous Feeding on fresh animal matter.
carrier wave The fundamental wave of a *resonant song* (Fig 5). (See also p. 28.)
cercus (plural: **cerci**) A paired structure at or near the tip of the *abdomen* (Figs 36, 37).
character displacement The enhancement of characters distinguishing two similar, partly *sympatric* species in the area(s) where they both occur.
clubbed Enlarged at the tip.
conspecific Belonging to the same species.

costal area An area of the wings, as shown in Fig. 37.

couplet A pair of mutually exclusive alternatives in a *dichotomous key*.

courtship song The special song produced by a male when close to a female.

crepitation (verb: **crepitate**) The production during flight (or by flight-like vibration of the wings while on the ground) of a rattling, whirring or buzzing sound; the sound so produced.

cross-vein A short wing-vein running across one of the wing-areas, connecting one longitudinal vein to another.

cryptic species *Sibling species* that are detected only when they are revealed by non-*morphological* studies.

cubitus-one (Cu$_1$) The *anterior* branch of the cubital vein, one of the main wing-veins (Fig. 37).

cubitus-two (Cu$_2$) The *posterior* branch of the cubital vein, one of the main wing-veins (Fig. 37).

denticulate With small teeth.

dichotomous key An identification key composed solely of *couplets*.

digital recorder An instrument for making *digital recordings* of sound.

digital recording A recording in which the sound is represented by numerous consecutive samples expressed in *binary digits* (as opposed to an *analogue recording*).

digitize To convert an *analogue recording* to a *digital recording*.

diplosyllable A *syllable* in which sound is produced by both opening and closing wing-strokes in the Ensifera, and both upward and downward leg-strokes in gomphocerine grasshoppers.

disc, pronotal The more or less flat *dorsal* part of the *pronotum* (Figs 36, 37).

distal Farthest from the centre of the body.

diurnal Active during the daytime.

DNA Deoxyribonucleic acid, a substance present in almost all living organisms and playing a key role in the transmission of hereditary characters.

dorsal Of or belonging to the upper surface of the body or one of its appendages.

dorsum The upper surface of the body.

echeme (pronounced eck´eem) A *first-order assemblage* of *syllables* (Figs 4–7). (See also p. 29.)

echeme-sequence A *first-order assemblage* of *echemes* (Figs 6, 7).

emarginate Concave or 'cut into' at the edge.

ethology The science of animal behaviour.

exponential relationship A relationship between two varying quantities in which one changes in proportion to the logarithm of the other.

F$_1$ First filial (generation); the immediate offspring produced by interbreeding (often between members of different species).

F$_2$ Second filial (generation); the immediate offspring produced by interbreeding between members of the first filial generation.

facultative Optional.

femur (plural: **femora**) The 'thigh' of the insect leg (Figs 36, 37).

filiform Thread-like, of even thickness.

first-order assemblage An assemblage one level in rank above its specified components.

foveolae A pair of shallow depressions, often quadrangular or triangular, on each side of the *vertex* in some grasshoppers.

frequency (of an oscillation) The repetition rate of the cycles of an oscillation (e.g. a sound wave). The frequency of audible sound waves is perceived by human ears as the pitch of the sound.

frequency spectrum The range of *frequencies* in a sound.

gene Each of the units of heredity by which characters are transmitted from parent to offspring.

gene flow The movement of *genes* from one population to another as a result of interbreeding.

gene pool The stock of *genes* in a freely interbreeding population.

genotype Genetic constitution.

harp An area of clear membrane in the male fore wings of true crickets and mole-crickets, similar in shape to the musical instrument (Figs 11, 12).

hemisyllable The sound produced by one unidirectional movement (opening, closing, upward or downward) of the fore wings or hind legs (Figs 4, 7). (See also p. 29.).

heterospecific Belonging to different species.

homologous Of the same origin and usually similar in basic structure or pattern.

homologue A counterpart of similar structure and development.

homonym Each of two or more identical scientific names for different *taxa* of the same rank.

intrasyllabic Within a *syllable*.

kilohertz (kHz) A unit of *frequency*: 1 kilohertz = one thousand cycles per second. 1 hertz (Hz) = one cycle per second.

knee The joint between the *femur* and *tibia*.

lateral carinae, pronotal Longitudinal *carinae* running along each side of the *pronotal disc*.

lateral lobes, pronotal The more or less vertical side-lobes of the *pronotum* (Figs 36, 37).

level, recording The strength of a recorded sound; the setting of the recording equipment control that determines this strength.

level, sound Loudness.

linear relationship A relationship between two varying quantities in which one changes in proportion to the other.

macropterous With fully developed wings.

macrosyllable In bush-cricket songs with *syllables* of contrasting duration, a longer (more normal) *syllable* (Fig. 4).

maxillary palps The more *anterior* of the two pairs of *palps*.

medial area An area of the wings, as shown in Fig. 37.

medial intercalary vein A longitudinal vein sometimes occurring in the *medial area* of grasshopper fore wings and often bearing stridulatory teeth or tubercles.

medial vein (M) One of the main wing-veins, as shown in Fig. 36.

median carina A longitudinal *carina* running along the mid-line.

metamorphosis The transformation that takes place in the development of some insects (e.g. caterpillar to butterfly).

metathorax The most *posterior segment* of the *thorax*.

metazona The rear part of the pronotal *disc*, often separated from the *prozona* by a transverse *sulcus*.

microsyllable In bush-cricket songs with *syllables* of contrasting duration, a shorter *syllable*, usually lasting less than 10 *ms* (Fig. 4).

mirror An area of clear membrane in the male fore wings of a bush-cricket or true cricket (Figs 8, 11).

mitochondria Structures occurring in most cells of living organisms, with important functions in respiration and energy exchange.

mitochondrial DNA (mtDNA) The *DNA* of the *mitochondria*.

monitor To listen to a sound picked up by a microphone and reproduced through headphones or loudspeaker, while making a recording of it.

morphology (The science of) the form or (usually external) structure of living organisms.

morphometry The measurement of external structures.

ms millisecond; one thousandth of a second.

nocturnal Active at night; occurring at night.

nymph The young stage of an insect that does not undergo a full *metamorphosis*, resembling an adult but without wings.

off-tape monitoring Listening to an instant play-back of a recorded sound at the same time as recording it.

olfactory Pertaining to the sense of smell.

oscillogram A visual record of the waveform (and hence *amplitude*) of a sound on a time scale (obtained from an *oscillograph*).

oscillograph Equipment for producing an *oscillogram*.

oscilloscope An instrument for displaying waveform on a screen.

over-modulation Recording at too high a *level*, thus causing distortion in the recorded sound.

ovipositor A structure at the tip of the female *abdomen* used for egg-laying.

palps Two pairs of small segmented *appendages* beneath the head (Figs 36, 37).

pheromone A substance produced by an animal to elicit a response from another, usually *conspecific*, animal.

phonotaxis (adjective: **phonotactic**) Directional movement in response to sound.

phonotaxonomy The *taxonomy* of animals as indicated by the sounds they produce.

phylogenetics (The science of) the evolutionary development (and hence relationships) of living organisms.

physiology (The science of) the function of living organisms and their organs.

phytophagous Feeding on plants.

plantula (plural: **plantulae**) A *ventral* flap at the base of the hind *tarsus* of some bush-crickets (Fig. 36).

plectrum A scraper on the fore wing of a bush-cricket, true cricket or mole-cricket, which is rubbed against a file on the other fore wing during singing (Fig. 8).

poikilothermic With a variable body temperature, affected by the temperature of the surrounding air and radiant heat.

polymorphic Occurring in several *sympatric* forms differing in genetically controlled characters.

posterior Behind; towards the hind end.

preoccupied name A scientific name already in use for a different *taxon* of the same rank; the junior (later in origin) of two *homonyms*.

pronotal disc See *disc, pronotal*.

pronotal lateral carinae See *lateral carinae, pronotal*.

pronotal lateral lobes See *lateral lobes, pronotal*.

pronotum The skeletal plate forming the top and sides of the *prothorax*, lying immediately behind the head (Figs 36, 37).

prosternum The *ventral* skeletal plate of the *prothorax*, lying between the bases of the fore legs.

prothorax The most *anterior segment* of the *thorax*.

proximal Nearest to the centre of the body.

prozona The front part of the pronotal *disc*, often separated from the *metazona* by a transverse *sulcus*.

radial area An area of the wings, as shown in Fig. 36.

radial vein (M) One of the main wing-veins, as shown in Fig. 37.

recording level See *level, recording*.

reproductive isolation The absence of interbreeding between populations, especially those of related *sympatric* species.

resonant song A song with an almost pure dominant *frequency*. If this frequency is sufficiently low, as is common in crickets, the song has a clear musical pitch. (See also p. 34.)

rivalry song The special song produced by two or more males reacting to one another.

rugose Roughened with minute wrinkles or similar irregularities.

s second.

scolopidium (plural: **scolopidia**) A *sensillum* specialized for the reception of sound or vibration.

segment A subdivision of the body or of an *appendage*.

sensillum (plural: **sensilla**) A small sensory receptor.

sibling species Species that are extremely similar to one another in *morphology*.

sister species Two species that are more closely related to each other (through recent common ancestry) than to any other species.

sonagram An *audiospectrogram* produced by a *sonagraph*.

sonagraph A particular kind of *audiospectrograph*, producing a *sonagram*.

sound level See *level, sound*.

spermatophore A capsule used for transferring sperm from the male to the female.

spiracle An external opening of the respiratory system.

stigma A spot, often rather indistinct, on the fore wings of some grasshoppers (Fig. 37).

stridulation (verb: **stridulate**) Sound production by rubbing one structure against another.

subcostal area An area of the wings, as shown in Fig. 37.

subgenital plate The *ventral* skeletal plate at the tip of the *abdomen* (Figs 36, 37).

subquadrate Approximately square in shape.

substrate The surface or material on which an organism moves or rests.

sulcus A groove.

supra-anal plate The *dorsal* flap at the tip of the *abdomen*, lying above the bases of the *cerci* (Fig. 36).

syllable The sound produced by one to-and-fro movement of the stridulatory apparatus (Figs 4–7). (See also p. 26.)

sympatric Occurring in the same area.

synonymize (Of scientific names) to treat as *synonymous*.

synonymous (Of scientific names) applying to the same *taxon*.

tarsus (plural: **tarsi**) The insect 'foot'.

taxon (plural: **taxa**) A taxonomic group of any rank (e.g. family, species).

taxonomy (The science of) classification.

tergite The *dorsal* skeletal plate of one *segment* of the body.

thorax The part of an insect body lying between the head and *abdomen*, bearing the legs and wings.

tibia (plural: **tibiae**) The 'shin' of the insect leg (Figs 36, 37).

titillators A pair of toughened, usually pointed structures at the tip of the abdomen of some male bush-crickets, lying near the bases of the *cerci* and often concealed.

trachea (plural: **tracheae**) An air-tube of the respiratory system (Figs 21, 22).

tremulation Rapid shaking.

truncate Ending in a straight edge as if cut off transversely.

tympanal aperture The external opening leading to the *tympanum*, when it is not completely exposed.

tympanum (plural: **tympana**) The membrane forming the outer part of the hearing organ (Figs 17–20).

type locality The locality of origin of the specimen chosen to bear the scientific name of a species or subspecies; usually the locality where a species or subspecies was originally found.

type species The species chosen to bear the scientific name of a genus or subgenus (thus determining the application of the name in any case of doubt); usually the species on which a genus was originally based.

ultrasound (adjective: **ultrasonic**) Sound above 20 *kHz* in *frequency*, and thus inaudible to human ears.

valve One of the paired components of the *ovipositor* or penis.

ventral Of or belonging to the under surface of the body or one of its appendages.

vertex The front part of the top of the head.

vestigial Almost completely absent.

viability Ability to live.

References

Ahlén, I. 1981. Ultraljud hos svenska vårtbitare. [Ultrasonics in songs of Swedish bush crickets (Orth., Tettigoniidae).] *Entomologisk Tidskrift* **102**: 27–41.

— 1982. *Songs of Swedish bush crickets (Orth., Tettigoniidae). Ultrasonics transformed to audible sounds.* Tape cassette. Uppsala (Swedish University of Agricultural Sciences).

Ahlén, I. & Degn, H.J. 1980. Lövvårtbitarens *Leptophyes punctatissima* sång. *Fauna och Flora* **75**: 265–312.

Ahmad, A. & Siddiqui, M.A. 1988. Physical processes and mechanism of sound production in field cricket *Gryllus bimaculatus* De Geer. *Entomon* **13**: 109–120.

Alexander, R.D. 1956. *A comparative study of sound production in insects, with special reference to the singing Orthoptera and Cicadidae of the eastern United States.* xviii + 529 pp. Ohio State University (unpublished thesis).

— 1957a. The song relationships of four species of ground crickets (Orthoptera: Gryllidae: *Nemobius*). *Ohio Journal of Science* **57**: 153–163.

— 1957b. The taxonomy of the field crickets of the eastern United States (Orthoptera: Gryllidae: *Acheta*). *Annals of the Entomological Society of America* **50**: 584–602.

— 1960. Sound communication in Orthoptera and Cicadidae [pp. 38–92]. *In* Lanyon, W.E. & Tavolga, W.N. [Eds], *Animal sounds and communication.* xiii + 443 pp. Washington (American Institute of Biological Sciences).

— 1961. Aggressiveness, territoriality, and sexual behavior in field crickets (Orthoptera: Gryllidae). *Behaviour* **17**: 130–223.

— 1962. The role of behavioral study in cricket classification. *Systematic Zoology* **11**: 53–72.

— 1967a. Acoustical communication in arthropods. *Annual Review of Entomology* **12**: 495–526.

— 1967b. *Singing insects. Four case histories in the study of animal species.* 86 pp. Chicago (Rand McNally).

Alexander, R.D., Pace, A.E. & Otte, D. 1972. The singing insects of Michigan. *Great Lakes Entomologist* **5**: 33–69.

Alexander, R.D. & Thomas, E.S. 1959. Systematic and behavioral studies on the crickets of the *Nemobius fasciatus* group (Orthoptera: Nemobiinae). *Annals of the Entomological Society of America* **52**: 591–605.

Allard, H.A. 1929a. Our insect instrumentalists and their musical technique. *Annual Report of the Board of Regents of the Smithsonian Institution* **1928**: 563–591.

— 1929b. The last Meadow Katydid; a study of its musical reactions to light and temperature (Orthoptera: Tettigoniidae). *Transactions of the American Entomological Society* **55**: 155–164.

— 1930. The chirping rates of the snowy tree cricket (*Oecanthus niveus*) as affected by external conditions. *Canadian Entomologist* **62**: 131–142.

Andrieu, A.-J. & Dumortier, B. 1963. *Chantes d'insectes.* 30 cm disc, 33 rpm. Neuilly-sur-Seine

(Pacific).

Andrieu, A.-J. & Dumortier, B. 1994. *Entomophonia. Chantes d'insectes.* Compact disc. Paris (Institut National de la Recherche Agronomique).

Audouin, [J.] V. & Brullé, [G.] A. 1835. *Histoire naturelle des insectes* 9, 415 pp. Paris (Pillot).

Azam, J. 1900. Description d'un Orthoptère nouveau de France. *Bulletin de la Société Entomologique de France* **1900**: 82–85.

— 1901. Catalogue synonymique et systématique des Orthoptères de France [fifth part]. *Miscellanea Entomologica* **9**: 145–160.

— 1907. Description d'un Orthoptère nouveau de France. *Bulletin de la Société Entomologique de France* **1907**: 262–264.

Baccetti, B. 1956. Notulae orthopterologicae. III. *Metrioptera caprai* n. sp. e *Chorthippus modestus* Ebner: Ortotteri endemici del Terminillo nuovi o poco noti. *Redia* **41**: 113–127.

Baier, L.J. 1930. Contribution to the physiology of the stridulation and hearing of insects. *Zoologische Jahrbücher (Zoologie und Physiologie)* **47**: 151–248.

Bailey, W.J. 1967. Further investigations into the function of the "mirror" in Tettigonioidea (Orthoptera). *Nature, London* **215**: 762–763.

— 1970. The mechanics of stridulation in bush crickets (Tettigonioidea, Orthoptera). I. The tegminal generator. *Journal of Experimental Biology* **52**: 495–505.

— 1972. The acoustic status of the species *Homorocoryphus nitidulus nitidulus* Scopoli (Tettigonioidea, Orthoptera) in southern Europe. *Zoologischer Anzeiger* **189**: 181–190.

— 1975. A review of the African species of *Ruspolia* Schulthess [Orthoptera Tettigonioidea]. *Bulletin de l'Institut Fondamental d'Afrique Noire* (A) **37**: 171–226.

— 1980. A review of Australian Copiphorini (Orthoptera: Tettigoniidae: Conocephalinae). *Australian Journal of Zoology* **27**: 1015–1049.

— 1990. The ear of the bushcricket [pp. 217–247]. *In* Bailey, W.J. & Rentz, D.C.F. [Eds], *The Tettigoniidae: biology, systematics and evolution.* ix + 395 pp. Bathurst, Australia (Crawford House Press).

— 1991a. *Acoustic behaviour of insects: an evolutionary perspective.* xv + 225 pp. London, etc. (Chapman & Hall).

— 1991b. Mate finding: selection on sensory clues [pp. 42–74]. *In* Bailey, W.J. & Ridsdill-Smith, J. [Eds], *Reproductive behaviour of insects.* [x +] 339 pp. London, etc. (Chapman & Hall).

Bailey, W.J. & Broughton, W.B. 1970. The mechanics of stridulation in bush crickets (Tettigonioidea, Orthoptera). II. Conditions for resonance in the tegminal generator. *Journal of Experimental Biology* **52**: 507–517.

Bailey, W.J. & Rentz, D.C.F. [Eds] 1990. *The Tettigoniidae: biology, systematics and evolution.* ix + 395 pp. Bathurst, Australia (Crawford House Press).

Bailey, W.J. & Robinson, D. 1971. Song as a possible isolating mechanism in the genus *Homorocoryphus* (Tettigonioidea, Orthoptera). *Animal Behaviour* **19**: 390–397.

Bauer, M. & Helversen, O. von. 1987. Separate localization of sound recognizing and sound producing neural mechanisms in a grasshopper. *Journal of Comparative Physiology* (A) **161**: 95–101.

Baumgartner, W.J. 1905. Observations on some peculiar habits of the mole-crickets. *Science, New York* **21** (New Series): 855.

— 1911. Observations on the Gryllidae: III. Notes on the classification and on some habits of certain crickets. *Kansas University Science Bulletin* **5**: 309–319.

Beier, M. 1955. Laubheuschrecken. *Neue Brehm-Bücherei* **159**, 48 pp.

— 1956. Feldheuschrecken. *Neue Brehm-Bücherei* **179**, 48 pp.

Bell, P.D. 1979. Acoustic attraction of herons by crickets. *Journal of the New York Entomological Society* **87**: 126–127.

Bellmann, H. 1985a. *Heuschrecken. Beobachten, bestimmen.* 210 pp. Melsungen (Neumann-Neudamm).

— 1985b. *Die Stimmen der heimischen Heuschrecken.* Tape cassette. Melsungen (Neumann-Neudamm).

— 1988. *A field guide to the grasshoppers and crickets of Britain and northern Europe.* 213 pp. London (Collins).

— 1993a. *Heuschrecken. Beobachten, bestimmen.* 349 pp. Augsburg (Naturbuch).

— 1993b. *Die Stimmen der heimischen Heuschrecken.* Tape cassette. Augsburg (Weltbild).

— 1993c. *Die Stimmen der heimischen Heuschrecken.* Compact disc. Augsburg (Weltbild).

Bellmann, H. & Luquet, G.C. 1995. Guide des sauterelles, grillons et criquets d'Europe occidentale. 383 pp. Lausanne & Paris (Delachaux et Niestlé).

Belwood, J.J. 1990. Anti-predator defences and ecology of neotropical forest katydids, especially the Pseudophyllinae [pp. 8–26]. *In* Bailey, W.J. & Rentz, D.C.F. [Eds], *The Tettigoniidae: biology, systematics and evolution.* ix + 395 pp. Bathurst, Australia (Crawford House Press).

Belwood, J.J. & Morris, G.K. 1987. Bat predation and its influence on calling behavior in neotropical katydids. *Science, Washington* **238**: 64–67.

Bennet-Clark, H.C. 1970a. The mechanism and efficiency of sound production in mole crickets. *Journal of Experimental Biology* **52**: 619–652.

— 1970b. A new French mole cricket, differing in song and morphology from *Gryllotalpa gryllotalpa* L. (Orthoptera: Gryllotalpidae). *Proceedings of the Royal Entomological Society of London* (B) **39**: 125–132.

— 1976. Le chant des insectes. *La Recherche* **70**: 731–740.

— 1984. Insect hearing: acoustics and transduction [pp. 49–82]. *In* Lewis, T. [Ed.], *Insect communication.* xvii + 414 pp. London, etc. (Academic Press).

— 1987. The tuned singing burrow of mole crickets. *Journal of Experimental Biology* **128**: 383–409.

Bentley, D.R. 1971. Genetic control of an insect neuronal network. *Science, Washington* **174**: 1139–1141.

— 1977. Control of cricket song patterns by descending interneurons. *Journal of Comparative Physiology* **116**: 19–38.

Bentley, D.R. & Hoy, R.R. 1972. Genetic control of the neuronal network generating cricket (*Teleogryllus Gryllus*) song patterns. *Animal Behaviour* **20**: 478–492.

Bentley, D. [R.] & Hoy, R.R. 1974. The neurobiology of cricket song. *Scientific American* **231**: 34–44.

Bentley, D.R. & Kutsch, W. 1966. The neuromuscular mechanism of stridulation in crickets (Orthoptera: Gryllidae). *Journal of Experimental Biology* **45**: 151–164.

Bérenguier, P. 1908. Notes orthoptérologiques. II. Biologie de l'*Isophya Pyrenæe* [sic], Serville. Variété *Nemausensis* (Nov.). *Bulletin de la Société d'Etude des Sciences Naturelles de Nîmes* **35**: 1–13.

Bertkau, P. 1879. Ueber den Tonapparat von *Ephippigera vitium. Verhandlungen des Naturhistorischen Vereines der Preussischen Rheinlande und Westfalens* **36**: 269–276.

Bessey, C.A. & Bessey, E.A. 1898. Further notes on thermometer crickets. *American Naturalist* **32**: 263–264.

Beukeboom, L. 1986. *De sprinkhanen van Nederland en Belgie.* 69 pp. Ghent (Jeugbondsuitgeverij).

Bigelow, R.S. 1960. Interspecific hybrids and speciation in the genus *Acheta* (Orthoptera, Gryllidae). *Canadian Journal of Zoology* **38**: 509–524.

Block, B.C. 1966. The relation of temperature to the chirp-rate of male snowy tree crickets, *Oecanthus fultoni* (Orthoptera: Gryllidae). *Annals of the Entomological Society of America* **59**: 56–59.

Blondheim, S.A. 1990. Patterns of reproductive isolation between the sibling grasshopper species *Dociostaurus curvicercus* and *D. jagoi jagoi* (Orthoptera: Acrididae: Gomphocerinae). *Transactions of the American Entomological Society* **116**: 1–65.

Bodson, L. 1976. La stridulation des cigales. Poésie grecque et réalité entomologique. *L'Antiquité Classique* **45**: 75–94.

Bolívar, I. 1873. Ortópteros de España nuevos ó poco conocidos. *Anales de la Sociedad Española de Historia Natural* **2**: 213–237.

— 1876. Sinópsis de los ortópteros de España y Portugal. Segunda parte. *Anales de la Sociedad Española de Historia Natural* **5**: 259–372.

— 1877. Sinópsis de los ortópteros de España y Portugal. Tercera parte. *Anales de la Sociedad Española de Historia Natural* **6**: 249–348.

— 1878. Analecta orthopterologica. *Anales de la Sociedad Española de Historia Natural* **7**: 423–470.

— 1884. [No title.] *Annales de la Société Entomologique de Belgique* **28**: cii–cvii.

— 1887. Especies nuevas ó críticas de ortópteros. *Anales de la Sociedad Española de Historia Natural* **16**: 89–114.

— 1893. Ad cognitionem Orthopterorum Europæ et confinium. *Actas de la Sociedad Española de Historia Natural* **22**: 22–26.

— 1897a. Insectos recogidos en Cartagena por D. José Sánchez Gómez. *Actas de la Sociedad Española de Historia Natural* **1897**: 166–174.

— 1897b. Catálogo sinóptico de los ortópteros de la fauna ibérica [second part]. *Annaes de Sciencias Naturaes* **4**: 203–232.

— 1898a. Catálogo sinóptico de los ortópteros de la fauna ibérica [third part]. *Annaes de Sciencias Naturaes* **5**: 1–48.

— 1898b. Catálogo sinóptico de los ortópteros de la fauna ibérica [fourth part]. *Annaes de Sciencias Naturaes* **5**: 121–152.

— 1908. Algunos ortópteros nuevos de España, Marruecos y Canarias. *Boletín de la Real Sociedad Española de Historia Natural* **8**: 317–334.

Bonnet, F.-R. 1995. *Guide sonore des sauterelles, grillons et criquets d'Europe occidentale*. Compact disc. Lausanne & Paris (Delachaux et Niestlé).

Borck, J.B. von. 1848. *Skandinaviens rätvingade insekters natural-historia*. xvi + 146 pp. Lund (Berling).

Bornhalm, D. 1991. Zur Biologie von *Bryodema tuberculata*. *Articulata* **6**: 9–16.

Bosc, [L.]. 1792a. *Acheta sylvestris*. *Actes de la Société d'Histoire Naturelle de Paris* **1**: 44.

— 1792b. *Locusta punctatissima*. *Actes de la Société d'Histoire Naturelle de Paris* **1**: 45.

Boyd, P. & Lewis, B. 1983. Peripheral auditory directionality in the cricket (*Gryllus campestris* L., *Teleogryllus oceanicus* Le Guillou). *Journal of Comparative Physiology* **153**: 523–532.

Brand, A.R. Kellogg, P.P. & Lutz, F.E. 1937. *Songs of insects*. 25 cm disc, 78 rpm. Issued privately in USA.

Brisout de Barneville, L. 1849. Catalogue des Acrididés que se trouvent aux environs de Paris. *Annales de la Société Entomologique de France* (2) **6** (1848): 411–425.

— 1856. Séance du 26 décembre 1855. *Bulletin de la Société Entomologique de France* (3) **3** (1855): cxiv–cxvi.

Brooks, M.W. 1881. Influence of temperature on the chirp of the cricket. *Popular Science Monthly* **20**: 268.

Broughton, W.B. 1952a. Recording of stridulation; possible responsiveness of female Acrididae (Orth.) to stridulation; stridulation of *Leptophyes punctatissima* (Bosc) (Orth., Tettigoniidae). *Entomologist's Monthly Magazine* **88**: 47.

— 1952b. Gramophone studies of the stridulation of British grasshoppers. *Journal of the South-west Essex Technical College and School of Art* **3**: 170–180.

— 1954. Oscillographe économique du schéma de modulation par un appareil de cout et de fonctionnement bon marché, utilisible avec les appareils enregistreurs à vitesse variable. *L'Onde Électrique* **34**: 204–211.

— 1955a. L'analyse de l'émission acoustique des Orthoptères à partir d'un enregistrement sur disque reproduit à des vitesses ralenties [pp. 82–88]. *In* Busnel, R.-G. [Ed.], Colloque sur l'acoustique des Orthoptères. *Annales des Épiphyties*, fascicule spécial de 1954, 448 pp.

— 1955b. Notes sur quelques caractères acoustiques de *Platycleis affinis* Fieber (Tettigoniidae) [pp. 203–247]. *In* Busnel, R.-G. [Ed.], Colloque sur l'acoustique des Orthoptères. *Annales des Épiphyties*, fascicule spécial de 1954, 448 pp.

— 1963a. Method in bio-acoustic terminology [pp. 3–24]. *In* Busnel, R.-G. [Ed.], *Acoustic behaviour of animals.* xx + 933 pp. Amsterdam, etc. (Elsevier).

— 1963b. Glossarial index [pp. 824–910]. *In* Busnel,R.-G. [Ed.], *Acoustic behaviour of animals.* xx + 933 pp. Amsterdam, etc. (Elsevier).

— 1964. Function of the 'mirror' in tettigonioid Orthoptera. *Nature, London* **201**: 949–950.

— 1965. Song learning among grasshoppers? *New Scientist* **27**: 338–341.

— 1972a. The grasshopper and the taxonomer. Part II. Acoustic aspects and the interpretation of results. *Journal of Biological Education* **6**: 333–340.

— 1972b. The grasshopper and the taxonomer. III. Keys to the species. *Journal of Biological Education* **6**: 385–395.

— 1976. Proposal for a new term 'echeme' to replace 'chirp' in animal acoustics. *Physiological Entomology* **1**: 103–106.

Broughton, [W.B.] & Lewis, [D.] B. 1979. Insect sound research at the Animal Acoustics Unit, City of London Polytechnic. *Recorded Sound* **74–75**: 77–80.

Broughton, W.B., Samways, M.J. & Lewis, D.B. 1975. Low-frequency sounds in non-resonant songs of some bush crickets (Orthoptera, Tettigonioidea). *Entomologia Experimentalis et Applicata* **18**: 44–54.

Brown, E.S. 1955. Mécanismes du comportement dans les émissions sonores chez les Orthoptères [pp. 168–174]. *In* Busnel, R.-G. [Ed.], Colloque sur l'acoustique des Orthoptères. *Annales des Épiphyties,* fascicule spécial de 1954, 448 pp.

Brown, L. [Ed.] 1993. *The new shorter Oxford English dictionary on historical principles.* xxviii + vii + 3801 pp. Oxford (Clarendon Press).

Brullé, A. 1832. Insectes. *Expédition scientifique de Morée* **3** (1) Zoologie, 2: 64–395.

Brunner von Wattenwyl, C. 1878. *Monographie der Phaneropteriden.* 401 pp. Vienna (Brockhaus).

Buchler, E.R. & Childs, S.B. 1981. Orientation to distant sounds by foraging big brown bats (*Eptesicus fuscus*). *Animal Behaviour* **29**: 428–432.

Bukhvalova, M.A. 1993a. Acoustic signals of *Arcyptera fusca fusca* (Pall.) and *A. fusca albogeniculata* Ikonn. (Orthoptera, Acrididae). *Vestnik Moskovskogo Universiteta* (Seriya 16, Biologiya) **1993** (1): 46–49. [In Russian with English summary.]

— 1993b. The song and morphological characters of some grasshoppers of the *Chorthippus biguttulus* group (Orthoptera, Acrididae) from Russia and adjacent territories. *Zoologicheskii Zhurnal* **72** (5): 55–65. [In Russian with English summary. English translation: 1995, *Entomological Review* **74** (1): 56–67.]

Bukhvalova, M.A. & Zhantiev, R.D. 1993. Acoustic signals in grasshopper communities (Orthoptera, Acrididae, Gomphocerinae). *Zoologicheskii Zhurnal* **72** (9): 47–62. [In Russian with English summary.]

Bull, C.M. 1979. The function of complexity in the courtship of the grasshopper *Myrmeleotettix maculatus. Behaviour* **69**: 201–216.

Burk, T. 1982. Evolutionary significance of predation on sexually signalling males. *Florida Entomologist* **65**: 90–104.

Burr, M. 1911. Orthoptères recueillis à Bagnoles-de-l'Orne. *Bulletin du Muséum d'Histoire Naturelle. Paris* **17**: 102–105.

— 1936. *British grasshoppers and their allies, a stimulus to their study.* xvi + 162 pp. London (Philip Allan; republished by Janson).

Burr, M., Campbell, B.P. & Uvarov, B.P. 1923. A contribution to our knowledge of the Orthoptera of Macedonia. *Transactions of the Entomological Society of London* **1923**: 110–169.

Burton, J.F. & Johnson, E.D.H. 1984. Insect, amphibian or bird? *British Birds* **77**: 87–104.

Burton, J.F. & Ragge, D.R. 1987. *Sound guide to the grasshoppers and allied insects of Great Britain and Ireland.* Tape cassette. Colchester (Harley Books).

Burton, M. 1969. *Insects. Sounds from the insect world introduced by Dr Maurice Burton.* 17·5 cm disc, 33 rpm. London (Macdonald Junior Reference Library, BPC Publishing).

Busnel, M.-C. 1953. Contribution à l'étude des émissions acoustiques des Orthoptères. 1ᵉʳ
Mémoire. Recherches sur les spectres de fréquence et sur les intensités. *Annales des Épiphyties* **4**:
333–421.

— 1955. Etude des chants et du comportement acoustique d'*Oecanthus pellucens* Scop. ♂ [pp. 175–
202]. *In* Busnel, R.-G. [Ed.], Colloque sur l'acoustique des Orthoptères. *Annales des Épiphyties*,
fascicule spécial de 1954, 448 pp.

— 1963. Caractérisation acoustique de populations d'*Ephippiger* écologiquement voisines
(Orthoptères; Ephippigeridae). *Annales des Épiphyties* **14**: 25–34.

Busnel, M.-C. & Busnel, R.-G. 1955. La directivité acoustique des déplacements de la femelle
d'*Oecanthus pellucens* Scop. [pp. 356–364]. *In* Busnel, R.-G. [Ed.], Colloque sur l'acoustique des
Orthoptères. *Annales des Épiphyties*, fascicule spécial de 1954, 448 pp.

Busnel, R.-G. 1955. Sur certains rapports entre le moyen d'information acoustique et le
comportement acoustique des Orthoptères [pp. 281–306]. *In* Busnel, R.-G. [Ed.], Colloque sur
l'acoustique des Orthoptères. *Annales des Épiphyties*, fascicule spécial de 1954, 448 pp.

Busnel, R.-G., Busnel, M.-C. & Dumortier, B. 1956. Relations acoustiques interspécifiques chez
les Ephippigères (Orthoptères, Tettigonidae). *Annales des Épiphyties* **7**: 451–469.

Busnel, R.-G., Busnel, M.-C. & Pasquinelly, F. 1954. Les constantes physiques des émissions
acoustiques du mâle d'*Oecanthus pellucens* Scop. *Comptes Rendus des Séances de la Société de
Biologie* **148**: 859–662.

Busnel, R.G. & Chavasse, P. 1951. Recherches sur les émissions sonores et ultra-sonores
d'Orthoptères nuisibles à l'agriculture: étude des fréquences. Note préliminaire. *Nuovo Cimento* **7**
(1950) (Supplemento): 470–486.

Busnel, R.-G., Dumortier, B. & Busnel, M.-C. 1956. Recherches sur le comportement
acoustique des Éphippigères (Orthoptères, Tettigoniidae). *Bulletin Biologique de la France et de la
Belgique* **90**: 219–286.

Busnel, R.-G. & Loher, W. 1953. Recherches sur le comportement de divers Acridoidea mâles
soumis à des stimuli acoustiques artificiels. *Comptes Rendus Hebdomadaires des Séances de
l'Académie des Sciences* **237**: 1557–1559.

Busnel, R.-G. & Loher, W. 1954*a*. Sur l'étude du temps de la réponse au stimulus acoustique
artificiel chez les *Chorthippus*, et la rapidité de l'intégration du stimulus. *Comptes Rendus des
Séances de la Société de Biologie* **148**: 862–865.

Busnel, R.-G. & Loher, W. 1954*b*. Mémoire acoustique directionelle du mâle de *Chorthippus
biguttulus* L. (Acrididae). *Comptes Rendus des Séances de la Société de Biologie* **148**: 993–995.

Busnel, R.-G. & Loher, W. 1954*c*. Recherches sur le comportement de divers mâles d'Acridiens à
des signaux acoustiques artificiels. *Annales des Sciences Naturelles* (Zoologie et Biologie Animale)
(11) **16**: 271–281.

Busnel, R.-G. & Loher, W. 1955. Recherches sur les actions de signaux acoustiques sur le
comportement de divers Acrididae mâles [pp. 365–394]. *In* Busnel, R.-G. [Ed.], Colloque sur
l'acoustique des Orthoptères. *Annales des Épiphyties*, fascicule spécial de 1954, 448 pp.

Busnel, R.-G. & Loher, W. 1956. Étude des caractères physiques réactogènes de signaux
acoustiques artificiels déclencheurs de phonotropismes chez les Acrididae [Orth.]. *Bulletin de la
Société Entomologique de France* **61**:52–60.

Busnel, R.-G., Pasquinelly, F. & Dumortier, B. 1955. La trémulation du corps et la transmission
aux supports des vibrations en résultant comme moyen d'information à courte portée des
Ephippigères mâle et femelle. *Bulletin de la Société Zoologique de France* **80**: 18–22.

Butlin, R. [K.] 1989. Reinforcement of premating isolation [pp. 158–174]. *In* Otte, D. & Endler, J.A.
[Eds], *Speciation and its consequences*. xiii + 679 pp. Sunderland, Massachusetts (Sinauer Associ-
ates).

Butlin, R.K. & Hewitt, G.M. 1985. A hybrid zone between *Chorthippus parallelus parallelus* and
Chorthippus parallelus erythropus (Orthoptera: Acrididae): behavioural characters. *Biological
Journal of the Linnean Society* **26**: 287–299.

Butlin, R.K. & Hewitt, G.M. 1986. Heritability estimates for characters under sexual selection in the grasshopper, *Chorthippus brunneus*. *Animal Behaviour* **34**: 1256–1261.

Butlin, R.K. & Hewitt, G.M. 1988. Genetics of behavioural and morphological differences between parapatric subspecies of *Chorthippus parallelus* (Orthoptera: Acrididae). *Biological Journal of the Linnean Society* **33**: 233–248.

Butlin, R.K., Hewitt, G.M. & Webb, S.F. 1985. Sexual selection for intermediate optimum in *Chorthippus brunneus* (Orthoptera: Acrididae). *Animal Behaviour* **33**: 1281–1292.

Cade, W. 1975. Acoustically orienting parasitoids: fly phonotaxis to cricket song. *Science* **190**: 1312–1313.

Canestrelli, P. 1981. Le casse di colmata della laguna media, a sud di Venezia. IX. La fauna ortotteroidea delle casse "A" e "B" (Ortotterofauna veneta: 3° contributo). *Lavori, Società Veneziana di Scienze Naturali* **6**: 13–31.

Cappe de Baillon, P. 1921. Note sur le mécanisme de la stridulation chez *Meconema varium* Fabr. [Orthopt. Phasgonuridae]. *Annales de la Société Entomologique de France* **90**: 69–80.

Caudell, A.N. 1906. The Cyrtophylli of the United States. *Journal of the New York Entomological Society* **14**: 32–45.

Cazurro y Ruiz, M. 1888. Enumeración de los Ortópteros de España y Portugal. *Anales de la Sociedad Española de Historia Natural* **17**: 435–513.

Charalambous, M. 1990. Is there heritable variation in female selectivity for male calling song in the grasshopper *Chorthippus brunneus* (Thunberg)? [pp. 255–268]. *In* Nickle, D.A. [Ed.], Proceedings of the 5th International Meeting of the Orthopterists' Society. *Boletín de Sanidad Vegetal. Plagas* (Fuera de Serie) **20**, xxx + 422 pp.

Charalambous, M., Butlin, R.K. & Hewitt, G.M. 1994. Genetic variation in male song and female song preference in the grasshopper *Chorthippus brunneus* (Orthoptera: Acrididae). *Animal Behaviour* **47**: 399–411.

Charpentier, T. de. 1825. *Horae entomologicae, adjectis tabulis novem coloratis.* xvi + 262 pp. Bratislava (A. Gosohorsky).

— 1841–45. *Orthoptera discripta et depicta.* iv [+ 116] pp. Leipzig (L. Voss).

Chavasse, P., Busnel, R.-G., Pasquinelly, F. & Broughton, W.B. 1955. Propositions de définitions concernant l'acoustique appliquée aux insectes [pp. 21–25]. *In* Busnel, R.-G. [Ed.], Colloque sur l'acoustique des Orthoptères. *Annales des Épiphyties*, fascicule spécial de 1954, 448 pp.

Chopard, L. 1922. Orthoptères et Dermaptères. *Faune de France* **3**, vi + 212 pp.

— 1923. Essai sur la faune des Orthoptères de la Corse. *Annales de la Société Entomologique de France* **92**: 253–286.

— 1939. Description d'une espèce nouvelle du genre *Omocestus* [Orth. Acrididae]. *Bulletin de la Société Entomologique de France* **44**: 172–173.

— 1952. Orthoptéroïdes. *Faune de France* **56** (1951), 359 pp.

— 1955. L'acoustique des Orthoptères. *La Nature* **83**: 425–429.

Clemente Espinosa, M.E. 1987. Revision de los generos *Stenobothrus* Fischer, 1853 *Omocestus* Bolívar, 1878 y *Myrmeleotettix* Bolívar, 1914 en la Península Ibérica (Orthoptera: Caelifera). [xii +] 349 pp. Universidad de Murcia (unpublished thesis).

Clemente, M.E., García, M.D. & Presa, J.J. 1986. Sobre la presencia de *Omocestus burri* Uvarov, 1936 (Orth. Gomphocerinae) en la Península Ibérica. *Boletín de la Asociación Española de Entomología* **10**: 189–197.

Clemente, M.E., García, M.D. & Presa, J.J. 1989a. Sobre la identidad taxonómica de *Omocestus burri* Uvarov, 1936, *O. knipperi* Harz, 1982 y *O. llorenteae* Pascual, 1978 (Orthoptera, Acrididae). *Boletín de la Asociación Española de Entomología* **13**: 99–108.

Clemente, M.E., García, M.D. & Presa, J.J. 1989b. Los Gomphocerinae de la Península Ibérica: I. *Stenobothrus* Fischer, 1853 y *Myrmeleotettix* Bolívar, 1914. *Graellsia* **45**: 35–74.

Clemente, M.E., García, M.D. & Presa, J.J. 1990. Los Gomphocerinae de la Península Ibérica: II.

Omocestus Bolívar, 1878. (Insecta, Orthoptera, Caelifera). *Graellsia* **46**: 191–246.

Clemente, M.E., García, M.D. & Presa, J.J. 1994. Descripción del canto de proclamación de *Chorthippus nevadensis* Pascual, 1978 (Orthoptera, Acrididae). *Boletín de la Real Sociedad Española de Historia Natural* (Sección Biológica) **91**: 199–202.

Corey, H.I. 1933. Chromosome studies in *Stauroderus* (an orthopteron). *Journal of Morphology* **55**: 313–347.

— 1937. Heteropycnotic elements of Orthopteran chromosomes. *Archives de Biologie* **49** (1938): 159–174.

Costa, A. 1863. Nuovi studii sulla entomologia della Calabria Ulteriore. *Atti dell'Accademia delle Scienze Fisiche e Matematiche* **1** (2), 80 pp.

Counter, S.A. 1976. Fourier and electrophysiological analyses of acoustic communication in *Acheta domesticus*. *Journal of Insect Physiology* **22**: 589–593.

Crankshaw, O.S. 1979. Female choice in relation to calling and courtship songs in *Acheta domesticus*. *Animal Behaviour* **27**: 1274–1275.

Currie, P.W.E. 1953. The 'drumming' of *Meconema thalassinum* Fabr. *Entomologist's Record and Journal of Variation* **65**: 93–94.

Dambach, M. & Huber, F. 1974. Perception of substrate-vibration in crickets. *Abhandlungen der Rheinisch-Westfälischen Akademie der Wissenschaften* **53**: 263–280.

Dambach, M. & Rausche, G. 1985. Low-frequency airborne vibrations in crickets and feed-back control of calling song [pp. 177–182]. *In* Kalmring, K. & Elsner, N. [Eds], *Acoustic and vibrational communication in insects*. viii + 230 pp. Berlin & Hamburg (Paul Parey).

Dambach, M., Rausche, H.-G. & Wendler, G. 1983. Perceptive feedback influences the calling song of the field cricket. *Naturwissenschaften* **70**: 417–418.

Dathe, H.H. 1974. Untersuchungen zum phonotaktischen Verhalten von *Gryllus bimaculatus* (Insecta, Orthopteroidea). *Forma et Functio* **7**: 7–20.

Davis, W.J. 1968. Cricket wing movements during stridulation. *Animal Behaviour* **16**: 72–73.

Defaut, B. 1979. Notes sur la morphologie et le chant des Orthoptères. 1. *Euthystira brachytera* [sic] (Ocsk.) et *Chrysochraon dispar* (Germ.) en Haut-Ariège. *Bulletin de la Société d'Histoire Naturelle de Toulouse* **115**: 232–241.

— 1981. Notes sur la morphologie et le chant des Orthoptères: 2. *Metrioptera buyssoni* (Saulcy) et *M. saussuriana* (Fr.-G.) dans les Pyrénées centrales. *Bulletin de la Société d'Histoire Naturelle de Toulouse* **117**: 117–124.

— 1983. Notes sur la morphologie et le chant des Orthoptères: 3. Relations entre température de l'air et stridulation ordinaire chez *Chorthippus p. parallelus* (Zett.). *Bulletin de la Société d'Histoire Naturelle de Toulouse* **119**: 53–57.

— 1984a. Sur la répartition d'*Uromenus rugosicollis* (Serville) [Ensifera, Ephippigerinae]. *L'Entomologiste* **40**: 119–122.

— 1984b. *Chorthippus yersini* Harz, espèce distincte de *Ch. biguttulus* (L.) (Caelifera, Acrididae). *Nouvelle Revue d'Entomologie* (Nouvelle Série) **1**: 299–301.

— 1987. *Recherches cénotiques et bioclimatiques sur les Orthoptères en région ouest paléarctique.* 522 pp. Université Paul Sabatier de Toulouse (unpublished thesis).

— 1988a. Notes sur la morphologie et le chant des Orthoptères: 4. Observations complémentaires sur les relations entre température de l'air et stridulation ordinaire chez *Chorthippus p. parallelus* (Zett.) (Orth., Acrididae). *Bulletin le la Société d'Histoire Naturelle de Toulouse* **124**: 25–28.

— 1988b. La détermination des Orthoptéroides ouest-paléarctiques. 6. Caelifera: Acrididae (suite). 7. Ensifera. 8. Mantodea. *Travaux du Laboratoire d'Ecobiologie des Arthropodes Edaphiques, Toulouse* **6** (1): 1–93.

De Geer, C. 1773. *Mémoires pour servir à l'histoire des insectes* **3**, viii + 697 pp. Stockholm (P. Hesselberg).

Deroussen, F. 1993. *Les chants de l'été.* Compact disc. Paris (Nashvert).

— 1994. *Sélection nature. I, France.* Compact disc. Paris (Nashvert).

Descamps, M. 1968. Notes sur le genre *Euchorthippus* [Orth. Acrididae]. Sa répartition dans le Vaucluse et les départements adjacents. *Annales de la Société Entomologique de France* (Nouvelle Série) **4**: 5–25.

Dethier, V.G. 1992. *Crickets and katydids, concerts and solos*. xiv + 140 pp. Cambridge, Massachusetts & London (Harvard University Press).

Doherty, J.A. 1985. Temperature coupling and 'trade-off' phenomena in the acoustic communication system of the cricket, *Gryllus bimaculatus* De Geer (Gryllidae). *Journal of Experimental Biology* **114**: 17–35.

Doherty, J.A. & Callos, J.D. 1991. Acoustic communication in the trilling field cricket, *Gryllus rubens* (Orthoptera: Gryllidae). *Journal of Insect Behavior* **4**: 67–82.

Doherty, J.A. & Huber, F. 1983. Temperature effects on acoustic communication in the cricket *Gryllus bimaculatus* DeGeer. *Verhandlungen der Deutschen Zoologischen Gesellschaft* **1983**: 188.

Dolbear, A.E. 1897. The cricket as a thermometer. *American Naturalist* **31**: 970–971.

Dubrovin, N.N. 1977. Frequency characteristics of sound signals in katydids (Orthoptera, Tettigoniidae). *Zoologicheskii Zhurnal* **56**: 380–385. [In Russian with English summary.]

Dubrovin, N.N. & Zhantiev, R.D. 1970. Acoustic signals of katydids (Orthoptera, Tettigoniidae). *Zoologicheskii Zhurnal* **49**: 1001–1014. [In Russian with English summary.]

Dufour, L. 1820. Description de dix espèces nouvelles ou peu connues d'insectes recueillis en Espagne. *Annales Générales des Sciences Physiques, Bruxelles* **6**: 307–331.

Duijm, M. 1990. On some song characteristics in *Ephippiger* (Orthoptera, Tettigonioidea) and their geographic variation. *Netherlands Journal of Zoology* **40**: 428–453.

Duijm, M. & Kruseman, G. 1983. *De krekels en sprinkhanen in de Benelux.* 186 pp. Amsterdam (Koninklijke Nederlandse Natuurhistorische Vereniging).

Duijm, M. & Oudman, L. 1983. Interspecific matings in *Ephippiger* (Orthoptera, Tettigonioidea). *Tijdschrift voor Entomologie* **126**: 97–108.

Duijm, M. & van Oyen, T. 1948. Het sjirpen van de zadelsprinkhaan. *Die Levende Natuur* **51**: 81–87.

Dumortier, B. 1956. A propos d'un film: Le monde sonore des Sauterelles. *La Nature* **84**: 224–230.

— 1963a. Etude expérimentale de la valeur interspécifique du signal acoustique chez les Éphippigères et rapport avec les problèmes d'isolement et de maintien de l'espèce (Orthoptères, Ephippigeridae). *Annales des Épiphyties* **14**: 5–23.

— 1963b. Morphology of sound emission apparatus in Arthropoda [pp. 277–345]. *In* Busnel, R.-G. [Ed.], *Acoustic behaviour of animals.* xx + 933 pp. Amsterdam, etc. (Elsevier).

— 1963c. The physical characteristics of sound emissions in Arthropoda [pp. 346–373]. *In* Busnel, R.-G. [Ed.], *Acoustic behaviour of animals.* xx + 933 pp. Amsterdam, etc. (Elsevier).

— 1963d. Ethological and physiological study of sound emissions in Arthropoda [pp. 583–654]. *In* Busnel, R.-G. [Ed.], *Acoustic behaviour of animals.* xx + 933 pp. Amsterdam, etc. (Elsevier).

Dumortier, B., Brieu, S. & Pasquinelly, F. 1957. Facteurs externes contrôlant le rythme des périodes de chant chez *Ephippiger ephippiger* Fiebig mâle (Orthoptère-Tettigonoidea). *Comptes Rendus Hebdomadaires des Séances de l'Académie des Sciences* **244**: 2315–2318.

Ebner, R. 1915. Zur Kenntnis der Orthopterenfauna der Abruzzen. *Deutsche Entomologische Zeitschrift* **1915**: 545–570.

— 1925. Biologische Beobachtungen an *Pycnogaster bolivari* Br.-W. (Orthoptera). *Zeitschrift für wissenschaftliche Zoologie* **125**: 357–363.

Edes, R.T. 1899. Relation of the chirping of the tree cricket (*Oecanthus niveus*) to temperature. *American Naturalist* **33**: 935–938.

Eiríksson, T. 1992. Density dependent song duration in the grasshopper *Omocestus viridulus*. *Behaviour* **122**: 121–132.

— 1993. Female preference for specific pulse duration of male songs in the grasshopper, *Omocestus viridulus*. *Animal Behaviour* **45**: 471–477.

Elliott, S. 1986. *The sounds of Britain's endangered wildlife.* Tape cassette. London (The British

Library National Sound Archive).

Elliott, C.J.H. & Koch, U.T. 1983. Sensory feedback stabilizing reliable stridulation in the field cricket *Gryllus campestris* L. *Animal Behaviour* **31**: 887–901.

Elliott, C.J.H. & Koch, U.T. 1985. The clockwork cricket. *Naturwissenschaften* **72**: 150–153.

Elsner, N. 1968. Die neuromuskalären Grundlagen des Werbeverhaltens der Roten Keulenheuschrecke *Gomphocerippus rufus* (L.). *Zeitschrift für vergleichende Physiologie* **60**: 308–350.

— 1970. Die Registrierung der Stridulationsbewegungen bei der Feldheuschrecke *Chorthippus mollis* mit Hilfe von Hallgeneratoren. *Zeitschrift für vergleichende Physiologie* **68**: 417–428.

— 1973. The central nervous control of courtship behaviour in the grasshopper *Gomphocerippus rufus* L. (Orthoptera: Acrididae) [pp. 261–287]. *In* Salánki, J. [Ed.], *Neurobiology of invertebrates. Mechanisms of rhythm regulation.* 494 pp. Budapest (Akadémiai Kiadó).

— 1974*a*. Neuroethology of sound production in gomphocerine grasshoppers (Orthoptera: Acrididae). I. Song patterns and stridulatory movements. *Journal of Comparative Physiology* **88**: 67–102.

— 1974*b*. Neural economy: bifunctional muscles and common central pattern elements in leg and wing stridulation of the grasshopper *Stenobothrus rubicundus* Germ. (Orthoptera: Acrididae). *Journal of Comparative Physiology* **89**: 227–236.

— 1975. Neuroethology of sound production in gomphocerine grasshoppers (Orthoptera: Acrididae). II. Neuromuscular activity underlying stridulation. *Journal of Comparative Physiology* **97**: 291–322.

— 1983*a*. Insect stridulation and its neurophysiological basis [pp. 69–92]. *In* Lewis, B. [Ed.], *Bioacoustics – a comparative approach.* x + 493 pp. London, etc. (Academic Press).

— 1983*b*. A neuroethological approach to the phylogeny of leg stridulation in gomphocerine grasshoppers [pp. 54–68]. *In* Huber, F. & Markl, H. [Eds], *Neuroethology and behavioural physiology.* xviii + 412 pp. Berlin, etc. (Springer-Verlag).

Elsner, N. & Hirth, C. 1978. Short- and long-term control of motor coordination in a stridulating grasshopper. *Naturwissenschaften* **65**: 160.

Elsner, N. & Huber, F. 1969. Die Organisation des Werbegesanges der Heuschrecke *Gomphocerippus rufus* L. in Abhängigkeit von zentralen und peripheren Bedingungen. *Zeitschrift für vergleichende Physiologie* **65**: 389–423.

Elsner, N. & Popov, A.V. 1978. Neuroethology of acoustic communication. *Advances in Insect Physiology* **13**: 229–355.

Elsner, N. & Wasser, G. 1995*a*. Leg and wing stridulation in various populations of the gomphocerine grasshopper *Stenobothrus rubicundus* (Germar 1817), I. Sound patterns and singing movements. *Zoology* **98**: 179–190.

Elsner, N. & Wasser, G. 1995*b*. Leg and wing stridulation in various populations of the gomphocerine grasshopper *Stenobothrus rubicundus* (Germar 1817), II. Neuromuscular mechanisms. *Zoology* **98**: 191–199.

Elsner, N. & Wasser, G. 1995*c*. The transition from leg to wing stridulation in two geographically distinct populations of the grasshopper *Stenobothrus rubicundus*. *Naturwissenschaften* **82**: 384–386.

Eversmann, E. 1848. *Additamenta quaedam levia ad Fischer de Waldheim celeberrimi Orthoptera Rossica.* 15 pp. Moscow (Semen).

— 1859. Orthoptera Volgo-Uralensia. *Bulletin de la Société Impériale des Naturalistes de Moscou* **32** (1): 121–146.

Ewing, A.W. 1984. Acoustic signals in insect sexual behaviour [pp. 223–240]. *In* Lewis, T. [Ed.], *Insect communication.* xvii + 414 pp. London, etc. (Academic Press).

— 1989. *Arthropod bioacoustics: neurobiology and behaviour.* x + 260 pp. Edinburgh (University Press).

Ewing, A.[W.] & Hoyle, G. 1965. Neuronal mechanisms underlying control of sound production in a cricket: *Acheta domesticus. Journal of Experimental Biology* **43**: 139–153.

Faber, A. 1928. Die Bestimmung der deutschen Geradflügler (Orthopteren) nach ihren Lautäusserungen. *Zeitschrift für Wissenschaftliche Insektenbiologie* **23**: 209–234.

— 1929*a. Chorthippus longicornis* Latr. (= *parallelus* Zett.) und *Chorthippus montanus* Charp. (bisher nach Finot als "*longicornis* Latr." bezeichnet). *Zoologischer Anzeiger* **81**: 1–24.

— 1929*b.* Die Lautäusserungen der Orthopteren. (Lauterzeugung, Lautabwandlung und deren biologische Bedeutung sowie Tonapparat der Geradflügler.) Vergleichende Untersuchungen I. *Zeitschrift für Morphologie und Ökologie der Tiere* **13**: 745–803.

— 1932. Die Lautäusserungen der Orthopteren II. (Untersuchungen über die biozönotischen, tierpsychologischen und vergleichend-physiologischen Probleme der Orthopterenstridulation. Methodik der Bearbeitung und Auswertung von Stridulationsbeobachtungen. Einzeldarstellungen.) *Zeitschrift für Morphologie und Ökologie der Tiere* **26**: 1–93.

— 1934. Neue Untersuchungen über die Lautäußerungen der Geradflügler (Orthopteren). *Der Biologe* **3**: 249–254.

— 1936. Die Laut- und Bewegungsäußerungen der Oedipodinen. *Zeitschrift für Wissenschaftliche Zoologie* **149**: 1–85.

— 1953*a. Laut- und Gebärdensprache bei Insekten. Orthoptera (Geradflügler). Teil I. Vergleichende Darstellung von Ausdrucksformen als Zeitgestalten und ihren Funktionen.* 198 pp. Stuttgart (Staatliches Museum für Naturkunde).

— 1953*b.* Ausdrucksbewegung und besonders Lautäußerung bei Insekten als Beispiel für eine vergleichend-morphologische Betrachtung der Zeitgestalten. *Zoologischer Anzeiger* Supplementband **17** (Verhandlungen der Deutschen Zoologischen Gesellschaft vom 2. bis 8. Juni 1952 in Freiburg): 106–115.

— 1957*a.* Über den Aufbau von Gesangsformen in der Gattung *Chorthippus* Fieb. (Orthoptera) und über phylogenetische Gemeinsamkeiten bei Stridulations- und anderen Bewegungsformen. *Stuttgarter Beiträge zur Naturkunde aus dem Staatlichen Museum für Naturkunde in Stuttgart* **1957** (1): 1–28.

— 1957*b.* Über parallele Abänderungen bei Lautäußerungen von Grylliden mit Bemerkungen zur Frage der Phylogenie von Stridulationsformen. *Stuttgarter Beiträge zur Naturkunde aus dem Staatlichen Museum für Naturkunde in Stuttgart* **1957** (2): 1–10.

— 1958. *Chorthippus erythropus* n. sp., ein nächster Verwandter der Gemeinen Grasschrecke [*Ch. longicornis* (Latr.); Orthopt., Acrid.]. *Stuttgarter Beiträge zur Naturkunde aus dem Staatlichen Museum für Naturkunde in Stuttgart* **1958** (16): 1–8.

— 1960. Form, Ableitung und Bedeutung von Stridulationsweisen im Verwandtschaftskreis um *Chorthippus longicornis* (Latr.) als grundsätzliches Beispiel der Vergleichung. *Chorthippus caffer* – eine selbständige Art (Orthoptera, Acridiidae). *Stuttgarter Beiträge zur Naturkunde aus dem Staatlichen Museum für Naturkunde in Stuttgart* **1960** (32): 1–12.

Fabricius, J.C. 1775. *Systema entomologiae, sistens insectorum classes, ordines, genera, species, adiectis synonymis, locis, descriptionibus, observationibus.* [xxxii +] 832 pp. Flensburg & Leipzig (Korte).

— 1787. *Mantissa insectorum sistens eorum species nuper detectas adiectis characteribus genericis, differentiis specificis, emendationibus, observationibus.* xx + 348 pp. Copenhagen (C.G. Proft).

— 1793. *Entomologia systematica emendata et aucta* **2**, viii + 519 pp. Copenhagen (C.G. Proft).

— 1798. *Supplementum Entomologiae Systematicae.* [iv +] 572 pp. Copenhagen (Proft & Storch).

Fieber, F.X. 1853. Synopsis der europäischen Orthoptera. *Lotos* **3**: 90–104, 115–129, 138–154, 168–176, 184–188, 201–207, 232–238, 252–261.

Fiebig, J. 1784. Beschreibung des Satelträgers (*Gryllus Ephippiger.*). *Schriften der Berlinischen Gesellschaft Naturforschender Freunde* **5**: 260–263.

Fischer, H. 1849. Beiträge zur Insekten-Fauna um Freiburg im Breisgau. Orthoptera. *Jahresbericht des Mannheimer Vereins für Naturkunde* **15**: 23–51.

— 1850. Beiträge zur Insekten-Fauna um Freiburg im Breisgau. (Erste Fortsetzung.) Orthoptera. *Jahresbericht des Mannheimer Vereins für Naturkunde* **16**: 25–40.

— 1853. *Orthoptera Europaea*. xx + 454 pp. Leipzig (Engelmann).

Fischer de Waldheim, G. 1833. Conspectus Orthopterorum Rossicorum. *Bulletin de la Société Impériale des Naturalistes de Moscou* **6**: 341–390.

— 1846. Orthoptères de la Russie. *Nouveaux Mémoires de la Société des Naturalistes de Moscou* **8**, iv + 413 pp.

Forrest, T.G. 1982. Acoustic communication and baffling behaviors of crickets. *Florida Entomologist* **65**: 33–44.

Fowler, H.G. & Garcia, C.R. 1987. Attraction to synthesized songs and experimental and natural parasitism of *Scapteriscus* mole crickets (Orthoptera: Gryllotalpidae) by *Euphasiopteryx depleta* (Diptera: Tachinidae). *Revista Brasileira de Biologia* **47**: 371–374.

Frey-Gessner, E. 1872. Orthopterologisches. *Mittheilungen der Schweizerischen Entomologischen Gesellschaft* **4**: 7–20.

Frings, H. & Frings, M. 1957. The effects of temperature on chirp-rate of male cone-headed grasshoppers, *Neoconocephalus ensiger*. *Journal of Experimental Zoology* **134**: 411–425.

Frings, H. & Frings, M. 1962. Effects of temperature on the ordinary song of the common meadow grasshopper, *Orchelimum vulgare* (Orthoptera: Tettigoniidae). *Journal of Experimental Zoology* **151**: 33–51.

Frings, M. & Frings, H. 1960. *Sound production and sound reception by insects – a bibliography*. 108 pp. Pennsylvania (State University Press).

Frivaldsky, J. 1867. A magyarországi egyenesröpüek magánrajza. (Monographia Orthopterorum Hungariae). *Ertekezések a Természettudományi osztály Köréböl* **1** (12), 201 pp.

Froehlich, C. 1989. Freilanduntersuchungen an Heuschrecken (Orthoptera: Saltatoria) mit Hilfe des Fledermausdetektors. Neue Erfahrungen. *Articulata* **4**: 6–10.

Fuessly, J.C. 1775. *Verzeichnis der ihm bekannten schweizerischen Insekten*. xii + 62 pp. Zürich & Winterthur (H. Steiner).

Fulton, B.B. 1925. Physiological variation in the Snowy Tree-cricket, *Oecanthus niveus* De Geer. *Annals of the Entomological Society of America* **18**: 363–383.

— 1930. Notes on Oregon Orthoptera with descriptions of new species and races. *Annals of the Entomological Society of America* **23**: 611–641.

— 1931. A study of the genus *Nemobius*. (Orthoptera: Gryllidae). *Annals of the Entomological Society of America* **24**: 205–237.

— 1932. North Carolina's singing Orthoptera. *Journal of the Elisha Mitchell Scientific Society* **47**: 55–69.

— 1933. Inheritance of song in hybrids of two subspecies of *Nemobius fasciatus* (Orthoptera). *Annals of the Entomological Society of America* **26**: 368–376.

— 1952. Speciation in the field cricket. *Evolution* **6**: 283–295.

Galvagni, A. 1955. Descrizione dello *Ephippiger Ruffoi* n. sp. (Orthoptera-Ephippigeridae). *Memorie del Museo Civico di Storia Naturale di Verona* **5**: 39–44.

García, M.D., Abellán, J., Clemente, M.E. & Presa, J.J. 1987. Consideraciones sobre el comportamiento acústico de *Arcyptera fusca fusca* (Pallas, 1773) y *A. microptera microptera* (Fischer Waldheim, 1833) (Orthoptera: Acrididae). *Anales de Biología* **11** (Biología Animal **3**): 81–89.

García, M.D., Clemente, M.E. & Presa, J.J. 1994. The acoustic behaviour of *Dociostaurus jagoi occidentalis* Soltani, 1978 (Orthoptera, Acrididae). *Zoologica Baetica* **5**: 79–87.

García, M.D., Clemente, M.E. & Presa, J.J. 1995. Manifestaciónes acústicas de *Chorthippus binotatus binotatus* (Charpentier, 1825) (Orthoptera: Acrididae). Su estatus taxonómico y su distribución en la Península Ibérica. *Boletín de la Asociación Española de Entomología* **19**: 229–242.

Gerhardt, U. 1914. Copulation und Spermatophoren von Grylliden und Locustiden. II. *Zoologische Jahrbücher* (Systematik) **37**: 1–64.

— 1921. Neue Studien über Copulation und Spermatophoren von Grylliden und Locustiden. *Acta*

Zoologica **2**: 293–327.

Germar, E.F. 1836. *Fauna insectorum Europae* **17**, 25 plates. Halle (Kümmel).

Girard, M. 1879. *Traité élémentaire d'entomologie* **2**, 1028 pp. Paris (Librairie J.-B. Baillière et fils).

Goeze, J.A.E. 1778. *Entomologische Beyträge zu des Ritter Linné zwölften Ausgabe des Natursystems* **2**, lxxii + 352 pp. Leipzig (Weidmanns Erben & Reich).

Götz, W. 1969. Beitrag zur Kenntnis einiger *Roeseliana*-Formen der Gattung *Metrioptera* (Orthoptera, Tettigoniidae). *Mitteilungen aus dem Zoologischen Museum in Berlin* **45**: 157–177.

— 1970. Zur Größenvariation im Formenkreis *Decticus verrucivorus* (Orthoptera, Saltatoria, Tettigoniidae). *Zoologische Abhandlungen Staatliches Museum für Tierkunde in Dresden* **31**: 139–191.

Goureau, [C.]. 1837*a*. Essai sur la stridulation des insectes. *Annales de la Société Entomologique de France* **6**: 31–75. [English translation: 1838, *Entomological Magazine* **5**: 89–102, 357–372.]

— 1837*b*. Note sur les sons insensibles produits par les insectes. *Annales de la Société Entomologique de France* **6**: 407–410.

Graber, V. 1871. Ueber den Ursprung und Bau der Ton-Apparate bei den Akridiern. *Verhandlungen der Zoologisch-Botanischen Gesellschaft in Wien* **21**: 1097–1102.

— 1872*a*. Ueber den Tonapparat der Locustiden, ein Beitrag zum Darwinismus. *Zeitschrift für Wissenschaftliche Zoologie* **22**: 100–119.

— 1872*b*. Anhang zu der Abhandlung über die Tonapparate der Locustiden. *Zeitschrift für Wissenschaftliche Zoologie* **22**: 120–125.

— 1873. Bemerkungen über die "Gehör- und Stimmorgane" der Heuschrecken und Cikaden. *Sitzungsberichte der Kaiserlichen Akademie der Wissenschaften, Mathematisch-Naturwissenschaftliche Classe, Wien* (Abteilung 1) **66** (1872): 205–213.

Graells, M. de la P. 1851. Descripcion de algunos insectos nuevos pertenecientes á la fauna central de España. *Memorias de la Real Academia de Ciencias Exactas, Físicas y Naturales de Madrid* (3. Ciencias Naturales) **1** (2): 109–166.

Grassé, P.P. 1924. Étude biologique sur *Phaneroptera 4–punctata* Br. et *Ph. falcata* Scop. *Bulletin Biologique de la France et de la Belgique* **58**: 454–472.

Gray, E.G. 1960. The fine structure of the insect ear. *Philosophical Transactions of the Royal Society of London* (Series B. Biological Sciences) **243**: 75–94.

Green, S.V. 1987. Acoustic and sexual behaviour in the grasshopper *Chorthippus brunneus* (Thunberg). [vii +] 189 + xxiv pp. University of East Anglia, Norwich (unpublished thesis).

— 1995. Song characteristics of certain Namibian grasshoppers (Orthoptera: Acrididae: Gomphocerinae). *African Entomology* **3**: 1–6.

Greenfield, M.D. 1997. Acoustic communication in Orthoptera [pp. 197–230]. *In* Gangwere, S.K., Muralirangan, M.C. & Muralirangan, M. [Eds], *The bionomics of grasshoppers, katydids and their kin.* xiii + 529 pp. Wallingford & New York (CAB International).

Grein, G. 1984. *Gesänge der heimischen Heuschrecken. Akustisch-optisch Bestimmungshilfe.* 30 cm disc, 33 rpm [with oscillograms]. Hannover (Niedersächsisches Landesverwaltungsamt).

Gribakin, E.G., Wiese, K. & Popov, A.V. [Eds]. 1990. *Sensory systems and communication in arthropods.* xviii + 423 pp. Basle, etc. (Birkhäuser).

Gross, S.W., Mays, D.L. & Walker, T.J. 1990. Systematics of *Pictonemobius* ground crickets (Orthoptera: Gryllidae). *Transactions of the American Entomological Society* **115**: 433–456.

Grzeschik, K.-H. 1969. Untersuchungen zur Systematik, Biologie und Ethologie von *Eugaster* Serville (Orthoptera, Tettigoniidae). *Forma et Functio* **1**: 46–144.

Gwynne, D.T. 1995. Phylogeny of the Ensifera (Orthoptera): a hypothesis supporting multiple origins of acoustical signalling, complex spermatophores and maternal care in crickets, katydids, and weta. *Journal of Orthoptera Research* **4**: 203–218.

Gwynne, D.T., Yeoh, P. & Schatral, A. 1988. The singing insects of King's Park and Perth gardens. *Western Australian Naturalist* **17**: 25–71.

Haes, E.C.M. & Else, G.R. 1976. Confirmation of *Metrioptera roeselii* (Hagenback [sic]) (Orth.,

Tettigoniidae) in Hampshire. *Entomologist's Monthly Magazine* **111** (1975): 183–184.

Hagenbach, J.J. 1822. *Symbola faunæ insectorum Helvetiæ exhibentia vel species novas vel nondum depictas* **1**, 48 pp. Basle (J.G. Neukirch).

Halfmann, K. & Elsner, N. 1978. Larval stridulation in acridid grasshoppers. *Naturwissenschaften* **65**: 265.

Hallenbeck, C. 1949. Insect thermometers. *Natural History, New York* **58**: 256–259, 285–287.

Hartley, J.C. 1990. Acoustic communication and phonotaxis in *Steropleurus nobrei* (Bolivar) [p. 408]. *In* Nickle, D.A. [Ed.], Proceedings of the 5th International Meeting of the Orthopterists' Society. *Boletín de Sanidad Vegetal. Plagas* (Fuera de Serie) **20**, xxx + 422 pp.

— 1993. Acoustic behaviour and phonotaxis in the duetting ephippigerines, *Steropleurus nobrei* and *Steropleurus stali* (Tettigoniidae). *Zoological Journal of the Linnean Society* **107**: 155–167.

Hartley, J.C. & Bugren, M.M. 1986. Colour polymorphism in *Ephippiger ephippiger* (Orthoptera, Tettigoniidae). *Biological Journal of The Linnean Society* **27**: 191–199.

Hartley, J.C. & Robinson, D.J. 1976. Acoustic behaviour of both sexes of the speckled bush cricket *Leptophyes punctatissima*. *Physiological Entomology* **1**: 21–25.

Hartley, J.C., Robinson, D.J. & Warne, A.C. 1974. Female response song in the ephippigerines *Steropleurus stali* and *Platystolus obvius* (Orthoptera, Tettigoniidae). *Animal Behaviour* **22**: 382–389.

Hartley, J.C. & Stephen, R.O. 1989. Temporal changes in the quality of the song of a bush cricket. *Journal of Experimental Biology* **147**: 189–202.

Hartley, J.C. & Warne, A.C. 1984. Taxonomy of the *Éphippiger ephippiger* complex (*ephippiger, cruciger* and *cunii*), with special reference to the mechanics of copulation (Orthoptera, Tettigoniidae). *Eos* **60**: 43–54.

Harz, K. 1955. Das Trommeln der Eichenschrecke *Meconema thalassinum* De Geer (Orthoptera, Ensifera). *Nachrichtenblatt der Bayerischen Entomologen* **4**: 91–92.

— 1957. *Die Geradflügler Mitteleuropas*. xxiv + 494 pp. Jena (G. Fischer).

— 1958a. Orthopterologische Beiträge (Schluß). *Nachrichtenblatt der Bayerischen Entomologen* **7**: 47–48.

— 1958b. Die Heideschrecke. *Entomologische Zeitschrift* **68**: 137–140.

— 1962. Orthopterologische Beiträge IV. *Nachrichtenblatt der Bayerischen Entomologen* **11**: 46–48, 50–56, 65–69.

— 1969. Die Orthopteren Europas I. *Series Entomologica* **5**, xx + 749 pp.

— 1971. Orthopterologische Beiträge IX. *Atalanta* **3**: 331–338.

— 1972. Orthopterologische Beiträge. *Atalanta* **4**: 129–132.

— 1975a. Eine neue *Platystolus*-Art aus Spanien. *Articulata* **1**: 17–18.

— 1975b. Die Orthopteren Europas II. *Series Entomologica* **11**, [viii +] 939 pp.

Harz, K. & Kaltenbach, A. 1976. Die Orthopteren Europas III. *Series Entomologica* **12**, 434 pp.

Haskell, P.T. 1953. The stridulation behaviour of the domestic cricket. *British Journal of Animal Behaviour* **1**: 120–121.

— 1955a. Intensité sonore des stridulations de quelques Orthoptères britanniques [pp. 154–167]. *In* Busnel, R.G. [Ed.], Colloque sur l'acoustique des Orthoptères. *Annales des Épiphyties*, fascicule spécial de 1954, 448 pp.

— 1955b. Vibration of the substrate and stridulation in a grasshopper. *Nature, London* **175**: 639–640.

— 1957. Stridulation and associated behaviour in certain Orthoptera. 1. Analysis of the stridulation of, and behaviour between, males. *British Journal of Animal Behaviour* **5**: 139–148.

— 1958. Stridulation and associated behaviour in certain Orthoptera. 2. Stridulation of females and their behaviour with males. *Animal Behaviour* **6**: 27–42.

— 1961. *Insect sounds*. viii + 179 pp. London (Witherby).

Hayward, R. 1901. The katydid's call in relation to temperature. *Psyche, Cambridge* **9**: 179.

Heath, J.E. & Josephson, R.K. 1970. Body temperature and singing in the katydid,

Neoconocephalus robustus (Orthoptera, Tettigoniidae). *Biological Bulletin, Marine Biological Laboratory, Woods Hole* **138**: 272–285.

Hedwig, B. 1986*a*. On the role in stridulation of plurisegmental interneurons of the acridid grasshopper *Omocestus viridulus* L. I. Anatomy and physiology of descending cephalothoracic interneurons. *Journal of Comparative Physiology* (A) **158**: 413–427.

— 1986*b*. On the role in stridulation of plurisegmental interneurons of the acridid grasshopper *Omocestus viridulus* L. II. Anatomy and physiology of ascending and T-shaped interneurons. *Journal of Comparative Physiology* (A) **158**: 429–444.

— 1990. Modulation of auditory responsiveness in stridulating grasshoppers. *Journal of Comparative Physiology* (A) **167**: 847–856.

Hedwig, B. & Elsner, N. 1981. A neuroethological analysis of sound production in the acridid grasshopper *Omocestus viridulus* [pp. 495–514]. *In* Salánki, J. [Ed.], *Neurobiology of invertebrates. Mechanisms of integration. [Advances in Physiological Science* **23**] ix + 581 pp. Budapest (Pergamon Press & Akadémiai Kiadó).

Hedwig, B. & Elsner, N. 1985. Sound production and sound detection in a stridulating acridid grasshopper (*Omocestus viridulus*) [pp. 61–72]. *In* Kalmring, K. & Elsner, N. [Eds], *Acoustic and vibrational communication in insects.* viii + 230 pp. Berlin & Hamburg (Paul Parey).

Hedwig, B., Gramoll, S. & Elsner, N. 1990. Stridulatory interneurons in the metathoracic ganglion of the grasshopper *Omocestus viridulus* L. [pp. 183–188]. *In* Gribakin, F.G., Wiese, K. & Popov, A.V. [Eds], *Sensory systems and communication in arthropods.* xviii + 423 pp. Basle, etc. (Birkhäuser).

Hedwig, B. & Meyer, J. 1994. Auditory information processing in stridulating grasshoppers: tympanic membrane vibrations and neurophysiology. *Journal of Comparative Physiology* (A) **174**: 121–131.

Heiligenberg, W. 1966. The stimulation of territorial singing in house crickets (*Acheta domesticus*). *Zeitschrift für Vergleichende Physiologie* **53**: 114–129.

— 1969. The effect of stimulus chirps on a cricket's chirping (*Acheta domesticus*). *Zeitschrift für Vergleichende Physiologie* **65**: 70–97.

Heinrich, R. & Elsner, N. 1997. Central nervous control of hindleg coordination in stridulating grasshoppers. *Journal of Comparative Physiology* (A) **180**: 257–269.

Heller, K.-G. 1984. Zur Bioakustik und Phylogenie der Gattung *Poecilimon* (Orthoptera, Tettigoniidae, Phaneropterinae). *Zoologische Jahrbücher* (Systematik) **111**: 69–117.

— 1986. Warm-up and stridulation in the bushcricket, *Hexacentrus unicolor* Serville (Orthoptera, Conocephalidae, Listroscelidinae). *Journal of Experimental Biology* **126**: 97–109.

— 1988. *Bioakustik der europäischen Laubheuschrecken [Ökologie in Forschung und Anwendung* 1]. 358 pp. Weikersheim, Germany (Margraf).

— 1990. Evolution of song pattern in east Mediterranean Phaneropterinae: constraints by the communication system [pp. 130–151]. *In* Bailey, W.J. & Rentz, D.C.F. [Eds], *The Tettigoniidae: biology, systematics and evolution.* ix + 395 pp. Bathurst, Australia (Crawford House Press).

Heller, K.-G. & Helversen, D. von. 1986. Acoustic communication in phaneropterid bushcrickets: species-specific delay of female stridulatory response and matching male sensory time window. *Behavioural Ecology and Sociobiology* **18**: 189–198.

Heller, K.-G. & Reinhold, K. 1992. A new bushcricket of the genus *Poecilimon* from the Greek islands (Orthoptera: Phaneropterinae). *Tijdschrift voor Entomologie* **135**: 163–168.

Heller, K.-G. & Willemse, F. 1989. Two new bush-crickets from Greece, *Leptophyes lisae* sp. nov. and *Platycleis (Parnassiana) tenuis* sp. nov. (Orthoptera: Tettigoniidae). *Entomologische Berichten, Amsterdam* **49**: 144–156.

Helluin, P. & Ribassin, P. 1961. *Francis de la nuit.* 17·5 cm disc, 33 rpm. Paris (Clartes).

Helversen, D. von. 1972. Gesang des Männchens und Lautschema des Weibchens bei der Feldheuschrecke *Chorthippus biguttulus* (Orthoptera, Acrididae). *Journal of Comparative Physiology* **81**: 381–422.

— 1993. 'Absolute steepness' of ramps as an essential cue for auditory pattern recognition by a grasshopper (Orthoptera; Acrididae; *Chorthippus biguttulus* L.). *Journal of Comparative Physiology* (A) **172**: 633–639.

Helversen, D. von & Helversen, O. von. 1975a. Verhaltensgenetische Untersuchungen am akustischen Kommunikationssystem der Feldheuschrecken (Orthoptera, Acrididae). I. Der Gesang von Artbastarden zwischen *Chorthippus biguttulus* und *Ch. mollis*. *Journal of Comparative Physiology* **104**: 273–299.

Helversen, D. von & Helversen, O. von. 1975b. Verhaltensgenetische Untersuchungen am akustischen Kommunikationssystem der Feldheuschrecken (Orthoptera, Acrididae). II. Das Lautschema von Artbastarden zwischen *Chorthippus biguttulus* und *Ch. mollis*. *Journal of Comparative Physiology* **104**: 301–323.

Helversen, D. von & Helversen, O. von. 1981. Korrespondenz zwischen Gesang und auslösendem Schema bei Feldheuschrecken. *Nova Acta Leopoldina* (Neue Folge) **54** (245): 449–462.

Helversen, D. von & Helversen, O. von. 1983. Species recognition and acoustic localization in acrid grasshoppers: a behavioral approach [pp. 95–107]. *In* Huber, F. & Markl, H. [Eds], *Neuroethology and Behavioral Physiology*. xviii + 412 pp. Berlin & Heidelberg (Springer-Verlag).

Helversen, D. von & Helversen, O. von. 1990. Pattern recognition and directional analysis: routes and stations of information flow in the CNS of a grasshopper [pp. 209–216]. *In* Gribakin, F.G., Wiese, K. & Popov, A.V. [Eds], *Sensory systems and communication in arthropods*. xviii + 423 pp. Basle, etc. (Birkhäuser).

Helversen, O. von. 1979. Angeborenes Erkennen akustischer Schlüsselreize. *Verhandlungen der Deutschen Zoologischer Gesellschaft* **72**: 42–59.

— 1986. Gesang und Balz bei Feldheuschrecken der *Chorthippus albomarginatus*-Gruppe (Orthoptera: Acrididae). *Zoologische Jahrbücher* (Systematik) **113**: 319–342.

— 1989. Bemerkungen zu *Chorthippus biguttulus hedickei* (Ramme 1942) und Beschreibung von *Chorthippus biguttulus euhedickei* n. ssp. *Articulata* **4**: 26–34.

Helversen, O. von & Elsner, N. 1977. The stridulatory movements of acrid grasshoppers recorded with an opto-electronic device. *Journal of Comparative Physiology* **122**: 53–64.

Helversen, O. von & Helversen, D. von. 1994. Forces driving coevolution of song and song recognition in grasshoppers [pp. 253–284]. *In* Schildberger, K. & Elsner, N. [Eds], Neural basis of behavioural adaptations. *Fortschritte der Zoologie* **39**, xiii + 284 pp.

Herbst, J.F.W. 1786. Fortsetzung des Verzeichnisses meiner Insektensammlung. Zweite Klasse. *Archiv der Insectengeschichte* **7–8**: 183–196.

Herrich-Schäffer, [G.A.W.]. 1840. *Nomenclator entomologicus. Verzeichniss der europäischen Insecten; zur Erleichterung des Tauschverkehrs mit Preisen versehen* **2**, viii + 244 pp. Regensburg (F. Pustet).

Hewitt, G.M. 1993. After the ice: *parallelus* meets *erythropus* in the Pyrenees [pp. 140–164]. *In* Harrison, R.G. [Ed.], *Hybrid zones and the evolutionary process*. x + 364 pp. New York & Oxford (Oxford University Press).

Hewitt, G.M., Butlin, R.K. & East, T.M. 1987. Testicular misfunction in hybrids between parapatric subspecies of the grasshopper *Chorthippus parallelus*. *Biological Journal of The Linnean Society* **31**: 25–34.

Hill, K.G., Loftus-Hills, J.J. & Gartside, D.F. 1972. Pre-mating isolation between the Australian field crickets *Teleogryllus commodus* and *T. oceanicus* (Orthoptera: Gryllidae). *Australian Journal of Zoology* **20**: 153–163.

Hodge, H. 1878. Further notes on *Acrida viridissima*. *Entomologist* **11**: 274–275.

Holst, K.T. 1970. Kakkerlakker græshopper og ørentviste XXVII. *Danmarks Fauna* **72**, 221 pp.

— 1986. The Saltatoria (bush-crickets, crickets and grasshoppers) of Northern Europe. *Fauna Entomologica Scandinavica* **16**, 127 pp.

Hörmann-Heck, S von. 1957. Untersuchungen über den Erbgang einiger Verhaltensweisen bei

Grillenbastarden (*Gryllus campestris* L.~*Gryllus bimaculatus* De Geer). *Zeitschrift für Tierpsychologie* **14**: 137–183.

Howse, P.E., Lewis, D.B. & Pye, J.D. 1971. Adequate stimulus of the insect tympanic organ. *Experientia* **27**: 598–600.

Hoy, R.R. 1974. Genetic control of acoustic behavior in crickets. *American Zoologist* **14**: 1067–1080.

Hoy, R.R., Hahn, J. & Paul, R.C. 1977. Hybrid cricket auditory behavior: evidence for genetic coupling in animal communication. *Science, Washington* **195**: 82–84.

Hoy, R.R. & Paul, R.C. 1973. Genetic control of song specificity in crickets. *Science, Washington* **180**: 82–83.

Huber, F. 1963. The role of the central nervous system in Orthoptera during the co-ordination and control of stridulation [pp. 440–488]. *In* Busnel, R.-G. [Ed.], *Acoustic behaviour of animals.* xx + 933 pp. Amsterdam (Elsevier).

— 1970. Nervöse Grundlagen der akustischen Kommunikation bei Insekten. *Vorträge. Rheinisch-Westfälische Akademie der Wissenschaften* **205**: 41–91.

— 1977. Lautäußerungen und Lauterkennen bei Insekten (Grillen). *Vorträge. Rheinisch-Westfälische Akademie der Wissenschaften* **265**: 15–66.

— 1983. Neural correlates of orthopteran and cicada phonotaxis [pp. 108–135]. *In* Huber, F. & Markl, H. [Eds], *Neuroethology and behavioral physiology.* xviii + 412 pp. Berlin & Heidelberg (Springer-Verlag).

Huber, F., Moore, T.E. & Loher, W. [Eds]. 1989. *Cricket behavior and neurobiology.* xvi + 565 pp. Ithaca & London (Cornell University Press).

Hudson, W.H. 1903. *Hampshire days.* xvi + 344 pp. London, etc. (Longmans, Green).

Hutchings, M. & Lewis, B. 1983. Insect sound and vibration receptors [pp. 181–205]. *In* Lewis, B. [Ed.], *Bioacoustics – a comparative approach.* x + 493 pp. London, etc. (Academic Press).

Inagaki, H. & Matsuura, I. 1985*a*. Discriminations génétique et acoustique de *Teleogryllus siamensis* n. sp. (Orthoptera, Gryllidae) et de son espèce jumelle *T. emma* Ohmachi et Matsuura. *Comptes Rendus de l'Académie des Sciences, Paris* (III) **300**: 247–250.

Inagaki, H. & Matsuura, I. 1985*b*. Proche parenté génétique de *Teleogryllus siamensis* Inagaki et Matsuura et de *T. taiwanemma* Ohmachi et Matsuura (Orthoptera, Gryllidae) et différence structurale de leurs signaux acoustiques. *Comptes Rendus des Séances de la Société de Biologie et de ses Filiales* **179**: 41–47.

Ingrisch, S. 1991. Taxonomy der *Isophya*-Arten der Ostalpen (Grylloptera: Phaneropteridae). *Mitteilungen der Schweizerischen Entomologischen Gesellschaft* **64**: 269–279.

— 1993. Taxonomy and stridulation of the Gomphocerinae and Truxalinae of Thailand (Orthoptera, Acrididae). *Revue Suisse de Zoologie* **100**: 929–947.

— 1995. Evolution of the *Chorthippus biguttulus* group (Orthoptera, Acrididae) in the Alps, based on morphology and stridulation. *Revue Suisse de Zoologie* **102**: 475–535.

Ingrisch, S. & Bassangova, N. 1995. Paarungswahlverhalten von *Chorthippus biguttulus* und *C. eisentrauti* (Orthoptera: Acrididae). *Mitteilungen der Schweizerischen Entomologischen Gesellschaft* **68**: 1–6.

Ingrisch, S., Willemse, F. & Heller, K.-G. 1992. Eine neue Unterart des Warzenbeißers *Decticus verrucivorus* (Linnaeus 1758) aus Spanien (Ensifera: Tettigoniidae). *Entomologische Zeitschrift, Frankfurt am Main* **102**: 173–182.

Isely, F.B. 1936. Flight-stridulation in American acridians. (Orthopt.: Acrididae). *Entomological News* **47**: 199–205.

Jacobs, W. 1944. Einige Beobachtungen über Lautäußerungen bei weiblichen Feldheuschrecken. *Zeitschrift für Tierpsychologie* **6**: 141–146.

— 1950*a*. Über den Weibchen-Gesang bei Feldheuschrecken. *Verhandlungen der Deutschen Zoologen* [vom 2. bis 6. August 1949 in Mainz] **1949**: 238–244.

— 1950*b*. Vergleichende Verhaltensstudien an Feldheuschrecken. *Zeitschrift für Tierpsychologie* **7**:

169–216.

— 1953a. Verhaltensbiologische Studien an Feldheuschrecken. *Zeitschrift für Tierpsychologie*, Beiheft **1**, vii + 228 pp.

— 1953b. Vergleichende Verhaltensstudien an Feldheuschrecken (Orthoptera, Acrididae) und einigen anderen Insekten. *Zoologischer Anzeiger* Supplementband **17** (Verhandlungen der Deutschen Zoologischen Gesellschaft vom 2. bis 8. Juni 1952 in Freiburg): 115–138.

— 1957. Über regionale Unterschiede des Verhaltens (Gesang) bei Feldheuschrecken. *Wanderversammlung Deutscher Entomologen* **8**: 15–21.

— 1963. Über das Singen der Feldheuschrecke *Chorthippus biguttulus* (L.) in verschiedenen Bereichen Mittel- und Westeuropas. *Zeitschrift für Tierpsychologie* **20**: 446–460.

Janiszewski, J. & Otto, D. 1989. Responses and song pattern copying of Omega-type I-neurons in the cricket, *Gryllus bimaculatus*, at different prothoracic temperatures. *Journal of Comparative Physiology* (A) **164**: 443–450.

Janssen, D. 1977. Some notes on *Conocephalus discolor* (Orthoptera: Tettigoniidae). *Bulletin of the Amateur Entomologists' Society* **36**: 43–44.

Jatho, M., Schul, J., Stiedl, O. & Kalmring, K. 1994. Specific differences in sound production and pattern recognition in tettigoniids. *Behavioural Processes* **31**: 293–300.

Jatho, M., Weidemann, S. & Kretzen, D. 1992. Species-specific sound production in three Ephippigerine bushcrickets. *Behavioural Processes* **26**: 31–42.

Jones, M.D.R. 1963. Sound signals and alternation behaviour in *Pholidoptera*. *Nature, London* **199**: 928–929.

— 1964. Inhibition and excitation in the acoustic behaviour of *Pholidoptera*. *Nature, London* **203**: 322–323.

— 1966a. The acoustic behaviour of the bush cricket *Pholidoptera griseoaptera*. 1. Alternation, synchronism and rivalry between males. *Journal of Experimental Biology* **45**: 15–30.

— 1966b. The acoustic behaviour of the bush cricket *Pholidoptera griseoaptera*. 2. Interaction with artificial sound signals. *Journal of Experimental Biology* **45**: 31–44.

Jones, M.D.R. & Dambach, M. 1973. Response to sound in crickets without tympanal organs (*Gryllus campestris* L.). *Journal of Comparative Physiology* **87**: 89–98.

Josephson, R.K. 1973. Contraction kinetics of the fast muscles used in singing by a katydid. *Journal of Experimental Biology* **59**: 781–801.

— 1984. Contraction dynamics of flight and stridulatory muscles of tettigoniid insects. *Journal of Experimental Biology* **108**: 77–96.

Kalmring, K. 1975. The afferent auditory pathway in the ventral cord of *Locusta migratoria* (Acrididae). II. Responses of the auditory ventral cord neurons to natural sounds. *Journal of Comparative Physiology* **104**: 143–159.

Kalmring, K., Kaiser, W., Otto, C. & Kühne, R. 1985. Coprocessing of vibratory and auditory information in the CNS of different tettigoniids and locusts [pp. 193–202]. *In* Kalmring, K. & Elsner, N. [Eds], *Acoustic and vibrational communication in insects*. viii + 230 pp. Berlin & Hamburg (Paul Parey).

Kalmring, K., Keuper, A. & Kaiser, W. 1990. Aspects of acoustic and vibratory communication in seven European bushcrickets [pp. 191–216]. *In* Bailey, W.J. & Rentz, D.C.F. [Eds], *The Tettigoniidae: biology, systematics and evolution*. ix + 395 pp. Bathurst, Australia (Crawford House Press).

Kalmring, K. & Kühne, R. 1983. The processing of acoustic and vibrational information in insects [pp. 261–282]. *In* Lewis, B. [Ed.], *Bioacoustics – a comparative approach*. x + 493 pp. London, etc. (Academic Press).

Kämper, G. & Dambach, M. 1981. Response of the cercus-to-giant interneuron system in crickets to species-specific song. *Journal of Comparative Physiology* **141**: 311–317.

Kämper, G. & Dambach, M. 1985. Low-frequency airborne vibrations generated by crickets during singing and aggression. *Journal of Insect Physiology* **31**: 925–929.

Karny, H.H. 1908. Ueber das Schnarren der Heuschrecken. *Stettiner Entomologische Zeitung* **69**: 112–119.

Kavanagh, M.W. & Young, D. 1989. Bilateral symmetry of sound production in the mole cricket, *Gryllotalpa australis. Journal of Comparative Physiology* (A) **166**: 43–49.

Kettle, R. 1984. A discography of insect sounds. *Recorded Sound* **85**: 37–61.

Keuper, A., Kalmring, K., Schatral, A., Latimer, W. & Kaiser, W. 1986. Behavioural adaptations of ground living bushcrickets to the properties of sound propagation in low grassland. *Oecologia* **70**: 414–422.

Keuper, A. & Kühne, R. 1983. The acoustic behaviour of the bushcricket *Tettigonia cantans*. II. Transmission of airborne-sound and vibration signals in the biotope. *Behavioural Processes* **8**: 125–145.

Keuper, A., Otto, C., Latimer, W. & Schatral, A. 1985. Airborne sound and vibration signals for bushcrickets and locusts; their importance for the behaviour in the biotope [pp. 135–142]. *In* Kalmring, K. & Elsner, N. [Eds], *Acoustic and vibrational communication in insects*. viii + 230 pp. Berlin & Hamburg (Paul Parey).

Keuper, A., Weidemann, S., Kalmring, K. & Kaminski, D. 1988*a*. Sound production and sound emission in seven species of European tettigoniids. Part I. The different parameters of the song; their relation to the morphology of the bushcricket. *Bioacoustics* **1**: 31–48.

Keuper, A., Weidemann, S., Kalmring, K. & Kaminski, D. 1988*b*. Sound production and sound emission in seven species of European tettigoniids. Part II. Wing morphology and the frequency content of the song. *Bioacoustics* **1**: 171–186.

Kevan, D.K.McE. 1952. A summary of the recorded distribution of British Orthopteroids. *Transactions of the Society for British Entomology* **11**: 165–180.

— 1955. Méthodes inhabituelles de production de son chez les Orthoptères [pp. 103–141]. *In* Busnel, R.-G. [Ed.], Colloque sur l'acoustique des Orthoptères. *Annales des Epiphyties*, fascicule spécial de 1954, 448 pp.

— 1974. Land of the grasshoppers. *Memoirs of the Lyman Entomological Museum and Research Laboratory* **2**, ix + 326 pp.

Khalifa, A. 1950. Sexual behaviour in *Gryllus domesticus* L. *Behaviour* **2**: 264–274.

Kirby, W.F. 1910. *A synonymic catalogue of Orthoptera. Vol. III. Orthoptera Saltatoria. Part II. (Locustidæ vel Acridiidæ)*. x + 674 pp. London (British Museum).

Kleukers, R., Nieukerken, E. van, Odé, B., Willemse, L. & Wingerden, W. van. 1997. *De sprinkhanen en krekels van Nederland (Orthoptera)*. 415 pp. Leiden & Utrecht (Nationaal Natuurhistorisch Museum & Koninklijke Nederlandse Natuurhistorische Vereniging).

Klingstedt, H. 1939. Taxonomic and cytological studies on grasshopper hybrids. I. Morphology and spermatogenesis of *Chorthippus bicolor* Charp. × *Ch. biguttulus* L. *Journal of Genetics* **37**: 389–420.

Klöti-Hauser, E. 1921. Insekten als Musikanten. Erster Teil. *Natur und Technik, Zürich* **3**: 49–57.

Kneissl, L. 1900. Die Lautäusserungen der Heuschrecken Bayerns. *Natur und Offenbarung* **46**: 41–55.

Koçak, A.Ö. 1984. On the nomenclatural status of two species group names in Orthoptera. *Priamus* **3**: 169–170.

Koch, L. & Hawkins, D. 1969. *A salute to Ludwig Koch and a selection of some of his finest recordings*. 30 cm disc, 33 rpm. London (British Broadcasting Corporation).

Koch, U.T. 1978. Analyse des Grillengesanges mit Miniatur-Bewegungsdetektoren. *Verhandlungen der Deutschen Zoologischen Gesellschaft* **71**: 174.

— 1980. Analysis of cricket stridulation using miniature angle detectors. *Journal of Comparative Physiology* **136**: 247–256.

Koch, U.T., Elliott, C.J.H., Schäffner, K.-H. & Kleindienst, H.-U. 1988. The mechanics of stridulation of the cricket *Gryllus campestris. Journal of Comparative Physiology* (A) **162**: 213–223.

Kollar, V. 1833. Systematisches Verzeichniss der im Erzherzogthume Oesterreich vorkommenden geradflügeligen Insekten. *Beiträge zur Landeskunde Oesterreich's unter der Enns* **3**: 67–87.

Komarova, G.F. & Dubrovin, N.N. 1973. A comparative study of the acoustic signals of two sibling-species of grasshoppers *Chorthippus dorsatus* Zett. and *Ch. dichrous* Ev. (Orthoptera, Acrididae). *Zhurnal Obshchei Biologii* **34**: 571–574. [In Russian with English summary.]

Korn-Kremer, H. 1963. Beiträge zur Analyse des Männchen-Gesangs und zur Biologie von *Chorthippus montanus* Charp. 1825 (Orthopt., Acrididae). *Zeitschrift für wissenschaftliche Zoologie* **168**: 133–183.

Korsunovskaya, O.S. 1978. Sound communication in tree cricket *Oecanthus pellucens* Scop. (Orthoptera, Oecanthidae). *Vestnik Moskovskogo Universiteta* (Biologiya) **1978** (4): 48–51. [In Russian with English summary.]

Krauss, H. 1873. Beitrag zur Orthopteren-Fauna Tirols mit Beschreibung einer neuen *Pterolepis*. *Verhandlungen der Kaiserlich-Königlichen Zoologisch-Botanischen Gesellschaft in Wien. Abhandlungen.* **23**: 17–24.

— 1879. Die Orthopteren-Fauna Istriens. *Sitzungsberichte der Kaiserlichen Akademie der Wissenschaften. Mathematisch-Naturwissenschaftliche Klasse.* Wien **78** (Abteilung I): 451–544.

Kreidl, A. & Regen, J. 1905. Physiologische Untersuchungen über Tierstimmen. (I. Mitteilung.) Stridulation von *Gryllus campestris*. *Sitzungsberichten der Kaiserlichen Akademie der Wissenschaften. Mathematisch-Naturwissenschaftliche Klasse.* Wien **114** (Abteilung III): 57–81.

Kruseman, G. & Jeekel, C.A.W. 1967a. *Stenobothrus (Stenobothrodes) cotticus* nov. spec., a new grasshopper from the French Alps (Orthoptera, Acrididae). *Entomologische Berichten* **27**: 1–7.

Kruseman, G. & Jeekel, C.A.W. 1967b. *Stenobothrus rubicundus* (Germar, 1817): an invalid name (Orthoptera). *Entomologische Berichten* **27**: 78–80.

Kutsch, W. 1969. Neuromuskuläre Aktivität bei verschiedenen Verhaltensweisen von drei Grillenarten. *Zeitschrift für Vergleichende Physiologie* **63**: 335–378.

— 1976. Post-larval development of two rhythmical behavioural patterns: flight and song in the grasshopper, *Omocestus viridulus*. *Physiological Entomology* **1**: 255–263.

Kutsch, W. & Schiolten, P. 1979. Analysis of impulse rate and impulse frequency spectrum in the grasshopper, *Omocestus viridulus*, during adult development. *Physiological Entomology* **4**: 47–53.

Kutsch, W. & Otto, D. 1972. Evidence for spontaneous song production independent of head ganglia in *Gryllus campestris* L. *Journal of Comparative Physiology* **81**: 115–119.

Laddiman, R. 1879. Observations on *Acrida viridissima*. *Entomologist* **12**: 21–23.

La Greca, M. 1974. Una nuova specie di *Yersinella* dell'Appennino Tosco-Emiliano (Orthoptera Tettigoniidae). *Bolletino della Società Entomologica Italiana* **106**: 60–64.

— 1986. Contributo alla conoscenza degli Ortotteri delle Alpi occidentali piemontesi con descrizione di una nuova specie di *Stenobothrus*. *Animalia* **12** (1985): 215–244.

Lakes, R. & Schikorski, T. 1990. Neuroanatomy of tettigoniids [pp. 166–190]. *In* Bailey, W.J. & Rentz, D.C.F. [Eds], *The Tettigoniidae: biology, systematics and evolution*. ix + 395 pp. Bathurst, Australia (Crawford House Press).

Lakes-Harlan, R. & Heller, K.-G. 1992. Ultrasound-sensitive ears in a parasitoid fly. *Naturwissenschaften* **79**: 224–226.

Landman, W., Oudman, L. & Duijm, M. 1989. Allozymic and morphological variation in *Ephippiger terrestris* (Yersin, 1854) (Insecta, Orthoptera, Tettigonioidea). *Tijdschrift voor Entomologie* **132**: 184–198.

Landois, H. 1867. Die Ton- und Stimmapparate der Insecten in anatomisch-physiologischer und akustischer Beziehung. *Zeitschrift für Wissenschaftliche Zoologie* **17**: 107–186.

— 1874. *Thierstimmen*. xi + 229 pp. Freiburg im Breisgau (Herder).

Latimer, W. 1980. Song and spacing in *Gampsocleis glabra* (Orthoptera, Tettigoniidae). *Journal of Natural History* **14**: 201–213.

— 1981a. Acoustic competition in bush crickets. *Ecological Entomology* **6**: 35–45.

— 1981b. Variation in the song of the bush cricket *Platycleis albopunctata* (Orthoptera,

Tettigoniidae). *Journal of Natural History* **15**: 245–263.

— 1981c. The acoustic behaviour of *Platycleis albopunctata* (Goeze) Orthoptera, Tettigoniidae). *Behaviour* **76**: 182–206.

Latimer, W. & Broughton, W.B. 1984. Acoustic interference in bush crickets; a factor in the evolution of singing insects? *Journal of Natural History* **18**: 599–616.

Latimer, W. & Schatral, A. 1986. Information cues used in male competition by *Tettigonia cantans*. *Animal Behaviour* **34**: 162–168.

Latimer, W. & Sippel, M. 1987. Acoustic cues for female choice and male competition in *Tettigonia cantans*. *Animal Behaviour* **35**: 887–900.

Latreille, P.A. 1804. *Histoire naturelle, générale et particulière, des crustacés et des insectes* **12**, 424 pp. Paris (F. Dufart).

Leroy, Y. 1962. Le signal sonore d'appel sexuel comme caractère spécifique chez les Gryllides. *Bulletin Biologique de la France et de la Belgique* **96**: 755–774.

— 1966. Signaux acoustiques, comportement et systématique de quelques espèces de Gryllides (Orthoptères, Ensifères). *Bulletin Biologique de la France et de la Belgique* **100**: 1–134.

— 1977. L'universe sonore animal. *Journal de Psychologie Normale et Pathologique* **1977**: 165–187.

— 1978. Signale sonore et systématique animale. *Journal de Psychologie Normale et Pathologique* **1978**: 197–230.

Leston, D. & Pringle, J.W.S. 1963. Acoustic behaviour of Hemiptera [pp. 391–411]. *In* Busnel, R.-G. [Ed.], *Acoustic behaviour of animals*. xx + 933 pp. Amsterdam, etc. (Elsevier).

Lewis, D.B. 1974. The physiology of the Tettigoniid ear. IV. A new hypothesis for acoustic orientation behaviour. *Journal of Experimental Biology* **60**: 861–869.

Lewis, D.B., Pye, J.D. & Howse, P.E. 1971. Sound reception in the bush cricket *Metrioptera brachyptera* (L.) (Orthoptera, Tettigonioidea). *Journal of Experimental Biology* **55**: 241–251.

Lewis, D.B., Seymour, C. & Broughton, W.B. 1975. The response characteristics of the tympanal organs of two species of bush cricket and some studies of the problem of sound transmission. *Journal of Comparative Physiology* **104**: 325–351.

Libersat, F., Murray, J.A. & Hoy, R.R. 1994. Frequency as a releaser in the courtship song of two crickets, *Gryllus bimaculatus* (de Geer) and *Teleogryllus oceanicus*: a neuroethological analysis. *Journal of Comparative Physiology* (A) **174**: 485–494.

Linnaeus, C. 1758. *Systema naturæ per regna tria naturæ, secundum classes, ordines, genera, species, cum characteribus, differentiis, synonymis, locis,* 10th edition, **1**, [iv +] 824 pp. Stockholm (L. Salvius).

— 1761. *Fauna Svecica sistens Animalia Sveciæ regni: Mammalia, Aves, Amphibia, Pisces, Insecta, Vermes,* editio altera. [xlviii +] 578 pp. Stockholm (L. Salvius).

— 1767. *Systema naturæ per regna tria naturæ, secundum classes, ordines, genera, species, cum characteribus, differentiis, synonymis, locis,* 12th edition, **1** (2), 533–1327 [+ xxxvi]. Stockholm (L. Salvius).

Loher, W. 1957. Untersuchungen über den Aufbau und die Entstehung der Gesänge einiger Feldheuschreckenarten und den Einfluss von Lautzeichen auf das akustische Verhalten. *Zeitschrift für Vergleichende Physiologie* **39**: 313–356.

Loher, W. & Broughton, W.B. 1955. Études sur le comportement acoustique de *Chorthippus bicolor* (Charp.) avec quelques notes comparatives sur les espèces voisines (Acrididae) [pp. 248–277]. *In* Busnel, R.-G. [Ed.], Colloque sur l'acoustique des Orthoptères. *Annales des Épiphyties,* fascicule spécial de 1954, 448 pp.

Loher, W. & Huber, F. 1964. Experimentelle Untersuchungen am Sexualverhalten des Weibchens der Heuschrecke *Gomphocerus rufus* L. (Acridinae). *Journal of Insect Physiology* **10**: 13–36.

Loher, W. & Huber, F. 1966. Nervous and endocrine control of sexual behaviour in a grasshopper (*Gomphocerus rufus* L., Acridinae). *Symposia of the Society of Experimental Biology* **20**: 381–400.

Lottermoser, W. 1952. Aufnahme und Analyse von Insektenlauten. *Acustica, Akustische Beihefte* **2**: 66–71.

Lucas, H. 1849. Histoire naturelle des animaux articulés. Cinquième classe. Insectes. (Suite.) Deuxième ordre. Les Orthoptères. *Exploration Scientifique de l'Algerie pendant les années 1840, 1841, 1842* (Sciences physiques, Zoologie) **3**: 1–39.

Lunau, C. 1952. Zum Vorkommen der Laubheuschrecke *Gampsocleis glabra* Herbst im Wilseder Heidepark. *Beiträge zur Naturkunde Niedersachsens* **5**: 12–14.

Luquet, G.C. 1978. La systématique des Acridiens Gomphocerinae du Mont Ventoux (Vaucluse) abordée par le biais du comportement acoustique [Orthoptera, Acrididae]. *Annales de la Société Entomologique de France* (Nouvelle Série) **14**: 415–450.

— 1991. Note sur la répartition et la raréfaction de quelques orthoptéroïdes de la faune française (Orthoptera). *Entomologica Gallica* **2**: 203–208.

— 1993. Les noms vernaculaires français, néerlandais, allemands et anglais des orthoptères d'Europe occidentale (Orth. Ensifera et Caelifera). *Entomologica Gallica* **4**: 97–124.

Lutz, F.E. 1924. Insect sounds. *Bulletin of the American Museum of Natural History* **50**: 333–372.

Lux, E. 1961. Biometrische und morphologische Studien an *Chorthippus longicornis* (Latr.) (= *parallelus* Zett.) und *montanus* (Charp.) unter Berücksichtigung regionaler Unterschiede. *Zoologische Jahrbücher* **88**: 355–398.

Malenotti, E. 1926. I canti delle Grillotalpe. *Atti e Memorie dell'Accademia di Agricoltura Scienze e Lettere di Verona* (5) **3**: 177–184.

Mangold, J.R. 1978. Attraction of *Euphasiopteryx ochracea, Corethrella* sp. and gryllids to broadcast songs of the southern mole cricket. *Florida Entomologist* **61**: 57–61.

Margoschis, R. 1977. *Recording natural history sounds*. [iv +] 109 pp. Barnet (Print & Press Services).

Marquet, [C.]. 1877. Notes pour servir à l'histoire naturelle des insectes Orthoptères du Languedoc. *Bulletin de la Société d'Histoire Naturelle de Toulouse* **11**: 137–159.

Masaki, S., Kataoka, M., Shirato, K. & Nakagahara, M. 1987. Evolutionary differentiation of right and left tegmina in crickets [pp. 347–357]. *In* Baccetti, B. [Ed.], *Evolutionary biology of orthopteroid insects*. 612 pp. Chichester (Ellis Horwood).

Mason, D.J. 1991. Genetic divergence in the *C. biguttulus* group (Orthoptera: Acrididae). 406 pp. University of Wales (unpublished thesis).

Matsuura, I. 1979. Japanese Gryllidae (6). *Nature Insects* **14** (8): 13–17. [In Japanese.]

Matthews, R.W. & Matthews, J.R. 1978. *Insect behavior*. xiii + 507 pp. New York, etc. (John Wiley & Sons).

McNeill, J. 1891. A list of the Orthoptera of Illinois. I–IV. *Psyche, a Journal of Entomology, Cambridge, Massachusetts* **6**: 3–9, 21–27, 62–66, 73–78.

Meissner, O. 1917. Beobachtungen an gefangenen Sattelschrecke. *Entomologische Zeitschrift, Frankfurt am Main* **31**: 37.

Menzies, I.S. & Shaw, H.K.A. 1947. Stridulation of *Leptophyes punctatissima* Bosc (Orth., Tettigoniidae). *Entomologist's Monthly Magazine* **83**: 95.

Messina, A., Arcidiacono, R., Failla, M.C., Lombardo, C. & Lombardo, F. 1980. Canto di richiamo in *Platycleis intermedia intermedia* (Serv.) di Sicilia (Orthoptera, Decticinae). *Animalia* **7**: 117–121.

Meyer, J. & Elsner, N. 1996. How well are frequency sensitivities of grasshopper ears tuned to species-specific song spectra? *Journal of Experimental Biology* **199**: 1631–1642.

Meyer, J. & Elsner, N. 1997. Can spectral clues contribute to species separation in closely related grasshoppers? *Journal of Comparative Physiology* (A) **180**: 171–180.

Michelsen, A. 1971. The physiology of the locust ear. I. Frequency sensitivity of single cells in the isolated ear. *Zeitschrift für Vergleichende Physiologie* **71**: 49–62.

Morris, G.K. & Walker, T.J. 1976. Calling songs of *Orchelimum* meadow katydids (Tettigoniidae). I. Mechanism, terminology, and geographic distribution. *Canadian Entomologist* **108**: 785–800.

Nadig, A. 1976. Beiträge zur Kenntnis der Orthopteren Marokkos: II. *Chorthippus (Glyptobothrus) biguttulus marocanus* ssp. n. (Orthoptera), ein Relikt "angarischer" Herkunft in den Gebirgen

Marokkos. *Revue Suisse de Zoologie* **83**: 647–671.

— 1980. *Ephippiger terrestris* (Yersin) und *E. bormansi* (Brunner v. W.) (Orthoptera): Unterarten einer polytypischen Art. Beschreibung einer dritten Unterart: *E. terrestris caprai* ssp. n. aus den ligurischen Alpen. *Revue Suisse de Zoologie* **87**: 473–512.

— 1986. Drei neue Gomphocerinae-Arten aus den Westalpen Piemonts. *Articulata* **2**: 213–233.

— 1987. Saltatoria (Insecta) der Süd- und Südostabdachung der Alpen zwischen der Provence im W, dem pannonischen Raum im NE und Istrien im SE (mit Verzeichnissen der Fundorte und Tiere meiner Sammlung). I. Teil: Laubheuschrecken (Tettigoniidae). *Revue Suisse de Zoologie* **94**: 257–356.

— 1991. *Stenobothrus nadigi* La Greca, 1985 und *St. ursulae* Nadig, 1986 sind synonym. *Articulata* **6**: 1–8.

Nadig, A. & Nadig, A. 1934. Beitrag zur Kenntnis der Orthopteren- und Hymenopterenfauna von Sardinien und Korsika. *Jahresbericht der Naturforschenden Gesellschaft Graubündens* **72**: 3–39.

Nagao, T. & Shimozawa, T. 1987. A fixed time-interval between two behavioural elements in the mating behaviour of male crickets, *Gryllus bimaculatus*. *Animal Behaviour* **35**: 122–130.

Nevo, E. & Blondheim, S.A. 1972. Acoustic isolation in the speciation of mole crickets. *Annals of the Entomological Society of America* **65**: 980–981.

Nickerson, J.C., Snyder, D.E. & Oliver, C.C. 1979. Acoustic burrows constructed by mole crickets. *Annals of the Entomological Society of America* **72**: 438–440.

Nickle, D.A. & Carlysle, T.C. 1975. Morphology and function of female sound-producing structures in ensiferan Orthoptera with special emphasis on the Phaneropterinae. *International Journal of Insect Morphology and Embryology* **4**: 159–168.

Nickle, D.A. & Castner, J.L. 1984. Introduced species of mole crickets in the United States, Puerto Rico, and the Virgin Islands (Orthoptera: Gryllotalpidae). *Annals of the Entomological Society of America* **77**: 450–465.

Nielsen, E.T. 1938. Zur Oekologie der Laubheuschrecken. *Entomologiske Meddelelser* **20**: 121–164.

— 1972. Precrepuscular stridulation in *Tettigonia viridissima* (Orthoptera ensifera). *Entomologica Scandinavica* **3**: 156–158.

Nielsen, E.T. & Dreisig, H. 1970. The behavior of stridulation in Orthoptera Ensifera. *Behaviour* **37**: 205–252.

Nocke, H. 1972. Biophysik der Schallerzeugung durch die Vorderflügel der Grillen. *Zeitschrift für Vergleichende Physiologie* **74**: 272–314.

— 1975. Physical and physiological properties of the tettigoniid ("grasshopper") ear. *Journal of Comparative Physiology* **100**: 25–57.

Ocskay, F. 1826. Gryllorum Hungariae indigenorum species aliquot. *Novo Acta Physico-Medica Academiae Caesareae Leopoldino-Carolinae Naturae Curiosorum* **13** (1): 407–410.

Odé, B. 1997. *De zingende sprinkhanen en krekels van de Benelux.* Compact disc. Leiden & Utrecht (Nationaal Natuurhistorisch Museum & Koninklijke Nederlandse Natuurhistorische Vereniging).

Oldfield, B.P. 1982. Tonotopic organisation of auditory receptors in Tettigoniidae (Orthoptera: Ensifera). *Journal of Comparative Physiology* **147**: 461–469.

Oldfield, B.P., Kleindienst, H.-U. & Huber, F. 1986. Physiology and tonotopic organization of auditory receptors in the cricket *Gryllus bimaculatus* De Geer. *Journal of Comparative Physiology* (A) **159**: 457–464.

Ossiannilsson, F. 1947. *Stritläten.* 30 cm disc, 78 rpm. Sweden (Radiotjänst). Produced for private circulation.

Otte, D. 1970. A comparative study of communicative behavior in grasshoppers. *Miscellaneous Publications of the Museum of Zoology, University of Michigan* **141**, 168 pp.

— 1972. Simple versus elaborate behavior in grasshoppers. An analysis of communication in the genus *Syrbula*. *Behaviour* **42**: 291–322.

— 1977. Communication in Orthoptera [pp. 334–361]. *In* Sebeok, T.A. [Ed.], *How animals communi-*

cate. xxi + 1128 pp. Bloomington & London (Indiana University Press).

— 1983. African crickets (Gryllidae). 2. *Afrogryllopsis* Randell and *Neogryllopsis* n. gen. of eastern and southern Africa (Gryllinae, Brachytrupini). *Proceedings of the Academy of Natural Sciences of Philadelphia* **135**: 218–235.

— 1985. African crickets (Gryllidae: Gryllinae). 7. The genus *Cryncus* Gorochov. *Proceedings of the Academy of Natural Sciences of Philadelphia* **137**: 129–142.

— 1987. African crickets. 9. New genera and species of Brachytrupinae and Gryllinae. *Proceedings of the Academy of Natural Sciences of Philadelphia* **139**: 315–374.

— 1992. Evolution of cricket songs. *Journal of Orthoptera Research* **1**: 25–49.

— 1994. *The crickets of Hawaii.* vi + 396 pp. Philadelphia (The Orthopterists' Society).

Otte, D. & Alexander, R.D. 1983. The Australian crickets (Orthoptera: Gryllidae). *Monographs of the Academy of Natural Sciences of Philadelphia* **22**, [v +] 477 pp.

Otte, D., Alexander, R.D. & Cade, W. 1987. The crickets of New Caledonia (Gryllidae). *Proceedings of the Academy of Natural Sciences of Philadelphia* **139**: 375–457.

Otte, D. & Cade, W. 1983*a*. African crickets (Gryllidae). 1. *Teleogryllus* of eastern and southern Africa. *Proceedings of the Academy of Natural Sciences of Philadelphia* **135**: 102–127.

Otte, D. & Cade, W. 1983*b*. African crickets (Gryllidae). 3. On the African species of *Velarifictorus* Randell (Gryllinae, Modicogryllini). *Proceedings of the Academy of Natural Sciences of Philadelphia* **135**: 241–253.

Otte, D. & Cade, W. 1984*a*. African crickets (Gryllidae). 4. The genus *Platygryllus* from eastern and southern Africa (Gryllinae, Gryllini). *Proceedings of the Academy of Natural Sciences of Philadelphia* **136**: 45–66.

Otte, D. & Cade, W. 1984*b*. African crickets (Gryllidae). 5. East and south African species of *Modicogryllus* and several related genera (Gryllinae, Modicogryllini). *Proceedings of the Academy of Natural Sciences of Philadelphia* **136**: 67–97.

Otte, D. & Cade, W. 1984*c*. African crickets (Gryllidae). 6. The genus *Gryllus* and some related genera (Gryllinae, Gryllini). *Proceedings of the Academy of Natural Sciences of Philadelphia* **136**: 98–122.

Otte, D., Toms, R.B. & Cade, W. 1988. New species and records of East and southern African crickets (Orthoptera: Gryllidae: Gryllinae). *Annals of the Transvaal Museum* **34**: 405–468.

Otto, D. 1971. Untersuchungen zur zentralnervösen Kontrolle der Lauterzeugung von Grillen. *Zeitschrift für Vergleichende Physiologie* **74**: 227–271.

Oudman, L., Duijm, M. & Landman, W. 1990. Morphological and allozyme variation in the *Ephippiger ephippiger* complex (Orthoptera, Tettigonioidea). *Netherlands Journal of Zoology* **40**: 454–483.

Oudman, L., Landman, W. & Duijm, M. 1989. Genetic distance in the genus *Ephippiger* (Orthoptera, Tettigonioidea) – a reconnaissance. *Tidschrift voor Entomologie* **132**: 177–182.

Pallas, P.S. 1773. *Reise durch verschiedene Provinzen des rußischen Reichs* **2**, [vi +] 744 pp. St Petersburg (Kaiserlichen Academie der Wißenschaften).

Panelius, S. 1978. The detailed distribution of *Tettigonia cantans* in Finland (Orthoptera, Tettigoniidae). *Notulae Entomologicae* **58**: 151–157.

Pantel, J. 1896. Notes orthoptérologiques. V. Les Orthoptères du "Sitio" dans la Sierra de Cuenca. *Anales de la Sociedad Española de Historia Natural* **25**: 59–118.

Panzer, G.W.F. 1796. *Faunae insectorum Germanicae initia oder Deutschlands Insecten.* Heft 33. 24 pp. Nürnberg (Felssecker).

Pascual, F. 1978. Descripción de una nueva especie de *Chorthippus* Fieber, 1852, de Sierra Nevada, España (Orth., Acrididae). *Eos, Madrid* **52** (1976): 167–174.

Pasquinelly, F. & Busnel, M.-C. 1955. Études préliminaires sur les mécanismes de la production des sons par les Orthoptères [pp. 145–153]. *In* Busnel, R.-G. [Ed.], Colloque sur l'acoustique des Orthoptères. *Annales des Épiphyties*, fascicule spécial de 1954, 448 pp.

Paul, R.C. 1976. Species specificity in the phonotaxis of female ground crickets (Orthoptera:

Gryllidae: Nemobiinae). *Annals of the Entomological Society of America* **69**: 1007–1010.

— 1977. Species-specific phonotaxis in *Gryllus* females. *Florida Entomologist* **60**: 67–68.

Pelletier, G. 1995. *Guide sonore et visuel des insectes chanteurs du Québec et de l'est de l'Amérique du Nord*. Booklet (61 pp.) and either tape cassette or compact disc. Quebec (Amateur Entomologists Association of Quebec & Editions Broquet).

Perdeck, A.C. 1951. Ervaringen met veldsprinkhanen. *Levende Natuur* **54**: 165–172.

— 1957. *The isolating value of specific song patterns in two sibling species of grasshoppers (Chorthippus brunneus Thunb. and C. biguttulus L.).* [viii +] 75 pp. Leiden (E.J. Brill).

Petrunkewitsch, A. & Guaita, G. von. 1901. Ueber den geschlechtlichen Dimorphismus bei den Tonapparaten der Orthopteren. *Zoologische Jahrbücher* (Systematik) **14**: 291–310.

Peyerimhoff, P de. 1908. Sur l'éclosion et la ponte d'*Ephippiger confusus* Finot [Orth.]. *Annales de la Société Entomologique de France* **77**: 505–516.

Pfau, H.K. 1984. Verbreitung und Gesänge kleiner korsischer *Omocestus*- (bzw. *Chorthippus*-) Arten. *Verhandlungen der Deutschen Zoologischen Gesellschaft* **77**: 267.

— 1986. Morphologie und Stridulation von *Metrioptera ambigua* nov. spec. aus Nordwestspanien, im Vergleich zu nahestehenden Arten (Insecta: Ensifera). *Stuttgarter Beiträge zur Naturkunde* (A) **389**: 1–10.

— 1988. Untersuchungen zur Stridulation und Phylogenie der Gattung *Pycnogaster* Graells, 1851 (Orthoptera, Tettigoniidae, Pycnogastrinae). *Mitteilungen der Schweizerischen Entomologischen Gesellschaft* **61**: 167–183.

— 1996. Untersuchungen zur Bioakustik und Evolution der Gattung *Platystolus* Bolívar (Ensifera, Tettigoniidae). *Tijdschrift voor Entomologie* **139**: 33–72.

Pfau, H.K. & Koch, U.T. 1994. The functional morphology of singing in the cricket. *Journal of Experimental Biology* **195**: 147–167.

Pfau, H.K. & Schroeter, B. 1983. Spinnenmimikry bei *Pycnogaster jugicola* Graells (Saltatoptera, Tettigoniidae)? *Articulata* **2**: 36–37.

Pfau, H.K. & Schroeter, B. 1988. Die akustische Kommunikation von *Platystolus martinezi* (Bolívar) – ein schnelles Antwort-Ruckantwort-System (Orthoptera, Tettigoniidae, Ephippigerinae). *Bonner Zoologischer Beiträge* **39**: 29–41.

Philippi, R.A. 1830. *Orthoptera Berolinensia*. [iv +] 46 pp. Berlin (F. Nietack).

Pierce, G.W. 1948. *The songs of insects, with related material on the production, propagation, detection, and measurement of sonic and supersonic vibrations*. vii + 329 pp. Cambridge, Massachusetts (Harvard University Press).

Pinedo, C. 1985. Contribucion al estudio del comportamiento acustico del Tetigonido *Platycleis intermedia* (Serville). *Actas do II Congreso Ibérico de Entomología* (*Boletim da Sociedade Portuguesa de Entomologia*, Suplemento 1) **1**: 27–36.

Pires, A. & Hoy, R.R. 1992. Temperature coupling in acoustic communication. I. Field and laboratory studies of temperature effects on calling song production and recognition in *Gryllus firmus*. *Journal of Comparative Physiology* (A) **171**: 69–78.

Pitkin, L.M. 1976. A comparative study of the stridulatory files of the British Gomphocerinae (Orthoptera: Acrididae). *Journal of Natural History* **10**: 17–28.

— 1977. A taxonomic study of the genus *Thyridorhoptrum* Rehn & Hebard (Orthoptera: Tettigoniidae), with a description of a new species. *Journal of Natural History* **11**: 645–659.

Poda, N. 1761. *Insecta Musei Græcensis, quæ in ordines, genera et species juxta Systema Naturæ Caroli Linnæi digessit*. [viii +] 127 [+ xii] pp. Graz (H. Widmanstad).

Popov, A.V. 1971. Sound-producing apparatus and structure of the call-song of the house cricket *Acheta domesticus*. *Zhurnal Evolyutsionnoi Biokhimii i Fiziologii* **7**: 87–95. [In Russian. English translation: 1971, *Journal of Evolutionary Biochemistry and Physiology* **7**: 67–74.]

— 1972. Sounds of crickets from southern regions of the European part of the U.S.S.R. *Entomologicheskoe Obozrenie* **51**: 17–36. [In Russian. English translation: 1972, *Entomological Review, Washington* **51**: 11–22.]

— 1975. Acoustic behaviour and migrations of field crickets *Gryllus campestris*. *Zoologicheskii Zhurnal* **54**: 1803–1809. [In Russian with English summary.]

— 1985. *Acoustic behaviour and hearing in insects.* 253 pp. St Petersburg (Nauka). [In Russian.]

Popov, A.V. & Shuvalov, V.F. 1974. Spectrum, intensity and directional characteristics of the calling song emission in the cricket *Gryllus campestris* under natural conditions. *Zhurnal Evolyutsionnoi Biokhimii i Fiziologii* **10**: 72–80. [In Russian with English summary. English translation: 1974, *Journal of Evolutionary Biochemistry and Physiology* **10**: 61–68.]

Popov, A.V. & Shuvalov, V.F. 1977. Phonotactic behavior of crickets. *Journal of Comparative Physiology* **119**: 111–126.

Popov, A.V., Shuvalov, V.F., Knyazev, A.N. & Klar-Spasovskaya, N.A. 1974. Communication calling songs of crickets (Orthoptera, Gryllidae) from south-western Tadjikistan. *Entomologicheskoe Obozrenie* **53**: 258–279. [In Russian with English summary. English translation: 1974, *Entomological Review, Washington* **53** (2): 11–24.]

Popov, A.V., Shuvalov, V.F. & Markovich, A.M. 1975. Spectrum of the calling signals, phonotaxis, and the auditory system in the cricket *Gryllus bimaculatus*. *Zhurnal Evolyutsionnoi Biokhimii i Fiziologii* **11**: 453–460. [In Russian with English summary. English translation: 1976, *Journal of Evolutionary Biochemistry and Physiology* **11**: 398–404.]

Popov, A.V., Shuvalov, V.F., Svetlogorskaya, I.D. & Markovich, A.M. 1974. Acoustic behaviour and auditory system in insects. *Abhandlungen der Rheinisch-Westfälischen Akademie der Wissenschaften* **53**: 281–306.

Poulton, E.B. 1896. On the courtship of certain European Acridiidae. *Transactions of the Entomological Society of London* **1896**: 233–252.

Prestwich, K.N. & Walker, T.J. 1981. Energetics of singing in crickets: effect of temperature in three trilling species (Orthoptera: Gryllidae). *Journal of Comparative Physiology* **143**: 199–212.

Princis, K. 1935. Zur Biologie von *Stauroderus pullus* Phil. (Orth. Loc.). *Internationale Entomologische Zeitschrift* **29**: 178–180, 183–186.

Prochnow, O. 1907. Die Lautapparate der Insekten. Ein Beitrag zur Zoophysik und Deszendenz-Theorie. *Internationale Entomologische Zeitschrift* **1**: 133–135, 141–143, 150–152, 157–159, 168–169, 173–174, 181–183, 190–191, 198–199, 207–208, 214–215, 221–223, 229–231, 237–239, 245–247, 253–255, 261–264, 269–271, 277–279, 285–287, 293–296, 301–304, 317–318, 333–334, 341, 349–350, 357–358, 368–370, 373–375, 377–379, 386–387.

Prozesky-Schulze, L., Prozesky, O.P.M., Anderson, F. & van der Merwe, G.J.J. 1975. Use of a self-made sound baffle by a tree cricket. *Nature, London* **255**: 142–143.

Pussard, R. 1942. Sur une pullulation d'*Orphania scutata* Br. (Orthoptère Phasgonuridae) dans les Alpes Maritimes. *Cahiers de Pathologie Végétale et Entomologie Agricole* **1942**: 16–24.

Pye, J.D. 1992. Equipment and techniques for the study of ultrasound in air. *Bioacoustics* **4**: 77–88.

Ragge, D.R. 1965. *Grasshoppers, crickets and cockroaches of the British Isles.* xii + 299 pp. London & New York (Warne).

— 1973. The British Orthoptera: a supplement. *Entomologist's Gazette* **24**: 227–245.

— 1974. A new species of *Tylopsis* from southern Israel (Orthoptera: Tettigoniidae). *Israel Journal of Zoology* **22** (1973): 63–66.

— 1976. A putative hybrid in nature between *Chorthippus brunneus* and *C. biguttulus* (Orthoptera: Acrididae). *Systematic Entomology* **1**: 71–74.

— 1981. An unusual song-pattern in the *Chorthippus mollis* group (Orthoptera: Acrididae): local variant or hybrid population? *Journal of Natural History* **15**: 995–1002.

— 1984. The Le Broc grasshopper population: further evidence of its hybrid status (Orthoptera: Acrididae). *Journal of Natural History* **18**: 921–925.

— 1986. The songs of the western European grasshoppers of the genus *Omocestus* in relation to their taxonomy (Orthoptera: Acrididae). *Bulletin of the British Museum (Natural History) (Entomology)* **53**: 213–249.

— 1987a. The songs of the western European grasshoppers of the genus *Stenobothrus* in relation to their taxonomy (Orthoptera: Acrididae). *Bulletin of the British Museum (Natural History)* (Entomology) **55**: 393–424.

— 1987*b*. Speciation and biogeography of some southern European Orthoptera, as revealed by their songs [pp. 418–426]. *In* Baccetti, B. [Ed.], *Evolutionary biology of orthopteroid insects*. 612 pp. Chichester (Ellis Horwood).

— 1990. The songs of the western European bush-crickets of the genus *Platycleis* in relation to their taxonomy (Orthoptera: Tettigoniidae). *Bulletin of the British Museum (Natural History)* (Entomology) **59**: 1–35.

Ragge, D.R., Burton, J.F. & Wade, G.F. 1965. *Songs of the British grasshoppers and crickets*. 17·5 cm disc, 33 rpm. London (Warne).

Ragge, D.R. & Reynolds, W.J. 1984. The taxonomy of the western European grasshoppers of the genus *Euchorthippus*, with special reference to their songs (Orthoptera: Acrididae). *Bulletin of the British Museum (Natural History)* (Entomology) **49**: 103–151.

Ragge, D.R. & Reynolds, W.J. 1988. The songs and taxonomy of the grasshoppers of the *Chorthippus biguttulus* group in the Iberian Peninsula (Orthoptera: Acrididae). *Journal of Natural History* **22**: 897–929.

Ragge, D.R., Reynolds, W.J. & Willemse, F. 1990. The songs of the European grasshoppers of the *Chorthippus biguttulus* group in relation to their taxonomy, speciation and biogeography (Orthoptera: Acrididae) [pp. 239–245]. *In* Nickle, D.A. [Ed.], Proceedings of the 5th International Meeting of the Orthopterists' Society. *Bolétin de Sanidad Vegetal. Plagas* (Fuera de Serie) **20**, xxx + 422 pp.

Rambur, P. 1838–1840. *Faune entomologique de l'Andalousie* **2**, 336 pp. Paris (A. Bertrand).

Ramme, W. 1921*a*. Orthopterologische Beiträge. *Archiv für Naturgeschichte* (A) **86** (12): 81–166.

— 1921*b*. Ein neuer *Chorthippus* aus Südtirol. *Deutsche Entomologische Zeitschrift* **3**: 246.

— 1923. Orthopterologische Ergebnisse meiner Reise nach Oberitalien und Südtirol 1921. *Archiv für Naturgeschichte* **89**: 145–169.

— 1927. Die Dermapteren und Orthopteren Siziliens und Kretas. *Eos, Madrid* **3**: 111–200.

— 1931. Beiträge zur Kenntnis der palaearktischen Orthopterenfauna (Tettig. et Acrid.). *Mitteilungen aus dem Zoologischen Museum in Berlin* **17**: 165–200.

Regen, J. 1902. Neue Beobachtungen über die Stridulationsorgane der saltatoren Orthopteren. *Zoologischer Anzeiger* **25**: 489–491.

— 1903. Neue Beobachtungen über die Stridulationsorgane der saltatoren Orthopteren. *Arbeiten aus den Zoologischen Instituten der Universität Wien und der Zoologischen Station in Triest* **14**: 359–422.

— 1908. Das tympanale Sinnesorgan von *Thamnotrizon apterus* Fab. ♂ als Gehörapparat experimentell nachgewiesen. *Sitzungsberichte der Kaiserlichen Akademie der Wissenschaften, Wien. Mathematisch-Naturwissenschaftliche Klasse* **117** (Abteilung III): 487–490.

— 1913. Über die Anlockung des Weibchens von *Gryllus campestris* L. durch telephonisch übertragene Stridulationslaute des Männchens. Ein Beitrag zur Frage der Orientierung bei den Insekten. *Pflügers Archiv für Gesammte Physiologie des Menschen und der Tiere* **155**: 193–200.

— 1914. Untersuchungen über die Stridulation und das Gehör von *Thamnotrizon apterus* Fab. ♂. *Sitzungsberichte der Kaiserlichen Akademie der Wissenschaften, Wien. Mathematisch-Naturwissenschaftliche Klasse* **123** (Abteilung I): 853–892.

— 1926. Über die Beeinflussung der Stridulation von *Thamnotrizon apterus* Fab. ♂ durch künstlich erzeugte Töne und verschiedenartige Geräusche. *Sitzungsberichte der Akademie der Wissenschaften in Wien. Mathematisch-Naturwissenschaftliche Klasse* **135** (Abteilung I): 329–368.

Renner, M. 1952. Analyse der Kopulationsbereitschaft des Weibchens der Feldheuschrecke *Euthystira brachyptera* Ocsk. in ihrer Abhängigkeit vom Zustand des Geschlechtsapparates. *Zeitschrift für Tierpsychologie* **9**: 122–154.

Renner, M. & Kremer, E. 1980. Das paarungsverhalten der Feldheuschrecke *Chrysochraon dispar* Germ. in Abhängigkeit vom Adultalter und vom Eiablagerrhythmus (Caelifera, Acrididae). *Spixiana* **3**: 25–32.

Rentz, D.C.[F.] 1963. Additional records of *Platycleis tessellata* (Charpentier) in California with biological notes (Orthoptera: Tettigoniidae). *Pan-Pacific Entomologist* **39**: 252–254.

— 1985. *A monograph of the Tettigoniidae of Australia. Volume 1. The Tettigoniinae.* ix + 384 pp. Australia (CSIRO) & Netherlands (Brill).

— 1988. The shield-backed katydids of southern Africa: their taxonomy, ecology and relationships to the faunas of Australia and South America (Orthoptera: Tettigoniidae: Tettigoniinae). *Invertebrate Taxonomy* **2**: 223–337.

— 1993. *Tettigoniidae of Australia. Volume 2. The Austrosaginae, Zaprochilinae and Phasmodinae.* x + 386 pp. Australia (CSIRO).

Reynolds, W.J. 1980. A re-examination of the characters separating *Chorthippus montanus* and *C. parallelus* (Orthoptera: Acrididae). *Journal of Natural History* **14**: 283–303.

— 1986. A description of the song of *Omocestus broelemanni* (Orthoptera Acrididae) with notes on its taxonomic position. *Journal of Natural History* **20**: 111–116.

— 1987. A description of the song of *Chorthippus cazurroi* (Orthoptera: Acrididae) with notes on its taxonomic position and distribution. *Journal of Natural History* **21**: 1087–1095.

Rheinlaender, J. 1976. Some aspects of neuronal adaptation in the CNS of crickets to conspecific songs [pp. 27–36]. *In* Broughton, W.B. [Ed.], *Insect neuroacoustics.* [ii +] 75 pp. London. (Proceedings of a seminar in the Animal Acoustics Unit of the City of London Polytechnic. Printed for private circulation.)

Rheinlaender, J. & Kalmring, K. 1973. Die afferente Hörbahn im Bereich des Zentralnervensystems von *Decticus verrucivorus* (Tettigoniidae). *Journal of Comparative Physiology* **85**: 361–410.

Rheinlaender, J., Kalmring, K., Popov, A.V. & Rehbein, H. 1976. Brain projections and information processing of biologically significant sounds by two large ventral-cord neurons of *Gryllus bimaculatus* DeGeer (Orthoptera, Gryllidae). *Journal of Comparative Physiology* **110**: 251–269.

Rheinlaender, J. & Römer, H. 1980. Bilateral coding of sound direction in the CNS of the bushcricket *Tettigonia viridissima* L. (Orthoptera, Tettigoniidae). *Journal of Comparative Physiology* **140**: 101–111.

Rheinlaender, J., Shuvalov, V.F., Popov, A.V. & Kalmring, K. 1981. Characteristics of movements in the female crickets *Gryllus bimaculatus* to the source of the calling song and dependence of accuracy of their orientation on the signal spectrum. *Zhurnal Evolyutsionnoi Biokhimii i Fiziologii* **17**: 25–33. [In Russian.]

Richards, T.J. 1952. *Nemobius sylvestris* in S. E. Devon. *Entomologist* **85**: 83–87, 108–111, 136–141, 161–166.

Riede, K. 1983. Influence of the courtship song of the acridid grasshopper *Gomphocerus rufus* L. on the female. *Behavioural Ecology and Sociobiology* **14**: 21–27.

— 1986. Modification of the courtship song by visual stimuli in the grasshopper *Gomphocerus rufus* (Acrididae). *Physiological Entomology* **11**: 61–74.

Riley, C.V. 1874. Katydids. *Annual Report of the State Entomologist on the Noxious, Beneficial and other Insects of the State of Michigan* **6**: 150–169.

Ritchie, M.G. 1991. Female preference for 'song races' of *Ephippiger ephippiger* (Orthoptera: Tettigoniidae). *Animal Behaviour* **42**: 518–520.

— 1992a. Behavioral coupling in tettigoniid hybrids (Orthoptera). *Behavior Genetics* **22**: 369–379.

— 1992b. Variation in male song and female preference within a population of *Ephippiger ephippiger* (Orthoptera: Tettigoniidae). *Animal Behaviour* **43**: 845–855.

Roberts, H.R. 1941. Nomenclature in the Orthoptera concerning genotype designations. *Transactions of the American Entomological Society* **67**: 1–34.

Robinson, D.J. 1980. Acoustic communication between the sexes of the bush cricket, *Leptophyes punctatissima*. *Physiological Entomology* **5**: 183–189.

— 1990. Acoustic communication between the sexes in bushcrickets [pp. 112–129]. *In* Bailey, W.J.

& Rentz, D.C.F. [Eds], *The Tettigoniidae: biology, systematics and evolution.* ix + 395 pp. Bathurst, Australia (Crawford House Press).

Robinson, D., Rheinlaender, J. & Hartley, J.C. 1986. Temporal parameters of male-female sound communication in *Leptophyes punctatissima. Physiological Entomology* **11**: 317–323.

Roché, J.-C. 1965. *Guide sonore du naturaliste, 1. Insectes.* 17·5 cm disc, 33 rpm. Sequedin (Edwards Records).

Ronacher, B. 1989. Stridulation of acridid grasshoppers after hemisection of thoracic ganglia: evidence for hemiganglionic oscillators. *Journal of Comparative Physiology* (A) **164**: 723–736.

— 1990. Contributions of brain and thoracic ganglia to the generation of the stridulation pattern in *Chorthippus dorsatus* [pp. 317–323]. *In* Gribakin, F.G., Wiese, K. & Popov, A.V. [Eds], *Sensory systems and communication in arthropods.* xviii + 423 pp. Basle, etc. (Birkhäuser).

— 1991. Contribution of abdominal commissures in the bilateral coordination of the hind legs during stridulation in the grasshopper *Chorthippus dorsatus. Journal of Comparative Physiology* (A) **169**: 191–200.

Ronacher, B. & Stumpner, A. 1988. Filtering of behaviourally relevant temporal parameters of a grasshopper's song by an auditory interneuron. *Journal of Comparative Physiology* (A) **163**: 517–523.

Rossi, P. de. 1790. *Fauna Etrusca sistens Insecta quae in provinciis Florentina et Pisana praesertim collegit* **1**, xxii + 272 pp. Livorno (T. Masi).

Rost, R. & Honegger, H.W. 1987. The timing of premating and mating behavior in a field population of the cricket *Gryllus campestris* L. *Behavioral Ecology and Sociobiology* **21**: 279–289.

Sakaluk, S.K. & Belwood, J.J. 1984. Gecko phonotaxis to cricket calling song: a case of satellite predation. *Animal Behaviour* **32**: 659–662.

Sales, G. & Pye, D. 1974. *Ultrasonic communication by animals.* xi + 281 pp. London (Chapman & Hall).

Samways, M.J. 1976a. The song of *Metrioptera azami* (Finot) (Orthoptera, Tettigoniidae), and new localities for the species. *Journal of Natural History* **10**: 469–473.

— 1976b. Song modification in the Orthoptera. I. Proclamation songs of *Platycleis* spp. (Tettigoniidae). *Physiological Entomology* **1**: 131–149.

— 1976c. Song modification in the Orthoptera. III. Microsyllabic echemes in *Platycleis* spp. (Tettigoniidae). *Physiological Entomology* **1**: 299–303.

— 1976d. Habitats and habits of *Platycleis* spp. (Orthoptera, Tettigoniidae) in southern France. *Journal of Natural History* **10**: 643–667.

— 1976e. Habitats and song of the five species of *Platycleis* (sensu stricto) (Tettigoniidae) from Montpellier, Hérault, southern France [pp. 334–335]. *In* Harz, K., Die Orthopteren Europas III. *Series Entomologica* **12**, 434 pp.

— 1977a. Bush cricket interspecific acoustic interactions in the field (Orthoptera, Tettigoniidae). *Journal of Natural History* **11**: 155–168.

— 1977b. Song modification in the Orthoptera. IV. The *Platycleis intermedia*/*P. affinis* interaction quantified. *Physiological Entomology* **2**: 301–315.

— 1989. Insect conservation and landscape ecology: a case-history of bush crickets (Tettigoniidae) in southern France. *Environmental Conservation* **16**: 217–226.

Samways, M.J. & Broughton, W.B. 1976. Song modification in the Orthoptera. II. Types of acoustic interaction between *Platycleis intermedia* and other species of the genus (Tettigoniidae). *Physiological Entomology* **1**: 287–297.

Samways, M.J. & Harz, K. 1982. Biogeography of intraspecific morphological variation in the bush crickets *Decticus verrucivorus* (L.) and *D. albifrons* (F.) (Orthoptera: Tettigoniidae). *Journal of Biogeography* **9**: 243–254.

Sarra, R. 1934. Notizie biologiche delle *Platycleis grisea* F. (Orth.-Phasgonuridae). *Bollettino del Laboratoria di Zoologia Generale e Agraria del R. Istituto Superiore Agrario in Portici* **27**: 197–207. [This paper appears to have been issued in separate form before the relevant volume of the journal; we have given the year cited in the publication date on the separate (27.xii.1934) in preference to that cited on the volume (1935).]

Saulcy, F. de. 1887. Encore trois nouveaux Orthoptères des Pyrénées. *Bulletin de la Société d'Histoire Naturelle de Metz* **17**: 189–191.

Saussure, H. de. 1898. Analecta entomologica. I. Orthopterologica. *Revue Suisse de Zoologie* **5**: 183–249.

Schäffner, K.-H. & Koch, U.T. 1987. Effects of wing campaniform sensilla lesions on stridulation in crickets. *Journal of Experimental Biology* **129**: 25–40.

Schatral, A., Latimer, W. & Kalmring, K. 1985. The role of the song for spatial dispersion and agonistic contacts in male bushcrickets [pp. 111–116]. *In* Kalmring, K. & Elsner, N. [Eds], *Acoustic and vibrational communication in insects.* viii + 230 pp. Berlin & Hamburg (Parey).

Schmidt, [F.I.] 1850. [No title.] *Bericht über die Mitteilungen von Freunden der Naturwissenschaften in Wien* **6**: 182–184.

Schmidt, G.H. 1978. Ein Beitrag zur Taxonomie von *Chorthippus* (*Glyptobothrus*) *biguttulus* L. (Insecta: Saltatoria: Acrididae). *Zoologischer Anzeiger* **201**: 245–259.

— 1987a. Adaptation of Saltatoria to various climatic factors with regard to their survival in different geographical regions [pp. 550–565]. *In* Baccetti, B. [Ed.], *Evolutionary biology of orthopteroid insects.* 612 pp. Chichester (Ellis Horwood).

— 1987b. Nachtrag zur biotopmäßigen Verbreitung der Orthopteren des Neusiedlersee-Gebietes mit einem Vergleich zur ungarischen Puszta. *Burgenländische Heimatblätter* **49**: 157–182.

— 1989. Faunistische Untersuchungen zur Verbreitung der Saltatoria (Insecta: Orthopteroidea) im tosco-romagnolischen Apennin. *Redia* **72**: 1–115.

— 1990. Notes on the *Chorthippus* (*Glyptobothrus*) species (Orthoptera: Acrididae) in Greece and the calling songs of their males [pp. 247–253]. *In* Nickle, D.A. [Ed.], Proceedings of the 5th International Meeting of the Orthopterists' Society. *Boletín de Sanidad Vegetal. Plagas* (Fuera de Serie) **20**, xxx + 422 pp.

— 1996. Biotopmäßige Verteilung und Vergesellschaftung der Saltatoria (Orthoptera) im Parco Nazionale del Circeo, Lazio, Italien. *Deutsche Entomologische Zeitschrift* **43**: 9–75.

Schmidt, G.H. & Baumgarten, M. 1977. Untersuchungen zur räumlichen Verteilung, Eiablage und Stridulation der Saltatorien am Sperbersee im Naturpark Steigerwald. *Abhandlungen des Naturwissenschaftlichen Vereins Würzburg* **15** (1974): 33–83.

Schmidt, G.H. & Schach, G. 1978. Biotopmäßige Verteilung, Vergesellschaftung und Stridulation der Saltatorien in der Umgebung des Neusiedlersees. *Zoologische Beiträge* (Neue Folge) **24**: 201–308.

Schmitz, B., Scharstein, H. & Wendler, G. 1982. Phonotaxis in *Gryllus campestris* L. (Orthoptera, Gryllidae). I. Mechanism of acoustic orientation in intact female crickets. *Journal of Comparative Physiology* **148**: 431–444.

Schroeter, B. & Pfau, H.K. 1987. Bemerkenswerte Sattelschrecken (Orthoptera, Ephippigerinae) aus Spanien und Portugal. *Articulata* **3**: 41–50.

Schroth, M. 1987. Nachweis der Plumpschrecke, *Isophya pyrenea* (Serville, 1839), für das Untermaingebiet mittels der Detektormethode (Saltatoria: Tettigoniidae). *Hessische Faunistische Briefe* **7**: 56–59.

Schwabe, J. 1906. Beiträge zur Morphologie und Histologie der tympanalen Sinnesapparate der Orthopteren. *Zoologica, Stuttgart* (**20**) 50, vi + 154 pp.

Scopoli, J.A. 1763. *Entomologia Carniolica exhibens Insecta Carnioliæ indigena et distributa in ordines, genera, species, varietates.* [xxxvi +] 423 pp. Vienna (J.T. Trattner).

— 1786. *Deliciae florae et faunae insubricae* **1**, viii + 85. Ticini (Monasterii S. Salvatoris).

Scudder, S.H. 1868a. The songs of the grasshoppers. *American Naturalist* **2**: 113–120.

— 1868b. Notes on the stridulation of some New England Orthoptera. *Proceedings of the Boston Society of Natural History* **11**: 306–313.

— 1893. The songs of our grasshoppers and crickets. *Report of the Entomological Society of Ontario* **1892**: 62–78.

Serville, J.G.A. 1838. *Histoire naturelle des insectes. Orthoptères.* xvii + 776 pp. Paris (Librairie Encyclopédique de Roret).

Sharov, A.G. 1968. Phylogeny of orthopteroid insects. *Trudy Paleontologicheskogo Instituta* **118**, 218 pp. [In Russian. English translation published as *Phylogeny of the Orthopteroidea* by Israel Program for Scientific Translations, Jerusalem, 1971, vi + 251 pp.]

Shull, A.F. 1907. The stridulation of the Snowy Tree-cricket (*Œcanthus niveus*). *Canadian Entomologist* **39**: 213–225.

Shuvalov, V.F. & Popov, A.V. 1973. The importance of the calling song rhythmic pattern of males of the genus *Gryllus* for phonotaxis of females. *Zoologicheskii Zhurnal* **52**: 1179–1185. [In Russian with English summary.]

Siebold, C.T. von. 1842. Beiträge zur Fauna der wirbellosen Thiere Preußens. Achter Beitrag: Preußische Orthoptera. *Preussische Provinzialblätter* **27**: 543–550.

— 1844. Über das Stimm- und Gehörorgan der Orthopteren. *Archiv für Naturgeschichte* **10**: 52–81.

Silver, S., Kalmring, K. & Kühne, R. 1980. The response of central acoustic and vibratory interneurones in bushcrickets and locusts to ultrasonic stimulation. *Physiological Entomology* **5**: 427–435.

Simmons, L.W. 1988. The calling song of the field cricket, *Gryllus bimaculatus* (De Geer): constraints on transmission and its role in intermale competition and female choice. *Animal Behaviour* **36**: 380–394.

Simms, E. 1979. *Wildlife sounds and their recording.* xvi + 144 pp. London (Paul Elek).

Sippel, M., Otto, C. & Kalmring, K. 1985. Significant parameters in conspecific signals for processing in vibratory-auditory neurons of bushcrickets and locusts [pp. 73–80]. *In* Kalmring, K. & Elsner, N. [Eds], *Acoustic and vibrational communication in insects.* viii + 230 pp. Berlin & Hamburg (Paul Parey).

Sismondo, E. 1979. Stridulation and tegminal resonance in the tree cricket *Oecanthus nigricornis* (Orthoptera: Gryllidae: Oecanthinae). *Journal of Comparative Physiology* **129**: 269–279.

— 1980. Physical characteristics of the drumming of *Meconema thalassinum. Journal of Insect Physiology* **26**: 209–212.

— 1993. Ultrasubharmonic resonance and nonlinear dynamics in the song of *Oecanthus nigricornis* F. Walker (Orthoptera: Gryllidae). *International Journal of Insect Morphology and Embryology* **22**: 217–231.

Skovmand, O. & Pedersen, S.B. 1978. Tooth impact rate in the song of a shorthorned grasshopper: a parameter carrying specific behavioral information. *Journal of Comparative Physiology* **124**: 27–36.

Skovmand, O. & Pedersen, S.B. 1983. Song recognition and song pattern in a shorthorned grasshopper. *Journal of Comparative Physiology* **153**: 393–401.

Soltani, A.A. 1978. Preliminary synonymy and description of new species in the genus *Dociostaurus* Fieber, 1853 (Orthoptera: Acridoidea; Acrididae, Gomphocerinae) with a key to the species in the genus. *Journal of Entomological Society of Iran*, Supplementum 2, 93 pp.

Spooner, J.D. 1964. The Texas bush katydid – its sounds and their significance. *Animal Behaviour* **12**: 235–244.

— 1968. Pair-forming acoustic systems of phaneropterine katydids (Orthoptera, Tettigoniidae). *Animal Behaviour* **16**: 197–212.

Stäger, R. 1930. Beiträge zur Biologie einiger einheimischer Heuschreckenarten. *Zeitschrift für Wissenschaftliche Insektenbiologie* **25**: 53–70.

Stearn, W.T. 1992. *Botanical Latin*, 4th edition. xiv + 546 pp. Newton Abbot (David & Charles).

Stephen, R.O. & Hartley, J.C. 1991. The transmission of bush-cricket calls in natural environments. *Journal of Experimental Biology* **155**: 227–244.

Stephen, R.O. & Hartley, J.C. 1995. Sound production in crickets. *Journal of Experimental Biology* **198**: 2139–2152.

Stiedl, O. & Bickmeyer, U. 1991. Acoustic behaviour of *Ephippiger ephippiger* Fiebig (Orthoptera,

Tettigoniidae) within a habitat of Southern France. *Behavioural Processes* **23**: 125–135.

Stiedl, O., Bickmeyer, U. & Kalmring, K. 1991. Tooth impact rate alteration in the song of males of *Ephippiger ephippiger* Fiebig (Orthoptera, Tettigoniidae) and its consequences for phonotactic behaviour of females. *Bioacoustics* **3**: 1–16.

Stiedl, O., Hoffmann, E. & Kalmring, K. 1994. Chirp rate variability in male song of *Ephippigerida taeniata* (Orthoptera: Ensifera). *Journal of Insect Behavior* **7**: 171–181.

Stiedl, O. & Kalmring, K. 1989. The importance of song and vibratory signals in the behaviour of the bushcricket *Ephippiger ephippiger* Fiebig (Orthoptera, Tettigoniidae): taxis by females. *Oecologia* **80**: 142–144.

Stout, J.F., DeHaan, C.H. & McGhee, R.W. 1983. Attractiveness of the male *Acheta domesticus* calling song to females. I. Dependence on each of the calling song features. *Journal of Comparative Physiology* **153**: 509–521.

Stumpner, A. & Helversen, O. von. 1992. Recognition of a two-element song in the grasshopper *Chorthippus dorsatus* (Orthoptera: Gomphocerinae). *Journal of Comparative Physiology (A)* **171**: 405–412.

Stumpner, A. & Helversen, O. von. 1994. Song production and song recognition in a group of sibling grasshopper species (*Chorthippus dorsatus, Ch. dichrous* and *Ch. loratus*: Orthoptera, Acrididae). *Bioacoustics* **6**: 1–23.

Sychev, M.M. 1979. Peculiarities of morphology and ecology of *Chorthippus biguttulus* L. and *Ch. mollis* Charp. (Orthoptera, Acrididae) in the mountains of the Crimea. *Entomologicheskoe Obozrenie* **58**: 78–87. [In Russian with English summary. English translation: 1980, *Entomological Review, Washington* **58** (1): 37–43.]

Targioni-Tozzetti, A. 1881. Orthopterorum Italiae species novae in collectione R. Musei Florentini digestae. *Bolletino della Società Entomologica Italiana* **13**: 180–187.

Taschenberg, [E.L.]. 1871. Orthopterologische Studien aus den hinterlassenen Papieren des Oberlehrers Carl Wanckel zu Dresden. *Zeitschrift für die Gesammten Naturwissenschaften Halle* **38**: 1–28.

Tenant, W.G. 1878. The green field-cricket (*Acrida viridissima*). *Entomologist* **11**: 183–185.

Thomas, E.S. & Alexander, R.D. 1957. *Nemobius melodius*, a new species of cricket from Ohio (Orthoptera, Gryllidae). *Ohio Journal of Science* **57**: 148–152.

Thomas, E.S. & Alexander, R.D. 1962. Systematic and behavioral studies on the meadow grasshoppers of the *Orchelimum concinnum* group (Orthoptera: Tettigoniidae). *Occasional Papers of the Museum of Zoology, University of Michigan* **626**, 31 pp.

Thompson, D.W. 1910. Historia animalium. *The works of Aristotle translated into English under the editorship of J.A. Smith M.A. [and] W.D. Ross M.A.* **4**, xv + [484] pp. Oxford (Clarendon Press).

Thorson, J., Weber, T. & Huber, F. 1982. Auditory behavior of the cricket. II. Simplicity of calling-song recognition in *Gryllus* and anomalous phonotaxis at abnormal carrier frequencies. *Journal of Comparative Physiology* **146**: 361–378.

Thunberg, C.P. 1815. Hemipterorum maxillosorum genera illustrata plurimisque novis speciebus ditata ac descripta. *Mémoires de l'Academie Impériale des Sciences de St. Pétersbourg* **5**: 211–301.

Tombs, D.J. 1974. Wildlife recording in stereo. *Recorded Sound* **54**: 278–289.

— 1980. *Sound recording, from microphone to master tape.* 222 pp. Newton Abbot, etc. (David & Charles).

Toms, R.B. 1992. Effects of temperature on chirp rates of tree crickets (Orthoptera: Oecanthidae). *South African Journal of Zoology* **27**: 70–73.

Tschuch, G. 1985. Temperaturgang der Parameter des Lockgesangs von *Gryllus bimaculatus* De Geer und *Acheta domesticus* L. *Zoologische Jahrbücher* (Physiologie) **89**: 115–118.

— 1986. Akustische Informationsübertragung bei *Gryllus bimaculatus* (Orthoptera). *Zoologische Jahrbücher* (Physiologie) **90**: 339–342.

Tschuch, G. & Köhler, G. 1990. Die Gesänge von *Chorthippus parallelus* (Zett.) und *Chorthippus montanus* (Charp.). *Articulata* **5** (1): 59–62.

Tümpel, R. 1901. *Die Geradflügler Mitteleuropas.* [ii +] 308 pp. Eisenach (Wilckens).

Tuttle, M.D., Ryan, M.J. & Belwood, J.J. 1985. Acoustic resource partitioning by two species of phyllostomid bats (*Trachops cirrhosus* and *Tonatia sylvicola*). *Animal Behaviour* **33**: 1369–1370.

Ulagaraj, S.M. 1976. Sound production in mole crickets (Orthoptera: Gryllotalpidae: *Scapteriscus*). *Annals of the Entomological Society of America* **69**: 299–306.

Ulagaraj, S.M. & Walker, T.J. 1973. Phonotaxis of crickets in flight: attraction of male and female crickets to male calling songs. *Science, New York* **182**: 1278–1279.

Uvarov, B.P. 1935. The Malcolm Burr collection of Palaearctic Orthoptera. *Eos, Madrid* **11**: 71–96.

— 1936. A new *Omocestus* from Spain (Orthoptera, Acrididæ). *Annals and Magazine of Natural History* (10) **18**: 378–380.

— 1966. *Grasshoppers and locusts, a handbook of general acridology* **1**, xi + 481 pp. Cambridge (University Press).

— 1977. *Grasshoppers and locusts, a handbook of general acridology* **2**, ix + 613 pp. London (Centre for Overseas Pest Research).

Vedenina, V.Yu. 1990. Responses of sympatric acridid species to natural and artificial sound signals [pp. 366–370]. *In* Gribakin, F.G., Wiese, K. & Popov, A.V. [Eds], *Sensory systems and communication in arthropods.* xviii + 423 pp. Basle, etc. (Birkhäuser).

Vedenina, V.Yu. & Zhantiev, R.D. 1990. Recognition of acoustic signals in sympatric species of locusts. *Zoologicheskii Zhurnal* **69** (2): 36–45. [In Russian with English summary. English translation: 1991, *Entomological Review, Washington* **69** (8): 41–50.]

Wadepuhl, M. 1983. Control of grasshopper singing behavior by the brain: responses to electrical stimulation. *Zeitschrift für Tierpsychologie* **63**: 173–200.

Waeber, G. 1989. Gesang und taxonomie der europäischen *Stenobothrus-* und *Omocestus*-Arten [Orthoptera, Acrididae]. [iv +] 156 pp. Friedrich-Alexander-Universität, Erlangen (unpublished thesis).

Walker, T.J. 1957. Specificity in the response of female tree crickets (Orthoptera, Gryllidae, Oecanthinae) to calling songs of the males. *Annals of the Entomological Society of America* **50**: 626–636.

— 1962*a*. The taxonomy and calling songs of United States tree crickets (Orthoptera: Gryllidae: Oecanthinae). I. The genus *Neoxabea* and the *niveus* and *varicornis* groups of the genus *Oecanthus. Annals of the Entomological Society of America* **55**: 303–322.

— 1962*b*. Factors responsible for intraspecific variation in the calling songs of crickets. *Evolution* **16**: 407–428.

— 1963. The taxonomy and calling songs of United States tree crickets (Orthoptera: Gryllidae: Oecanthinae). II. The *nigricornis* group of the genus *Oecanthus. Annals of the Entomological Society of America* **56**: 772–789.

— 1964*a*. Experimental demonstration of a cat locating orthopteran prey by the prey's calling song. *Florida Entomologist* **47**: 163–165.

— 1964*b*. Cryptic species among sound producing ensiferan Orthoptera (Gryllidae and Tettigoniidae). *Quarterly Review of Biology* **39**: 345–355.

— 1969*a*. Systematics and acoustic behavior of United States crickets of the genus *Orocharis* (Orthoptera: Gryllidae). *Annals of the Entomological Society of America* **62**: 752–762.

— 1969*b*. Systematics and acoustic behavior of United States crickets of the genus *Cyrtoxipha* (Orthoptera: Gryllidae). *Annals of the Entomological Society of America* **62**: 945–952.

— 1969*c*. *Oecanthus jamaicensis*, n. sp.: a *Cecropia*-inhabiting cricket (Orthoptera: Gryllidae). *Florida Entomologist* **52**: 263–265.

— 1971. *Orchelimum carinatum*, a new meadow katydid from the southeastern United States (Orthoptera: Tettigoniidae). *Florida Entomologist* **54**: 277–281.

— 1973. Systematics and acoustic behavior of United States and Caribbean short-tailed crickets (Orthoptera: Gryllidae: *Anurogryllus*). *Annals of the Entomological Society of America* **66**: 1269–1277.

— 1974. Character displacement and acoustic insects. *American Zoologist* **14**: 1137–1150.

— 1975a. Stridulatory movements in eight species of *Neoconocephalus* (Tettigoniidae). *Journal of Insect Physiology* **21**: 595–603.

— 1975b. Effects of temperature, humidity and age on stridulatory rates in *Atlanticus* spp. (Orthoptera: Tettigoniidae: Decticinae). *Annals of the Entomological Society of America* **68**: 607–611.

— 1975c. Effects of temperature on rates in poikilotherm nervous systems: evidence from the calling songs of meadow katydids (Orthoptera: Tettigoniidae: *Orchelimum*) and reanalysis of published data. *Journal of Comparative Physiology* **101**: 57–69.

— 1977. *Hapithus melodius* and *H. brevipennis*: musical and mute sister species in Florida (Orthoptera: Gryllidae). *Annals of the Entomological Society of America* **70**: 249–252.

— 1979. Calling crickets (*Anurogryllus arboreus*) over pitfalls: females, males, and predators. *Environmental Entomology* **8**: 441–443.

— 1993. Phonotaxis in female *Ormia ochracea* (Diptera: Tachinidae), a parasitoid of field crickets. *Journal of Insect Behavior* **6**: 389–410.

Walker, T.J., Brandt, J.F. & Dew, D. 1970. Sound-synchronized, ultra-high-speed photography: a method for studying stridulation in crickets and katydids (Orthoptera). *Annals of the Entomological Society of America* **63**: 910–912.

Walker, T.J. & Dew, D. 1972. Wing movements of calling katydids: fiddling finesse. *Science, New York* **178**: 174–176.

Walker, T.J. & Greenfield, M.D. 1983. Songs and systematics of Caribbean *Neoconocephalus* (Orthoptera: Tettigoniidae). *Transactions of the American Entomological Society* **109**: 357–389.

Walker, T.J. & Gurney, A.B. 1972. Systematics and acoustic behavior of *Borinquenula*, a new genus of brachypterous coneheaded katydids endemic to Puerto Rico (Orthoptera, Tettigoniidae, Copiphorinae). *Annals of the Entomological Society of America* **65**: 460–474.

Walker, T.J., Whitesell, J.J. & Alexander, R.D. 1973. The robust conehead: two widespread sibling species (Orthoptera: Tettigoniidae: *Neoconocephalus* "*robustus*"). *Ohio Journal of Science* **73**: 321–330.

Wallin, L. 1979. *Svenska gräshoppors och vårtbitares sångläten.* 9 pp. Uppsala (Uppsala Universitet). [With tape cassette of recorded songs.]

Weber, H.E. 1984. Bestimmungsschlüssel für Heuschrecken und Grillen in Westfalen nach akustischen Merkmalen. *Natur und Heimat* **44**: 1–19.

Weidemann, S. & Keuper, A. 1987. Influence of vibratory signals on the phonotaxis of the gryllid *Gryllus bimaculatus* DeGeer (Ensifera: Gryllidae). *Oecologia* **74**: 316–318.

Weih, A.S. 1951. Untersuchungen über das Wechselsingen (Anaphonie) und über das angeborene Lautschema einiger Feldheuschrecken. *Zeitschrift für Tierpsychologie* **8**: 1–41.

Weissman, D.B. & Rentz, D.C.F. 1977. Feral house crickets *Acheta domesticus* (L.) (Orthoptera: Gryllidae) in southern Calif. *Entomological News* **88**: 246–248.

White, G. 1789. *The natural history and antiquities of Selborne, in the county of Southampton: with engravings, and an appendix.* v + 468 [+ xiv] pp. London (B. White & Son).

Willemse, C. 1943. Het sjirpen van *Mecostethus grossus* L. de groote weidesprinkhaan. *Nutuurhistorisch Maandblad* **32**: 7–8.

Willemse, F. 1979. A review of the species of *Acrometopa* Fieber, 1853 (Orthoptera, Tettigonioidea, Phaneropterinae) with special reference to the Greek fauna. *Bijdragen tot de Dierkunde* **49**: 135–152.

Willemse, F. & Heller, K.-G. 1992. Notes on the systematics of Greek species of *Poecilimon* Fischer, 1853 (Orthoptera: Phaneropterinae). *Tijdschrift voor Entomologie* **135**: 299–315.

Willey, R.L. 1970. Sound location of insects by the dwarf weasel. *American Midland Naturalist* **84**: 563–564.

Wolf, H. 1985. Monitoring the activity of an auditory interneuron in a freemoving grasshopper [pp. 51–60]. In Kalmring, K. & Elsner, N. [Eds], *Acoustic and vibrational communication in insects.* viii + 230 pp. Berlin & Hamburg (Paul Parey).

Xi, R., Liu, J., He, Z. & Chen, N. 1992. Studies of sound structure of the acridoids. *Sinozoologia* **9**: 35–42. [In Chinese with English summary.]

Yersin, [A.]. 1852. Séance du 18 février 1852. *Bulletins des Séances de la Société Vaudoise des Sciences Naturelles* **3**: 100–104.

— 1853. Séance annuelle et publique du 29 juin 1853. *Bulletins des Séances de la Société Vaudoise des Sciences Naturelles* **3**: 239–242.

— 1854*a*. Sur quelques Orthoptères nouveaux ou peu connus, du midi de la France. *Bulletins des Séances de la Société Vaudoise des Sciences Naturelles* **4**: 63–70.

— 1854*b*. Mémoire sur quelques faits relatifs à la stridulation des Orthoptères et leur distribution géographique en Europe. *Bulletins des Séances de la Société Vaudoise des Sciences Naturelles* **4**: 108–128.

— 1857. Sur les Orthoptères et quelques Hémiptères des environs d'Hyères en Provence. *Annales de la Société Entomologique de France* (3) **4** (1856): 737–748.

— 1858. Note sur un Orthoptère nouveau. *Annales de la Société Entomologique de France* (3) **6**: 111–122.

— 1860. Note sur quelques Orthoptères nouveaux ou peu connus d'Europe. *Annales de la Société Entomologique de France* (3) **8**: 509–537.

— 1863. Description de deux Orthoptères nouveaux d'Europe. *Annales de la Société Entomologique de France* (4) **3**: 285–292.

Young, A.J. 1971. Studies on the acoustic behaviour of certain Orthoptera. *Animal Behaviour* **19**: 727–743.

Zanon, V. 1926. Contributo alla conoscenza degli Ortotteri del dintorni di Roma. *Memorie dell'Accademia Pontificia dei Nuovi Lincei* (2) **9**: 173–194.

Zaretsky, M.D. 1972. Specificity of the calling song and short term changes in the phonotactic response by female crickets *Scapsipedus marginatus* (Gryllidae). *Journal of Comparative Physiology* **79**: 153–172.

Zeller, P.C. 1849. Ueber *Decticus tessellatus* Charp., *D. Philippicus* Zell. und *D. strictus* Zell. *Entomologische Zeitung Herausgegeben von dem Entomologischen Vereine zu Stettin* **10**: 113–116.

Zetterstedt, J.W. 1821. *Orthoptera Sveciæ, disposita et descripta.* 152 pp. Lund (Berling).

Zeuner, F. [E.]. 1931. Beiträge zur deutschen Orthopterenfauna. *Mitteilungen der Deutschen Entomologischen Gesellschaft* **2**: 75–78.

— 1941. The classification of the Decticinae hitherto included in *Platycleis* Fieb. or *Metrioptera* Wesm. (Orthoptera, Saltatoria). *Transactions of the Royal Entomological Society of London* **91**: 1–50.

Zhantiev, R.D. 1981. *Bioacoustics of insects.* 256 pp. Moscow (University Press). [In Russian.]

Zhantiev, R.D. & Chukanov, V.S. 1972. Frequency characteristics of tympanal organs of the cricket *Gryllus bimaculatus* Deg. (Orthoptera, Gryllidae). *Vestnik Moskovskogo Universiteta* (Series 6, Biologiya, Pochvovedenie) **1972** (2): 3–8. [In Russian with English summary.]

Zhantiev, R.D. & Dubrovin, N.N. 1971. On the sensitivity of tympanal organs of katydids (Orthoptera, Tettigoniidae) to sounds of different frequencies. *Proceedings of the XIII International Congress of Entomology* **2** (1968): 45–46.

Zhantiev, R.D. & Dubrovin, N.N. 1974. Acoustic signals of crickets (Orthoptera, Oecanthidae, Gryllidae). *Zoologicheskii Zhurnal* **53**: 345–358. [In Russian with English summary.]

Zhantiev, R.D. & Korsunovskaya, O.S. 1973. Sound communication and some characteristics of the auditory system in mole crickets (Orthoptera, Gryllotalpidae). *Zoologicheskii Zhurnal* **52**: 1789–1801. [In Russian with English summary.]

Zhantiev, R.D. & Korsunovskaya, O.S. 1978. Morphofunctional organization of tympanal organs in *Tettigonia cantans* (Orthoptera, Tettigoniidae). *Zoologicheskii Zhurnal* **57**: 1012–1016. [In Russian with English summary.]

Zhantiev, R.D. & Korsunovskaya, O.S. 1986. Sound communication in bush crickets (Tettigoniidae, Phaneropterinae) of the European part of the U.S.S.R. *Zoologicheskii Zhurnal* **65**: 1151–1163. [In Russian with English summary.]

Zhantiev, R.D. & Korsunovskaya, O.S. 1990. Sound communication of Phaneropteridae (Orthoptera) [pp. 402–406]. *In* Gribakin, F.G., Wiese, K. & Popov, A.V. [Eds], *Sensory systems and communication in arthropods.* xviii + 423 pp. Basle, etc. (Birkhäuser).

Zhantiev, R.D., Korsunovskaya, O.S. & Byzov, S.D. 1995. Acoustic communication in desert bushcrickets (Orthoptera, Bradyporidae). *Zoologicheskii Zhurnal* **74** (9): 58–72. [In Russian with English summary. English translation: 1996, *Entomological Review, Washington* **75** (6): 91–105.]

Zimmermann, U., Rheinlaender, J. & Robinson, D. 1989. Cues for male phonotaxis in the duetting bushcricket *Leptophyes punctatissima. Journal of Comparative Physiology* (A) **164**: 621–628.

Zippelius, H.-M. 1949. Die Paarungsbiologie einiger Orthopteren-Arten. *Zeitschrift für Tierpsychologie* **6**: 372–390.

Index to vernacular names

The page references given are normally for the main treatment of each species in the text. References to the colour plates are given only for the English vernacular names. A full list of page references is given under the Latin name of each species in the General Index.

General index

Principal page references are in **bold** type. Vernacular names are indexed separately (p. 579).

NOTES

Printed in the United States
by Baker & Taylor Publisher Services